NARROW ROADS OF GE

The Collected Papers of
W. D. HAMILTON

NARROW ROADS OF GENE LAND

The Collected Papers of
W. D. HAMILTON

VOLUME 3

Last Words

Edited by
Mark Ridley

This book has been printed digitally and produced in a standard specification in order to ensure its continuing availability

OXFORD
UNIVERSITY PRESS

Great Clarendon Street, Oxford OX2 6DP

Oxford University Press is a department of the University of Oxford.
It furthers the University's objective of excellence in research, scholarship,
and education by publishing worldwide in

Oxford New York

Auckland Cape Town Dar es Salaam Hong Kong Karachi
Kuala Lumpur Madrid Melbourne Mexico City Nairobi
New Delhi Shanghai Taipei Toronto
With offices in
Argentina Austria Brazil Chile Czech Republic France Greece
Guatemala Hungary Italy Japan South Korea Poland Portugal
Singapore Switzerland Thailand Turkey Ukraine Vietnam

Oxford is a registered trade mark of Oxford University Press
in the UK and in certain other countries

Published in the United States
by Oxford University Press Inc., New York

Reprinted 2010

ISBN 978-0-19-856690-8

Contents

Editorial Preface

The Narrow Roads of Gene Land is W. D. Hamilton's title for his collected papers. This book is Volume 3, and completes the set. In the previous two volumes, however, Bill (as I'll refer to W. D. Hamilton here) did much more than simply reprint his scientific papers. For each paper, he also wrote an autobiographical introduction, and those amazing introductions were for many readers the most immediate attraction of the books, notwithstanding the scientific fire-power of the papers themselves. Bill died in the year 2000. He had seen Volume 1 through publication (it came out in 1996). He left almost-publishable manuscripts for Volume 2, together with all the decisions about which papers were to be included. After some editorial work, and guess-work, Volume 2 was published posthumously in 2001; it included papers published up to 1990. Bill intended to produce a Volume 3, but when he died he had done no work on the introductions nor left any indication about which papers would be included.

The editor at Oxford University Press, Michael Rodgers, who had dealt with the publication of Volumes 1 and 2, spoke round and corresponded with a number of people about how to publish Volume 3. I am not sure who exactly invented the form that the book has taken; Luisa Bozzi and Marlene Zuk may have been particularly influential, along no doubt with Michael Rodgers himself. Anyhow, a plan was somehow devised in which Volume 3 would include the papers from Bill's final years, together with more-or-less personal introductions written by Bill's co-authors. The introductions would work rather like Bill's own autobiographical introductions in Volumes 1 and 2, taking the reader somewhat closer to Bill's extraordinary personality and intellect. I was subsequently (though by then Michael Rodgers had retired from the OUP) invited to edit the volume, according to that plan. I accordingly encouraged Bill's collaborators, in their chapters, to write personally about Bill (if they wished to do so) as well as introducing the science. These introductory sections now provide, I believe, another way to get to know one of the great scientist's of the twentieth century, through the eyes of his collaborators. And the papers themselves enable readers to find out about, or remind themselves of, Bill's scientific output for 1990–2000 (though papers continued to be published until 2003). In addition to the co-authored papers, Volume 3 also contains several papers of which Bill was the sole author, and these are reprinted without introduction.

Volumes 1 and 2 both had leading themes. Most of the papers in Volume 1 were about social behaviour, and most of those in Volume 2 were about the evolution of sex. The themes are identified in the subtitles that Bill gave the volumes. By the time Volume 3 begins, Bill's research was moving into a more diversified phase. Bill continued to be interested in sex, and particularly its relation with parasitic disease; several of the chapters in this book are on this topic or something close to it. But he was also thinking about a huge range of topics, and often collaborating with someone else who worked on a particular topic in more detail. Some of the co-authors who have contributed introductions here have remarked how they had little idea that Bill was also working with half a dozen other people on disparate research topics at the same time as he was collaborating with them. I initially hoped to provide a subtitle for Volume 3 that would link its diversified papers into an identifiable theme; but I failed to find one and fell back on chronology. Olivia Judson invented the particular subtitle I have used, 'Last words'.

Bill's thinking ranged from highly imaginative abstract theory, to exact mathematical and computer modeling, and he liked to relate the theory to abstruse natural history, particularly from entomology. Volume 3 shows him at work in all these ways. He began the decade doing parasitically revved up computer simulations of genetic algorithms with Brian Sumida (Chapter 1); he ended it doing simulations of 'pacemakers' in spatial models of host–parasite coevolution, with Akira Sasaki and Francisco Ubeda (Chapter 18). In-between, he helped some astonishing work on antiobiotically cured parthenogenesis into print (Chapter 2); wrote about gender with Laurence Hurst (Chapter 4), the weird habits of Strepsiptera with Jeya Kathirithamby (Chapter 6), virulence with Dieter Ebert (Chapter 10), and diversity with Pete Henderson (Chapter 16). He backed the controversial hypothesis that the AIDS pandemic had accidentally originated in the polio vaccination campaign in Africa (Chapter 14). He had some wonderful flights of Bill-style theorizing about Gaia (Chapter 15) and the colours of autumn leaves (Chapter 17). Moreover, the co-authors who have contributed introductions here are by no means the only who collaborated with Bill in these years, though they are a good sample. Bill's sole authored papers look at models of sex (Chapter 7) and—again with some characteristically imaginative thinking—at inbreeding (Chapter 8).

Volume 3 contains almost all the published papers that appeared with Bill's name on after the end of Volume 2. Bill excluded a few minor publications from Volumes 1 and 2—publications such as letters to the editor, and short book reviews (though one more substantial book review made it into Volume 1); he also excluded at least one co-authored paper from the chronological period covered by Volume 2. For Volume 3, I have followed similar principles, though Bill's changing work-mode has suggested some slight modifications. I have again excluded short book reviews, letters to the editor, and minor abstract-length publications, though I encouraged co-authors to quote from and cite sources of this kind (as well as correspondence) in their introductions if they thought it appropriate. I also excluded a posthumous paper that had Bill's name on but that he knew nothing of—the posthumous papers included here are ones that Bill had worked on, contributed to, and knew were destined for submission. Finally, I excluded one or two manuscripts, of conference lectures, that Bill had worked on before he died, and probably would have been published; they seemed to me to be too incomplete for most readers to be able to follow.

On the other hand, I have included some papers that Bill might just not have included—either because he made only small contributions to them, or because he might have judged them too minor. Bill made little contribution to the Wolbachia paper (Chapter 2) or the second Gaia paper (Chapter 15), but they provide interesting sidelights on the way Bill was working now that he was famous. I also included a couple of lecture-addresses (Chapters 11 and 12), given when Bill received major prizes. They only just make it past the 'published' criterion—technically, they were published, but privately by the foundations concerned. Part of the reason to include them, along with Bill's bravura personal eschatology (Chapter 3), a bibliographical piece (Chapter 9), and a preface to a book on paper wasps (Chapter 13), is their autobiographical interest. Volumes 1 and 2 were rich in Bill's autobiography, and I inclined to stretch the net to admit some autobiography here too. In the end, about 90% of the decisions about inclusion and exclusion were straightforward, but there was a residue that was inevitably arbitrary. The book also includes a chapter by Jeremy Leighton John on the Hamilton archive—'Bill's last great work'—complete with irresistible pictures (Chapter 19), and Alan Grafen's biographical memoir (Chapter 20) by way of overview of Bill's life and work.

Finally, the book is necessarily missing the largest part of Bill's writings from his final decade: the autobiographical introductions to Volumes 1 and 2 of *Narrow Roads of Gene Land.* They amount to more than the length of the papers included in this volume. As Alan Grafen says (Chapter 20), Bill had invented an original way of writing autobiography, and one that is peculiarly appropriate for a scientist. Any one who wants a full picture of Bill's activities in the 1990s will need to add them to the publications reprinted here. If, by some paradox, those autobiographies had been included in this volume, I'd have offered as subtitular theme for Bill's final decade 'collaborations and autobiographies.'

Mark Ridley
Oxford, December 2004

List of Contributors

Brown, Sam. Section of Integrative Biology, University of Texas at Austin, 1 University Station C0930, Austin, TX 78712-0253, USA. E-mail: sam@biosci.utexas.edu

Ebert, Dieter. Departement de Biologie, Ecologie, et Evolution, Université de Fribourg, Chemin du Musee 10, 1700 Fribourg, Switzerland. E-mail: dieter.ebert@unifr.ch

Grafen, Alan. Department of Zoology, South Parks Road, Oxford OX1 3PS, UK. E-mail: alan.grafen@sjc.ox.ac.uk

Henderson, Peter. PISCES Conservation Ltd, IRC House, The Square, Pennington, Lymington, Hants SO41 8GN, UK. E-mail: pisces@irchouse.demon.co.uk

Hooper, Ed., Ludwell's Farm, Stathe, Bridgwater, TA7 0JL, UK. E-mail: edhooper@withy.f9.co.uk

Hurst, Laurence D. Department of Biology and Biochemistry, University of Bath, Claverton Down, Bath, Somerset BA2 7AY, UK. E-mail: l.d.hurst@bath.ac.uk

John, Jeremy Leighton. Department of Manuscripts, Directorate of Scholarship and Collections, 96 Euston Road, The British Library, London NW1 2DB, UK. E-mail: jeremy.john@bl.uk

Kathirithamby, Jeyaraney. Department of Zoology, South Parks Road, Oxford OX1 3PS, UK. E-mail: jeyaraney.kathirithamby@zoology.oxford.ac.uk

Lenton, Tim. Centre for Ecology and Hydrology—Edinburgh, Bush Estate, Penicuik, Midlothian EH26 0QB, UK. E-mail: tlent@ceh.ac.uk

Ridley, Mark (editor) Department of Zoology, South Parks Road, Oxford OX1 3PS, UK. E-mail: mark.ridley@zoo.ox.ac.uk

Sasaki, Akira. Department of Biology, Faculty of Science, Kyushu University, Hakozaki 6-10-1, Higashi-ku, Fukuoka 812-8581, Japan. E-mail: asasascb@mbox.nc.kyushu-u.ac.jp

Stouthamer, Richard. Department of Entomology, University of California, Riverside, CA 92521, USA. E-mail: richard.stouthamer@ucr.edu

Sumida, Brian. 5534 Merriewood dr., Oakland, CA 94611, USA. E-mail: pszyk2001@yahoo.com

CHAPTER 1

OKU NO HOSOMICHI

Roads to Hamilton's 'Wrightian' Digital Parasites in Geneland

BRIAN SUMIDA

> The title, too, is a *tour de force*. Bashō borrowed the name of a road he came across in his northbound odyssey. Laden with meaning, his inspired title is difficult to translate. Some of the word *oku's* meanings are: 'the far recesses, the inner reaches, the hinterland, the interior.' *Oku* is also an abbreviation of Michinoku, or Michi-no-Oku, 'the road's far recesses.' Japanese nouns are at once both singular and plural. In taking Oku-no-Hosomichi for his title, Bashō must have had in mind its wider meaning, embracing all the desolate, rugged footways penetrating the remote frontier. The original 'Narrow Road' is now a busy thoroughfare in Iwakiri on the outskirts of the bustling city of Sendai[1].

I first met Bill Hamilton after he arrived at the Animal Behaviour Research Group (ABRG) in the department of Zoology, as a Royal Society Research Professor. At that time, the ABRG held weekly meetings in a building named Holywell Manor, which belonged to Bailiol College, on Thursday evenings. Bill's presence was memorable as he sat, hunched over, in a chair that seemed too small for his large frame: his tweed jacket, voluminous mane of grey hair and reading glasses (secured around his neck with nylon fishing line) adding to the professorial air.

Shortly thereafter, while I was working in the Zoology computer room, Bill walked up asking me to help him logon to the Oxford University mainframe. While showing Bill the login procedure, he said that, as part of the agreement with the Royal Society, he had purchased a computer workstation to use in his research. Bill had selected the Whitechapel MG-1 because of its high-resolution graphics hardware. A serious disadvantage of this choice, however, from Bill's point of view, was that the Whitechappel's operating system was UNIX, which he had no experience with. I had worked with UNIX at the University of California and offered to help set up his workstation. Bill was encouraged with this knowledge, asking me whether I would consider working as his research assistant to write computer programs for his workstation and to do the systems administration. Although I was knowledgeable on the subject of computer programming, I did not have any background in evolution. Bill did not see this as a hindrance and I accepted his offer since this was a fantastic, and unexpected, opportunity to work with a great scientist.

So I began my working relationship with Bill, occupying a corner of his office where he had chosen to set up the Whitechappel. I began my apprenticeship by translating the computer simulations that he had written at the University of Michigan with Robert Axelrod and Rieko Tanese.[2]

While I was translating Bill's programs, we discussed the sorts of computer simulations that interested him. Bill said that John Holland, who he met at the University of Michigan, had invented an interesting class of computer programs that John called genetic algorithms (GA). Since Holland was a computer scientist, his GAs did not accurately imitate the mechanisms of population genetics, although it was possible to modify his GAs to make them more realistic. In essence, GAs are a kind of optimization algorithm that uses simulated evolution to search for better solutions. In contrast to the analytical models that Bill had previously programmed and that had unique solutions, GAs were fuzzy in that they evolved groups of possible solutions by trial and error. GAs appealed to Bill because they provided him with a rich framework to model virtual populations. Populations were composed of individual entities, not frequencies, as was more common in other models. Bill was the sole architect of the GA projects, devising the specifications for our versions of Holland's GA. The evolutionary features (e.g. truncation selection) that Bill selected to include in our models were interesting to me, as these choices reflected which elements he felt were the most influential mechanisms. Since Bill was an advocate of Sewall Wright, he modeled the GA populations that contained a number of structural units that Sewall Wright described as demes, each deme-unit comprising a small (10–20) number of individuals. The demes were isolated from one another but could receive migrants from other demes. Migration of individuals between demes introduced genetic variability. In the simulated GA world that Bill constructed, demes would inhabit an environment with the geometry of a rectangular grid whose sides were reflecting barriers (like Kimura's stepping-stone model), or, alternately the world may have a toroidal surface.

Bill's major GA innovation was to introduce coevolving parasites.[3] As in real life, the simulated parasites reduced the host's fitness proportional to the closeness of their match to the host genotype. Coevolving parasites prevent a population from becoming fixed on a particular solution, forcing the hosts into a constant flight from the parasites. The dynamics of host–parasite GAs exhibited the following pattern. A host population ascends toward a 'fitness' peak as selection favours the genetic combinations that have higher fitnesses. Once a peak has been reached, the host population loses genetic variability, as it becomes fixed for the currently best genotype. As host variability decreases, the intensity of parasite attack increases, which results in the host population being forced off the peak. Hosts are pushed down, by parasite pressure, into the valleys between peaks where exploration eventually leads the host population to the attraction zone of another peak. Selection then operates to draw the host population to this peak. Using this cycle of hill climbing, peak descent and exploration, the GA is able to continuously sample the environments.

Not one to take the cautious approach, Bill decided to use our GAs to solve NP-complete problems. These sorts of problems are very difficult due to the large numbers of possible solutions, generally not solvable in finite time. Our first project simulated the NP-complete travelling salesman problem (TSP). Bill's goals were usually unconservative. His objective in any project was to create the best in the world. He expressed this thought aloud when we were discussing a project to evolve optimal binary sorts. On a figure plotting the theoretical limit of binary sorts, Bill pointed to a region well beyond the best-known sorts and said, 'that is where we will want to be'. Bill was always aggressive in setting goals. When deciding on the number of cities for our GA travelling salesman tour, Bill felt that 100 cities would present a worthy challenge for the host–parasite GA. The number of possible tours for 100 cities is 100 factorial, which is a number greater than all of the atoms in the visible universe. To further enhance the tour's complexity, Bill positioned the cities of the tour so as to replicate a fractal tour, such that there were a great many nearly optimal tours. Each nearly optimal tour was equivalent to a fitness peak in the fitness landscape.

Bill had a foray into optimization theory, collaborating with Alasdair Houston, John McNamara and myself on a GA model of bird song. I had previously collaborated with Alasdair Houston, who was a member of the Edward Grey Institute in the department of Zoology. At first glance, Alasdair appeared an imposing figure, resembling a fierce Celtic warrior, being in excess of six feet in stature, bearing wild curly hair and threatening moustache. On the contrary, however, Alasdair is the most gentle person you could meet. I remember having tea in his well-tended garden where he would often interrupt our conversation to point out the birds that entered his visual and aural beacons. Alasdair's area of expertise was in melding experimental psychology with mathematical optimization. I was in awe of his writing skill, with the apparent ease and speed with which he produced a working manuscript. Through Alasdair I met his long-time collaborator, John McNamara, a first-rate mathematician who is on the faculty at Bristol University. The four of us constructed a GA model of bird song, which evolved a dawn chorus (after finding the right parameter settings!). Although the paper presented some novel results, Bill felt that the model was too far divorced from real genetics to motivate him to further explore this kind of model, despite the enthusiasm of both Alasdair and John.

Bill defined the problems and scope for our projects. My primary responsibility was to write the computer code for Bill's models. Bill's approach was to include lots of parameters into the model, so that a large range of settings could be explored. Debugging such complex programs is often very difficult. Bill and I debugged programs by laying out 132 column fan-fold paper printouts across the length of his office floor. Armed with different coloured highlighter pens, we then proceeded to trace the logical flow of the program, annotating the code as we moved from the top to the bottom of the printout often in excess of 10 feet in length. Bill's debugging powers were frequently required because his coding resembled the way he thought. If you picture Bill's finely detailed figures in his geometry for the selfish herd paper, then you can get an impression of the intricacy of his coding. Simplicity and (programming) safety was not one of the guidelines Bill followed when writing FORTRAN. Bill used goto statements and common blocks liberally—most software engineers avoid using these programming elements. Worst of all (to me), Bill would code with the execrable *computed goto*. As any programmer knows, debugging is part science and part art. Many debugging strategies exist, mostly centred around following a single logic thread through the program. Bill's method, however, was more macroscopic. He would look at the entire data output to work out what sort of logical error would produce the results, then back reference to the code to find the section that violated the logic. Most programmers (including myself) work in the reverse fashion, debugging by examining the code, tracking the values of a single variable or a data structure. Bill's method was like taking a picture of a freeway system from a helicopter, and by examining the patterns, deduce the position and timing of the stop lights from first principles.

Bill used lots of visual models when he was documenting or expressing his ideas. He drew meticulously detailed geometric figures on A4 paper, annotating these with his extremely tiny writing (using a 0.5-mm mechanical pencil). Bill also liked to construct physical models. In Bill's office there was a piece of pine that Bill had carved a 3-dimensional fitness landscape, complete with gridlines cut into the surface at regular intervals. Another of Bill's 3D models, which he used to illustrate host–parasite coevolution was a perspex cube with holes bored into each corner. Each corner of the cube represented a combination of alleles. A stainless-steel rod was threaded through two opposing corners of the cube. Bill held the ends of the steel rods and rotated one of the non-threaded corners upwards. He instructed the viewer to imagine that the genotype at the top apex had, at the moment, the highest fitness. As this genotype spread throughout the population, however, the variability of genotypes decreased, giving the parasites a target to evolve towards. As the parasites evolved better matches to the host genotype their hosts' finesses steadily diminished. Bill rotated the perspex cube to cause the uppermost corner to sink downwards reflecting the increasing pressure from parasites. Bill thought this model nicely illustrated the dynamics of the host–parasite chase and flight. When Bill purchased a colour workstation he asked me to write a 3D-computer graphic that emulated his perspex cube model, plotting the trajectories of host and parasite genotypes in different colours. The cover of *Narrow Roads of Gene Land*, Volume 2, displays screenshots of the program's output displayed as a hypercube.

When writing a paper with Bill, the manuscript was divided into sections where each of us had the greater expertise. Bill's writing style was so unique that I rarely edited his sections except to correct some technical point. On the other hand, I was happy to let Bill edit my text to suit his sense of style and content, for I felt the manuscript would be a better product after his highly focused editing. Even when writing e-mail messages he would revise the text many times, focusing as intently on its contents as if it were a scientific paper. Not only was Bill's handwriting very small it was also very difficult to read, even by himself. Nevertheless, when you look at his 3″ × 5″ index card library of notes of scientific papers (Bill referred to these cards as his 'brain cells') you cannot help but marvel at the sheer quantity of information that Bill managed to pack into these little pieces of paper.

Bill enjoyed ideas (and people) that were unusual, or outside the conventional thinking, especially those in science whose ideas were not accepted. I think that he felt a sympathy for the underdog because he had also experienced rejection of his ideas earlier in his career, a sentiment that he still felt strongly about many years later. Bill supported Ed Hooper's efforts to publicize a theory that AIDS emerged as a result of the early live-polio vaccine trials in Africa by writing to Science in protest against the journal's refusal to publish Hooper's theory. During this time Ed Hooper and Bill were corresponding using e-mail. Bill said that Hooper was virtually in hiding, fearing persecution and that he was instructed by Ed to encrypt all his messages. To encrypt Bill's messages, I installed Phil Katz's Pretty Good Encryption (PGP) program on Bill's computer. After installing the software, I asked Bill to give me an encryption phrase to use to encode his message. After a short pause Bill began, 'A ship, an isle, a sickle moon, With few but with how splendid stars'. Another pause then, 'the mirrors of the sea are strewn between their silver bars . . . '. Bill continued to recite James Elroy Flecker's poem, 'A Ship, An Isle A Sickle Moon' in its entirety. I waited for Bill to complete his recitation (I enjoyed the poem) before commenting that there was a 255-character limit for the key. Bill often quoted from literature like this. He was fond of Shakespeare's sonnets and selecting Shakespeare, could play the UNIX first line–next line game, scoring respectably. One of the rituals that I copied from Alasdair Houston was to insert a literary quote at the beginning of a paper. When Bill and I were writing Section 1.2 on GAs that follows this introduction, I quoted a passage from T. S. Eliot's 'Sweeney Agonistes' that alluded to (pursuit by) parasites. Returning the next day, after reading a draft of the manuscript, Bill brought in his faded Penguin paperback copy of Aeschylus's *Oresteia trilogy* (in English) with a post-it marking the page with the original passage. While handing me the book, Bill commented that after rereading Aeschylus, he felt the T. S. Eliot version worked better for our purposes.

An aspect of Bill's vision, which I most admired, was his magician-like ability to discover something extraordinary within the ordinary. In the

Zoology coffee area there were a number of glass-fronted exhibition cases containing various animals on display. These displays were generally passed unnoticed by the myriad of students and staff on their way to coffee or tea since their contents, such as sticklebacks, could not be described as exotic. One morning, however, I spied Bill perched upon a chair, staring intensely into a display case. When I asked him what it was that had his attention, Bill said that there was a rare ant species (not an official component) living in a mossy portion of the display. Although everyone shared the same view of the displays, only Bill was able to 'see' something special. It was one of his special gifts. Bill had a total awareness of his surroundings. But, his Zen-like focus did not seem to be functional in ordinary everyday activities. Bill was often unaware of the day of the week, for example.

Another instance of Bill's insight that comes to my mind is his explorations of a simple limit-cycle algorithm, which he called Cycesh. I wrote a program in the C language, with a 10-line recursive algorithm for Bill so that he could simulate limit cycles on his Whitechappel MG-1. Bill spent many hours exploring different values for the three or four parameters in the limit-cycle equation. Bill's intuition operated in multidimensions, allowing him to instinctively select a number of parameters values simultaneously. John Holland would say that Bill used hyperplane sampling. Instead of points, Bill selected *sets* of points. I remember seeing Bill, bathed in the fading evening light of the late afternoon, hunched over the keyboard in near darkness while watching the orbital dance of Cycesh. Over time, Bill patiently collected a large collection of limit cycles cataloging the printouts in black 2-ring binders.[4] Many of the limit cycles that Bill discovered were quite exquisite in form. When simulating limit cycles, the algorithm can easily blow up (go to infinity), which often happened to me when I ran Cycesh. There is a delicate balance of settings that result in stability. Bill's adroitness in selecting parameter values resembled, to me, the kind of skill displayed by a musician who, while playing an electric guitar on the verge of total feedback is still able to coax beautiful music from the instrument. The analogy is somewhat inappropriate, however, since Bill was (by his own admission) completely tone deaf.

I would like to conclude by adding a short note to the story of Bill's famous cycling episode where he collided with a moving automobile and went through the back window.[5] Bill and I had arranged to meet at his office in the Zoology department to work on a computer simulation. Due to the accident, Bill arrived at his office late, as was not uncommon, appearing with multiple lacerations in his face where the glass shards had imbedded themselves. He was very irritated because the doctor who extracted the glass fragments had made an appointment for Bill to have his skull X-rayed for possible fractures. This irritated Bill immensely since he was completely unconcerned about his injuries. What *did* concern Bill was the fact that he had to leave his bicycle on the pavement of Woodstock Road. Bill asked me to fetch the bike with urgency since it was quite precious to him. He said that the bicycle had belonged to his brother who had been killed in a climbing accident. Bill said that whilst riding on the bicycle he would often think of his brother.

Bill's scientific genius is without question. But, after having worked with Bill for many years, what impressed me most about Bill was his humanity.[6] Bill was always respectful to everyone; always open to new ideas. He was uncommonly humble, gentle and honest. Bill was a great classical scientist, knowledgeable in Latin and Greek mythology and poetry. Like many others I shall miss his company.

My collaboration with Bill ended after our proposal for evolving optimal binary sorts using a host–parasite GA was rejected. Following this event, I made the decision to begin a career outside academics as a software engineer at Oxford Molecular designing bioinformatics software.

In this slim volume, you will find prose of spare simplicity, and phrases of beautifully polished elegance. You will find writing of robust, masculine strength, as well as touches of delicate, feminine grace.

As you read *The Narrow Road to a Far Province*, at times you will find yourself rising up to applaud. At other times you will quietly hang your head with emotion.[7]

Soryō
Early summer of the Seventh
Year of Genroku [1694]

References

1. From the introduction in *A HAIKU JOURNEY Bashō's Narrow Road to a Far Province*. Translated by Dorothy Britton. Kodansha International. Tokyo. Bill always traveled through the far recesses of Geneland.

2. W. D. Hamilton, Robert Axelrod, and Reiko Tanese, Sexual Reproduction as an Adaptation to Resist Parasites (A Review), reprinted in *Narrow Roads of Gene Land*, vol. 2, Chapter 16.

3. Daniel Hillis had previously designed GAs with components that he called parasites. His parasites, however, were not parasites in the biological sense. In Hillis's model GAs were evolving optimal sorts. His 'parasites' were the lists of data that were being sorted. The lists that impeded the best sorts were favoured and used to create the next generation of lists. Further, I believe that the lists did not reproduce sexually.
4. I believe these are in the Hamilton archive, British Library, that is described by Jeremy John's Chapter 19 of this volume.
5. Eulogy delivered by Richard Dawkins in the Foreward to *Narrow Roads of Gene Land*, Vol. 2.
6. On one occasion I was bedridden for three days with a nasty virus. Concerned, Bill visited me at my flat on 15 Kingston Road to see how I was and to ask if I needed any provisions! Bill was always doing little favours for people.
7. *A HAIKU JOURNEY Bashō's Narrow Road to a Far Province*. Translated by Dorothy Britton. Kodansha International. Tokyo.

GENETIC ALGORITHMS AND EVOLUTION†

B. H. SUMIDA, A. I. HOUSTON, J. M. MCNAMARA
AND W. D. HAMILTON

Abstract

The genetic algorithm (GA) as developed by Holland (1975, *Adaptation in Natural and Artificial Systems*. Ann Arbor: University of Michigan Press)[8] is an optimization technique based on natural selection. We use a modified version of this technique to investigate which aspects of natural selection make it an efficient search procedure. Our main modification to Holland's GA is the subdividing of the population into semi-isolated demes. We consider two examples. One is a fitness landscape with many local optima. The other is a model of singing in birds that has been previously analysed using dynamic programming. Both examples have epistatic interactions. In the first example we show that the GA can find the global optimum and that its success is improved by subdividing the population. In the second example we show that GAs can evolve to the optimal policy found by dynamic programming.

Introduction

Darwin stressed that the action of natural selection could result in animals that appear to be well-designed. This view of natural selection has been developed into the application of optimality theory to the study of the morphology and behaviour (e.g. refs 1–4; see also refs 5, 6). Although the use of optimality to study evolution is widespread, it is also criticized (e.g.).[7]

In this paper we make no attempt to review these arguments. Instead we consider recent attempts to use the evolutionary process to provide optimization techniques (e.g. refs 8–10; for some related ideas based on neurobiology and physics see ref 11). In particular we concentrate on Holland's genetic algorithms, and use them to study biological problems. We begin with a historical background to genetic algorithms. The genetic algorithm kernel (GAK), based upon

† *J. Theoret. Biol.* **147**, 59–84 (1990).

the original model[8] is described. The basic GAK model is modified to include additional biological features. We are interested in investigating how various aspects of natural selection influence the efficiency of the GA as a search procedure. In as much as we are following the spread of genes through a population, we are doing population genetics. We are, however, looking at fitness landscapes which are highly complex, with many local optima or strong epistatic effects. It is a whole research programme to explore how evolution depends on all aspects of an organism's biology. Of necessity, we must restrict attention to a few features. A modified GAK is developed to incorporate features of the Shifting Balance Theory of evolution as proposed by Wright.[12] Two case studies, both involving epistatic effects, are described. The first examines the evolution of an optimal genotype in a multiple-peaked fitness landscape. We investigate the effects of population structure, mutation rate, selection strength and migration frequency on evolution's ability to reach the global optimum. The second case study describes the evolution of daily routines of singing and foraging in a small bird. There has been considerable discussion about the evolutionary advantages of daily routines of singing and foraging (see ref. 13 for a review). McNamara *et al.*[14] found optimal routines using dynamic programming. We show that these GAs can evolve to produce these optimal routines with relative ease.

Background and History

The foundation of genetic algorithms (GAs) is contained in the monograph *Adaptation in Natural and Artificial Systems* by Holland.[8] Holland[8] is the acknowledged founder of the field of GAs. Other computer simulations with a genetic algorithm flavour (e.g. refs 15, 16) had appeared in the literature before Holland's[8] monograph, but his was the first to formalize the theory and apply GAs to artificial systems. The GA has its roots in the field of artificial intelligence as an optimization technique that generates adaptive responses to changing environments. What then is a GA? GAs are exceptionally powerful heuristic search algorithms which use biological evolution as the framework for their search process, executing a structured yet randomized information exchange to produce fitter individuals through time. The algorithms are *genetic* in the sense that their primary operations, mating, mutation and crossover (see below for details), emulate Mendelian populations (see, e.g. ref. 17). Individuals or organisms are represented by their chromosomes and are implemented in the GA by vectors of data structures. An ensemble of individuals, called the population, generates a new set of individuals at the end of a generation, by mating the fittest individuals from the current generation. Occasionally, a chromosomal mutation occurs, introducing new information and preventing fixation or loss of genetic information.

GAs, when compared to other optimization methods, offer many advantages. GAs are computationally simple despite their efficiency in tackling difficult problems. Two features that may make an optimization problem difficult are a very large search space of possible solutions and the existence of many local optima. On such problems ennumerative procedures such as dynamic programming would require prohibitively large amounts of computer time. GAs are, however, highly efficient techniques for such problems.[18,19] Optimization analyses which rely on gradient techniques (i.e. 'hill climbing') assume the existence of derivatives, thus imposing constraints on the characteristics of the search space. Such techniques work well when there is a single optimum, but fail when there are many local optima. The GA does not require derivatives, only a payoff, or fitness measure is necessary. Even if the fitness function has many local optima the GA can perform well.[20] These factors contributed to the success which has resulted from applying the GA method to a host of problems (e.g. Travelling salesman problem, see ref 18).

The applications which followed Holland's[8] original work were centred around engineering problems. Since then the method has attracted a wider audience resulting in an increased diversity of applications in other fields (see refs 21, 22). For example, Axelrod[23] effectively used genetic algorithms to study the evolution of co-operation in the Prisoner's Dilemma of game theory. In observing the emergence of game strategies through simulated evolution. Axelrod[23] discovered many interesting and surprising behaviours, including the evolution of a strategy that rivalled the celebrated 'Tit for Tat' strategy. He went on to suggest that genetic algorithms may be used profitably to investigate the evolution of other sorts of behaviours. The studies described in this paper affirm his prediction.

The Genetic Algorithm Kernel (GAK)

The GA programme is flexible and many variations have been written according to specific requirements. In general, however, GAs will contain the following essential components:

(a) A chromosomal representation of possible solutions to the problem.

(b) A fitness (i.e. objective) function to evaluate the performance of individual solutions.

(c) A reproduction function to produce a new set of offspring.

(d) A mutation operator.

A rule for solving the problem is encoded by an ordered sequence of numbers. By analogy to biology, we refer to this ordered sequence as a chromosome. In biology, the chromosomes are sequences of genes that contain the information for producing an individual. In the context of GAs a single (haploid) chromosome contains all the information for specifying a rule. Again by analogy to biology, we refer to the position in the sequence of numbers as a locus and the value of the number at a position as an allele. Holland[8] used bit-string (i.e. elements from the set {0, 1}) representations in his work. Although much less is known about other forms of encoding in GAs, it will be shown in this paper that alternative schemes can be successful.

The success or failure of any given GA may depend on the encoding scheme. At the most basic level, the encoding scheme will determine the set of possible solutions, and thus imposes an evolutionary constraint. We illustrate this later in the paper in the context of the model for singing vs. foraging. For some problems, there will be a natural coding in which different chromosome segments correspond to functionally independent aspects of the problem. Recombination can then bring together the best segments for each aspect of the problem. In many cases, however, there will be a complex interdependence between the performance of any segment of the chromosome and the values coded on other parts of the chromosome. This is likely to be true when attempts are made to code daily routines or life histories; the model of singing vs. foraging is a clear example.

Having fixed a code, a chromosome represents a particular attempt at a problem-solving rule and the performance of such a rule is measured by a single number which we refer to as the fitness of the chromosome. One of the strengths of the GAs, as originally envisaged by Holland,[8] is the flexibility that results from the fact that the chromosomes are defined independently of the performance criterion for the problem. Thus, if the parameters of the problem, and hence the performance criterion, changes it is possible to utilize the same chromosomes and encoding for the new circumstances. Holland[8] saw this ability to adapt to changes as an important attribute of an artificial intelligence system. There is a direct biological analogy in that the same method of encoding (genes) are used when the environment changes and hence the fitness of various phenotypes change.

The GA is based on the evolution of a population of chromosomes. In the simplest version, the population has discrete non-overlapping generations. In each generation, the fitness of each chromosome is determined and a new generation of chromosomes is produced. The composition of this new population is determined by mating, recombination and mutation, where the mating success of each chromosome is determined by its fitness.

In the GA sexual reproduction is the process by which two parent chromosomes give rise to offspring by hybridizing the parental chromosomes via recombination. The recombination procedure begins by choosing a random interlocus interval on the chromosome. This point on the chromosome forms the crossover chiasma, or the point of interchange between the two parent chromosomes. Figure 1.1 illustrates the crossover operation.

Parents are chosen in pairs from the original population according to some probability distribution. These probabilities are chosen in such a way that the

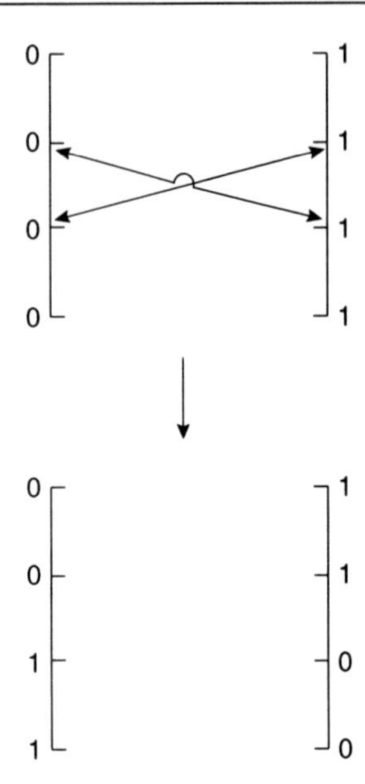

Figure 1.1. The crossover operation in the genetic algorithm. Offspring are formed, with a process analogous to sexual recombination, by hybridizing genetic material from two parents. In the upper half of the figure, the crossover point, where genetic material is exchanged between the parents, is indicated by a chiasma. The lower half of the figure illustrates the offspring resulting from a crossover at the indicated position. Although two offspring logically result from recombination of two parents, only one offspring is kept.

expected number of offspring that a parent has is an increasing function of its fitness. The total number of offspring produced is usually constrained so that the total number of chromosomes in a generation is kept constant.

The offspring, constructed from elements of the fittest chromosomes in the previous generation, form the new population. Such selective pressure, over generations, tends to raise mean fitness. Holland[8] argues that the power of the GA—and of evolution by natural selection—stems from two separate processes that work in concert to direct progress towards a better solution. Selection isolates and preserves the best building blocks—subsets of coadapted alleles. Crossover effects a highly efficient search, testing subsets of information rather than single points. He made this observation of GAs and applied the term 'implicit parallelism' to describe the process. A consequence of subset recombination emerges: crossover enables the bringing together of two or more beneficial mutations which occur on different homologous chromosomes. The probability of a single chromosome acquiring two beneficial and rare mutations by chance would be exceedingly low.

Mutation is a necessary component of GAs. Without mutation the population will lose alleles through fixation or deletion. Mutation rates in GAs, as in nature, are usually quite low. Because mutation is undirected a high mutation rate is more likely to disrupt the GA than to improve its performance. Mutation in the GAK is carried out after a new population has been formed and involves a random change at selected loci.

In our version of the GA, each locus on each chromosome has a given small probability of undergoing a mutation. If occurence of a mutation is determined, the resulting allele is chosen at random from the set of all possible alleles at that locus.

The Modified Genetic Algorithm Kernel

The GAK outlines the basic plan for any genetic algorithm simulation. In order to be able to investigate which features of natural selection result in an effective search procedure, we modify the GAK to incorporate certain additional features.

Only a single crossover point is used in the GAK. This does not coincide with observations of nuclear division in cells, where crossover may occur repeatedly along the chromosome.[24] Multiple crossover, with an independent probability of interlocus crossover, is possible in the modified GAK.

Populations in the GAK, like annual plants, are entirely replaced after each generation, forming non-overlapping generations. Our modifications allow overlapping generations, i.e. a fraction of the population up to the entire population can be replaced in the next generation.

Selection was simulated using a form of soft (see ref. 25 for a definition) threshold, or truncation, selection: a fixed percentage of least fit members of the population was 'killed' every generation. Each individual in the current population that 'dies' is replaced by an offspring formed from randomly selected parents. Only one offspring is produced from each mating. Offspring are assigned to replace current least fit individuals and the replacement proceeds in an ascending fitness order. The mortality fraction in each generation determines the strength of selection and, in all but one experiment, 16% mortality was used. Truncation selection was used here because this type of selection has been demonstrated[26–28] to be the most effective procedure for bringing about directional gene-frequency changes.

Our most drastic modification of the GAK is to make the GA compatible with Wright's shifting balance model of evolution (e.g. ref. 29, and later) as follows.

The Search Problem and Wright's Shifting Balance Model

Deterministic hill-climbing is a method commonly used in optimization procedures. This term is derived from the use of the gradient (i.e. the slope or derivative) to locate the optimum. A positive (uphill) slope indicates progress toward the optimum; a negative (downhill) slope indicates a retreat from the optimum. If movement in any direction results in a negative slope then the optimum has been reached. A hill-climbing algorithm, having discovered an optimum, cannot distinguish whether what has been achieved is a global optimum or a local optimum. Thus, hill-climbing optimization procedures are only appropriate when the solution set is known to contain a single peak.

Unfortunately, many problems in nature may contain multiple peaks (see, e.g. refs 20, 30, 31), resembling a surface of peaks and valleys. Wright[32] used the metaphor of an adaptive landscape to describe a hypersurface where relative fitness is plotted as a function of the various gene frequencies (see also, refs 33–35). A population subject to selection, like hill-climbing optimization, is attracted towards the nearest peak in the adaptive landscape. This need not be the highest. The population eventually becomes stuck on a peak, through local adaptation. Wright considered what might cause a population to move away from a locally stable equilibrium (*viz* peak), and descend into the surrounding valleys and so reach the domain of attraction of another, higher peak. In particular, Wright studied the consequences of random changes in allele frequencies caused by genetic drift believing it to be an important force in evolution. Fisher[36] did not agree with Wright

'The views of Fisher and Wright contrast strongly on the significance of random changes in the population. Whereas, to Fisher, random change is essentially noise in the system that renders the deterministic processes somewhat less efficient than they would otherwise be, Wright thinks of such random fluctuations as one aspect whereby evolutionary novelty can come about by permitting novel gene combinations.'[37]

Whether random genetic drift is a way of creating favourable gene combinations, or is useless, evidence suggests that it is prevalent in nature.[38–40] Counterbalancing the rewards gained from random genetic drift is the loss in mean fitness as the population drifts away from the highest peak. But Wright[41] argued that this is a necessary cost for letting evolution find new gene combinations.

To enable populations to move from local to global optima, Wright introduced his shifting balance model. The details of the model's processes can be outlined as follows. The population is partitioned into small and somewhat isolated subpopulations. These subpopulations undergo largely random differentiation due to random genetic drift since the effect of genetic drift is greater in small and isolated populations. Now and then a subpopulation acquires a genotype which posesses superior fitness, whereupon mass selection overpowers genetic drift and fixes this genotype in the sub-population. In other words the subpopulation crosses a saddle to another peak. Because of its increased fitness the subpopulation then disperses greater numbers of emmigrants than its neighbours thereby attracting surrounding subpopulations to follow across the valley to its (currently) higher peak. This process of 'intergroup selection' is repeated, in an ever widening circle to draw the entire population to this peak.

In this manner there is a chance that virtually the whole of the adaptive landscape can be explored by continual shifting of control by one adaptive peak to control by a superior one until the global optimum is found.

The SBGA Structure

Motivated by Wright's 'continent-island' model, we modify the GAK so that (i) the population is divided into semi-isolated subpopulations and (ii) there is migration between the subpopulations. The subpopulations are referred to as demes, and constitute a locally interbreeding group. Our simulations usually used a deme size of six individuals, or, since the model is haploid, six chromosomes.

Migration between demes provides for the constitution of a single metapopulation. The form and amount of migration in actual populations is important. Our model is an 'island' (see ref. 42 for definition) model in which migrants are drawn from a random sample of the entire population. In nature, migrants may come primarily from nearby populations. To the extent that nearby populations will have similar allele frequencies, the effects of migration will be smaller than predicted by the island model. The amount of migration sets the limit to how much genetic divergence can occur through random genetic drift within a deme. A high migration rate negates the effect of genetic drift, and the population behaves as a single large unit, its deme structure being irrelevant. Low migration rates increase the strength of local genetic drift but this may (a) obstruct progress to new found peaks, and (b) prevent a deme on a near peak 'infecting' others near it with its discovery. It is clear that migration level needs to be carefully chosen. Some guidance about how to achieve this balance can be obtained from studying the quantitative relationship between migration and drift.

The work of Crow & Kimura[37] following Wright suggests that if the number of migrants is much more than one per deme per generation, there is little local differentiation. In fact remarkably little migration is required to prevent significant genetic divergence among subpopulations due to genetic drift. In line with this result we usually use a mean migration rate of one migrant per deme every two generations.

Wright's theory also requires (in the absence of deme extinctions) greater numbers of emigrants from subpopulations at higher peaks[43,44] this being the expression of interdeme advantage. The SBGA emulated this feature in the following way. Migration occurred every second generation. The total number of migrants in the population was equal to the number of demes. Emigrants were selected from the population until the required number of migrants was obtained. The probability of an emigrant being drawn from a given subpopulation was proportional to that subpopulation's fitness.

The interplay between migration and random genetic drift in the absence of selection was investigated with a one locus two allele SBGA to check both the programming and the correspondence of the discrete model to the limiting continuous distribution of Wright. Small differences are expected for two reasons: the discrete situation vs. the idealized continuous distribution, and the finite size of our population. Figure 1.2 displays histograms of the equilibrium distribution of allele frequencies that result from three different migration rates with random mating and no selection after 1000 generations. The histograms

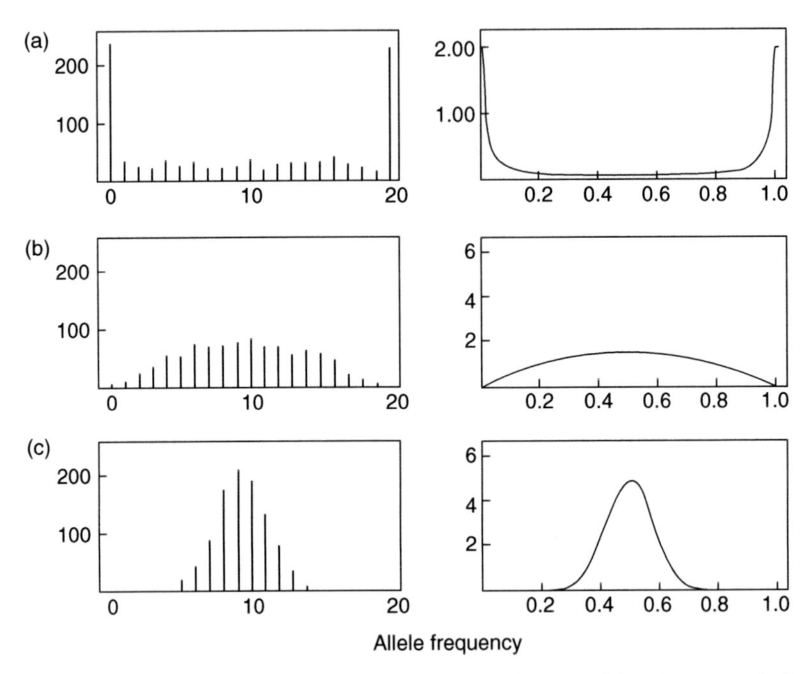

Figure 1.2. The equilibrium allelic frequency distributions resulting from simulation drift and migration with the SBGA. The population consists of 20 000 individuals divided into 1000 demes of 20 individuals. Individuals carrying the '*A*' allele were counted in each deme after 1000 generations and cumulated over the 21 possible counts. Three migration rates are plotted: (a) 0.01, (b) 0.5 and (c) 5 migrants per deme per generation. When migration rate is low, (a), drift is considerable and loss of genetic diversity results from fixation and deletion of the allele (see as accumulations in outer histogram bars). At an intermediate migration rate, (b), the effect of drift is balanced by migration and genetic diversity is preserved, as is seen on the graph, by the wide distribution of the allele frequency in the population. A high migration rate, (c), effectively turns the subpopulations into a single population and the frequency of the allele in the population converges toward the mean value. To the right of each empirical distribution is shown the beta distribution obtained when $Nm = 0.01, 0.5$ or 5.0 are applied in formula 6.15 of Moran.[45] Index coefficients of Moran's rather than Wright's are appropriate here because Moran's derivation is for a haploid model like ours. Correspondence between the empirical distributions and their limiting continuous counterparts is good, confirming theory and the programming of our model.

were compiled by measuring the allele frequency in a population composed of 1000 demes. We can observe from the figure that a low migration rate increases the effect of random genetic drift resulting in fixation and loss of the allele in a large proportion of the population. Intermediate migration rates maintain allelic diversity within demes. High migration rates cause the subpopulations to remain very similar, in effect to fuse into a single population. The figure also shows theoretical curves based on the analyses in Wright[32] and more specifically on formula 6.15 in Moran[45] because Moran treats as we do a haploid population. It can be seen that there is very good qualitative agreement between the simulation and theoretical predictions.

Experiment 1: Optimization in a Multiple-peaked Fitness Landscape

REPRESENTATION

The SBGA was studied with a variation of Wright's[12] *Model with Six Homo-allelic Selective Peaks*. The primary difference is that Wright's[12] model is diploid whereas the SBGAs model uses haploid chromosomes.

The population consisted of chromosomes with 16 loci. Each locus had two alleles, An allele had either the value of zero (0) or a one (1). This configuration gives 2^{16} possible genotypes. We defined the optimal genotype as a series of eight consecutive zeros followed by eight consecutive ones. A multi-peaked environment was formed using the following algorithm to determine the fitness scores:

- For each chromosome, sum the value, ν_i, at each locus, i, to give

$$x = \sum_{i=0}^{15} \nu_i.$$

- Calculate the preliminary fitness function

$$\omega = 64 - (x - 8)^2.$$

If an optimum is defined as a position in the chromosome state space such that a single step in any direction decreases fitness, the above function specifies $(8!)^2/16! = 12\ 870$ equal optima, which correspond to all chromosomes where numbers of ones and zeros are equal.

- Differentiate peaks to create a global optimum

$$\hat{\omega} = \omega + k \sum_{i=0}^{15} i\nu_i,$$

where k is a small constant.

This new fitness function gives moderately higher values to strings with more 1s in the string. The number of peaks produced by this algorithm with 16 loci and k small remains 12 870, but one, that of eight zeros followed by eight ones, is now a global optimum.

The SBGAs task was to evolve the global optimum genotype. This is a sensible task given a haploid population, but it should be noted that in a diploid population selection will tend to maximize mean fitness, and this does not necessarily produce any single optimal genotype. For the problem considered there are 65 536 possible genotypes of which 12 869 are local optima. Preliminary studies indicated that the problem was far too simple. Even scaled up versions of the same problem were easily solved. For example the SBGA consistently evolved the optimum genotype within 1000 generations from 2^{64} possible genotypes in an environment with $1.83 \cdot 10^{18}$ peaks using a population of only 36 chromosomes! Rather than increasing the search space and the number of peaks further, which did not seem to impede the efficiency of the search process, a device inhibiting the building sequences of ones or zeros which could then be recombined, was added to the fitness algorithm to raise the level of difficulty. A sequence of ones flanked at both ends by a zero is called a block of ones. A parameter, C, giving the length of a standard block is defined. Every block of ones of length G in the chromosome is given a score

$$b = K(G-1)(G-C)(C-1)^2,$$

where K is a constant. This score is zero if $G=1$ or $G=C$, is negative if $1 < G < C$ and is positive for $G > C$. The score, b, for every sequence of 1s is added to w to give the chromosome fitness. The parameter C can be imagined as representing the size of the valley that separates the global optimum from most of the local optima—from all those that do not already have blocks larger than C. Larger values of C would then represent a wider, more difficult to cross, valley.

To make the search problem difficult, the occurrence of blocks of ones in the initial population was minimized by setting the probability of a '1' being assigned to a chromosomal locus to 0.005. The scoring system used to measure performance was based on the population mean modulus error after 1000 generations. The modulus error lies between 0 (perfect match) and 16 (all bits wrong) with a mean modulus error of eight corresponding to a random choice of zeros and ones. The distribution of mean modulus error that we obtained was bimodal, with modes occuring at 0 and 8 and very few values between 3 and 6. We therefore categorized the data into two discrete classes. A modulus error between 0 and 4.5 inclusive was classed a 'hit' and a score greater than 4.5 was classed as a 'miss'.

The effect of deme structure in the case of a population of total size 72 is explored in Fig. 1.3, which plots the number of 'hits' against deme structure for $C=3$, 4, 5, 6 and 7. Taking an arcsine transformation of the proportion of hits,

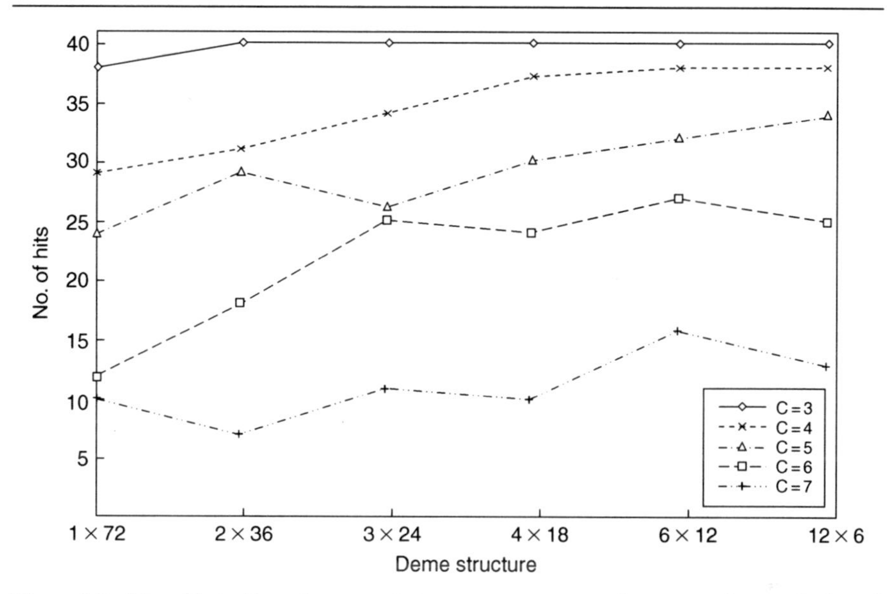

Figure 1.3. The effect of varying population structure on performance. A population of 72 individuals was subdivided into one deme of 72, two demes of 36, three demes of 24, four demes of 18, six demes of 12 and 12 demes of 6. The performance, number of 'hits', of each population structure is plotted for $C = 3$, 4, 5, 6 and 7. The improvement in performance as the population is partitioned into more numerous demes is statistically significant (see text).

a multiple regression reveals a significant effect of the number of demes ($P < 0.05$, df. $= 11$, $F = 43.59$, $r^2 = 0.9224$). As we might expect, the effect of C is highly significant; as C increases, fewer hits occur. Although increasing the number of demes makes search more efficient in this case, the effect seems to depend on total population size. Table 1.1 looks at two examples in which total population size is 72 and two examples which total population size is 36. Subdividing the population of size 36, Table 1.1(a), does not significantly increase the number of 'hits'. Increasing the population size from 36 to 72 chromosomes, Table 1.1(b), resulted in significant differences in the scores obtained from an undivided and divided population when $C = 4$ ($\chi^2 = 7.44$, $0.001 < P < 0.01$), $C = 5$ ($\chi^2 = 6.27$, $0.01 < P < 0.05$) and $C = 6$ ($\chi^2 = 8.498$, $0.001 < P < 0.01$).

In only one comparison out of the ten does the undivided population score more 'hits'. However, rather surprisingly that case was more severe than average and overall the difference was not greatest when the problem set is of maximum severity: in neither case was it significant for $C = 7$.

The overall advantages of both large and subdivided populations, however, are clear. Cumulating the 'hit' scores across all values of C, for a population size

Table 1.1. Comparison between performances of subdivided and undivided populations in a multi-peaked environment for various lengths of penalty block (C). Migrations between the subpopulations were proportional to fitness (see text) and the rate of migration was 0.5 migrants per deme per generation. Rank-order soft selection was used for both types of population structure with the mortality rate set to 16% of the population. Each run lasted 1000 generations. The other parameters were K = 0.8, v = 0.5, mutation = 0.005, probability of crossover = 0.5. Results from a 2×2 χ^2 contingency test is given in the far-right column. Significance is indicated by a star (*)

(a) Total population size was 36 organized as either a single population of 36 individuals or six subpopulations of six individuals

	6×6		1×36		
C	Hit	Miss	Hit	Miss	χ^2 (significance)
3	34	6	33	7	0.092 N.S.
4	30	10	23	17	2.739 N.S.
5	24	16	18	22	1.8 N.S.
6	15	25	19	21	0.818 N.S.
7	7	33	6	34	0.092 N.S.

(b) Total population size was 72, organized as either a single population of 72 individuals or 12 sub-population of six individuals

	12×6		1×72		
C	Hit	Miss	Hit	Miss	χ^2 (significance)
3	40	0	38	2	2.051 (N.S.)
4	38	2	29	11	7.44 ($0.001 < P < 0.01$)*
5	34	6	24	16	6.27 ($0.01 < P < 0.05$)*
6	25	15	12	28	8.498 ($0.001 < P < 0.01$)*
7	13	27	10	30	0.549 (N.S.)

N.S. = Not significant.

of 36, reveals that the overall success of the subdivided population (= 110) is greater than the undivided population (= 99). Increasing the population size to 72 results in an increased number of 'hits' for both the subdivided (= 150) and the undivided population (= 113). But the magnitude of improvement is greater for the subdivided population (+40 hits) as compared to the undivided population (+14 hits). Thus, a larger population size improves performance but the degree of improvement is greater when the population is subdivided.

Tables 1.2–1.4 investigate the effects of various biological features on the number of 'hits' obtained. In each case the population is divided into 12 demes of six individuals, and each datum is based on 40 runs of 1000 generations.

Migrations between subpopulations were proportional to fitness. In each analysis, the proportion of 'hits' was subjected to an arcsine transformation in order to normalize the variance.

Table 1.2. Performances of a subdivided population (12 demes of six individuals) under different migration frequencies for various lengths of penalty block (C). The migration frequencies were varied from four generations to 32 generations between migrations. Rank-order soft selection was used with the mortality rate set to 16% of the population. The other parameters were K = 0.8, v = 0.5, mutation = 0.005, probability of crossover = 0.5

Generations between migrations	C	Hit	Miss
	3	40	0
	4	39	1
4	5	34	6
	6	30	10
	7	22	18
	3	40	0
	4	33	7
8	5	29	11
	6	21	19
	7	8	32
	3	37	3
	4	16	24
16	5	14	26
	6	7	33
	7	2	38
	3	23	17
	4	1	39
32	5	1	39
	6	2	38
	7	0	40

In Table 1.2 we investigate the effect of migration frequency. A multiple regression based on C^2 and the number of generations between migrations showed that both these factors were significant ($P < 0.0005$, df. = 16, $F = 44.82$, $r^2 = 0.89$).

Table 1.3 shows a similar analysis based on selection strength, expressed as percentage mortality. The effect of C is significant ($P < 0.005$, df. = 11, $F = 43.59$, $r^2 = 0.92$) but the effect of mortality (i.e. selection strength) is not.

The final example (Table 1.4) shows that mutation rate and C both have a significant effect ($P < 0.0005$, df. = 11, $F = 57.22$, $r^2 = 0.94$).

The examples given above illustrates that, at least in some circumstances, the SBGA is able to find the global optimum more efficiently than an algorithm based on a single population. To look at a specifically biological problem, we now consider a second example.

Table 1.3. Performances of a subdivided population (12 demes of six individuals) under varying levels of mortality. Rank-order soft selection was used with mortality rates set to 33, 50 and 67% of the population. The other parameters were $K=0.8$, $v=0.5$, mutation $=0.005$, probability of cross-over $=0.5$

C	Mortality (%)	Hit	Miss
	33	40	0
3	50	40	0
	67	40	0
	33	33	7
4	50	34	6
	67	28	12
	33	27	13
5	50	32	8
	67	30	10
	33	17	23
6	50	18	22
	67	19	21
	33	11	29
7	50	12	28
	67	16	24

Table 1.4. Performance of a subdivided (12 demes of six individuals) population under different mutation rates. Rank-ordeer soft selection was used with mortality rates set to 16% of the population. The other parameters were $K=0.8$, $v=0.5$ probability of crossover $=0.5$

Mutation rate	*C*	Hit	Miss
	3	40	0
	4	40	0
0.01	5	38	2
	6	32	8
	7	15	25
	3	28	12
	4	16	24
0.001	5	17	23
	6	18	22
	7	10	30

Experiment 2: Daily Routines of Singing and Foraging—the Evolution of a Dawn Chorus

In this section we describe an application of the SBGA to the problem of how a male bird should allocate his time between singing and foraging. The SBGA evolves policies which are used during the course of a day to decide whether to sing or feed based upon energy reserves. A high level of singing at dawn is referred to as the dawn chorus (see refs 13, 46, 47, for discussion). From previous work,[14] the optimum policy is known: a dawn chorus should evolve when overnight energy expenditure is variable.

THE MODEL AND REPRESENTATION

The model is the same as that developed by McNamara *et al.*[14] The male is characterized by his level of energy reserves, x. If x falls to zero, then the male dies. McNamara *et al.*[14] considered a male singing over a period of 5 days. Each day is divided into a period of daylight and a period of darkness. There are 24 intervals during the daylight period. At the start of each interval, the male can decide either to sing during the interval or to forage during the interval. If the male sings, then it uses energy (thus decreasing its reserves) but has a probability of attracting a mate. If the male forages, then it also uses energy, but it obtains a random amount of food (the amount may be zero). The parameter *IP* controls the variability of the food by determining the probability that the bird is interrupted while feeding. During the night the male uses a normally distributed amount of energy with mean μ_N and standard deviation σ_N (for further details see ref. 14).

To find the policy of singing and foraging that maximizes a male's fitness, McNamara *et al.*[14] introduced a terminal reward at the end of the fifth day that specifies the male's expected future reproductive success as a function of its state (see ref. 48 for a detailed discussion of this approach). The reward is zero if the male is dead, one if the male is alive without a mate and two if the male is alive and has obtained a mate. A sensitivity analysis showed that the exact values of this reward function were not crucial. Using dynamic programming[49] it was found that at time t on day n it was optimal to sing if reserves were above $C^*(t, n)$ and to forage if reserves were below $C^*(t, n)$. Apart from the last day ($n = 5$), the policies were virtually independent of n. We will denote this day-independent critical level by $C^*(t)$.

In the investigation of the GA, all animals start at dawn on day 1 with $x = 8$. Each bird then follows the policy given by its chromosome until dusk at the end of the second day. We calculate the probability distributions of energy states at the following dawn under this policy. This series of calculations is referred to as the forward procedure. The animal's state at this dawn determines the animal's reward in the following way. Amimals which are dead at this time receive no reward. Animals that are alive and have a mate receive a reward of 2. Animals that are alive without a mate receive a reward equal to the mean reward they would have obtained from following the optimal policy for the three remaining days of a 5 day period, as given by McNamara *et al.*[14] For each chromosome we calculate the expected total reward obtained and equate this to the fitness of the chromosome.

The chromosome determines behaviour as a function of energy reserves and time of day. The optimality analysis carried out by McNamara *et al.*[14] indicates that the optimal policy always involves a critical level above which the bird should sing and below which it should forage. Thus out of the enormously large range of possible policies, we restrict attention to policies of this form: i.e. a chromosome will specify a critical level $C(t)$.

Within this class of policies, there are various ways to code for $C(t)$. The results given below are based on the following coding. Each chromosome has six loci, labelled $i=0$ to $i=5$. Locus i contains an integer c_i between 0 and $C_{\max}$ ($C_{\max}=32$ in all computations reported here). The integers c_i determine a function $C(t)$ as follows

$$C(24i/5) = c_i, \quad i = 0, \ldots, 5,$$

$C(t)$ at other values is given by linear interpolation.

The male's behaviour is taken to be the same on each day of a 2-day period and is given by the rule forage if $x(t) < C(t)$ and sing otherwise.

The values at a given locus only influences the policy around that time of day. It is important to note, however, that the fitness consequences of the value at a given locus depend on the values at all other loci. In other words, there are strong epistatic effects. One of the main points made by McNamara *et al.*[14] is that the timing of singing cannot necessarily be understood by considering the environmental conditions at the precise time when singing occurs. One may only be able to understand the timing of singing by considering the bird's daily routine as a whole.

We also tried a different form of coding based on a polynomial equation for $C(t)$. The equation was of the form $C(t) = k_0 + k_1 t + k_2 t^2 + k_3 t^3 + k_4 t^4$. There were five loci with locus i coding for k_i. This worked less well than the coding described above. A possible reason for this is that the value at each loci influences the policy at all times of the day. Another possibility is that the optimal critical level $C^*(t)$ is essentially a piece-wise linear function of t and there are limits to the accuracy with which a fourth-order polynomial can approximate a piece-wise linear function. The polynomial coding may thus have imposed considerable constraints on the set of possible phenotypes.

A population of 36 chromosomes was partitioned into six demes of six chromosomes. The following SBGA parameter settings were used for all runs. The probability of mutation was set to 0.01. When a mutation occurred at a locus, it resulted in the assignment of a random integer between 0 and $C_{\max}$ to that locus. Migration rate was 0.5 migrants per deme per generation. Selection within demes was by mortality with two deaths per deme per generation. The interlocus crossover probability during reproduction was set to 0.5.

The chromosomes were initialized by setting each locus to a random integer between 0 and C_{max}. Preliminary runs indicated that fixation of the chromosomes occurred at $\approx$100 generations. After fixation only mutation can alter the chromosomal structure. Since we were primarily interested in selection effects, runs were halted after 100 generations and the evolved policies were saved.

Results

NO OVERNIGHT LOSS

We begin by looking at the case in which there is no overnight energy expenditure, i.e. μ_N and σ_N are zero. In these circumstances the optimal policy does not depend on time of day, and is given by the rule 'sing if reserves are above C^*, otherwise forage'.

The evolved policies resulting from six typical runs of the SBGA are plotted in Fig. 1.4. It can be seen that the fit between the evolved policies and the optimum is very close. Figure 1.5 shows the effect of increasing the interruption probability (*IP*). Increasing *IP* makes the food supply less reliable, and hence the animal should carry more reserves, i.e. C^* is increased. It can again be seen that the fit between the evolved policies and the optimum policy is very close. When there is no overnight loss, the proportion of unmated birds that are singing is constant throughout the day under the optimal policy. The behaviour that results from the SBGA policies evolves to this pattern in both cases that we have considered.

OVERNIGHT LOSS

When a variable amount of energy is expended overnight, an animal must build up its reserves before dusk. Consequently, the optimal policy involves a threshold $C^*(t)$ which increases as dusk approaches. Figure 1.6 compares six typical evolved policies with the optimal policy. Figure 1.7 shows the progressive evolution of singing behaviour over 100 generations, as well as the singing behaviour that results from following the optimal policy. It can be seen that the optimal routine involves a dawn chorus, a mid-afternoon dip in singing followed by a rise in singing at dusk (cf. ref. 14). After 100 generations, the evolved routine has all of these features.

Increasing the interruption probability when there is overnight expenditure means that the animal must not only carry higher reserves but start to build up reserves earlier in the day. Figure 1.8 shows the optimal policy together with six typical evolved policies and Fig. 1.9 shows the evolution of singing behaviour over 100 generations together with the singing behaviour under the optimal policy. It can again be seen that the evolved daily routine of singing has the same qualitative form as the optimal singing routine. The policies evolved by the GA achieved close to optimal fitness as can be seen from Table 1.5.

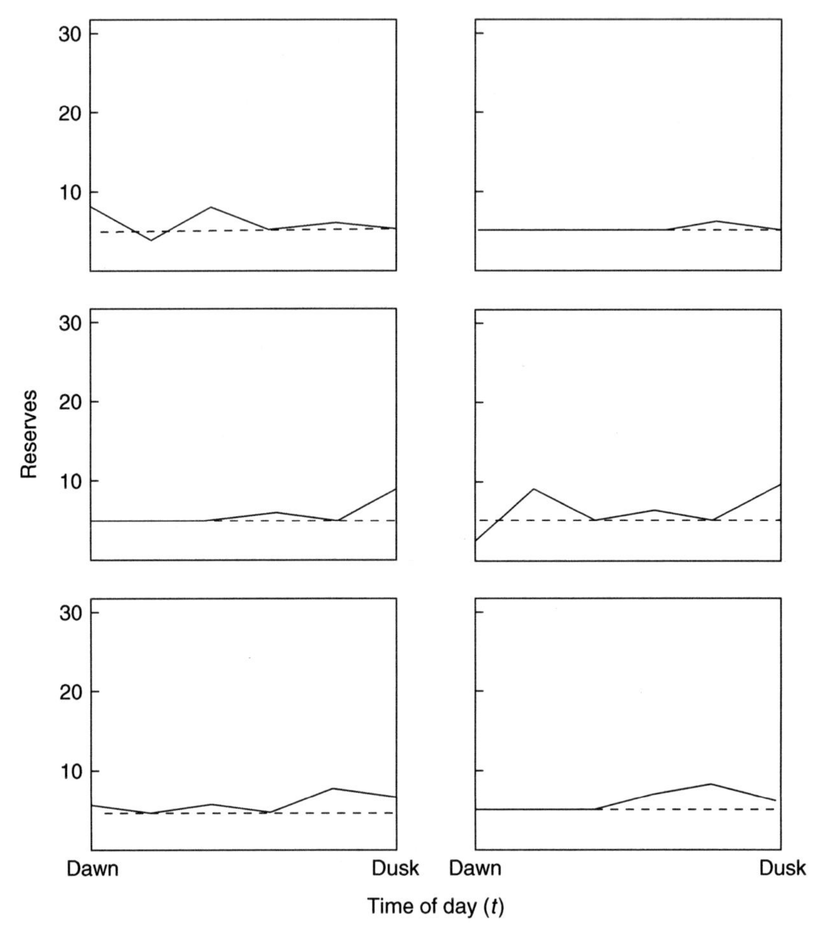

Figure 1.4. Evolved policies for singing vs. foraging after 100 generations from six runs of the SBGA. The evolved policy $C(t)$ is given by a solid line and the optimal policy C^* (which is the same for each graph) is given by the broken line. The parameter values were: $\mu_N = 0$; $\sigma_N = 0$; and $IP = 0.0$. The SBGA parameters were: mutation rate = 0.01; crossover probability = 0.5; migration rate = 0.5 migrants per deme per generation. A population of 36 chromosomes was divided into six demes of six chromosomes. Mortality selection, with two deaths per deme per generation, was used for all runs.

Discussion

Viewed as an optimization procedure, GAs have certain advantages over usual methods of optimization. Holland[8] and De Jong[19] argue that GAs perform better than many standard optimization procedures when the payoff function is discontinuous or has multiple peaks. GAs are flexible in that using the same encoding, they can track changes in the payoff function. Holland[8] saw this as

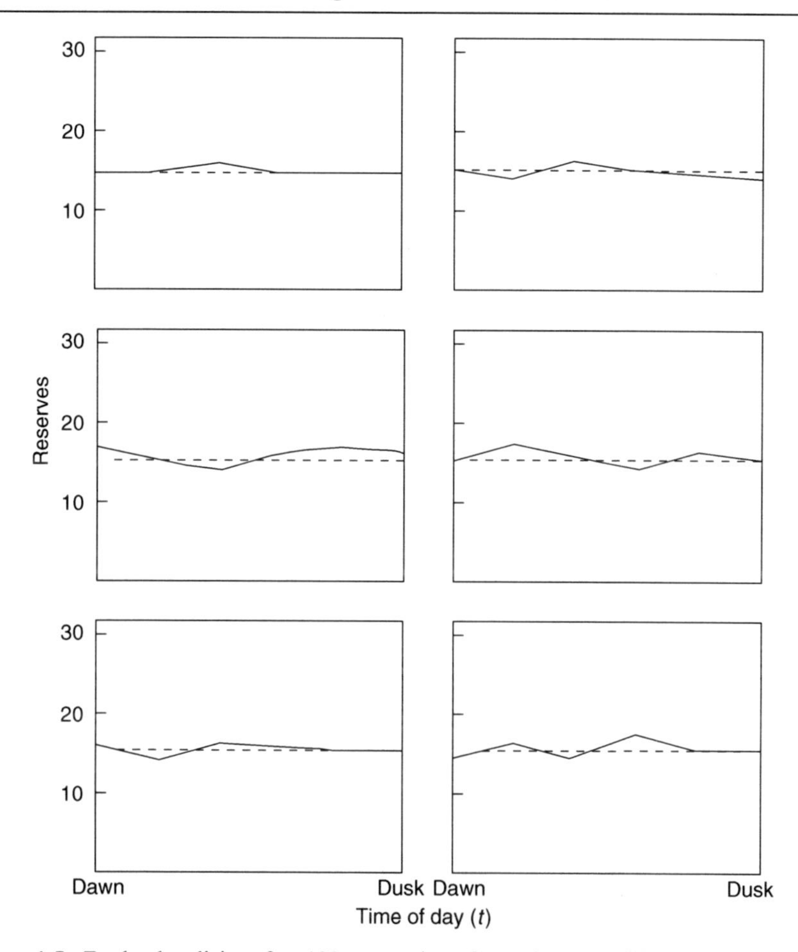

Figure 1.5. Evolved policies after 100 generations from six runs of SBGA. The evolved policy $C(t)$ is given by a solid line and the optimal policy C^* (which is the same for each graph) is given by the broken line. The parameter values were: $\mu_N = 0$; $\sigma_N = 0$; and $IP = 0.4$. The SBGA parameters are as for Fig. 1.4.

a necessary feature of an adaptive search procedure. The GAs have their basis in biology, and one can use biological insight to improve their performance. Motivated by Wright's ideas, we have introduced a modified GA in which the population is divided into semi-isolated demes. We have shown that there are circumstances in which this SBGA performs better than a single population GA. Further work needs to be done to investigate the general effect of this modification and the other modifications that we have introduced (e.g. non-overlapping generations, crossover frequency and proportional vs. non-proportional migration) on the search efficiency of GAs.

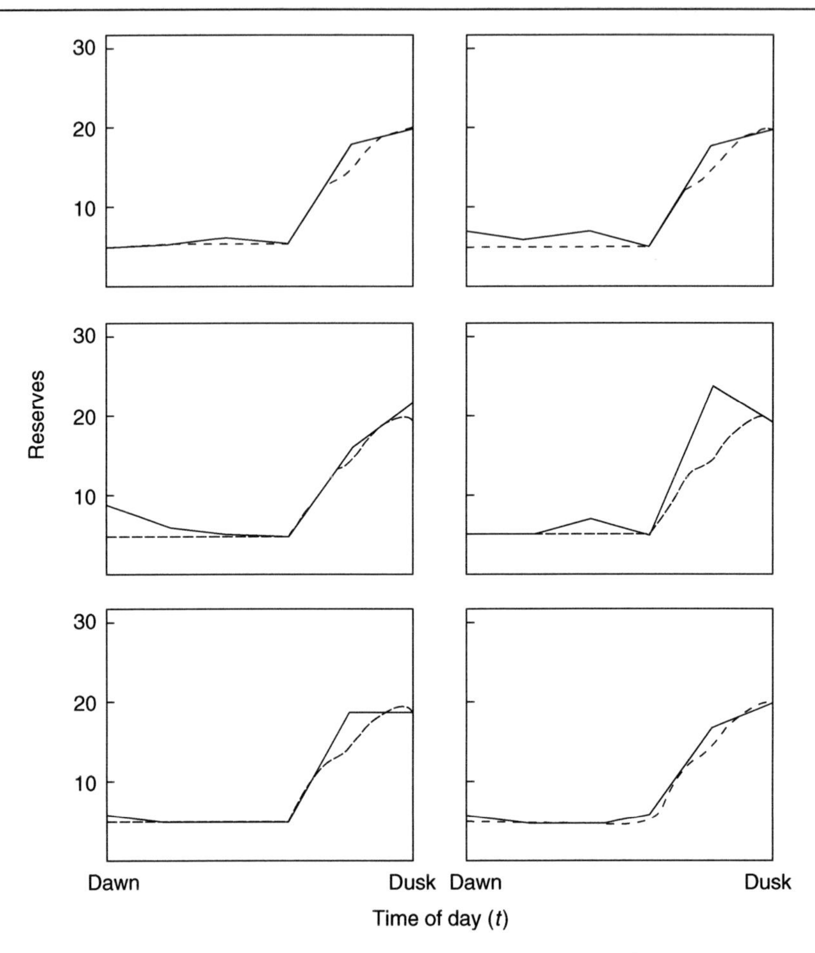

Figure 1.6. Evolved policies after 100 generations from six runs of SBGA. The evolved policy $C^*(t)$ is given by a solid line and the optimal policy C^* (which is the same for each graph) is given by the broken line. The parameter values were: $\mu_N = 12$; $\sigma = 4$; and $IP = 0.0$. The SBGA parameters are as for Fig. 1.4.

We applied the SBGA to cases in which dynamic programming had already been used.[14] The SBGA found near-optimal solutions within a short period of time (within 100 generations). This performance is impressive compared to a procedure based on exhaustively searching the $32^6 = 1\ 073\ 741\ 824$ possible chromosomes. But in this one-dimensional case, dynamic programming is a more efficient procedure. The computational costs of dynamic programming rise exponentially as the dimensionality of a problem increases ('the curse of dimensionality', Bellman,[49]), whereas for the GA, an increase in the

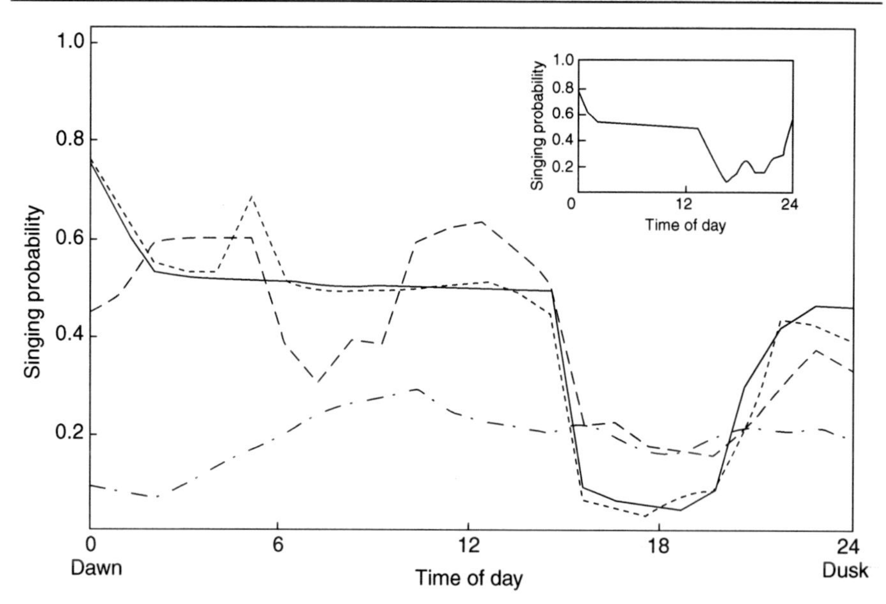

Figure 1.7. Dynamics of the evolution of behaviour. The ordinate plots the behaviour, as probability of singing, and the abscissa gives the time of day. Behaviour, calculated using the forward procedure, was plotted after 1 (------); 20 (— · —); 50 (– – –); and 100 (——) generations of evolution. Environmental parameters were $\mu_N = 12$, $\sigma_N = 4$ and $IP = 0.0$. The inset shows the routine of singing under the optimal policy.

dimensionality of the problem need not substantially increase the computational costs. It follows that GAs are likely to be far more efficient than dynamic programming in high-dimensional problems.

As well as using GAs as efficient search procedures, they can also be used to shed light on evolution in biological systems. Not only do they provide a convenient way to find an optimum, but they also make it plausible to assume that evolution could reach this optimum too. It must be remembered, however, that what the GA can evolve is constrained by the scheme of encoding. On the one hand, this can be seen as a disadvantage, in that the optimum is dependent on the genetic system. On the other hand, it can be seen as getting around a criticism of optimality techniques in biology. It has been pointed out that finding optimal solutions ignores genetic constraints; GAs make it possible to investigate the effects of genetic constraints on optimal solutions.

The GAs provide a natural setting for investigating evolution in systems with epistatic interactions between genes. Such interactions are present in both the examples discussed in this paper. In the case of singing and foraging, there are epistatic effects because the fitness consequences of an action at one time of day depend on the actions performed at other times of day.

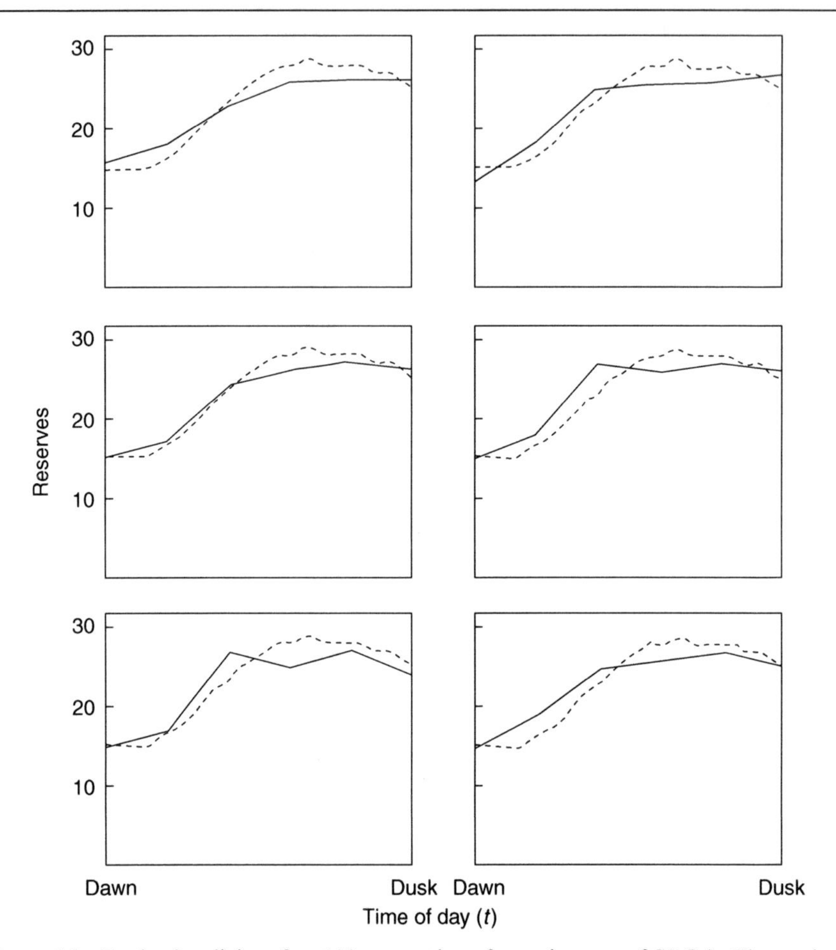

Figure 1.8. Evolved policies after 100 generations from six runs of SBGA. The evolved $C(t)$ is given by a solid line and the optimal policy $C^*(t)$ (which is the same for each graph) is given by the broken line. The parameter values: $\mu_N = 12$; $\sigma_N = 4$; and $IP = 0.4$. The SBGA parameters are as for Fig. 1.4.

At a higher level the GA allows one to explore the efficiency of evolution by natural selection as a search mechanism, and in particular to discover which features make the search effective and which features are detrimental.[50]

We now give some examples of potential applications of GAs in biology.

(i) Game theoretic interactions between animals. Houston & McNamara[51] modified the model of McNamara *et al.*[14] (1987) to include interactions between singing males. By considering an infinite population limit, they were able to restrict attention to policies that depended only on a male's level of reserves and time of day. Even so, this dynamic game was computationally much more

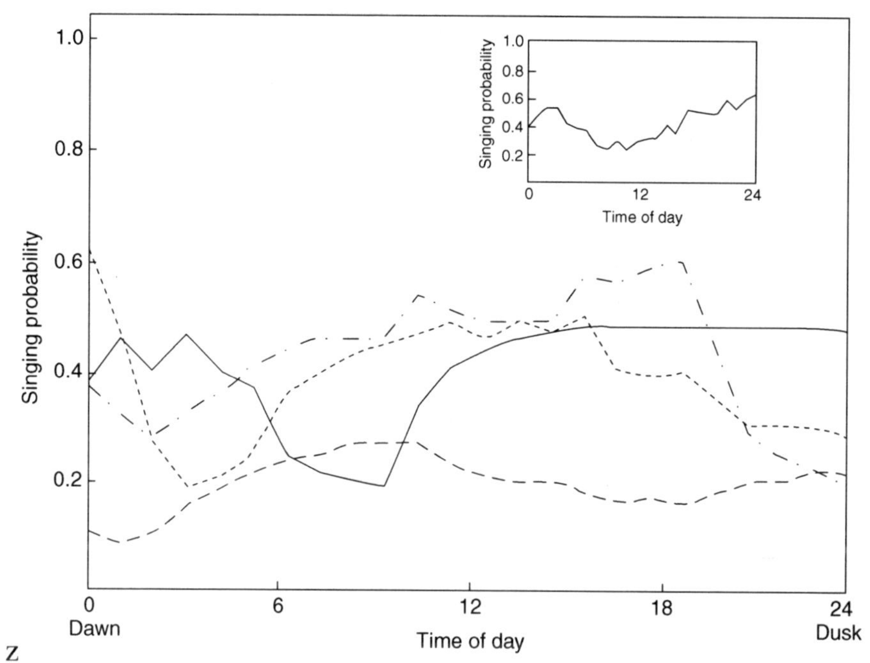

Figure 1.9. Dynamics of the evolution of behaviour. The ordinate plots the behaviour, as probability of singing, and the abscissa gives the time of day. Behaviour, calculated using the forward procedure, was plotted after 1 (---), 20 (— · —), 50 (– – –) and 100 (———) generations of evolution. Environmental parameters were $\mu_N = 12$, $\sigma_N = 4$ and $IP = 0.4$. The inset shows the routine of singing under the optimal policy. Key as for Fig. 1.7.

expensive than the optimization problem faced by a single male. Any reasonable model based on a finite population would have to allow a male's behaviour to depend on the previous behaviour of the other males in the population. Incorporating additional state variables representing this information would render an approach based on dynamic programming completely impractical. One could, however, use GAs to encode the way in which a male responds to the previous behaviour of other males and hence, analyse the action of evolution on such systems.

(ii) Learning rules. There has been considerable discussion of rules which are efficient at exploiting a particular environment.[52–55] Such analyses usually assume that the parameters of the environment are known to the animal. Many animals are constantly exposed to new environments, and so a full account of behaviour needs to consider a higher level of rule which is capable of both learning about, and exploiting a range of *a priori* unknown environments. The GAs provide an obvious and powerful framework for investigating the evolution of such learning rules.

Table 1.5. Evolved policies from replicate runs along with their final fitnesses under four environmental conditions. Genetic algorithm parameters for all runs are: mutation probability = 0.01, mortality = 2, deme number = 6, local population = 6, migrants per generation = 0.5, crossover probability = 0.5 and number of generations = 100

Optimal fitness	Evolved fitness	Evolved policy					
(a) Condition: $\mu_N = 0$, $\sigma = 0$ and $IP = 0.0$							
1.5180	1.5178	8	4	8	5	6	5
	1.5180	5	5	5	5	6	5
	1.5179	5	5	5	6	5	9
	1.5178	2	9	5	6	5	9
	1.5178	6	5	6	5	8	7
	1.5178	5	5	5	7	8	6
(b) Condition: $\mu_N = 0$, $\sigma = 0$ and $IP = 0.4$							
1.5100	1.5100	15	15	16	15	15	15
	1.5100	15	14	16	15	15	14
	1.5100	17	15	14	16	17	16
	1.5100	15	17	15	14	16	15
	1.5100	16	14	16	16	15	15
	1.5100	14	16	14	17	15	15
(c) Condition: $\mu_N = 12$, $\sigma = 4$ and $IP = 0.0$							
1.4488	1.4486	7	6	7	5	18	20
	1.4487	5	5	6	5	18	20
	1.4486	5	5	7	5	24	19
	1.4485	9	6	5	5	16	22
	1.4487	6	5	5	6	17	20
	1.4487	6	5	5	5	19	19
(d) Condition: $\mu_N = 12$, $\sigma = 4$ and $IP = 0.4$							
1.4367	1.4364	13	18	25	26	26	27
	1.4364	16	18	23	26	26	26
	1.4364	15	18	27	26	27	26
	1.4364	15	17	24	26	27	26
	1.4364	15	19	25	26	27	25
	1.4360	15	17	27	25	27	24

The GAs were originally developed as an artificial intelligence technique. A simple view of evolution by natural selection was used to provide an efficient search procedure. We have now come full circle, where what has been learnt about the GA can be used, in a defined type of strategy-analytic simulation, to improve our understanding of the evolution of natural populations.

Acknowledgements

B. H. Sumida was supported by S.E.R.C.; A. I. Houston was supported by S.E.R.C., N.E.R.C. and King's College, Cambridge. We thank the following for helpful comments and criticisms: Andy Barto, Sean Nee and Andrew Pomiankowski. Joy Bergelson provided advice on statistical analysis.

References

1. R. Alexander, McN. *Optima for Animals*. London: Edward Arnold (1982).
2. J. Maynard Smith, *Ann. Rev. ecol. Syst.* **9**, 31–56 (1978).
3. N. S. Goel, and R. L. Thompson, *Int. Rev. Cytol.* **103**, 1 (1986).
4. G. F. Oster, and E. O. Wilson, *Caste and Ecology in the Social Insects*. (Princeton University Press, Princeton, NJ, 1978).
5. B. Borstnik, D. Pumpernik, and G. L. Hofacker, *J. theor. Biol.* **125**, 249–268 (1987).
6. M. Eigen, and P. Schuster, *The Hypercycle—A Principle of Natural Self-organization*. (Springer-Verlag, Berlin, 1979).
7. S. J. Gould, and R. C. Lewontin, *Proc. R. Soc., Lond. B*, **205**, 581–598 (1979).
8. J. Holland, *Adaptation in Natural and Artificial Systems*. (University of Michigan Press, Ann Harbor, 1975).
9. Q. Wang, *Biol. Cybern,* **57**, 95–101 (1987).
10. R. Galar, *Biol. Cybern,* **60**, 357–364 (1989).
11. D. G. Bounds, *Nature, Lond.* **329**, 215–219 (1987).
12. S. Wright, *Evolution and Genetics of Populations*, Vol. **3**. *Experimental Results and Evolutionary Deductions*. (University of Chicago Press, Chicago, 1977).
13. A. Kacelnik, and J. R. Krebs, *Behaviour* **83**, 287–309 (1982).
14. J. M. Mcnamara, R. H. Mace, and A. I. Houston, *Behav. ecol. Socio.* **20**, 399–405 (1987).
15. A. S. Fraser, *J. theor. Biol.* **2**, 329–346 (1962).
16. H. J. Bremmerman, In: *Self-Organizing Systems* (Yovits, J. and Goldstein, G. D., eds) (Spartan Books, D.C. Washington, 1962).
17. A. Jacquard, *The Genetic Structure of Populations*. (Springer-Verlag, Berlin, 1974).
18. R. M. Brady, *Nature, Lond.* **317**, 804–806 (1985).
19. K. De Jong, *IEEE Trans. on Sys. Man and Cyber.* **9**, 566–574 (1980).
20. P. Schuster, Optimization and complexity in molecular biology and physics. In: *Optimal Structures in Heterogenous Reaction Systems* (Plath, P. J., ed.) (Springer-Verlag, Berlin, 1989).
21. L. Davis, *Genetic Algorithms and Simulated Annealing*. (Pitman, London, 1987).
22. D. S. Goldberg, *Genetic Algorithms*. Addison-Wesley, Massachusetts (1989).
23. R. Axelrod, In: *Genetic Algorithms and Simulated Annealing* (Davis, L., ed.) Pitman, London (1987).
24. J. D. Watson, N. H. Hopkins, J. W. Roberts, J. A. Steitz, and A. M. Weiner, *Molecular Biology of the Gene*, 4th edn. Benjamin/Cummings, Menlo Park (1987).
25. B. Wallace, *Evolution* **29**, 465–473 (1975).
26. M. Kimura, and J. F. Crow, *Proc. natn. Acad. Sci. U.S.A.* **76**, 396–399 (1979).
27. M. Kimura, and J. F. Crow, *Proc. natn. Acad. Sci. U.S.A.* **75**, 6168–6171 (1978).
28. A. S. Kondrashov, *J. theor. Biol.* **107**, 249–260 (1984).
29. S. Wright, *Anat. Rec.* **44**, 287 (1929).
30. W. M. Schaffer, and M. L. Rosenzweig, *Ecology* **58**, 60–72 (1977).

31. T. D. Price, P. R. Grant, H. L. Gibbs, and P. T. Boag, *Nature, Lond.* **309**, 787–791 (1984).
32. S. Wright, *Evolution and Genetics of Populations*, Vol. 2. *The Theory of Gene Frequencies*. (University of Chicago Press, Chicago, 1969).
33. J. R. G. Turner, *Am. Nat.* **105**, 267–278 (1971).
34. M. Conrad, *Biosystems* **11**, 167–182 (1979).
35. W. B. Provine, *Sewall Wright and Evolutionary Biology*. (University of Chicago Press, Chicago, 1986).
36. R. A. Fisher, *The Genetical Theory of Natural Selection*. Clarendon Press, Oxford. 2nd edn. 1958: Dover Publications (1930).
37. J. F. Crow and M. Kimura, *An Introduction to Population Genetics*. Burgess, Minneapolis (1970).
38. M. Kimura, *Gen. Res.* **11**, 247–269 (1968).
39. J. L. King, and T. H. Jukes, *Science* **164**, 788–798 (1969).
40. J. F. Crow. In: *Proceedings of the XII International Congress on Genetics* **3**, 105–113 (1969).
41. S. Wright, In: *Mathematical Topics in Population Genetics* (Kojima, K., ed.) (Springer-Verlag, New York, 1970).
42. J. Felsenstein, *Ann. Rev. Genet.* **10**, 253–280 (1976).
43. S. Wright, *Ecology* **26**, 415–419 (1945).
44. S. Wright, *Am. Nat.* **131**, 115–123 (1988).
45. P. A. P. Moran, *The Statistical Process of Evolutionary Theory*. (Clarendon Press, Oxford, 1962).
46. J. R. Krebs, In: *Evolutionary Ecology* B. Stonehouse, and C. M. Perrins, (ed.) (McMillan Press, London, 1977).
47. R. Mace, *Anim. Behav.* **34**, 621–622 (1986).
48. J. M. Mcnamara and A. I. Houston, *Am. Nat.* **127**, 358–378 (1986).
49. R. Bellman, *Dynamic Programming*. (Princeton University Press, New Jersey, 1957).
50. J. J. Grefenstette, *IEEE Trans. on Sys. Man and Cybern.* **16**, 122–128 (1986).
51. A. I. Houston, and J. M. Mcnamara, *J. theor. Biol.* **129**, 57–68 (1987).
52. R. F. Green, *Theor. pop. Biol.* **18**, 244–256 (1980).
53. Y. Iwasa, M. Higashi, and N. Yamamura, *Am. Nat.* **117**, 710–723 (1981).
54. J. M. McNamara, *Theor. pop. Biol.* **21**, 269–288 (1982).
55. J. M. McNamara and A. I. Houston, In: *Quantitative Analyses of Behavior* Vol. **6**, *Foraging* Commons, M. L., Kacelnik, A. K. and Shettleworth, S. J., (ed.) pp. 23–39, (Lawrence Erlbaum, New Jersey, 1987).

BOTH WRIGHTIAN AND 'PARASITE' PEAK SHIFTS ENHANCE GENETIC ALGORITHM PERFORMANCE IN THE TRAVELLING SALESMAN PROBLEM†

BRIAN H. SUMIDA AND WILLIAM D. HAMILTON

Orestes: You don't see them, you don't—but I see them:
they are hunting me down, I must move on.
Choephoroi (from *Sweeney Agonistes* (T. S. Eliot))

Background

A genetic algorithm (GA) is a heuristic optimization technique based on natural evolution.[1] Although GAs are executed with the aid of a computer, the GA itself is inspired by principles derived from both computer science and biology. Hence there are two potential sources from which to seek insights towards improving the performance of GAs. A particular GA implementation may be improved in the first instance by seeking more efficient algorithms (e.g. for sorting) and secondly, by deriving more appropriate evolutionary models.[2] In this chapter we describe a GA based on a new evolutionary model which, by adding parasites and structured populations in the model, greatly improves the performance on a particular highly multi-peaked, adaptive landscape, that of the travelling salesman problem (TSP).

Parasites as potential stimulants of evolution have been the theme of recent theoretical[3–5] and computational[6,7] Collins, in this volume) studies. Parasites

† In R. Paton (ed), Computing with Biological Metaphors, pp. 264–279. (Chapman and Hall, London 1994).

have also been credited as a key factor for the maintenance of sex,[6,8] and, as Holland realized much earlier, sex, apart from natural selection itself, is the most important process that GAs use.

Populations of organisms which inhabit the natural world are usually clustered into groups of individuals. For a population of a given size, having its members compete and mate largely within groups greatly alters its properties under natural selection. Sewall Wright[9] was the first to give serious attention to this fact and developed a body of theory describing what he called a 'shifting balance' referring to alternative adaptations and levels of adaptation among semi-isolated populations. In the important case for this theory local populations, or demes, interbreed freely within themselves but only occasionally transfer individuals from one deme to another. Such partial isolation is highly preservative of global population variation[10] and from the point of view of evolution becomes most creatively so if the demes are small because isolation facilitates random genetic drift.[11] Since drift is a non-directional process, in the ensemble all directions of drift are experienced, including, at times, different allele and genotype fixations; hence the ensemble stores variability. Adding natural selection Wright saw that (a) different adaptive peaks would tend to be attained in different demes, and (b) by a combination of drift and migration, demes finding superior peaks could transform neighbouring demes to their discovery. Independently, Holland[1] conceived of evolutionary 'building blocks' arising in GAs which, in a context of demes, means that one deme might discover one part of a superior adaptation (a strong opening in chess, say) while another deme might discover another complementary part (a strong end game). In the meta-population (i.e., the aggregate of demes) interdemic migration and subsequent intrademic recombination could bring the two parts together.

The Travelling Salesman Problem (TSP)

The TSP, although difficult to solve, is easy to describe. A travelling salesman provided with a list of cities must visit each city once only and end the tour at the starting city. The cost of a journey between any two cities is known and the salesman's objective is to organize a tour permutation which incurs the least cost. Given a tour containing n cities, the number of possible permutations is $n!$. It is obvious that an exhaustive search for the optimal permutation is impossible for even moderately sized (~50) values of n. The problem is inherently extremely multi-peaked and at many levels, so eliminating classical optimization methods. Such 'ruggedness' of landscape arises because there are many radically different approaches to touring cities, both within any cluster and also for touring between the clusters. Thus, should Oxford, Banbury and Aylesbury be visited while touring up the east side of England or when touring down the west? Are they indeed best treated as a cluster, neighbours to be visited in some close sequence? If a salesman decides to change his western tour to include these, it is almost certain he will need to adjust other parts of both eastern and western limbs of his total tour to say nothing of changing order within the three mentioned cities themselves. Once a fairly efficient tour has been devised all simple 'cut and paste' operations of segments of tours are most likely to be harmful even though they may suggest, to an intelligent viewer, dramatically improved tours that could follow. More explicit examples of such inherent ruggedness peaks will appear later. Non-GA optimization methods that improve on blind exhaustive search by eradicating blocks of alternatives that have no chance to be successful are available. These methods, however, are still extremely tedious. Using them the current record for the longest solved TSP of 3038 cities[12] was obtained. The unsolved and perhaps unsolvable difficulty of the TSP provides a challenging test bed for the GA to prove its effectiveness. A concise review of the application of GAs to the TSP can be found in ref. 13.

Tour Representation, Host Sexuality and Host Fitness

A novel chromosomal representation of travelling salesman tours is presented. A given tour's representation comprises two arrays: a next city array and a previous city array. In each the array index corresponds to the city name. Actually one of the two arrays is sufficient to define the tour but there is an advantage of clarity in the algorithm as so far developed to maintain both.

Both arrays are used in sexual reproduction to form hybridized offspring tours. In each deme, for each offspring to be produced, the individual with the best tour becomes one parent and a random deme member the other parent. If there are to be further offspring, this procedure is repeated exactly—the best tour becomes in effect polygamous. An offspring tour is created from a pair of parents by an act resembling bacterial conjugation: the best tour parent donates a subtour (a sequence of city visits) from itself to be inserted into a copy of the tour of the other parent. The transformed copy is the offspring.

The first step in the insertion procedure is to pick a random start city, SD, in the donor and a random subtour length, l (i.e., number of cities, $0 < l < n$ where n is total number of cities). The second step is to pick a random city, SR, in the recipient tour. SD is to be the first novel city visited in the transformed recipient tour, immediately following city SR. Thus the 'next' of SR in the recipient tour is changed to SD and the 'previous' of SD is changed to become SR. What about the next of SD itself? Soon it will be the next city, if there is to be one, from the donated segment; but in case there is not to be one, and to keep tours legal at every point of the insertion procedure, what is done immediately is to make the next of SD into what was the next of SR and to make the previous of that next into SD.

An example may help to clarify this procedure. Suppose that one route being tried in a particular deme, let us call it R (denoting 'recipient'), around British cities includes the sequence: → Manchester, Shrewsbury, Gloucester →. Another route, D (denoting 'donor'), in the same deme has a sequence: → Leicester, Wolverhampton, Birmingham, Nottingham, Sheffield → and let us suppose that this tour is the best (i.e., shortest) tour yet. The R and D tours happen to be selected as parents for a 'mating' and in the random city drawings for the mated pair SR chances to be Shrewsbury and SD to be Wolverhampton. The procedure described above declares that for the second tour R, in two steps, the next of Shrewsbury and previous of Gloucester are to become Wolverhampton (a change that happens to look rather good on the map). But the R tour, of course, was already visiting Wolverhampton at some other stage. Suppose it was in a sequence: → Coventry, Wolverhampton, Oxford →. Since R must not visit Wolverhampton twice, this section of its tour is now made to be simply: → Coventry, Oxford →. (This 'bypass' also happens to look good, but that is only because we have chosen it to show how improvements can occur.) Reverting to the language of the arrays, the final step involves the next of SD's former previous and the previous of SD's former next which must be changed to specify each other. Once all this is complete city SD has been 'inserted' to follow SR and the recipient still has a legal tour encompassing all the cities. The algorithm can now proceed in the same way to insertion of the second city of the donated subtour (in the example it would be Birmingham) and so on onward until all l cities have been inserted.

For clarity in programming the algorithm, it was found convenient to think in terms of operations that were called 'arrows', an arrow being the creation of a step from one city to another. An arrow may itself be thought of as comprising two 'half arrows', these being simply the outward bound 'next' of the arrowtail city and a corresponding inward bound 'previous' of the arrow-head city. In these terms, the insertion of one novel city as described above can be accomplished (given also a few storage additions to make sure one is not using values already altered) by calling a function of the form 'arrow(A,B)' three times, which effects a step A → B. The three steps can be described as 'moving in', 'continuing' and 'bypassing old position' as above.

For simplicity the description so far has assumed that the inserted segment of donor tour keeps its original direction. We saw no biological reason why this order should be kept, however, and saw certain advantages for the TSP that could be gained if it were not. Therefore we arranged that, with $p = 0.5$, the original direction would be used, otherwise the segment of l cities would be read 'backwards' along the donor tour from the start at city SD. This is easily done by referring at a certain point of the program to elements of the 'previous' array instead of the elements of the 'next' array. The obvious advantage for the TSP is that inserting self-segments backwards can 'untwist' geometrical crossovers in the tour since the presence of such sequences can never be optimal. Insertion of sequences already present in the recipient (i.e., self-segments) of course occurs with increasing frequency as demes tend to monomorphism.

The set of values in, say, an offspring's next array is always a permutation of the cities. But to be a TSP tour it has to be a permutation of a restricted kind, otherwise detached subtours and immobile salesmen will occur. This however is easily arranged at the start of the GA by creating a set of differing full random tours and then ensuring, as in the three-step insertion procedure above, that every transformation effected on a tour creates a full tour.

A host's primary fitness is obtained by summing the distance travelled whilst traversing the tour specified by the chromosome—fitness increases with decreasing tour length. Primary fitnesses yield secondary fitnesses by increasing host tour length (i.e., decreasing fitness) by a quantity reflecting the success of parasitism. Secondary fitnesses are used to allocate reproductive parents and to determine who dies. As already stated the individual with the highest secondary fitness in the deme is always selected as one member of each mating pair. This mating advantage effects part of the Darwinian selection driving the evolutionary process. Mortality, the other common aspect of selection, is represented by overwriting the individuals with the lowest fitnesses with newly created offspring.

Parasite Tours and Parasite Sex

Parasite chromosomes are the same length as host chromosomes but a parasite tour differs in that not every city need be visited. Indeed it is common for a parasite tour to visit only a few cities. However, a single-city parasite would always match and we felt that even a two-city matching parasite would too easily evolve. The GA needs time to consolidate its discoveries and to export them from the demes making the discovery before being forced by parasitism to 'doubt' and so to change them. Small matching parasites may reduce the value of new features almost from the moment of their arrival. From these considerations we decided that a parasite tour could be any legal closed TSP tour or subtour subject to $n \geq 3$. Cities not visited are marked on the chromosome by having their next and previous array values set to a null value for which we used -1. Since these null parts of the parasite's chromosome do not need to be handled in most of the operations involving parasites their effective chromosomes are shorter and their replications are carried out speedily. Parasites can become small or large in the course of evolution. As it turns out most of the time they are on average small and we feel that this is life-like and potentially beneficial to the GA (see below). In contrast, antagonistic organisms of equal size to the hosts such as are used by Hillis[7] and Collins (this volume) seem to us more aptly described as coevolving predators.

At the stage of applying parasites, random hosts and random parasites from their respective demic populations are paired and out of the pairing a parasite fitness is calculated which is then deducted from host fitness, as mentioned above. A parasite's fitness bears no relation to the geometrical length of its tour. Instead it depends upon the sequences of cities that the parasite manages to imitate in the tour of the host it is paired with. All 'arrows' of the host tour duplicated by the attached parasite are noted in the program and each such arrow is a potential contribution to fitness; but in the version we have used so far we count only the longest unbroken sequence of corresponding arrows as the parasite's primary fitness. The parasite's secondary fitness was taken to be this count divided by the total number of cities. In such a scoring system, 'short' parasites should find it relatively easy to mimic a host along most of their length; however, such parasites do not contain as much potential for pay-off as longer parasites have. We wanted short parasites to be present for speed in the early and very changeable stages of host evolution but also to allow for the possibility of a slow acquisition by the parasites of long mimetic sequences that reflected major elements of TSP strategy. These long sequences would correspond to the chromosomes of hosts perched on the well-stabilized adaptive hilltops. The essence of using parasites in the GA is that their effect should be gently to lower and flatten achieved peaks so that eventually the slopes of neighbouring peaks can be discovered.[14,15]

In order to allow parasite size to evolve we arranged that when parasites mated two off-spring would result. The procedure of selecting parents, one best and the other random, was the same as for hosts and the procedure of donating a segment of tour from this donor to create offspring by modifying a copy of the recipient's tour was also the same except that the 'bypass' operation could often be skipped, an omission keyed in the algorithm by discovery of '-1' values in the arrays. Each time the bypass operation is omitted, the recipient subtour grows in length by one city. While this was going on, differently than with hosts, a copy of the donor tour was simultaneously modified by elimination of the cities it donated. 'Next' and 'previous' array elements of all the eliminated cities were set to -1 and the cities bounding the gap connected by an arrow. This is a 'bypass' operation similar to that already described for a recipient host tour but nevertheless is quite distinct from it. That former kind of bypass may be made in a recipient parasite tour also, as already mentioned, but (a) it never involves more than one city at a time, and (b) it does not shorten the recipient tour. The present operation on a parasite donor is an elimination rather than a bypass and has the result that every act of mating usually produces one augmented tour and always produces one diminished tour, as compared to what had existed before. Either of the new tours may turn out to be an improvement in the context of the current host population. Thus mean parasite length might evolve up or down. With this procedure we felt we provided both speed when variability was high in the host population and the parasite's target very changeable, and effective discovery and harassment of hosts when subpeak host adaptations were becoming stabilized: parasites as long as necessary to damage hosts seriously would evolve to discourage any spreading conviction that a best tour had been found.

An illustration of the parasite pursuit arising from our variable parasite length model is given in Fig. 1.10. In this figure we can observe the evolving

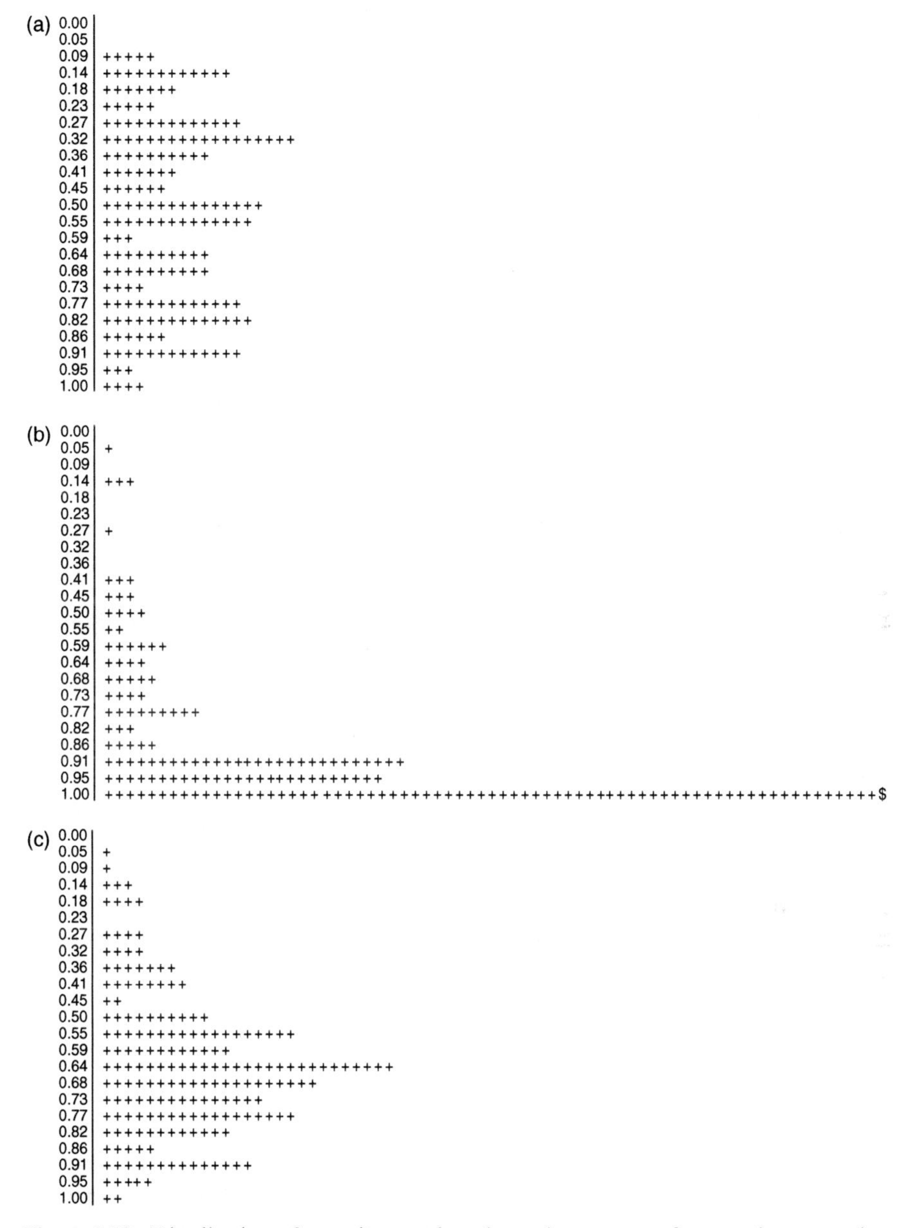

Figure 1.10. Distribution of parasite tour lengths at three stages of a run where parasites are present in a single deme. When a parasite tour length is equal to a host tour length it receives a value of '1.0'. The 0th bin represents parasite tours of lengths up to 5% of the host tour. Each '+' symbol represents an individual of the total population; when the limit of a bin is reached (72 individuals) an overflow symbol, '$', is printed (Fig. 1.10(b)). Population size for both hosts and parasites was 195 individuals in a single deme. Host mortality and parasite mortality was 16 and 40 deaths per generation respectively. (a) gives distribution at the start of the simulation; (b) is the distribution after 500 generations; and (c) is the distribution after 1500 generations.

distribution of parasite chromosome lengths as the parasite attempts to mimic the host chromosome. The distribution of parasite chromosome lengths is initially flat (Figure 1.10(a)) due to the random start. As the hosts evolve, selection finds and preserves the fittest individuals, thus carrying the population towards the nearest peak in the fitness landscape. As the peak is approached the population becomes more uniform. A uniform population presents the parasites with a static target and, consequently, parasites quickly evolve to imitate the chromosome characterizing the peak and in doing so grow long (Figure 1.10(b)). Increased success by parasites is matched by decreasing host fitness which in turn eventually gives advantage to residual variants and new suboptimal mutants. These residuals, mutants, and their combinations with each other, along with the current peak genotypes, explore the valleys and saddles around the current peak. When the upward slope of another peak is discovered and that peak begins to be 'colonized', a peak shift follows accompanied by a large increase of both host and parasite variability. The latter manifests itself in part as a decrease in mean length of parasites and an increase in their length variance. Both result from the fact that during the peak shift long mimetic sequences in the parasite are fractured and there is therefore no longer an advantage in being long. Figure 1.10(c) illustrates a final state where the best peak has been flattened by successful parasitism but no next ascending slope has been found. In this terminal state considerable variability is again to be expected. Indeed to identify the final best tour/peak genotype it is necessary to switch off parasite action and continue the run in order to eliminate the suboptimal, opportunistically probing genotypes the parasites are still causing to be created. Such a clearance of parasites probably sometimes occurs in the real world when a host species colonizes an oceanic island or a new continent. Populations on islands are sometimes very innovative; an example of this might be the finches that uniquely use cactus spines as tools to extract insect larvae from dead trees in the Galapagos Islands.[16]

Deme Structure and Migration

As already implied, matings, selection and reproduction occur between individuals within demes. In our GA simulation the demes are spatially arranged as a toroidal world forming a continuous lattice with no barriers. Movement of individuals between demes is possible but constrained to create the condition of semi-isolation. In our model, individuals are dispersed through the toroidal start world by two classes of migration. In nearby migration an individual may migrate into any one of the eight adjacent 'near' demes with equal probability. Far migration differs from this in that an individual is allowed to migrate to any deme in the toroidal world with equal probability as in Wright's 'Island' model.[10] As in the natural world, migration to and from nearby demes in the toroidal world was made much more likely than a distant migration. The frequency of interdemic migration strongly affects gene distribution in the population.[17] If set too high or too low, we regress either to one large deme or to many isolated small ones. In both cases the benefit to the GA of having population structure is nullified.[18,19]

The population structure of our model superficially resembles the models of Hillis[7] and Collins, although in the absence of precise detail of their population handling we cannot say how functionally or demographically equivalent they are. Our implementation discussed in this chapter is clearly distinguished from theirs by our use of much smaller population size (hundreds of individuals as compared to tens of thousands) and by the presence of far migration. Obviously GAs themselves should evolve to be efficient 'organisms' in their intended niche and if a certain degree of difficulty of TSP can be overcome by a program that is economical in its code, storage requirements and CPU time, then this bodes well for its success on bigger problems when more time and storage are available. Such extension of efficiency probably applies most strongly, however, if merely tours of a given standard of efficiency and not ultimate optimal tours are being sought. There seems to be no escape from the rapid increase of difficulty of the strict optimal tour problem as the number of cities increases. For the strict TSP applying to very many cities, some degree of 'concept formation' and of 'appreciation of the pattern of the whole' seem necessary. Our program is certainly not yet at a stage of such abilities.

Outcomes using Parasites and Population Structure

An experiment was run to determine whether adding parasites and partitioning the population into demes aided the GA. We tested four conditions: (1) no population structure and no parasites; (2) a structured population but no parasites; (3) no population structure but parasites present; and (4) a structured population with parasites.

A population of 192 individuals was divided into 16 demes each containing 12 individuals. Truncation selection (see ref. 17 for definition) at the lower end of the secondary fitness ordering was used with the mortality in each deme set to two deaths for the hosts and six deaths for the parasites in every generation. This implies that in each host's deme the two worst individuals are killed and the best host individual is twice made a donor parent whereas in each parasite deme the six worst individuals are killed with the best parasite individual made parent three times (recall that parasite matings give two offspring each). This different mortality level for the parasite population is one of two ways in which selection on parasites is made stronger. The other is that parasites are given a generation time half of that of the host. In detail the last assumption means that after a generation is run in which both parasites and hosts undergo selection, parasites are run again in a generation in which hosts do not change; then both change again, and so on throughout.

Nearby migration and distant migration probabilities for both hosts and parasites per individual per generation were set at 0.01 and 0.001 respectively. Ten replicates for each condition were run, each run lasting 5000 generations.

Results

The evolution of the population towards the optimal tour is shown in a series of tour graphs (Fig. 1.11–1.14). The best tour from each of the sixteen demes is plotted in a subdivided square, with the best overall tour outlined. Figure 1.11 shows the population at the start of the simulation. Evolution has not begun at generation zero so the best tour here is merely the best of the random initial tours. After 1500 generations, in Fig. 1.12, we can see several strategies for visiting cities emerging. It can also be observed from the figure that tours in adjacent demes, due to the mainly local migration, are beginning to exhibit

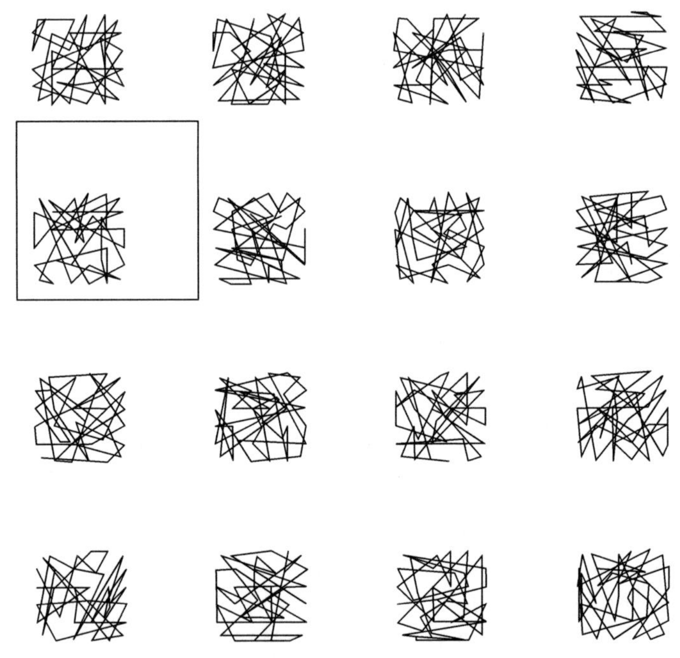

Figure 1.11. Best host tours from each of 16 demes. The overall best tour is surrounded by a white box. The population of 192 was arranged into 16 demes of 12 individuals and there were two parasite generations per host generation. This figure is for the 0th and consequently the tours here are random.

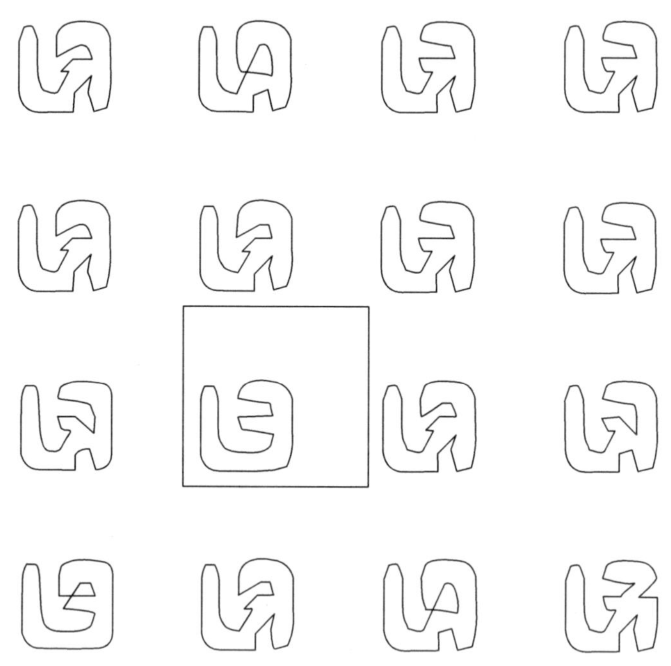

Figure 1.12. Best host tours from each of 16 demes after 1500 generations. Alternative general approaches can be seen. It can be seen from the figure that tours in adjacent demes share similar subtour elements due to the mainly local migration.

common subtour elements. At generation 4500, Fig. 1.13, we can see that most of the population variance has been lost, but an example of an optimal tour has evolved in the upper left-hand corner of the figure. At the run's conclusion at 5000 generations, Fig. 1.14 the optimal tour (the length of which is $(11+3\sqrt{10})1.5=30.7302$) is present in nearly every deme in the population. Notice, however, that even at this late stage three other variants remain. Given the presence of the best tour in the demes surrounding these variants, their existence must certainly be due to the influence of the parasites. In other words, in terms of secondary fitnesses, these host variants have become locally better in secondary fitness than the globally best TSP tour because they have broken the mimicking sequences of their local parasites.

The results of runs are shown, as means of the best evolved tour lengths over the ten replicates, in Table 1.6.

The full data set was analysed with a twoway analysis of variance ($DF=39$). Subdividing the population into the toroidal arrangement of demes gave a highly significant improvement ($F=20.407, p\leq 0.0002$). A presence of parasites also gave a highly significant improvement ($F=13.09$, $p\leq 0.0012$), but this factor as shown in the means and significance levels was slightly less effective

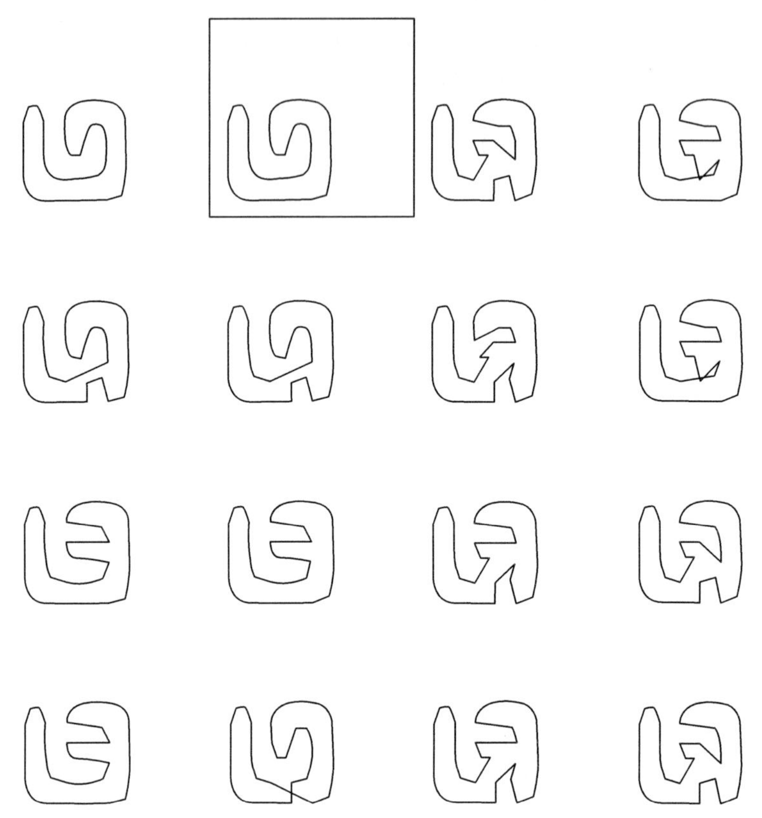

Figure 1.13. Best host tours from each of 16 demes after 4500 generations. An optimal tour has been discovered, seen in the upper left-hand corner. Much of the variability in the population has been lost.

Table 1.6.

	No parasites	*Parasites*
No demes	34.152	32.187
Demes	31.879	31.369

than deme structure. The interaction between deme structure and parasites was significant ($F = 4.524$, $p \leq 0.0381$) but comparatively weak.

Discussion

Variation must come from somewhere if evolution is to be effective. Subdividing the population into small demes increases the power of random genetic drift and

Figure 1.14. Best host tours from each of 16 demes after 5000 generations. The optimal tour discovered in generation 4500 (Fig. 1.13) has been dispersed throughout the population. Only three of the 16 demes give a non-optimal tour, these deviations having almost certainly arisen secondarily as the result of the local advantage of tours which escape attacking parasites.

this counters the loss of variation in the population. In addition, in our model it is likely, especially early on, that different demes will be attracted to different peaks. This, of course, also helps to maintain variation although it is not in obviously useful form. How it can be made useful depends on the effects of migration.

The role of migration is somewhat subtle. In one sense, in unifying the demes into a meta-population, even if only barely, a loss of variation may result because the metapopulation now has a chance to fix all demes at the same peak. On the other hand when the attraction of a new and better peak is found in any deme, migration enables the discovery to disseminate to other demes and in the course of this dispersal, assisted by recombination, a new wave of variation becomes liberated for global experimentation (Fig. 1.10(c)).

We have not had time yet to examine the relative effects of distant versus local migration. It may have appeared that the evolutionary process of mutation (often treated as equivalent to far migration in population genetical accounts) was omitted from our simulation. Mutation-like effects, however, are implicit in the way that the subtours are inserted during reproduction and therefore explicit mutation was deemed unnecessary. It should be recalled that even with a uniform population, because SRs and SDs are picked at random and tour parts are being inserted in normal or reversed orientation, sexual reproduction is continually causing changed tours to be tried. It is worth pointing out here that the potential of migration to prevent (almost) any local fixation was seen clearly by Wright[20]:

Moreover, the short range means of dispersal that have been postulated are likely to be supplemented by occasional long range dispersal. All of these tend to prevent fixation of one type even locally.

The same passage goes on:

On the other hand, selection may favor one allele in some places and others in other places. This would tend to increase local differentiation.

In our model this is true in two ways, one of which Wright may have had in mind, while the other he probably did not. Firstly, differences of selection arise through different demes being approximated to different peaks. Secondly, hosts from different demes coevolve with different sets of parasites. In the real world both factors must occur along with the even more obvious one of selection differing locally due to heterogeneity of the physical environment.

Evolutionary pressures acting on our host population stem from the advantage of better tour permutations and, at the same time, from avoidance of parasites. The action of these pressures can be described geometrically using the adaptive landscape metaphor.[9] Mapping the possible tour permutations onto a fitness surface results in the TSP case in an adaptive landscape of innumerable peaks of varying heights, representing fitter tours, and valleys of various depths representing the less fit tours. As one example demonstrating the existence of peaks, note that our chosen map of cities actually has four equivalent optimal peaks (Fig. 1.15). It so happens that the run we illustrate in Figs. 1.11–1.14 finds the one we intended to be optimal but it could equally well have found one of the other three optimal tours of whose existence the GA quickly informed us. (To be more precise the observation informed us of the existence of at least two forms and further analysis suggested a third and fourth which were subsequently proven to complete the optimal set. The fourth (top right), which has not yet been found by our GA in any run, presumably has a smaller 'catchment' in the total set of tours. From frequent observations of all the alternative optimal tours, catchment sizes appear to decrease counter-clockwise from the top left (most common) tour.)

It is obviously impossible to go from any one of the four optimal tours to any other by making any single insertion from the tour into an identical copy. It is

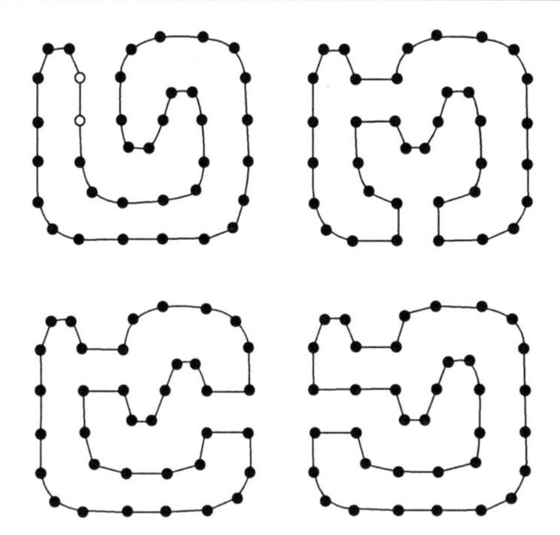

Figure 1.15. Optimal tour permutations for the 36 city TSP experiment. Each of the four illustrated tours have the same tour length. The chosen city pattern is based on a 6×6 square grid of points. Twelve points are moved inwards half-way towards the centres of lattice squares that they bound so as to suggest the 'worm' shape at the top left. This pattern could be made the only optimal tour (as originally intended) by moving the lower of the two lattice points represented by unfilled circles (in top-left tour) toward the one above it by any small distance, thus lengthening slightly the 'bridge' to the right that appears from the worm's neck in the three other tours.

even more obvious that no minimal modification (transfer of one city to another place) will effect the transition. Therefore it must involve steps down into a valley. Minor 'foothills' are also very easily illustrated. We are dealing with a landscape with four equal peaks and a very great number of foothills.

Over many generations populations of hosts move about on the landscape surface, blindly but effectively,[21] seeking the highest peak and the fittest tour. When the hosts are in a valley but within the basin of attraction of any given local peak, selection will drive the population towards the summit. This hill-climbing is a local optimization. The best or peak tour, once found by any individual, will spread by means of migration throughout the population, eventually resulting in the population becoming homogeneous. Selection in the absence of any other agency will never produce other than the local peak tour, implying that if the local population is isolated and large then, as Wright pointed out, the population is trapped on its peak. However, if there is (a) small deme size and/or (b) mutation/migration the population may be able to move away from its adapted peak towards other basins of attraction according to the Wrightian scheme.

The other evolutionary pressure moving populations away from current peaks is provided by parasites. This will work in opposition to the ultimate landscape selection even in large isolated populations. With or without the presence of (a) and/or (b) above, parasite pressure may be able to dislodge hosts and move them onward very much like random genetic drift in Wright's theory. As the best tour's genotype increases in frequency in the population, parasites, with their shorter generation time, quickly evolve tours which, by successfully matching a now common best host tour, drive the fitness of its parasitized host downwards from its peak onto adjacent saddles—or, as a more accurate image, flattens the subpeak itself. The best of the depressed hosts is temporarily liberated from parasites by having broken, by recombination, the matching parasite sequences. When such individuals attain the slopes of neighbouring peaks, selection becomes the stronger influence again: hill-climbing then recommences and the process begins anew. Moving from peak to peak in this manner (ref. 22) the population has a better chance to discover a global optimum.

In the present study a new evolutionary model of the GA incorporating both parasites and structured populations was tested with a 36 city travelling salesman tour using a desktop microcomputer (33 MHz 80486). Finding the optimal tour of this shortish problem cannot be considered trivial yet is not difficult either. As it turns out our GA finds the solution quite easily. What the experiment demonstrates is that significant improvements are gained both by adding parasites and by altering the population structure. Each factor alone was significant: two-sample T-test, population variances unknown and not assumed equal. Significance level was $p \leq 0.0015$ for the test of no population and no parasites versus no population structure but with parasites. For the test of no population structure and no parasites versus population structure but no parasites, the significance level was $p \leq 0.0008$. Effort and data are insufficient so far to assess the degree of interaction between the two factors, however. Probably the moderately significant antisynergistic interaction observed in Table 1.6 is due to nothing more than that each factor by itself was proving so effective that there was not much left that a combination could achieve.

Of course our model need not be limited to short tours. To demonstrate this our GA was run with a much more difficult tour of 100 cities randomly placed within a square. To compensate for the increased complexity, population size was increased slightly to 384 (32 demes of 12 individuals) and the number of generations was increased from 5000 to 20 000. Mortality and migration for hosts and parasites were set to the parameter values used in the previously described experiment. Figure 1.16 shows an evolved best tour for this 100 city tour. As can be seen in more than one place the evolved tour is certainly not quite optimal. But it is, at least to the eye or to a not-too-fastidious salesman, a satisfyingly efficient tour beyond which probably only a small percentage improvement can be made. Indeed some fairly obvious local algorithmic attention could polish this evolved tour into a form that would be quite difficult to improve at all. That such polishing has not been done by the GA is probably due to the presence of the parasites as indicated by the earlier discussion. If so,

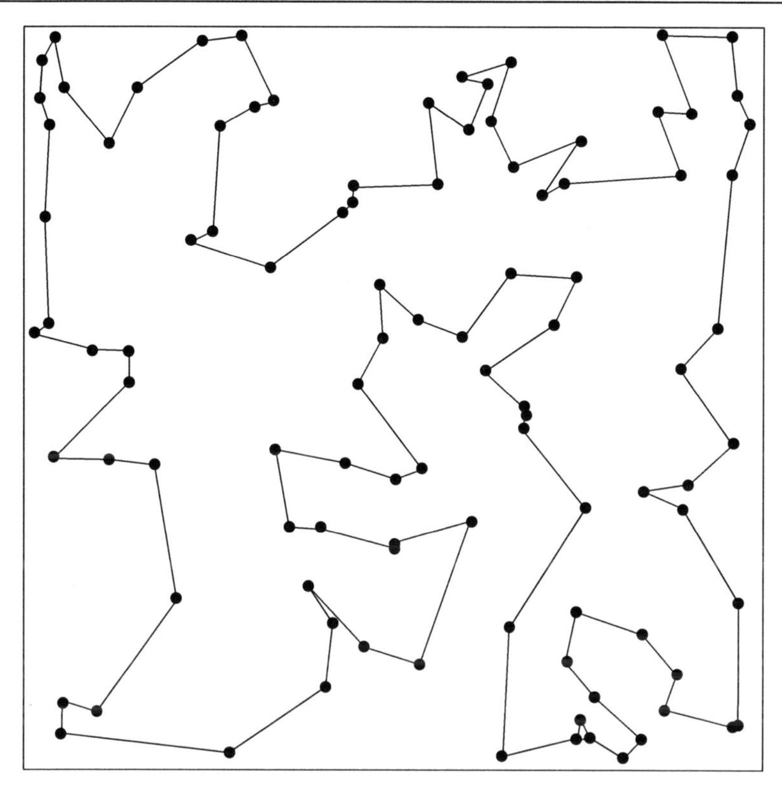

Figure 1.16. Evolved best tour for visiting 100 randomly placed cities. Population sizes for host and parasites were 384 individuals arranged into 32 demes of 12 individuals. Parasite and host mortalities were two and six deaths per generation respectively. Parasites were allowed two generations for every host generation. Total number of generations was 20 000. The optimal tour for this configuration of cities is not known but the evolved tour appears an efficient route unlikely to be far from optimal in terms of distance. Minor local imperfections, at least one of which is easily seen, are likely to be due to the continuing influence of parasites (see text).

then their final slightly adverse effect can be eliminated, as already mentioned, by continuing the simulation in a terminal optimizing period with parasite influence switched off.

In this chapter we have demonstrated how Nature's own software solutions can be transformed and transplanted into the language of the GA to provide results which can rival or even exceed the improvements that could be obtained by the use of larger populations and faster computers. By restricting ourselves to small GAs and, for now, small problems we hope we have shown that parasites as well as population structure have potential to help with far more substantial problems.

Acknowledgements

B. H. Sumida was supported by a SERC postdoctoral research grant. The authors wish to thank A. I. Houston for helpful comments.

References

1. J. H. Holland, *Adaptation in Natural and Artificial Systems.* The University of Michigan Press, Ann Arbor, MI (1975).
2. B. H. Sumida, Genetics for genetic algorithms. *SIGBIO Newsletter* **12**, 44–6 (1992).
3. W. D. Hamilton, Seething genetics of health and the evolution of sex. In *The Proceedings of the International Symposium on the Evolution of Life*, March 25–28 1990. Springer, Tokyo, pp. 225–52 (1990*a*).
4. W. D. Hamilton, Mate choice near and far. *American Zool.* **30**, 341–52 (1990*b*).
5. W. D. Hamilton, Memes of Haldane and Jayakar in a theory of sex. *J. Genet.* **69**, 17–32 (1990*c*).
6. W. D. Hamilton, R. Axelrod, and R. Tanese, Sexual reproduction as an adaptation to resist parasites (A review). *Proceedings of the National Academy of Sciences, USA* **87**, 3566–73 (1990).
7. W. D. Hillis, Co-evolving parasites improve simulated evolution as an optimization procedure. In *Artificial Life II* (ed. C. G. Langton, C. Taylor, J. D. Farmer and S. Rasmussen) Addison-Wesley, Reading, MA, pp. 313–24 (1991).
8. W. D. Hamilton, Sex versus non-sex versus parasite. *Oikos* **35**, 282–90 (1980).
9. S. Wright. *Evolution and Genetics of Populations*, Vol. 2. *The Theory of Gene Frequencies.* (University of Chicago Press, Chicago, (1969).
10. S. Wright, Evolution in Mendelian populations. *Genetics* **16**, 97–159 (1931).
11. D. J. Futuyma, *Evolutionary Biology.* Sinauer, Sunderland, MA (1986).
12. A. Sangalli, Short circuiting the travelling salesman problem. *New Scientist* **134**, 16 (1992).
13. Z. Michalewicz, *Genetic Algorithms + Data Structures = Evolution Programs.* Springer-Verlag, Berlin (1992).
14. S. Wright, *Evolution and Genetics of Populations*, Vol. 3. *Experimental Results and Evolutionary Deductions.* (University of Chicago Press, Chicago, 1977).
15. R. E. Lenski, Experimental studies of pleiotropy and epistasis in *Escherichia coli.* II. Compensation for maladaptive effects associated with resistance to virus T4. *Evolution* **42**, 443–40 (1988).
16. P. R. Grant, *Ecology and Evolution of Darwin's Finches.* (Princeton University Press, Princeton, NJ, 1986).
17. J. F. Crow, and M. Kimura, *An Introduction to Population Genetics Theory.* (Harper and Row, New York, 1970).
18. R. Tanese, Parallel genetic algorithms for a hypercube. In *Genetic Algorithms and Their Applications: Proceedings of the Second International Conference on Genetic Algorithms* (ed. J. J. Grefenstette). (Lawrence Erlbaum, New Jersey, 1987).
19. B. H. Sumida, A. I., Houston, J. M. McNamara, and W. D. Hamilton, Genetic Algorithms and Evolution. *J. Theoret. Biol.* **147**, 59–84 (1990).

20. S. Wright, Isolation by distance. *Genetics* **28**, 114–38 (1943).
21. R. Dawkins, *The Blind Watchmaker*. Longman, Essex (1986).
22. S. Wright, Statistical genetics in relation to evolution. In *Actualités scientifiques et industrielles, 802: Exposés de Biometrie et de la statistique biologique XIII*. (Hermann and Cie, Paris, 1939).

CHAPTER 2

MANIPULATING MICROBE PROCEEDINGS

Cytoplasmic Bacteria That Cause Parthenogenesis

RICHARD STOUTHAMER

I have to admit that Bill had little to do with the actual research that was reported in this paper but, his work formed the inspiration for this study and without his help—both editorial and conceptual—this paper would not have been published in the Proceedings of the National Academy of Sciences. Here I will describe how Bill got involved.

As a graduate student, under the guidance of Bob Luck in the Department of Entomology of the University of California Riverside, I had become very interested in the quite common occurrence of thelytoky in parasitic hymenoptera. In Hymenoptera the common mode of reproduction is called arrhenotoky in which fertilized (diploid) eggs develop into females and unfertilized (haploid) eggs develop into males. Thelytoky is used for the mode of reproduction where females are capable of producing daughters from unfertilized eggs. The Department of Entomology at UCR has a long tradition in studying parasitic hymenoptera with the goal of applying them in biological pest control.[1] From this applied perspective these thelytokous forms could, in principle, be much more efficient in reducing pest populations, because their population growth rate would be higher than that of sexual forms; thelytokous females do not 'waste' any of their eggs for the production of males.

The system I worked on for several years was a wasp species where both thelytokous and arrhenotokous forms were present. According to the literature the arrhenotokous form could be rendered thelytokous by keeping the wasps at constant rearing temperatures. After trying—in vain—for many generations to verify this effect,[2] I switched to working with *Trichogramma* wasps. *Trichogramma* are very small parasitic wasps that lay their eggs in the eggs of moths and butterflies. The larval wasps consume the butterfly egg, develop and then emerge as adult wasps from the shell of the butterfly egg. They have a short generation time of about 10 days at 26 °C. In the department a large collection of *Trichogramma* lines was present, assembled by John Pinto, who was doing a revision of the North American species of this genus. As a basis for the revision, he established hundreds of isofemale lines from field-collected wasps. Among these were a substantial number of thelytokous lines.[3]

It had been reported several times in the literature that these thelytokous lines, which normally would only produce females, could be made to produce some male offspring when they were reared under higher than normal temperatures (>28 °C). Also, one report from a species (*T. pretiosum*) with both thelytokous and arrhenotokous lines showed that the 'thelytokous' males (derived by temperature treatment) could father offspring with the arrhenotokous females.[4] Pinto's collections allowed me to verify that high rearing temperatures indeed led to the production of male offspring in thelytokous lines, and that these males were capable of fathering offspring with conspecific females of arrhenotokous lines. This enabled me to determine if the trait 'thelytoky' could be passed on in a simple Mendelian manner. I did this by introgressing the thelytokous genome into the arrhenotokous line and testing if the trait 'thelytoky' would express itself. This was not the case and it left two possible explanations: 1) the trait was chromosomally based but had many genes involved or was caused by epistatic interactions that could not be recreated through introgression, or 2) that the trait was extrachromosomally inherited.

I was well aware of Bill's work on extraordinary sex ratios and the prediction that cytoplasmic genes would favour a 100% female-biased sex ratio.[5] In addition, several papers had been published by Jack Werren showing that bacteria inherited through the egg cytoplasm in the wasp species *Nasonia vitripennis* could be killed by antibiotic treatment.[6] This opened up the possibility that micro-organisms might be involved in causing the thelytoky trait. It took me a while to figure out how to get the antibiotics into the wasps. Initially I was focused on trying to feed the antibiotic to the wasp larvae, but I could not think of a way of injecting the antibiotic into the moth eggs (= wasp larval food) without destroying the moth eggs. After a number of weeks I finally got the idea of feeding the antibiotic to the mothers. I mixed up tetracycline with honey, and to be sure that they would get enough antibiotics, I made strong solutions of 10% antibiotic in honey.

Ten days after the feeding of the antibiotics, the offspring of the treated females started to emerge, and to my big disappointment the offspring consisted of females only. Apparently, I could not cause the thelytokous females to produce male offspring by antibiotic treatment. I decided to continue the treatment for one more generation before ending the experiment. The offspring of the second generation consisted of females plus a large number of males. The reason that we did not get males in the first generation was later discovered. The 10% antibiotic concentration was too high. It killed the treated females before they were able to lay eggs that had been completely cured. This all took place in the summer of 1986. That same year, we showed that we indeed could cure thelytokous lines of many *Trichogramma* species of their infection and establish lines that had been permanently reverted to the normal sexual mode of reproduction (arrhenotoky). We did this by mating the males from an infected line with infected females from the same line and feeding these mated females antibiotics. The daughters produced from fertilized, cured eggs, then formed the basis for establishing new arrhenotokous lines. Later, it appeared that only in *Trichogramma* was it possible to establish sexual lines. In many other species of infected parasitoids, antibiotic treatment also results in male offspring, but these newly produced males are not able to father offspring in the females they mate with. Therefore sexual lines can not be established in these species.

By 1988 we had written up the findings of our research and, figuring that this was a huge breakthrough, we submitted a manuscript entitled: 'Extrachromosomal inheritance of uniparental reproduction in wasps' to *Nature*. We got it back very fast; it had not been sent out for review. Next, we reworked it for *Science* with the same result. By then I was very disappointed, and I was getting sick and tired of looking at the manuscript. Later that year, I presented this work as a ten-minute talk 'Microorganisms implicated as a cause of thelytokous parthenogenesis in *Trichogramma* species', at the International Congress of Entomology in Vancouver, Canada. For some reason I had been placed in the section 'Insect Pathology and Biological Control', where I got to present my paper in between two talks on the use of *Bacillus thuringiensis* to kill insects.

Before my talk Bob Luck had attended a presentation by Bill, 'Dead tree arthropods: advanced or primitive?', and he had asked Bill to come and see my presentation. After my presentation we talked for a while, and Bill expressed his enthusiasm about the findings. He mentioned that he had predicted the involvement of cytoplasmic genes in causing extremely female-biased sex ratios, but he had not thought that bacterial infections could cause the ultimate female bias of 100% females and could induce thelytoky. We explained to him our difficulty in trying to get this work published in one of the prominent journals. Bill suggested that we send him the manuscript and he would look at it to see if he could suggest changes. Within a couple of weeks, he send the manuscript back with a lot of changes, and in the accompanying letter he wrote: 'I think the approach I suggest would make it sound more topical news (which those journals want) and put it more thoroughly in the context of current discussion.' He also added some papers to be cited at different places in the manuscript and went on to write: 'Citation 7 is my paper 'Wingless and flightless males . . .' in the book edited by Blum and Blum[7] on sexual selection in insects. Figure 6 in that paper, in fact, will show you why I was excited to hear of your result: I can't remember whether I discuss my expectation that such effects would be found in that paper, or indeed whether I have discussed it properly in print, but I think that my paper 'Diversity under bark'[8] in the Royal Ent. Soc. Symposium volume has at least some further comments on the potential importance of symbionts in meddling with sex'.[9] After the paper had been rewritten, largely based on his comments, we invited him to be the third author on the paper. In his letter back he wrote regarding the authorship; 'Although I don't think I really deserve to be, for the little that I have sent, I would be happy and flattered to become a third author, and think that just possibly it might help it to be reconsidered by *Nature*, which seems to me obviously where it should go. What we might try is that, if you sent me a final draft, I would send it to the editor, explaining my enthusiasm, how I think it is likely to be an expanding theme for the future, and so on. I have in the past managed to get *Nature* (and *Science*) to reverse decisions about papers this way. However, I admit I didn't become an author on any of them, so I don't think that is any substantial reason to include me now—you should consider primarily how it will seem on your c.v.'.[10] I considered how it would seem on my c.v. and he submitted the manuscript to *Nature*, now under the title: 'Antibiotics cure parthenogenesis in *Trichogramma* wasps'. The first two reviews that came back were very positive, and it looked like the paper would be accepted. However, after the editor had asked the third reviewer several times to send in his or her review, it finally came back with the following objections to the paper: '1. They have not isolated the microorganism they suspect to be involved in causing parthenogenesis. 2. They have not shown that the 'factor' can be experimentally transmitted from a thelytokous strain to normally arrhenotokous individuals of the same species thereby producing thelytokous individuals, . . .' and ending the review with 'In short while I am inclined to agree that thelytoky in these *Trichogramma* strains is due to a microorganism, I do not feel that the evidence they present is sufficiently convincing.'[11] Bill tried to argue with the editor but the paper was refused by Nature, and in his letter to me he wrote: 'To look on the bright side, I seem to have seen this reaction to truly novel (and eventually validated) papers so often now. I see your troubles as an index of the importance of your results. It also makes me personally feel less of a back-number as a scientist that I still have the mental flexibility to notice and to want to support results like yours—even if making a mess of doing so. My own very first paper ever, on the kin selection criterion, was also rejected by Nature (actually much faster than this)—so in a way it makes me feel young! More seriously, I am sure that applying antibiotics to parthenogenetic lines and species is going to be commonly tried once your result is well known, and in many cases I bet it is going to work.'.[12]

Next, we rewrote it a bit and submitted it to *Science*, the reviews we got back from *Science* were all negative. Here is one example:

'As has been previously pointed out, under certain circumstances, thelytoky may confer a reproductive advantage on females practicing it and on genes causing it. It has also been pointed out that bacteria and other genetic entities which are transmitted only through the egg cytoplasm will gain an advantage if they can shift the sex ratio toward more females. It has not been predicted, however, that the two phenomena will be linked. But this is what the authors of this paper have found. Moreover, they found the phenomenon in several species and demonstrated that it is possible to restore sexuality by treating the females with certain antibiotics or with high temperature. However, while the discovery of this phenomenon has not been predicted, it could have been. (*How do you respond to this? R.S.*) Moreover, I believe that so far it has not provided us with new insights into the origin of thelytoky, haplodiploidy, or the evolution of selfish genetic entities, because the authors still do not know whether the agents are bacteria; how the agents are transmitted; how they cause the egg to begin to develop; whether the egg is haploid or diploid, and if it is haploid how the diploidy is restored: whether the agents kill the males, etc.'[13]

By then I was ready to submit this paper to the *Annals of the Entomological Society of America* and hope they would accept it. However, Bob did not think that was a good idea and Bill suggested we would send it to the *Proceedings of the National Academy of Sciences*, now under the title: 'Antibiotics cause parthenogenetic *Trichogramma* to revert to sex'. We changed the title because in the reviews we got back from *Nature* and *Science*, some reviewers objected to the term 'cure' in the title. The reviews we got back from the PNAS were a lot more favorable and the paper was accepted by the end of 1989. When it was published, *Nature* printed a News and Views article by Laurence Hurst, Charles Godfray and Paul Harvey (1990), ironically under the title 'Antibiotics cure asexuality'.[14]

Since this first publication on the topic, we have been able to overcome many of the objections the reviewers of this paper had. We have identified the bacteria associated with the parthenogenesis (*Wolbachia*),[15] we know how diploidy is restored in the unfertilized infected eggs,[16] and we have shown that they can sometimes be transmitted from infected individuals to uninfected individuals and induce parthenogenesis.[17] After we got the paper published in the PNAS, I saw Bill only once more at a meeting in Florence in 1996, where we discussed some models on the evolution of resistance to the parthenogenesis inducing *Wolbachia*. His enthusiasm about our results, and his willingness to help with getting our initial findings published have meant a lot to me, and in retrospect, his involvement did not hurt my c.v. Finally, many of the predictions he made in his papers regarding the genes located on elements in the cytoplasm have come true; the study of how these elements are manipulating the reproduction of many organisms has taken a huge leap over the last 20 years.[18,19]

References

1. R. Stouthamer, *Wolbachia*-induced parthenogenesis Chapter 4, pp. 102–124, *in* S. C. O'Neill, J. H. Werren, and A. A. Hoffmann, (ed.) *Influential Passengers.* (Oxford University Press, Oxford, 1997).
2. R. Stouthamer and R. F. Luck, Transition from bisexual to unisexual cultures in *Encarsia perniciosi* Tower (Hymenoptera; Aphelinidae): New data and a reinterpretation, *Ann. Entomol. Soc. Am.* **84**, 150–157 (1991).
3. J. D. Pinto, Systematics of the North American species of *Trichogramma* Westwood Hymenoptera: Trichogrammatidae). *Mem. Entomol. Soc. Wash.* **22**, 1–287 (1999).

4. G. M. Orphanides and D. Gonzalez, Identity of a uniparental race of *Trichogramma pretiosum*. *Ann. Entomol. Soc. Am.* **63**, 1784–1786 (1970).
5. W. D. Hamilton, Extra-ordinary sex ratios. *Science* **156**, 477–488 (1967).
6. J. H. Werren, S. W. Skinner and A. M. Huger, Male-killing bacteria in a parasitic wasp. *Science* **231**, 990–992 (1986).
7. W. D. Hamilton, Wingless and fighting males in fig wasps and other insects. p. 167–220 in M. S. Blum and M. A. Blum (ed.), *Sexual Selection and Reproductive Competition in Insects* (Academic Press, NY, 1979).
8. Letter from W. D. H. to R. S. strangely enough dated January 27th, 1986. The date on his computer must have been wrong, the letter was postmarked August 1st, (1988).
9. W. D. Hamilton, Evolution and diversity under bark, pp. 154–175, in L. A. Mound and N. Waloff (ed.), *Diversity of insect faunas*. Symposia of the Royal Entomological Society of London, no. 9 (Blackwell Scientific, Oxford, 1978).
10. Letter W. D. H to R. S. August 25th, 1988.
11. Review back from Nature November 30th, 1988.
12. Letter W. D. H to R. S, December 7th, 1988.
13. Review back from Science, Spring 1989.
14. L. D. Hurst, H. C. J. Godfray and P. H. Harvey, Antibiotics cure asexuality. Nature **346**, 510–511 1990.
15. R. Stouthamer, J. A. J. Breeuwer, R. F. Luck, and J. H. Werren, Molecular identification of microorganisms associated with parthenogenesis. *Nature* **361**, 66–68 (1993).
16. R. Stouthamer, and J. D. Kazmer, Cytogenetics of microbe-associated parthenogenesis and its consequences for gene flow in *Trichogramma* wasps. *Heredity* **73**, 317–327 (1994).
17. M. E. Huigens, R. F. Luck, R. G. H. Klaassen, F. M. P. M. Maas, M. J. T. N. Timmermans, and R. Stouthamer, Infectious parthenogenesis *Nature* **405**, 178–179 (2000).
18. S. C. O'Neill, J. H. Werren, and A. A. Hoffmann, ed. *Influential Passengers*. (Oxford University Press, Oxford, 1997).
19. K. Bourtzis, and T. A. Miller (ed.) *Insect Symbionts*, (CRC Press, Boca Raton, FL, 2003).

ANTIBIOTICS CAUSE PARTHENOGENETIC *TRICHOGRAMMA* (HYMENOPTERA/ TRICHOGRAMMATIDAE) TO REVERT TO SEX†

RICHARD STOUTHAMER, ROBERT F. LUCK,
AND W. D. HAMILTON

Abstract

Completely parthenogenetic *Trichogramma* wasps can be rendered permanently bisexual by treatment with three different antibiotics or high temperatures. The evidence strongly suggests that maternally inherited microorganisms cause parthenogenesis in these wasps. Theories predict female-biased sex ratio in offspring under the influence of maternally inherited symbionts, but extreme sex ratios of 100% females were never considered because the lack of males would prevent the host's reproduction.

In recent discussion of the reasons for sex or its abandonment,[1–3] surprisingly little attention has been given to possible roles of the biological agents for which sex creates the starkest alternatives. Agents transmitted solely in the female line, which includes mitochondria, plasmagenes, and, in some species, microbial symbionts, have encountered biological disaster when they find themselves passed to a male zygote: they then leave no descendants beyond the lifetime of their bearer. Mitochondria are considered to be under almost total control of the host genome; but what of the rest? The possibility that such agents might bias the sex ratio of sexual species in favor of females has sometimes been

---

† *Proc. Nat. Acad. Sci.* USA. **87**(7), 2424–2427 (1990).

theorized[4,5] and shown,[6,7] and the point has been made that their optimum is no males at all if this is still consistent with fertility.[8] However, the idea that the agents might somehow cause uniparental reproduction has not been considered. We now report results strongly suggestive that this can happen. Treatment of adults of four species of thelytokous *Trichogramma* wasps with antibiotics specific for prokaryotes, or their subjection to high temperatures, reverts their thelytoky to arrhenotoky.

Thelytoky is the term used to describe female-to-female parthenogenesis in the Hymenoptera. This potentially permanent mode of reproduction is to be contrasted to arrhenotoky, which is the normal (and essentially sexual) system of reproduction in the Hymenoptera. Under arrhenotoky, males arise from unfertilized eggs and females from fertilized ones. The males are haploid, and the underlying cytogenetic system is called male haploidy. The offspring produced by virgin females is used to distinguish between thelytokous and arrhenotokous reproduction: thelytokous virgins produce female offspring, while arrhenotokous virgins produce male offspring. The 'reverted' *Trichogramma* cultures mentioned above have returned permanently to arrhenotoky from thelytoky.

The genus *Trichogramma* has a cosmopolitan distribution and consists of >100 described species.[9] These minute parasitic wasps (about 1 mm long) lay their own eggs within the eggs of moths and butterflies. At 24 °C they complete their development from egg to adult in about 10 days. Among and within host species, the number of wasp eggs laid in a host egg increases with its size. From hosts parasitized by arrhenotokous females, one to several females may emerge together with usually just one male. As a rule *Trichogramma* species reproduce by arrhenotoky; however, thelytokous reproduction also occurs. Some species are entirely thelytokous, others consist of both thelytokous and arrhenotokous populations, while in several arrhenotokous species an occasional thelytokous female is found. Males are normally rare in the offspring of thelytokous females, but if such females are exposed to temperatures >30 °C during their larval development, some males will usually appear in their offspring.[10,11] The same phenomenon has been found in thelytokous forms of many different Hymenoptera species.[12–17] Generally such temperature-induced males are considered to be nonfunctional.[10,12,14] However, in several species the males are functional, and their sperm can successfully fertilize eggs of conspecific arrhenotokous females[17,18] or thelytokous females.[19]

Materials, Methods, and Results

The material in our study came from parasitized host eggs that were collected throughout North America and Hawaii for a taxonomic survey of *Trichogramma* species. Infrequently thelytokous females emerged from these field-collected eggs, and they were used to initiate thelytokous laboratory cultures of *Trichogramma pretiosum, Trichogramma deion, Trichogramma platneri*, and *Trichogramma chilonis* by using cabbage looper (*Trichoplusia ni*) eggs as hosts. Each culture was started with one thelytokous female. At the locations where thelytokous females of *T. deion* (Irvine, CA; Mountain Center, CA; Sanderson, TX; and Belle Fourche, SD), *T. platneri* (Riverside, CA), and *T. chilonis* (Kauai, HI) were collected, arrhenotokous females of the same species were also found. The collection of *T. pretiosum* (Kauai, HI) consisted of only thelytokous females, while *T. pretiosum* (Nuevo Leon, Mexico) was collected together with arrhenotokous conspecifics. Finally, the collection of *T. pretiosum* (Lara State, Venezuela) consisted of only one thelytokous female.

BACKCROSSING EXPERIMENTS

To determine if chromosomal factors might affect the inheritance of thelytoky in *Trichogramma*, we created females that had the genome of a thelytokous strain and the cytoplasm of a field-collected arrhenotokous strain of the same species and determined their mode of reproduction (i.e., arrhenotoky vs. thelytoky). If thelytoky were inherited chromosomally, these females would be expected to produce only female offspring. To create such females, we crossed males from a thelytokous strain (produced by exposing thelytokous females to temperatures >30 °C) with females from an arrhenotokous strain kept at 24 °C. The resulting hybrid females (averaging 7 females per generation for *T. deion* and 8 for *T. pretiosum*) were mated again with males from the thelytokous strain. We continued these backcrosses for five generations with *T. deion* and for nine generations with *T. pretiosum*. Since cytoplasm is inherited through the maternal line, all of the cytoplasm in these backcrossed females arose from the arrhenotokous strain. On average, the percentage of the genome stemming from the thelytokous line in females of the *n*th generation of backcrossing is given by the relationship $1 - (0.5)^n$; this percentage in generations 5 and 9 is 96.9% and 99.8%, respectively. We tested the mode of reproduction of the backcrossed females every generation by supplying virgin backcrossed females (averaging 23 virgins for *T. deion* and 17 for *T. pretiosum*) with host eggs: all female offspring indicated the females reproduced thelytokously, whereas all male offspring indicated they reproduced arrhenotokously.

The backcrossed females remained arrhenotokous, showing that thelytoky is not inherited in a simple way through the chromosomes of the wasp. (In generation 4 of *T. deion*, 45 virgins were tested, and the offspring of 1 of these females consisted of 2 males and 2 females. This female was probably not a virgin when used in this test. The all-male offspring of 36 virgins in the next generation of backcrossing corroborates this.) Therefore, extrachromosomal inheritance is probable.

ANTIBIOTIC EXPERIMENTS

To test if thelytoky is caused by extrachromosomally inherited microorganisms, we allowed newly emerged wasps to feed on honey mixed with one of several antibiotics for 1 day before providing them with host eggs. Populations were exposed to antibiotics for several generations by repeating this procedure. In those females fed honey mixed with the antibiotics tetracycline hydrochloride, sulfamethoxazole, and rifampicin (100 mg of antibiotic per 1 ml of honey), males appeared in the offspring of generation 1 or 2 and were present in all subsequent generations. Mixtures of 1% tetracycline in honey proved effective in other experiments. Male offspring did not occur where thelytokous females were fed the antibiotics gentamycin, pennicillin G, and erythromycin. We initiated three to five isofemale lines (lines started with one female) with mated females from each population that had been exposed for several generations to tetracycline or high temperatures (Table 2.1). The new lines were maintained without antibiotics at the normal rearing temperature of 24 °C. We tested the mode of reproduction of these lines between generations 2 and 5 after they were initiated. Those reproducing by arrhenotoky in this test were maintained and remained arrhenotokous when tested again around generations 10 and 20 (Table 2.1). Therefore, we had permanently changed the mode of reproduction from thelytoky to arrhenotoky by exposing thelytokous females of these four *Trichogramma* species to either antibiotics or high temperatures. The lines where the treatment failed to revert thelytoky to arrhenotoky often produced

Table 2.1. Species name and collection site of initially thelytokous *Trichogramma* that were treated for several generations to either temperatures >30 °C (T) or the antibiotic tetracycline hydrochloride (A)

Species and collection site	Treatment	Generations treated, no.	Lines started, no.	Arrhenotokous lines in 1st test, no.	Tested in generation
T. pretiosum					
Nuevo Leon, Mexico	T	23	5	5	4, 12, 24
	A	4	4	4	3, 12, 28
Lara State, Venezuela	A	3	4	3	1, 17
Hawaii	T	13	4	2	4
	A	7	4	2	3, 10, 22
T. deion					
Irvine, CA	A	4	3	3	5, 15
Mountain Center, CA	T	9	4	4	2, 12, 22
	A	6	4	4	2, 12, 23
Belle Fourche, SD	A	5	4	1	2, 7, 10, 13, 22
Sanderson, TX	T	4	4	3	2, 11, 24
	A	5	5	5	5, 22
T. platneri					
Riverside, CA	A	7	4	4	2, 10, 26
T. chilonis					
Hawaii	A	8	4	4	5, 10, 23

At the end of the treatment, lines were started with individual mated females. These lines were tested for their mode of reproduction after several generations. After the first test, only the arrhenotokous lines were maintained. In subsequent tests they remained arrhenotokous.

some male offspring in the first few generations after the treatment had ended, but by generation 5, only female offspring were produced. One exception to this were two lines of *T. deion* (SD) that consisted of a mixture of thelytokous and arrhenotokous individuals when tested in generation 7 after the treatment had ended. Three new isofemale lines were subsequently started with mated arrhenotokous females from one of the two original lines tested again for their mode of reproduction 12 generations later and were found to be arrhenotokous.

Discussion

The backcrossing experiments showed that thelytoky was not inherited as a simple chromosomal trait; therefore, extrachromosomal inheritance of thelytoky became likely. Recently, several sex ratio-distorting factors have been discovered in insects—for instance, the parasitic wasp *Nasonia vitripennis*[4] and various *Drosophila*[5] species. Females carrying these factors produce abnormal sex ratios in their offspring. These sex-ratio distorters are inherited extrachromosomally and are in some cases microorganisms that are passed on from mother to offspring through the cytoplasm of the egg.[5] These microorganisms could either be killed by treatment with tetracycline or by exposure to heat. We exposed our thelytokous cultures to several antibiotics and high rearing temperatures. The three antibiotics that caused a reversion from thelytoky to arrhenotoky all differ in their mode of action.[20] Rifampicin inhibits prokaryotic DNA-dependent RNA polymerase, tetracycline affects the protein synthesis on microbial ribosomes, whereas sulfamethoxazole inhibits the folate synthesis in microorganisms. Microorganisms need to synthesize their own folate and cannot use preformed folate, while animals only use preformed folate. The mode of action of the ineffective antibiotics cannot be used to better identify the presumed microorganism for several reasons. We could not determine (*i*) if the honey mixed with the ineffective antibiotics was ingested by the wasps and (*ii*) if ingested, whether the antibiotic reached the microorganisms.

The treatment failed to lead to arrhenotoky in a few lines started from a treated culture (Table 2.1). The males that were produced in these 'failed' lines during the first few generations after the treatment ended can be explained by an incomplete removal of the microorganisms. With a lower microorganism titer once the treatment has ended, the females are still thelytokous but produce some male offspring. Over generations, the microorganism titer would increase again and result in only female offspring.

Two alternative hypotheses can be posed to explain these results. First, high temperatures and certain antibiotics may signal to the female that the environment is about to change and that arrhenotokous reproduction may be favored. A situation similar to that has been shown in aphids, where there is a transition from asexual to sexual reproduction induced by low temperatures and day length.[21] The permanently arrhenotokous cultures that were established after treatment are difficult to explain by this hypothesis. Why would these cultures remain arrhenotokous when they are put under conditions that previously favored thelytoky? A second alternative is that the exposure to high temperatures or certain antibiotics may cause some chromosomal damage that induces a change from thelytoky to arrhenotoky. What kind of mechanism would cause such a change is not known, but if such a change occurs, it does not lead to gross chromosomal rearrangements because the males from 'reverted' lines are completely compatible in crosses with females from field-collected arrhenotokous lines.[22] Additionally, experiments[23] in which the three effective antibiotics were tested for their effects on longevity and fecundity showed that whereas 10% mixtures of tetracycline had a negative influence on both longevity and fecundity, possibly consistent with the chromosomal-damage hypothesis, neither sulfamethoxazole nor rifampicin had a significant influence on these traits. Therefore, if some chromosomal change occurs, this change must be very specific and does not have any negative influence on these life-history traits.

These two alternative hypotheses are less well supported by the data than the microbial hypothesis. Symbiontic microbes are often extremely fastidious, and they have only been successfully cultured in a few cases, so that Koch's postulates could be satisfied.[24,25] Although we have not yet satisfied Koch's postulates and therefore have not proven the microbial involvement in thelytoky, the backcrossing and antibiotic experiments strongly suggest that microbial symbionts are involved in causing thelytoky.

The advantage to vertically transmitted symbionts of causing a transition to either spanandrous arrhenotoky (arrhenotoky in which few males are produced) or thelytoky was pointed out above. Symbionts with ovarial transmission or other elements (excluding mitochondria) that pass only in the female line are known to be associated with a majority of cases of male haploidy and parahaploidy,[26–30] but their possible role in causing transition to these modes of reproduction has not been reviewed. Here (sic) is merely pertinent to note that induction of male haploidy opens the way, via arrhenotokous control of the sex of offspring by the mother, to strong female biases of sex ratio, especially under conditions of local mate competition. These biases are common[31] and, while they are fairly well explained by current evolutionary stable strategy (ESS) theory concerning the interests of the host genome,[32–34] they also converge with the adaptations of a symbiont when inbreeding is close. When broods are large and sex ratios are very female-biased,[8] further selection to eliminate the male will be weak. However, this is hardly the case with *Trichogramma*, where brood sizes are constrained by host egg sizes and are often small (one to two eggs per host egg): a change in *Trichogramma* from arrhenotoky to thelytoky gains a substantial fraction of immediate fitness. Data on the offspring production by thelytokous and 'reverted' arrhenotokous cultures show[23] that thelytokous females produce the same number of off-spring per host egg as arrhenotokous individuals. Therefore, thelytokous females produce more female offspring per parasitized egg than arrhenotokous females.

There are also other possible interesting evolutionary connections of our phenomenon. In many species of thelytokous wasps, males are induced by exposing females to temperatures >30 °C.[10–17] We suspect that microorganisms cause thelytoky in these wasps also, although permanent reversion to arrhenotoky was not shown. In some cases thelytokous females produce male offspring in response to exposure to high temperatures during their larval development; however, when such adult females are subsequently exposed to lower temperatures, they will start to lay female offspring again,[13] indicating that the high temperatures might affect the transmission of the microorganism or its effect on thelytoky, without actually killing it. The temperature sensitivity of thelytoky suggests that microorganisms may be the agents involved in the cyclic thelytoky characteristic of many invertebrates.[1] These invertebrates alternate biparental and uniparental generations. The correlation of sex with season in alternation suggests a possibility of the inactivation of the microorganisms' influence, at least in the origin of these systems. While the evidence in our case suggests elimination of a maternally inherited symbiont, evolution of a temporary controlled incapacition without elimination in a potentially powerful adaptation for the host, affording entry to facultative or cyclical parthenogenesis. In two of the best known groups of insects showing such modes, coccids and aphids, maternally transmitted symbionts, sometimes of several kinds, are already known to be universal in the females. Although they have not yet been shown to have causative roles in gametogenesis or sex determination, they are closely associated with the events and, in at least two cases, fail to enter male eggs or embryos.[35,36]

Acknowledgements

We thank E. R. Oatman and J. D. Pinto for supplying us with the thelytokous cultures from their field collections, Gary Platner and Barbara Pfrunder for their assistance, and J. H. Werren for stimulating discussions.

References

1. G. Bell, *Masterpiece of Nature* (Univ. of California Press, Los Angeles 1982).
2. R. E. Michod and B. R. Levins, *The Evolution of Sex: An Examination of Current Ideas* (Sinauer, Sunderland, MA 1987).
3. S. C. Stearns, ed, *The Evolution of Sex and Its Consequences* (Birkhauser, Basel 1987).
4. L. M. Cosmides and J. Tooby, *J. Theor. Biol.* **89**, 83–129 (1981).
5. J. H. Werren, *J. Theor. Biol.* **124**, 317–334 (1987).

6. J. H. Werren, S. W. Skinner, and A. M. Huger, *Science* **231**, 990–992 (1986).
7. D. L. Williamson and D. F. Poulson, in *The Mycoplasmas III* (eds.) M.F. Barile and S. Razin (Academic, New York), pp. 175–208 (1979).
8. W. D. Hamilton, in *Sexual Selection and Reproductive Competition in Insects*, (eds.) M. S. Blum and N. A. Blum (Academic, New York), pp. 167–220 (1979).
9. B. Pintureau, Thesis (l'Université Paris VII 1987).
10. W. R. Bowen and V. M. Stern, *Ann. Entomol. Soc. Am.* **59**, 823–834 (1966).
11. G. T. Cabello and P. P. Vargas, *Z. Ang. Entomol.* **100**, 434–441 (1985).
12. S. E. Flanders, *Ecology* **23**, 120–123 (1942).
13. M. Laraichi, *Entomol. Exp. Appl.* **23**, 237–242 (1978).
14. F. Wilson and L. T. Woolcock, *Nature (London)* **186**, 99–100 (1960).
15. S. E. Flanders, *Am. Nat.* **94**, 489–494 (1965).
16. G. Gordh and L. Lacey, *Proc. Entomol. Soc. Washington* **78**, 132–144 (1976).
17. E. F. Legner, *Can. Entomol.* **117**, 385–389 (1985).
18. G. M. Orphanides and D. Gonzalez, *Ann. Entomol. Soc. Am.* **63**, 1784–1786 (1970).
19. Y. Rössler and P. Debach, *Hilgardia* **42**, 149–175 (1973).
20. L. S. Goodman and A. Gilman. *The Pharmacological Basis of Therapeutics* (Macmillan, New York 1975).
21. D. F. Hales and T. E. Mittler, *Genome* **29**, 107–109 (1987).
22. R. Stouthamer, J. D. Pinto, G. R. Platner, and R. F. Luck, *Ann. Entomol. Soc. Am.*, **83**, 475–481 (1990).
23. R. Stouthamer, Ph.D. thesis (Univ. of California, Riverside 1989).
24. K. J. Hackett, D. E. Lynn, D. L. Williamson, A. S. Ginsberg, and R. F. Whitcomb, *Science* **232**, 1253–1255 (1986).
25. N. L. Somerson, L. Ehrman, J. P. Kocka, and F. J. Gottlieb, *Proc. Natl. Acad. Sci. USA* **81**, 282–285 (1984).
26. B. Peleg and D. M. Norris, *Nature (London) New Biol.* **236**, 111–112 (1972).
27. A. Bournier, *Ann. Soc. Entomol. Fr.* (N.S.) **11**, 415–435 (1966).
28. P. Büchner, *Z. Morphol. Oekol. Tiere* **50**, 1–80 (1961).
29. P. Büchner, *Endosymbiosis of Animals with Plant Microorganisms* (Wiley-Interscience, New York 1965).
30. W. P. J. Overmeer and R. A. Harrison, *N. Z. J. Science* **12**, 920–928 (1969).
31. B. H. King. *Qt. Rev. Biol.* **62**, 367–396 (1987).
32. L. Nunney and R. F. Luck, *Theor. Pop. Biol.* **33**, 1–30 (1988).
33. S. A. Frank, *Evolution* **39**, 949–964 (1985).
34. S. A. Frank, *Theor. Pop. Biol.* **29**, 312–342 (1986).
35. P. Büchner, *Z. Morphol. Oekol. Tiere* **43**, 262–312 (1954).
36. G. Lampel, *Z. Morphol. Oekol. Tiere* **48**, 320–348 (1959).

CHAPTER 3

MY INTENDED BURIAL AND WHY*[†]

WILLIAM D. HAMILTON

There seem to be people incurably fascinated by insects. I am one of them. The interest arises untaught in infancy and harsh experiences of childhood seem to do little to change it. I remember a walk to a wild blackberry patch just after the war and my mother teaching one of my sisters not to be scared of the spiders that were common on huge orb webs among the blackberries. Our big garden Araneus loves to build right in front of the best, ripest berries, as if it knows that these are good places to catch flies. Of course, the webs often intercepted our hands as well. The spiders, however, were never aggressive; when disturbed they would run and hide among leaves. But they were fat and looked—well, spidery, and quite off putting even to me. My mother on this occasion set a spider to walk on her hand; then she encouraged my sister to do the same. I was impressed. Shortly, I too picked one out of its web and placed it on my hand: it walked a few paces, then bit me quite painfully. My old caution returned, but I think my admiration for the courage of the tiny creature only increased: what human would dare stab a giant of equal proportion, even if the giant had just ruined your home? Another time it was not curiosity that caused me to be hurt but simply trying to be kind. A bumblebee stung the finger on which I lifted it, perhaps clumsily, out of a pond where it was drowning. That must have been earlier than the spider incident because we were evacuated to Edinburgh, avoiding the German flying bombs that fell around our home in Kent. From this event, I began to understand that the real world of animals might be fairly different from the world of Aesop's fables; but the huge furry and stripy insects going to and from their hole in the back masonry wall of the garden pond continued to fascinate me. I smell the stale water and the damp and dark

* In honour of Bill Hamilton (1936–2000), one of the greatest biologists of the 20th century and member of the Advisory Board of EEE, we publish—with great emotion—this article, which originally appeared in the Japanese entomological journal 'The Insectarium' in 1991.

[†] *Ethology Ecology and Evolution* **12**, 111–122 (2000). English translation of paper originally published in Japanese in *The Insectarium* (1991).

leaf-sodden corner of that Edinburgh garden to this day: the corner held the first bumblebee nest that I ever found.

Foolishly regardless of risks she was creating to my economic success in later life (for though rich men sometimes later become good amateur entomologists, those drugged too early by the fascination of insects seldom become rich), or more likely simply contemptuous of the risks, my mother encouraged my interests. Tirelessly she provided the jam jars for the caterpillars that I brought in and as tirelessly with her hairpins punctured air holes in their paper lids. Tirelessly also she tried not to mind my regular overturning of the stones of her rock garden in my search for beetles and whatever other strange creepies might lurk beneath them.

Stone turning—that, as it now occurs to me, is a trait that might almost define us compulsive juvenile entomologists. We are turners over of junk in waste places, pullers of loose bark from rotting logs. You never know what you may find, the rosy or greenish worm, the spilled-rice confusion of pupae of an ant nest, the sinister dark and yet glorious iridescence of a ground beetle {*Carabus violaceus*}. Other more vertebrate prizes were not to be neglected either. Under a piece of old corrugated iron half buried in grass may lurk not only these but a mouse nest full of babies or best of all a glossy, sinuous slow worm {*European legless lizard*}. In my case stone turning accidentally led also in a more dangerous direction. This began when a big flint pulled out of a bank revealed not the expected scurry of insects but rusting tins that had been pushed down a rabbit burrow. Real treasure at last or at least hidden food, I thought, even better than insects; but the tins had only sugary and salty substances in them, and one had aluminium powder which coloured everything it touched silvery grey. They turned out to be the ingredients for explosives that my father, who had experimented new designs for hand grenades for the British Home Guard in the war, had hidden there. The tins were of course taken away from me but I did not fail to notice where they went—this time to a high shelf of our shed. Later I took up the question of what the mysterious 'ingredients' could be mixed to make. The answers to this question very soon nearly ended my life and I came back to entomology with shorter fingers and, for a long time, a dull pain in the back and front of my chest.

Did baboon-like hominid ancestors on the high plains of Africa already take advantage of a diversity of temperament which made some members of each group specialise in spotting the insect meals, the best stones to turn, the best logs to pull apart in the search for the grubs that made part of the ape-humans' normal diet? Were these specialists supposed to learn and to teach which creatures bit or stung or were poisonous, and which could be grabbed and eaten? Going farther with such speculation, might one suggest not only that this natural variation was already present but also that, as the more advanced ancestors of all mankind spread out from Africa over the rest of the world, some sifting of such temperaments, perhaps by chance, happened to send a rather higher proportion of the stone-turning biotype to the East than to the West? It seems to me that the very fact of existence of the magazine for which I write this article indicates a higher proportion of the stone-turning fraternity to be now living upon Asia's Eastern fringes than on her Western. In other words it is my impression that in Japan there is less of a human divergence in attitudes about insects and fewer examples of the almost hysterical dislike and fear which is quite common in Britain. Here most people wince and draw back at what is seen under a stone, they stamp on the stag beetle that flies uninvited into the garden party. From books, from the live insects, and from the cages for them that are sold in Japanese shops, as well as from the insect-related incidents that I have seen in the streets of Japan and could never imagine in Britain—a man catching cicadas from a tree for a small girl, using a long net on a busy suburban street, small children beside a shrine netting dragonflies, for which they have brought with them a plastic and manufactured cage—I judge that most or all Japanese children must be brought up (or may simply grow up) with a moderate liking for insects. The rather common portrayal of insects and insect-damaged leaves in the art of the Far East, rare to see in Western art, seems to point the same way.

In Europe, to be fated to join the sparse company of stone turners, or to grow up to be a runner after butterflies, is to become an oddity among your schoolfellows, almost a figure of ridicule. In spite of this I doubt that many of the afflicted will express regret over their eccentricity, for the rewards are also great. Perhaps foremost is the instant admission to a world where, first as child, then as adult, and finally even as a veteran right until the very failing of sight, no one can be bored.

The world contains so many insect lives and so few are known. As to sheer diversity, by far the majority of the kinds of objects known and carefully described upon the surface of our planet are insects. Kinds of rock, kinds of soil, kinds of plant still more, these are indeed many and their classification complex; but it is certain that their numbers and complexity are quite small matters compared to the numbers and complexity of insects. Consider for example one very arbitrary class of animals, all that have longish snouts. Other animals than insects are perhaps more elegantly or more hugely snouted—we think here of the anteaters, tapirs, elephants and kiwis. But there are merely two edentate anteaters (decently snouted, that is), two elephants, two tapirs, and a mere trio of kiwis... in contrast looking for snouted beings among arthropods, there are far more weevils than there are of all species of the mammals put together. One could toss in all birds, long-beaked or not, upon the side of the mammals and still the weevils—a mere single family among the Coleoptera—would greatly exceed them all. Out of such multitudes, very few have their lives recorded. Yet the habits of weevils, when known, again taking these as example, are often fascinating. I could mention the leaf-rollers, Byctiscus and allies, whose careful endowment of food and protection to their offspring would put many birds and mammals to shame. Or, did you know that some weevils (and some bearing extra horns too) seem to be beginning to play fig wasp—pollen missiles that home to their target—to certain plants? Some are bred up in flowers by *palms* (Elaeis) for whose pollination they seem to be subsequently nearly essential. With all this going on and most of it 'unknown' there is, consequently, always something for the entomologist to be doing. Possibly this is why so many continue into an unusually scientifically productive old age. Perhaps such a living interest can for a time resist the degenerative chemistry of our bodies or can stop even the museum napthalene, which is intended slowly to kill or to drive out the invading live secondary insects that ruin collections, from killing or driving out the human curators, seeming on the contrary to make addicts of them for the tarry smell and to leave them juvenescent.

I am digressing: I meant to write of my early interest in insects. I am writing instead what are, no doubt, wishful thoughts for my future. To these, however, in more extreme and perhaps sombre vein, justifying my title, I must come back at the end.

Returning to about my thirteenth year I see myself now labouring on the steep hills of the North Downs of Kent, riding or pushing my bicycle. The stick of my butterfly net fits neatly in the bicycle's frame, my net itself and my killing bottle are in the saddlebag: a fanatical butterfly collector is out hunting. The brass shrapnel of my bomb-making period (my other boyhood passion) is well bedded now in scar tissue far back in my chest and my lung hardly hurts as I puff on the hills. And if it does hurt I hardly notice, for new chalky fields, rich ground for my 'blues' {*Polyommatini*}, are rising into my view, and new mounds of roadside blackthorn {Prunus spinosa}, perhaps hiding the rare British 'hairstreaks' {*Theclini*} that this hunter has never found, are passing.

I started my collecting with moths as well as butterflies and became no stranger to the special excitement of collecting at night. But I was discouraged by the sheer numbers, especially of all the 'micros' which, to be consistent, I ought to include but which proved such a tedious impossibility to 'set' nicely for my collection. I had soon realised moreover I couldn't afford all the boxes I needed for storage, leave alone could afford that ideal—magnificent polished cabinets with glass-topped drawers. Yet, still with a love perhaps the more pure because it was less destructive, I continued to observe the velvety beauties of the night: always I kept lights around me as bright as possible on summer nights and would open all curtains and windows of any room where I was doing school homework, bathing or the like, in order to welcome for inspection and identification before release all moths that deigned to enter. One group, the hawk moths (Sphingidae), remained so attractive to me by their name and shape and sheer size and dashing vigour that I gave them honorary status as butterflies and kept a place for them in my collection. Collecting occupied a large part of my spare time. I worked from books my mother had given me and from others I occasionally bought for myself in secondhand shops. The books praised collecting as an end in itself, treated it almost as an art. I accepted this wholeheartedly. I had at first virtually no scientific aims. My carefully selected and shaped sticks for my nets came from the local woods and, oddly enough, the steel frames for the nets came from the woods also: I developed a line of springy but effective frames that I made from the strands of a steel cable of a barrage balloon which had come down to trail over the trees after an air raid. My mother sewed and re-sewed the net bags which the thorny scrub of the chalk hills so quickly wore out. My hunting range ever increased and came to include occasional one-night and lonely camps in abandoned chalk quarries and under hedgerows. My books also encouraged the rearing of caterpillars, advising me that this was the way to obtain perfect specimens. Progressing from my mother's jam jars, I therefore hammered together larger and larger wooden net-sided cages. Soon they evolved to airy palaces in which even 'difficult' species, I hoped, would feel at ease and would mate and lay eggs; but when my palaces began at last to achieve this, the tedium of months of attention to the brood that followed was often great compared to the final reward, so soon I decided that only very rare ones would be reared from the egg. Unconstructive as all this activity was scientifically, I often think that the familiarity I gained for a particular group of animals has been an advantage to me as a biologist. A good hunter has to become a naturalist, and a collector, almost as certainly, has to become a systematist and to acquire ability to spy out natural relationships. Many of the ideas of evolutionary biology I became interested in later turned out to be open to illustration by examples from my chosen group, just as many initially attractive theoretical ideas, whether my own or someone else's, could be dismissed or else favoured out of the same stock of knowledge. Looking for food plants on which to find caterpillars I learned field botany, while from the broad landscapes of my wanderings I acquired a general eye for habitats, and through this, some inklings of community ecology, this becoming fixed in me by the intensity of my desire to guess what species each field or wood might be going to contain.

About 1946, a big new book turned up in the book shop of the town where I went to school. It had a suspiciously popular and simple title: 'Butterflies'. Was this quite the thing for a Collector, I wondered? It had lots of colour photographs of living butterflies but rather few of the customary symmetrical arrays of pinned specimens: I sensed a possibility of some ambivalence or heresy in its author towards my fine and tested ideals. Nevertheless, every time I went to the book shop I found myself handling the book with longing. After my parents had given it to me for a birthday present about a year after its appearance, it began to open new worlds. Its subject was indeed butterflies, but it was also about genetics, and beyond that, about principles that shape all life, about evolution by natural selection. I had already had hints of Darwinism from my mother and a rough idea of how natural selection worked. I was ready for a change, and E. B. Ford, the strange, learned and misogynous batchelor, author of the book (and not much stranger, probably, than the fanatic who ached to follow his words) was my guide to lead me out of my old temple and into another. I perceived that he had been (perhaps was still?) a collector himself and yet he had gone on . . . where? Through Ford I began to be fascinated by variation, and beyond that I became even more incurably drawn in by the study of evolution itself than I had been by my hitherto brilliant instances of its working, that which underlay the orderly spread of the insects of my collection. Soon I used a school prize to buy 'The Origin of Species' by Darwin. This was enthralling to me also, but of course didn't help at all with what had become for me the most challenging side of Ford's book—, his mysterious genetics—, which weren't invented in Darwin's time. Some mysteries, for example Ford's curious triple ratio, p^2: 2pq: q^2, took me about five years to understand. They only became resolved thoroughly when I discovered works of Ford's more brilliant and innovative colleague, R.A. Fisher, in the St John's College library in Cambridge. Other mysteries were never unravelled, but with these I came slowly to the conclusion that neither Ford, Fisher, nor anyone else, had succeeded with them either. They are what I still work on. As to the more immediate effects of the book, it rekindled, if anything, my collecting mania but brought a new hybrid, more scientific spirit into my work. The first fruits were still bigger and better cages in which I now intended *to cross varieties*. I wanted to duplicate some of the marvellous Mendelian ratios Ford described, to make them believable to me through my own experience; at the same time I wanted to obtain *wholly new varieties which, following his recipes, I would create*. I hoped for new rare *'Vars'* and *'Aberrs'*, as the books called them: I would create among them cases so rare that they would be mine alone! But I found Ford had made it sound too easy. Two arduous years of breeding and two years of *not* revealing even any genes of the old pale 'peppered' variety out of the local melanic population of the moth Biston betularia discouraged me. Perhaps too, I was beginning to be stirred by Ford's obvious contempt for the mere amassing and describing of varieties. By the time I left school, I had almost ceased to collect.

What remains from all that fanatical activity? An odd and socially underdeveloped personality is probably the most conspicuous consequence as far as my friends and family are concerned. On the positive side, there undoubtedly has also been the gradual induction of a rather vague and illogical brain into the endless fascination of science. But above both good and bad legacies, from my own point of view it has helped me to carry on from child to adult a deep and never-ending gasp of wonder. First induced in me by the shining violet of a carabid's cuticle beneath my earliest stones, there lighting dark earth where woodlice crept and showers of collembola leapt under my breath, then induced again and intensified by the deep velvet black and brilliant red and the white of a red admiral's {*Vanessa atalanta*} wing close to my face on the flowers of my mother's Michaelmas Daisies {*Aster*}, a long, long indrawn breath at the beauty of insects has stayed with me. I can still almost hear the hiss of the movement of those great wings, more beautiful in pattern on the underside than a Persian carpet, as they are raised and lowered and the butterfly basks in the sunshine on my mother's purple flowers. How clearly in imaginative memory I can still watch a silver-washed fritillary {*Argynnis paphia*} glide to a bramble flower in the heavy larch-scented air of our summer woods! How well I understood how Wallace, the co-discoverer with Darwin of the principle of Natural Selection, could describe himself almost fainting with joy when he first captured the Rajah Brooke Swallowtail in Borneo! How similar indeed have seemed my own emotions to this when, one evening in early September, I chanced on a colony of Britain's most brilliant lycaenid, the Adonis Blue {*Lysandra bellargus*} on a grassy hill slope amid the old (and also rare) juniper bushes within two miles of my home! I remember even the hour of this event, 6 o'clock, and how, still flying in the late warm sun, the butterflies flashed their wings as if displaying for me alone, like fallen flakes of the blue sky, and how when probing for nectar with closed wings on the purple heads of devil's bit scabious {Succisa pratensis} or settling to sleep on the grass heads, even their undersides seemed more clear cut, much more of every thing, than those of the Common and the Chalkhill Blues {*Polyommatus icarus & Lysandra coridon*} that I knew well before! So I had found it at last!—the Adonis! Never again would I be fooled by an unusually bright Common Blue hoping it was this! From that day halfway to the present in settings become utterly different (vertical sunlight breaching the deep forest in air heavy and warm as a bathroom) I have watched the flash of the same clear blue on a Morpho's wing that is perhaps fifty times the expanse of the whole lycaenid of my boyhood. Glimpsed first in a patch of light far down a path ahead, I have waited for these new vast sky flakes come bouncing by with the same awe and excitement that I knew first standing by the Michaelmas Daisies, in the larch woods, or among the junipers of that grassy evening slope. Yet, for all the size and strangeness of exotic splendours, I have never known the excitement of the early experiences to be exceeded.

Such were my early insect aims and experiences. On the whole, it was only insects that were large, eye-catching, or possibly painful, adding the thrill of danger, that became my prey. A maiden aunt of my mother who long ago had collected herself, gave me her cabinet still full of her beetles. These were all neatly arranged as to species but had no locality or date labels on them. I wanted the space badly for my own collection and with the self righteousness of a True Collector following principles stated in my Books, which held that a specimen without locality and date label was Worthless, I threw all her collection out, keeping only a few spectaculars and a still smaller few that had been fully labelled. I have long regretted this vandalism: as my scientific interest in the marvellous diversity of beetles increased I have often found myself wishing to know whether my aunt did in fact collect this or that one of the British list, and whether a species was more common in her time than now, and what might have happened in fifty years in the Berkshire, which not only was her home county, but which became that county where I too have lived most of my life. But perhaps even more important for me than her cabinet, and never to be treated in such barbarian spirit, were my aunt's books. First they were borrowed from her and later inherited. A special new interest for me lay in her copies of popular translations from the works of J. H. Fabre, the French naturalist, from whose Victorian writings and quaint style I learned again, as from Ford, the lesson that there is far more interest latent in insects than mere collecting. Fabre's new wonder for me was the beginning of an objective study of behaviour.

At the same time that they created new ideals for me, these author mentors had made new confusion. Later at Cambridge University, and afterwards as a graduate student for the MSc and PhD at University of London, beginning to think more for myself, I began to see that the matters that all my revered Fs (Ford and Fabre), to say nothing of my revered D (Darwin), had left confusing might just possibly not be beyond comprehension. With this in mind I began a new onslaught on the mystery of p^2: 2pq: q^2 and its related topics. The endeavour brought me as if by a winding underground tunnel to pop up eventually into the sciences of genetics of populations and of domestic animal breeding as well as into a small smattering of applied mathematics. Finally I came to that field of evolutionary study that is now called 'Kin Selection'. Simultaneously, my small fund of knowledge of the Hymenoptera (started through Fabre and by my part time help with my mother's honeybees) plus a prodding towards True Science that had been applied during my zoology course at Cambridge (Wigglesworthian ramifications of juvenile hormone—not exciting but I could see their point) were jointly pushing me towards my next brush with insects. My ideas about kin selection were at last written down and submitted to a journal. I was pretty sure they were right—that is, that they were correctly argued. If right in this way, it was clear that no amount of evidence from nature would make them wrong; or, if it did, then at least for my comfort, Darwin's and Fisher's evolution versions would have to crash along with mine. Nevertheless, as an aspirant now of True Science I was acutely aware—was made to be so by scientists around me—that I had somehow by-passed (or as I said tunnelled under) an essential stage of scientific development. I ought now to make up, to fact-pile something somewhere, test something, become respectable. Besides which, anyway, I badly needed a change from life in my bed sitter, from Waterloo Station and from Senate House Library. Without even waiting for the editor's verdict on my paper, I set off for Brazil, hoping at least to pretend to test my principles by restudying some of the social and unsocial bees which C.D. Michener, W.E. Kerr and S.F. Sakagami had worked on. At the same time, I was also hoping to solve a less touched-on mystery of how the Brazilian social wasps managed to keep their colonies cohesive and cooperating in spite of highly multiple egg-laying queens. I am ashamed to say that I didn't get far with either objective during the year that I was there. The reason is probably that, almost compulsively following my boyhood self-training, I started to Collect social wasps and their nests wherever I went. The doing of this properly (as my aunt had not not done it in her travels) and labelling everything took up a great amount of time. I am further ashamed to say I didn't personally go on even to describe what I collected and observed, even though I realised that quite a bit of it must be 'new to science', even if most of it was not new to the Brazilian countrymen who led me to the nests; but closely observing the wasps and their ways in order to hunt them in this fashion added another intensively known group to the body of my knowledge. After my return my scientific conscience was to be relieved somewhat when my collections were quite eagerly taken over by Professor O.W. Richards and used in his major taxonomic work on the South American wasps.

In the course of my own work I observed many marvellous nest forms. My efforts to obtain nests or even just to get a standard sample of the wasps as specimens usually resulted in one or two stings from each nest, but luckily I never developed any of the venom allergy that not uncommonly impedes students of social insects. Rather, in my case, the opposite happened; I became less sensitive to stings. It was lucky because the wasps were often very fierce. Confronted by some magnificent new nest structure, which I thought perhaps had never been seen by scientists before, my old hunting instincts were aroused, just as in former days they had been by butterflies, but with the addition that here before me was big game, an army of Lilleputians that might, by their extreme numbers or subtlety of venom, leave me unconscious or even dead in the forest or savannah. In fact, my only nearly serious such accident in the course of the year was not with the wasps at all, but with a colony of the newly introduced African honey bee, and then there was a farmer and his family nearby to help.

This farmer, like me, had not realised what kind of bees he had. He had seen my one glove and the veil that I used for the wasps and he asked me to help him take honey from a 'strong' hive which, as it turned out, he kept in an upturned wooden packing case perched on top of a tree stump! Within a second or two of trying to prize the first board from the packing case with my machete I began to guess what sort the bees might be. Within about thirty seconds and with just one board half off from the box everything had clarified beyond any doubt: everyone anywhere near to the hive was going to have to flee for their lives. My hands, in so far as I could still see them through a brown fog of bees, had become boxing gloves: they appeared as large brown moving balls. One hand under its furious covering had the leather glove and the other only a sock which the farmer had provided. The bees were very quickly reminding me which hand was which while at the same time others all over sought successfully to remind me that I was wearing only tropical clothing. The farmer rushed to his house; I to my jeep. A mile up the mountain road when there were more bees inside my vehicle than were accompanying it outside, I jumped out and ran further on foot into the forest, and finally, when there seemed to be more bees inside my veil than outside, I flung that off as well and ran further. Half an hour later, daring to come back and dreading to find what had happened to the family, my first ominous encounter was with a pig dashing grunting up the road, evidently bound on the same errand as mine and trailing behind it a small but still furious retinue. But when I reached the farm, however, I found that no human or even large animal seemed to have been seriously stung. The farmer indeed was again in the open, engaged in breaking apart the remainder of his pig-sties to let his pigs escape. His chickens and ducks which could have escaped, however, had failed to do so, and all of them were already laid out and dead under the bushes where they had crept unavailingly for shelter. Yet somehow, as I said at the beginning, unlike what was the probable effect of my bomb explosion on my love of chemistry, events of this kind despite their pain and their aftermath (in this case one day of fever), do not seem deter me. I keep or help keep honeybees, and I still must walk up to any new nest of a social insect that I have never seen as near as I can and do so still with a joyous rather than falling heart.

Between the hunting of wasps and wasp nests, I was not neglecting my old inclinations for turning stones, rolling logs, and pulling loose bark. I was beginning to find, in the tropical strangeness of the fauna of the latter two places, one of my next main insect interests. This came to be in the lives, the sex ratios and, finally, in the remarkable major evolutionary potentials, of all kinds of tiny arthropod *inbreeders*. Examples of habitual inbreeders turned out to be fairly common in the confined spaces such as occur, for instance, under still-attached old bark, or in new or disused beetle galleries or similar places. I was shortly to realise, however, that they were even more common still in other enclosed sites, such as in galls, in the strange microcosm of the interior of figs, in enclosed flowers and in parasitised pupal cases of other insects. As had happened with the ideas about kin selection I had worked on earlier, identification of the evolutionary advantages that determine sex ratios turned out to have much wider applications than the insect examples that I had first been puzzled by and had used to illustrate and develop them. In general, however, although some of the new cases were not insects, the insects continued to provide the best examples of all; indeed it was as if in return for the affection and curiosity I paid to them, they kindly agreed to fit my quantitative theories more closely than a population or evolutionary biologist normally dares hope for. But then perhaps it is just that there are so many insects so diverse that if you look far enough you can find examples to fit any theory, however bad...

By no means all insects living under bark inbreed, however. It was only when I returned to this habitat many years later that I began to realise the full range of the extraordinary lives being lived there—lives surely that would have delighted even Fabre's taste for the bizarre and for quasi-moralistic reflections that they could cast towards humankind. As one example, consider that most extreme of all inbreeders that I know. It is not an insect I admit. It has eight legs rather than six as an adult and belongs where this feature suggests, among the Acari, but it is at least a member of the same great exoskeletoned and segmented supergroup or phylum, the Arthropoda. I had pulled a piece of white-rotten wood from a dead branch and had noticed some odd pearly droplets adhering to the velvety broken surface of fungal mycelium. Under a microscope the drops proved to be even more strange than I had imagined: each possessed a head and legs. After immersing them in water on a slide and using a microscope I was able see what was going on. Each drop was the swollen body of a female mite distended with eggs and young. Tiny fully formed adults were bursting from eggs inside their mother. I saw the hatchlings joining a throng that was already floating or slowly swimming about in the maternal fluid and soon made out that there were three kinds, two of them female (both numerous overall, one kind a lobster-clawed dispersal morph that clings to the hairs of larger emigrant insects and the other more plain) and one type of male. The males were very uncommon, one or two to a brood, but they were always first to hatch and once hatched were very active. They spent all their time searching for new sisters in the crowd and right there, within their mother's womb, copulated with them. Their genitalia were very large, so large that it was at first hard to guess from terminal protuberances of the body which end of a male mite was front and which back, the whole animal suggesting to me a physical embodiment of a Freudian Id. Incestuous love-making by babies within the womb: this is a very strange sexual and reproductive life, you may think. But surely it is hardly stranger than that of a half millimetre beetle that I found living and breeding under bark around and amidst exactly the same fungi that supported the mites. One peculiarity of this insect is in its sometimes being winged and sighted, and sometimes wingless and blind. It was an additional character of the wingless form, however, that of being almost devoid of pigment and therefore transparent, that enabled a student working with me, Victoria Taylor, to spot this beetle's truly most extraordinary physical characteristic. This is the transmission of a spermatozoon longer than the male animal producing it! Needless to say, the male cannot carry more than a few of his conger-eel-like gametes at one time. Nor can the female possibly accept more than a very few into the curiously sculptured (and doubtless and necessarily strong) receptacle of her sperm storage organ; of the formation and the behaviour of such sperm in both sexes much remains to be worked out. The beetles are of the genus *Ptinella* and are to be found in rotting logs all over the world (undoubtedly including Japan). How many other strange lives are in the same habitat! Within the same soft-rotten log as the Ptinella and the mites, leaping in size from the scale of such minute arthropods to the middle range, we may find the grooved and jet black Passalids that live in families and which hold together in the darkness of their galleries, in all probability, through their chorus of chirps. Both the adults, which are normally a male—female pair, and the larvae, who are long, white, fairly active grubs, up to some 10 or so in number, can make sounds. For the larvae, chirping must be quite important, for they have actually sacrificed one-third of an insect heritage for the purpose. Their third pair of legs have become atrophied to tiny organs that now have no function except to scratch on a tympanum and, by this means, probably they call to their parents like young birds wanting to be fed; it is known that the parents prepare a partially digested mash of the wood in which they live to regurgitate to the young ones. But the passalids, varying themselves between species a great deal in size but little in shape, are but part way to the opposite extreme of size of insects inhabiting a log. Just as the Ptinella are almost the smallest of beetles (and the very smallest of all in fact live in the sporulating tubes of polypore fungi that are usually nearby), so the giants of the beetle families Cetoniinae, Dynastinae and Prioninae, which include the very largest insects, also breed in this habitat. Many of the giants are dramatically horned in the males and indeed, whatever the size, horns and fighting with horns is a repeating theme in dead wood inhabitants in various families. I have not personally seen much of the truly exceptional giant insects of dead wood, but I have interested myself occasionally in the middling giants, the 'stag' beetles of the family Lucanidae, which are in fact quite closely related to the much less ornate Passalids just mentioned. If the horns, which are actually mandibles in the lucanids, are enlarged in a given species, the males are usually fierce fighters. Some, like the British stag beetle, seem to need the presence of a female to inspire combat, but others, like the extreme Chilean lucanid, Chiasognathus granti, engage any male on sight and fight until one is either thrown completely from the victor's vicinity or runs away. Charles Darwin (who encountered them but seemingly did not see fighting) was, unusually, wrong about the effectiveness of the seeming ridiculously long, tong-like mandibles of this species, thinking them to be used mainly for display and for charming the females. In nature, the males rather fight for command of the sap flows on the southern beech trees (Nothofagus) where the females come to feed, or else around the massed flowers of an ivy-like tree-climbing Hydrangea where females similarly assemble. The less strong or expert males are wrenched from their footing by their opponents and, by the combination of the long tongs and the long legs of the victor, are held far out over space and simply dropped away from the scene of rivalry. I have not witnessed quite natural tournaments of this kind with my own eyes but I confidently reconstruct them from accounts given by Chileans and from the remnants of a tournament by Chiasognathus and a replay of it that I did witness in Chile.

I had made a hurried trip from the town of Coyhaique, just eastward across the Andes, to Puerto Aisen on its western fiord of the Pacific. I had come to the more forested coastal zone, hopeful to find Chiasognathus as well as other insects. My whole day's walk and search in the area had been unlucky, but returning towards my bus stop with only few minutes to spare, I at last came upon a tree clothed in the climbing and flowering hydrangea under which a dozen or more male beetles were crawling about on the grass. I had no time to look for the real champion, the triumphant caster-down of the animals I was seeing. He doubtless must have been just out of sight among the bunched flowers above my head. I simply scooped all the beetles I could see from the grass into a polythene bag and ran for my bus. All the way in the bus back to Coyhaique a crackling from the bag in my back pack drew some odd stares from my neighbours. Listening to it myself I felt already that I knew Darwin must be wrong: it did not sound like any mere symbolic fighting. In fact I had seen some of the beetles fall to some quite vicious struggles even as I dropped them into the bag. By the time the bus had brought me to my hotel near the cordilleran pass, my captives had, however, quietened somewhat and outside, a chilly night with threat of frost was coming on. My room in the hotel was small and ill-lit and I was tired so I decided to leave the bag outside where cold would hopefully immobilise everything until morning. I was not hopeful that even when warmed up I would see much now: surely after all that noise on the bus there could not be much fighting spirit left, perhaps not even many live beetles.

By the end of my breakfast next day a brilliant sun was clearing the last traces of frost from the grass of the lawn outside. Taking my bag to the sunniest spot, I tipped all the beetles out on the grass to see what the remnants were and could do. To my pleasure, all the beetles were intact and alive. What they did was very simple and, in hindsight, quite understandable. Sluggish at first with the cold, but soon warming, they all turned on the grass and crawled towards me. Evidently I, looming above, was for them the tree they had fallen from last night, and they had every intention to go up it and to continue their battle for the flowers where doubtless, the females were soon due to arrive. Where they met as they converged towards me across the grass, they fought; as they climbed onto my boots and met there, they fought some more; and finally the few that reached the level of my calf or knee on my trousers fought most fiercely and desperately of all. With loud crackling of their serrate tongs rasped on cuticle they grabbed at one another. Having secured a grip, they endeavoured to wrench their opponent free of his holds and into the air. As I already described, when one succeeded he would reach out with his long legs, hold his struggling victim over empty space, and drop him. Here of course, they fell only a few inches into the grass and undeterred, the larger beetles would quickly repeat the journey to my boots and re-ascend: but when a small one was dropped he seemed to sense the degree of his outmatching, and he would instead run over the grass in some random direction and attempt to fly. What still interests me out of this tournament, and out of a rather similar one that I have conducted with British stag beetles {*Lucanus cervus*}, is that the overall victor turned out to be not the largest of the beetles, but second largest. Is there then scope for bluffing? Does it sometimes pay, perhaps in the more hopeful early days of mating, when discretion may indeed be a better part of valour, as the proverb puts it, to have more outward size and more awesome mandibles than your opponent, even when these features outreach your strength?

From such questions about these most macho of insects, the horned lucanids and dynastines, I have gone recently (following perhaps trends of the human world) to an interest in those rare and exceptional cases where it is the *female* that wears the horn, sings the song, or in general, takes the more demonstrative sexual part. What is so different about these species? Perhaps it would be most sensible for me to study songs rather than the horns, and to approach those tiny singers that trill in our damp and neglected rooms from *female* abdomens vibrated against wood or paper. These are the so-called book lice {*Psocoptera*}.

It is a strange fact that even in the common genus Lepinotus, where the female has been known to be the singer, for a long time there has been no study to determine the causes of the strange reversal, even though the animals are very easy to keep (fed upon yeast) in small containers. But alas, the songs that I used to hear so clearly from dusty book shelves or from woodpiles in an outhouse when I was a boy now pass by me unheard, just as, rather similarly, the two-millimetre bodies of the insects themselves begin to defeat my eyes. I can make up, of course, for the latter with spectacles and lenses, but this is much less convenient. I have therefore left such interesting tinies to others and have turned to study progress of the 'female insect liberation movement' at its other extreme of size. So instead of rapt and hearkening in the wood shed at home in Britain, you find me prone on the forest floor in Brazil, in the dusk, closely watching something entirely different. What is it? Well, at the moment it seems to be just a dead chicken. What am I doing? You will see in a moment: a heaving has started somewhere beneath the breast feathers of the chicken, as if from being dead, it had suddenly started to try to breathe. This movement, however, works its way down towards the crop. A mound grows bulbous, the feathers spread out. Suddenly bursting from a ragged hole at the base of the chicken's neck comes something shiny and green, grotesquely curvaceous, messy—a huge rotund insect, big as a golf ball, sparkling its vivid cuticle from under the blood and flesh and worse kinds of muck that cover its surface in the most glorious gold, yellow and green, as shown in the light of my torch. It has a long black horn sweeping back from its head across a thorax scooped as concave as most beetles have them convex, a full bull-dozer blade of hard body, glossy-metallic chitin warped as if into a phantasy sculpture by Henry Moore. As he—for this one is the male—pushes swiftly down through the feathers to the ground, his messy covering seems to slip lightly from his shoulders, like the disguise of crude clothing slipping from a king who has pretended to be a workman. In a moment, most of him is as resplendent and perfect as a cetoniine chafer daintily climbing down from a flower. There is, however, one thing that perhaps spoils his beauty and makes him a workman still—but also makes him far more interesting in my eyes because such male behaviour is so rare—the big pink ball of flesh, a good human mouthful and half his own size, that he is still carrying in his arms. Where is he taking it? I hold my breath and hardly notice the first heavy drops of the evening storm that are beginning to thump on my back. It is not often I get a chance to see all this, and usually the transactions about to be described take place in darkness beneath the chicken. My eye catches another movement; it is in the dead leaves and the mounds of loose fresh clay piled at the side, a pile thrown up as if by a mole. A moment later and the female is coming out to meet him, head and thorax bursting from her own less bloody hole in the soil. Is she the smaller, rounder, duller being which, as female, whether of horned lucanid, dynastine or scarab, she ought to be, her horns mere shadows and lumps and bumps of her surface, in comparison to those of her mate? No! This one is as large—in fact larger—than the male. A superb horn sweeps even further back over her body, which itself is glinting in the light of my torch, quite as brilliantly as the other. Only perhaps at the extremities of the raised and forked boss of the thorax where the horn can rest, may one judge it to be a little less flared and extravagant. But see, she's in a hurry, or else she doesn't like my torch: she almost snatches at the load of meat that the male holds out to her and backs with it instantly into the dark hole she came from. Who does she fight with that horn of hers? What does she fight for? We don't yet exactly know; so far the many minor skirmishes prior to such (temporary?) pairings as I describe seem to be quite unisex, all against all. Once again, the beetles are abreast of the fashion.

But now the male too has gone under the feathers again and the cold rain comes down on me in a torrent.

Soaked, I hurry to my dinner in the open iron-roofed canteen of Reserva Ducke. It is chicken here too—fried, of course—but I am thinking more of the mysteries of the forest undertakers I have been watching, and *their* chicken. Of how they sometimes bury their carcass entire, as a team, several pairs together, while other times, even though quite as many are present, they work to remove its flesh on the surface in the way I have described. When they work together, are they all siblings or cousins? Could they know about this?... Shivering a little, I think of how, by the time I am old, all these secrets of their work will be known, of how easily, then, we will super-attract beetles, if we care to, from large areas of forest by means of foetid chemicals... I think how, by that time, I will confidently arrange what I have thought of. I will leave a sum in my last will for my body to be carried to Brazil and to these forests. It will be laid out in a manner secure against the possums and the vultures, just as we make our chickens secure; and this great Coprophanaeus beetle will bury me. They will enter, will bury, will live on my flesh; and in the shape their children and mine, I will escape. No worm for me nor sordid fly, I will buzz in the dusk like a huge bumble bee. I will be many, buzz even as motorbikes, be borne, body by flying body out into the great Brazilian wilderness beneath the stars, lofted under those beautiful and un-fused elytra which we will all hold over our backs. So, finally, I too will shine like a violet ground beetle under a stone.

CHAPTER 4

SEX, SEXES AND SELFISH ELEMENTS

LAURENCE D. HURST

It was a chance conversation that led to me doing a DPhil (as they call their PhDs in Oxford) under Bill's supervision. In the summer of 1986, just prior to my final undergraduate year studying Zoology at Cambridge, I happened to talk to Nick Davies, our lecturer in behavioural ecology. We chatted about what I found interesting. The idea that male birds show off their parasite status to females was a really fascinating idea, I remarked. He suggested that I should therefore do a DPhil with Bill Hamilton in Oxford. Nick kindly wrote a letter of introduction and, sometime later that year, I found myself in Bill's office in the Zoology department, being interviewed by him and Alan Grafen for a DPhil position that they would jointly supervise.

At that first meeting I didn't know what to expect. Bill came over as quiet and thoughtful, but also more than a little bumbling. His hair was rather unkempt, his face rugged, his hands large and hardened. He was 50 that year but, not knowing this, I guessed he must be nearer 70. I was proud just to say that I had met him.

At one point the conversation turned to the question of what I might like to research. Parasites, birds and the MHC came up, but we moved off that quite fast. What else did I find interesting, they asked? *Chlamydomonas* and uniparental inheritance of chloroplasts, I said. Like humans, zygotes of this single-celled green alga inherit all their cytoplasmic organelles (chloroplasts, mitochondria) from just one of the two parents. Unlike humans, however, the gametes of *Chlamydomonas* are all the same size. I had met this in genetics lectures and thought it bizarre. Why, when you have made a zygote with cells the same size, each contributing equally in terms of chloroplasts, is one of the parental cell's organelles destroyed? This seemed to me no different from any of the other paradoxes that I had been introduced to in behavioural ecology lectures. We had a discussion about the relevant facts, but none of us were especially well up on the details. Bill and Alan agreed that these were interesting questions. There was some talk of mating types and I confused matters by bringing up relative sexuality, the still poorly understood ability for some species to make up their minds what sex they might be after meeting a potential partner.

Anyone might think that it was then a straight line from this interview to the paper Bill and I published together,[1] relating the control of uniparental inheritance to the evolution of mating types. Nothing could be further from the truth. Soon after my arrival Bill and Alan sat me down to decide what I should do. The parasite and showy birds story was being heavily worked on and neither thought this a good area to go into. I suggested looking for phylogenetic inertia in behavioural traits. Alan squashed that one early on (to my eternal gratitude). Bill had an idea. He was interested in the wood formation of members of the *Compositae* and suggested that I see what literature there was on the subject. I asked why this was interesting and got back an answer, but not one that I understood nor, consequently, one that I can now relate. I guess it was something to do with parasites. Alan looked somewhat sceptical, but, as I had no better idea, I hit the libraries. I would report back in a month and they would see what progress had been made. There was, it turned out, quite a bit of literature on wood formation in the *Compositae*, although for the most part, it is rare within the group. I copied all the papers for Bill and duly handed them over. I remained none the wiser. Alan suggested this topic wasn't going anywhere. I could only agree, and Bill was happy for me to look in a different direction.

In the end, I fumbled my way onto a project. Much inspired by Hamilton and Axelrod's ideas,[2] I had been thinking what might happen when parasites mix. I thought sperm might be small to avoid such mixing of vertically transmitted parasites.[3] Quite by accident, uniparental inheritance of cytoplasmic factors was back on my horizons.

Bill was very encouraging and commented extensively on early drafts of this, my first, paper.[4] In one such draft, I had noted that pollen infected with viruses compete less well than uninfected ones. Bill picked up on this and added that this suggests '*a possible advantage to long styles*'.[5] In the margin he scribbled: '*It might be interesting to consider whether plants with exceptionally long styles are unusually plagued with vertically transmitted viruses. One thinks immediately of the long flowers of Nicotiana and tobacco mosaic virus—if that really is a special disease of Nicotiana*'.[6] Later he notes that one of the few cases that I found of efficient pollen transfer (Alfalfa Mosaic virus) is in a plant with very short flowers. Bill might have thought immediately of the long flowers of *Nicotiana* and the short ones of alfalfa, but I had no idea what they looked like. I was, nonetheless, happy to add the conjecture to the paper.

So where should my studies go next? Bill at the time had been working with Richard Stouthamer on the problem of asexuality induced by vertically transmitted microbes.[7] That maternally transmitted microbes should distort sex ratios was clear in Bill's early papers.[8] I asked Bill if anyone had ever written a review of the incidences of these as, if this literature had not been trawled, it seemed like an obvious next step for me: from why have uniparental inheritance, to what happens if you have it. Bill said he knew of no such review and thought it an excellent idea. Alan agreed and so, while not knowing it at the time, I had started using Bill's own research method: sit in a library and read.

I spent the next two years solidly sifting through any paper that might even mention a strange sex ratio, anything to do with cytoplasmic factors and anything that took my fancy just because it looked odd. When I found a particularly good report of this or that I would often go into Bill's room and announce that I had found a really intriguing paper and present him with a copy. This I discovered was a good way to start a conversation and to pick his brains. Rich Ladle, who joined to do a DPhil with Bill after me, solved the same problem by 'bribing' him with interesting beetles that he found while out sampling. His greatest success were some metallic green chrysomelids that had sex continuously. These prompted, from Bill, some energetic impromptu speculations on the efficacy of mate-guarding and sperm competition.

My offerings of interesting papers rarely raised Bill's interest as much as Rich's beetles. Nonetheless, if not berating me for pronouncing coccid like psocid and thereby getting him very confused, Bill would helpfully say that he had met something similar and go in hunt for the references. After much searching through his extensive card index, he would pull out a dozen or so cards and hand them to me, insisting that I return them when finished. On each was scribbled the details of a paper and usually some notes on its contents. Many of these feature as some of the more obscure references in his papers.

These cards were, he often told me, his 'extra-bodily grey matter'. He couldn't remember all the papers he had read, so this was his only record of many of them. The note cards were extremely valuable to me as a way into the literature. I don't recall ever finding great insights in Bill's notes on them, but that was not what I was looking for. I thought that Bill had contributed a lot to the review that subsequently appeared several years later.[9] He, didn't want to be co-author, saying that I had done just about all of the work. I did, however, make a point of providing a special acknowledgement of the extra-bodily grey matter.

While Bill certainly did rely on these note cards, I am sure he also had a special place in his real grey matter where he left a compendium of matters outstanding. One day while Alan and I were having coffee, Richard Dawkins and Bill joined us. The coffee room was just outside the library and that morning Dan Promislow came out and announced that he had a just read a paper about a coccid (definitely not a psocid) in which the males were minute appendages on the females' legs. Bill was ever so animated. Which species was it, he asked? Dan and Bill rushed into the library. Clearly a missing piece of some Hamiltonian jigsaw puzzle was about to be put in place. Richard and Alan asked me to remind them just what a coccid was.

It was a similar episode that led to the shorter of the two pieces I wrote with Bill.[10] I had, as usual, been reading that day's new journals when they arrived in the department's library. There was a very interesting paper on the discovery of a sexual representative of a group otherwise thought to be asexual.[11] I chatted it over with Rich Ladle who was doing his thesis with Bill on the evolution of sex. We decided that it would be worth writing up as a news piece for TREE. I handed Bill a copy of the paper and asked him what he thought of the idea of a short commentary. At the same time, I had been working on cytoplasmic sex-ratio distorters, not least of which were the parthenogenesis inducers that Bill had written about earlier.[12] While the literature was patchy, it seemed clear that certain lineages had more of one type of sex-ratio distorter than others. Inbred wasps, for example, seemed much more commonly to have asexuals than other species, and at least in some cases this was due to the vertically transmitted microbes (now known mostly to be Wolbachia). Bill and I had talked about this and it was clear that it tallied with his view that there might be lineages that, for whatever reason, produce asexuals at a higher rate than others. What we then see in such clades is mostly asexuals, doomed to failure probably, with a central core mother species spawning off her asexual descendents. The paper by

Pernin and colleagues[13] seemed to fit with such a pattern. Bill was happy to use this paper as a hook for the idea.

Bill's contribution came mostly in the latter third of the published paper. The references to *Rubus* and *Taraxum* possibly matching the pattern were Bill's insertion. It was Bill's idea to put in the reference to martians observing (or rather failing to observe) human sex. The close reader might also note a change of style in the last paragraph. This was all Bill's work. I remember when I first read it, it seemed almost poetical ('the softest, moistest of moss cushions where life is most benign'). Bill, as I now discover,[14] had an inordinate fondness for moss.

Bill's contribution to the other paper we published together was more substantive. It was an obvious jump for me to go from asking about uniparental inheritance and its relationship to the sexes in anisogamous organisms (i.e. why have small sperm), to asking about mating types in isogamous ones and the control of organelle inheritance. I came up with a simple verbal model in which biparental inheritance is bad, as it permits the spread of organelles that are aggressive to others. This would lead to the spread of a nuclear enforcer of uniparental inheritance, which in turn makes the conditions for the evolution of choice and hence of mating types. I was very much thinking at this time in behavioural ecological ways and only later reworked the model to have greater genetical relevance[15–17] (see also[18]). A very similar idea had been put forward, but dismissed, by Rolf Hoekstra.[19] I never understood why he dismissed it. I had reviewed as much of the mating type and organelle inheritance literature as I could find and things seemed to fit very nicely. That ciliates and fungi, which didn't allow cytoplasm to mix, often had very many sexes was just as expected. I talked the evidence through with Bill. He remarked that, in his mind, he would know that an idea was right when the exceptions started to fit into place. This, no doubt, informed our emphasis in the brief introduction.

The only reason he did maths, Bill once told me, was to get his papers past the journal editors. I was never quite sure if he was joking. Nonetheless, for publication something more than a verbal model was needed, so I put together a mathematical model. I had come upon a technical problem that I couldn't see the way out of. Somewhere in my recursions I had generated a quartic equation. I just couldn't see how to provide a neat analytical solution. I had a chat with Bill. He asked me to leave my notes on the maths and let him think it over for the weekend. On Monday he returned with hand written set of equations (in pencil) and a printout from Mathematica, with prototypes for figures a and b. My own model was fine and tractable, but my nomenclature was utterly dreadful, he remarked. Coming from the author of the paper with the most impossible nomenclature,[20] this was pretty severe criticism.

Bill suggested we try *Nature*. Someone had told me that *Nature* papers were by necessity short but you could cram loads into figure legends without them noticing, hence the size of the figure legend. A few weeks after submission we got referees' reports back and a letter saying that, while the manuscript was interesting, one of the referees thought the piece was too speculative. The other two reviewers had been positive. I told Bill and showed him the letter. Bill commented how *Nature* had it in for him, that his PNAS sex review[21] should have been published there, but it had a bad review from a former colleague. Bill was a bit down at the time. He shrugged his shoulders and advised that we submit elsewhere.

The paper sailed through the system at *Proc R Soc B*, and *Science* picked up on the story, doing a two-page spread on it.[22] This was very much a case of the left hand not knowing what the right was doing, as after *Nature*, we sent the paper off to *Science* only to have it returned a few days later unrefereed, with the usual by-line that it was 'not of adequate interest to the readers'. Bill shrugged his shoulders at that one as well.

After the spread in *Science* the story became news for a short while (sex and selfishness are a good mix for science journalists). Bill was very gracious and refused to take any of the credit. On the back of it, I had the honour of presenting, along with Paul Sherman, the Crafoord lectures, at the time when Bill received the Crafoord prize in Stockholm, in 1993. A few days before the ceremony, the Zoology department in Uppsala invited Bill and me to visit and to present a talk. Bill gave a strikingly novel lecture on sphagnum bogs and Gaia. He concluded that while sphagnum bogs may be a level of selection, Gaia couldn't work. I had not the slightest inkling that he was thinking about such things.

Prior to this, I hadn't much socialized with Bill; that wasn't his style. He was always a little difficult in conversation and pregnant awkward silences were not uncommon, both socially and in lectures. On one occasion, presenting a talk in the department of Zoology in Oxford, when attempting to remember the name of a coccid (yes, them again), he took his hands to his face and stood in silence for what must have been many minutes, punctuated only by the strange humming noise of Bill musing. I doubt that anyone in the room would have heard about the species in question, which, on this occasion, Bill couldn't remember anyway.

In Uppsala, Bill and I both stayed in the same hotel and one day after dinner we took the lift up together. In the lift I thought I would try and lighten the mood, so asked him a playful question. It was what is sometimes called the good fairy question. What if a good fairy could answer any question you really wanted answering. Any question, no matter how big—a 'life the universe and everything' sort of question. He asked me to go first. A phylogeny of everything I thought would be fascinating, not least because we will not get it without neo-divine intervention. Bill's turn. Pregnant silence. Was I going to get an answer before the lift let us out? I pushed him for a reply. Not quite a question, he replied. What he really wanted was to be taken back in time. To the time of the dinosaurs. His gift from the good fairy, was to really know how big the dinosaurs were. Bill wanted to understand to the bottom of his boots, what Tyrannosaurus rex was really like.

Not long after, I left Oxford and my interactions with Bill became relatively rare as our interests diverged and Bill was spending more and more time in the tropics. Our paths crossed again when, in 1998, I edited a review by Stuart West and colleagues[23] that advocated the development of mixed models to understand the evolution of sex. Bill's parasite model, they argued, may be part of the explanation, but you also needed to factor in the sorts of forces that Alexey Kondrashov had discussed, relating to mutational decay. It was my responsibility to find authors to comment on the review, comments that would be published alongside the review.

As prime developer of the parasite hypothesis, Bill was an obvious candidate to write such a critique. At around this time Bill and I were both speaking at a meeting organised by G.C. Williams at Stonybrook, New York, so I broached the issue then. I followed up with a letter and then a phone call. He apologized but said he was not keen on writing the piece. Not only was he very busy preparing for a trip to the tropics, but he didn't see the point of exploring mixed models until someone could show him that he was wrong. Alexey Kondrashov also thought the enterprise premature until someone could show him that he was wrong. In the end Kondrashov did contribute a critique[24] but Bill, to my great regret, stayed silent.

References

1. L. D. Hurst and W. D. Hamilton, Cytoplasmic fusion and the nature of sexes. *Proc. R. Soc. Lond. B* **247**, 189–194 (1992).

2. R. Axelrod and W. D. Hamilton, The evolution of cooperation. *Science* **211**, 1390–1396 (1981).

3. L. D. Hurst, Parasite diversity and the evolution of diploidy, multicellularity and anisogamy. *J. theor. Biol.* **144**, 429–443 (1990).

4. See no 3.

5. The quotes are from Bill Hamilton's manuscript comments on a draft of L. D. Hurst. *J. theor. Biol.* **144**, 429–443 (1990). The manuscript draft is in the possession of the paper's author.

6. See no 5.

7. R. Stouthamer, R. F. Luck, and W. D. Hamilton, Antibiotics cause parthenogenetic *Trichogramma* (Hymenoptera/Trichogrammatidae) to revert to sex. *Proc. Natl Acad. Sci. USA* **87**, 2424–2427 (1990).

8. W. D. Hamilton, in *Reproductive Competition, Mate Choice and Sexual Selection.* Blum, M. S. and Blum, N. A. (ed.) 167–220 (Academic Press, 1979).

9. L. D. Hurst, The incidences, mechanisms and evolution of cytoplasmic sex ratio distorters in animals. *Biol. Rev.* **68**, 121–193 (1993).

10. L. D. Hurst, W. D. Hamilton, and R. J. Ladle, Covert sex. *Trends Ecol. Evol.* **7**, 144–145 (1992).

11. P. Pernin, A. Ataya, and M. L. Cariou, Genetic-Structure of Natural-Populations of the Free-Living Ameba, Naegleria-Lovaniensis—Evidence for Sexual Reproduction. *Heredity* **68**, 173–181 (1992).

12. See no 7.

13. See no 11.

14. W. D. Hamilton, *The Narrow Roads of Gene Land Volume 2 Evolution of Sex* (Oxford University Press, Oxford, 2001).

15. L. D. Hurst, Selfish genetic elements and their role in evolution: the evolution of sex and some of what that entails. *Phil. Trans. R. Soc. B* **349**, 321–332 (1995).

16. L. D. Hurst, Why are there only 2 sexes? *Proc. R. Soc. Lond. B* **263**, 415–422 (1996).

17. J. P. Randerson and L. D. Hurst, Small sperm, uniparental inheritance and selfish cytoplasmic elements: a comparison of two models. *J. Evol. Biol.* **12**, 1110–1124 (1999).

18. V. Hutson and R. Law, Four steps to two sexes. *Proc. R. Soc. Lond. B* **253**, 43–51 (1993).

19. R. F. Hoekstra, in *The Evolution of Sex and its Consequences*, Stearns, S. C. (ed.) 59–91 (Birkhauser, Basil, 1987).

20. W. D. Hamilton. The genetical evolution of social behaviour I and II. *J. theor. Biol.* **7**, 1–16 and 17–52 (1964).

21. W. D. Hamilton, R. Axelrod, and R. Tanese, Sexual reproduction as an adaptation to resist parasites (a review). *Proc. Natl. Acad. Sci. USA* **87**, 3566–3573 (1990).

22. A. Anderson, The Evolution of Sexes. *Science* **257**, 324–5 (1992).

23. S. A. West, C. M. Lively, and A. F. Read, A pluralist approach to sex and recombination. *J. Evol. Biol.* **12**, 1003–1012 (1999).

24. A. S. Kondrashov, Being too nice may be not too wise—Commentary. *J. Evol. Biol.* **12**, 1031–1031 (1999).

CYTOPLASMIC FUSION AND THE NATURE OF SEXES[†]

LAURENCE D. HURST AND WILLIAM D. HAMILTON

Summary

Binary mating types are proposed to arise in a three-stage process through selection of nuclear genes to minimize cytoplasmic gene conflict at the time of gamete fusion. In support of this view we argue that: (i) in systems with fusion of gametes, the mating type genes are typically binary and regulate cytoplasmic inheritance; (ii) binary sexes have evolved several times independently associated with fusion, although at least twice binary types have been lost, associated with a loss of fusion; further, in accordance with the theory are findings for isogamous species that (iii) close inbreeding may correlate with less than two sexes and biparental inheritance of cytoplasmic genes; and (iv) species with more than two sexes may have uniparental inheritance of cytoplasmic genes, be rare and be afflicted by deleterious cytoplasmic genes which attempt to pervert normal cytoplasmic genetics. Such facts and their rationale support a new and unified definition of sexes based on the control of the inheritance of cytoplasmic genes. For the common cases, the male sex is that which resigns attempts to contribute cytoplasmic genes to the next generation. We differentiate between sexes and the incompatibility types of ciliates, basidiomycetes, some angiosperms and a few other organisms which are independent of organelle contribution.

Introduction

Why in a sexual population are there different sexes? In the most rudimentary condition the gametes are isogamous but are still differentiated into mating

[†] Proc. Roy. Soc. Lond. B 247 (1320), 189–194 (1992).

types. Following Hoekstra,[1] we extend the view that sperm evolved to prevent the mixing of cytoplasmic genes from different parents,[2–6] and further propose that the fundamental asymmetry of the sexes evolved as a means to minimize the damage caused by conflict between cytoplasmic genes in the zygote. A new definition of sexes based on the control of the inheritance of cytoplasmic genes clearly excludes the incompatibility types of ciliates, basidiomycetes, some angiosperms and a number of other species. With very few exceptions, sexes as newly defined are not multiple. They are either non-existent or binary, and these classes correlate well with the presence or absence of cytoplasmic fusion of gametes. The rare outliers to these generalities illustrate unusual conditions that render them also compatible with the theory.

The Model

Consider first a population of haploid sexual chlamydomonad-like algae. The gametes of the algae are isogamous and in their primitive condition are not differentiated into mating types. Any gamete can mate with any other. Let us assume that each gamete carries one chloroplast (any similar cytoplasmic organelle would be equivalent for the following argument), and that at first these chloroplasts are all passive, by which is meant they do not react to the presence of other chloroplasts introduced from the partner's cytoplasm during gametic fusion. Now consider the fate of a mutant chloroplast gene which destroys the chloroplast of the partner. Such a destroyer type will go to fixation so long as any costs incurred by its actions do not outweigh the twofold transmission advantage obtained by eliminating the competitor chloroplast. 'Homozygous' destroyer zygotes will be especially prone to costs due to balanced but still damaging activities. At worst, mutual destruction and death of the new zygote might occur.

Imagine next a nuclear Suppressor gene which leaves its organelle both unable to destroy its opponent and vulnerable to annihilation by that opponent. Its opposite allele in the genome is the wild-type Non-suppressor. Let the relative fitness of cells produced from a zygote formed by fusion of two unsuppressed cells be β, where $0 < \beta < 1$. Let there also be a cost to suppression such that the relative fitness of cells produced by the fusion of two suppressor cells is α, where $0 < \alpha < 1$. Cells which are the product of fusions between Suppressors and Non-suppressors suffer only half the cost of suppression and avoid the potential damage inflicted by two Non-suppressors. They have relative fitness unity.

The two costs specify a situation of heterozygote advantage which has a well-known equilibrium when, if p is the gene frequency of Suppressor and q that of Non-suppressor (with $p + q = 1$),

$$p(1 - \alpha) - q(1 - \beta) = 0. \qquad (1)$$

Consider now an unlinked choosy gene which prefers its haploid cell to fuse only with one of the opposite suppressor type and thus tends to avoid non-optimal matings. We assume that Suppressor and Non-suppressor cells are somehow recognizable before irreversible fusion. Suppression is likely to be mediated by the production of a protein which might be directly assessed or which might have side consequences which reveal its presence. Alternatively, substances related to the preparedness of the chloroplast for aggressive fusion might be detected. Let the haploid cells or gametes mix and meet at random. For convenience we will consider a situation in which all cells have only two opportunities for mating. After the first encounter, if both of the cells accept the mating then it proceeds. If one or both of the cells reject union then the pair breaks apart and all those cells which did not mate in the first round make another attempt to find a mate. In this second phase, no cells demonstrate choosiness, hence mating is random. Entering this second round of mating has a cost that applies a fitness factor, ϕ.

The choosy gene starts to spread if its fitness in the average zygote that it forms when rare is greater than fitness in the non-choosing population at large. As in the standard theory above, this is

$$W = p^2\alpha + 2pq + q^2\beta. \tag{2}$$

Assuming the Choosers are so uncommon that their meetings with each other are negligible, and that they occur by mutation in random association with Suppressors and Non-suppressors, their mean fitness is

$$V = p^2(p\alpha + q)\phi + 2pq + q^2(p + q\beta)\phi. \tag{3}$$

It is easily seen that for any internal equilibrium of the suppressor type by heterozygote advantage, if $\phi = 1$ then $V > W$ and the Chooser gene can enter. Lowering ϕ, however, eventually brings a condition where invasion is barred. The critical ϕ^* at which entry is just prevented, obtained by solving $V = W$, is shown as a function of α and β in Fig. 4.1*a*.

Assuming that ϕ is above the critical value and Chooser enters, does it go to fixation? As Chooser invades it always moves q towards 0.5, hence reducing the frequency of necessity for repairing, although never to below 0.5. Thus if invasion is possible it proceeds to fixation. This point is illustrated by the lower sheet in Fig. 4.1*a* which shows the critical ϕ' for invasion by Non-chooser if Chooser is universal: clearly also a threshold is implied such that if $\phi' < \phi < \phi^*$, Chooser can still enter and be established if sufficiently assisted initially by genetic drift or hitch-hiking.

If Chooser occurs inseparably linked to Suppressor or to Non-suppressor we have, replacing (3),

$$V = p^2(p\alpha + q)\phi + 2pq + q^2\beta$$

or

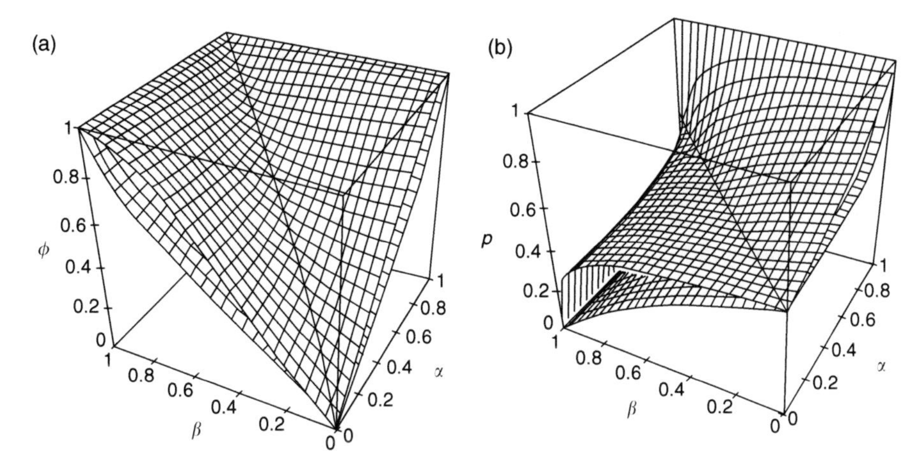

Figure 4.1. (*a*) Invasion and counter-invasion conditions for a gene causing choice of mating type, and (*b*) stable frequencies of a gene causing suppression of cytoplasmic conflict under choice and non-choice. (*a*) Critical ϕ^* for entry of a nuclear Chooser gene into a population of a random mating gene is given by the solution for ϕ of $V = W$ (equations (1) and (3), see text). This is

$$\phi* = (p^2\alpha + q^2\beta)/[p^2(p\alpha + q) + q^2(p + q\beta)], \qquad (4)$$

and is shown as the upper sheet. When Choosers are universal, (2) and (3) still give the mean fitness of Non-Choosers and Choosers but equilibrium is no longer as in (1); instead, frequency-dependent selection as in (3) holds, giving equilibrium condition

$$p(1 - p\alpha\phi) - q(1 - q\beta\phi) = 0. \qquad (5)$$

Thus for a critical ϕ' for re-entry of Non-chooser, (5) instead of (1) must be solved with $V = W$ for ϕ. The resulting quartic solution, obtained iteratively, is shown in the lower sheet of (*a*). Sheets coincide along $\{\alpha = 1, \phi = 1\}$, $\{\beta = 1, \phi = 1\}$ and $\{\alpha = \beta, \phi = 2\alpha/(\alpha + 1)\}$. (*b*) shows equilibrium frequencies p of a nuclear gene suppressing cytoplasmic conflict under two conditions: the more uniformly slanted sheet applies to random mating (absence of Chooser) and displays the classical solution of (1): $p = (1 - \beta)/(2 - \alpha - \beta)$. The more terrace-like sheet emerging into full view to the left applies to universal presence of choice for mating-type choice and displays $p = (A - \sqrt{AB})/(A - B)$, where $A = 1 - \beta\phi$ and $B = 1 - \alpha\phi'$ (by quadratic solution of (5)). Equilibrium frequencies under the two conditions have identical fixations along $\alpha = 1$, and $\beta = 1$, and are also identical along $\{\alpha = \beta, p = q = 0.5\}$.

$$V = p^2\alpha + 2pq + q^2(p + q\beta)\phi.$$

These give $V > W$ under similar conditions. Chooser invades if ϕ is large enough but now does not go to fixation even if $\phi = 1$, for clearly Chooser cannot carry the frequency of either linked suppression type to above 0.5. However, it is noteworthy that Chooser mutations of all three linkage conditions—unlinked, linked to Suppressor, linked to Non-suppressor—converge in building a bisexual system in which Fisherian selection of sex ratio holds. For the case of non-linkage, the onset of a stable sex ratio situation is shown in Fig. 4.1*b* in the wide 'terrace' around $q = 0.5$ for the states of completed invasion. Still wider, flatter terraces are observed if more than one round of choice is allowed. Other simpler but related choice systems that begin the same tendency to a typical sex ratio situation are mentioned below.

Realistic values of the three parameters are unknown. It is, however, reasonable to presume that α and β will be less than unity. In a parallel indication of cost, the gene determining insensitivity to the meiotic driver Segregation Distorter in *Drosophila melanogaster* is known to inflict a cost relative to the wild-type Non-suppressor.[7] Similar costs have been conjectured to be involved in the suppression of various sex ratio distorters.[8] The cost of choice would be dependent on the ecology, and could possibly be anywhere in the range zero to one, but under numerous circumstances is likely to be closer to zero.

The model can be elaborated for three or more rounds of mating for more complex choosy genes which have a strength of preference which declines with time. By extension of the above result, there is a trade-off such that the decline over time of the strength of preference balances the cost of choice. Unless there is no cost to choice a number of residual cells which have failed to find an opposite type mate will pair without the exercise of preference.

As detection that the host cell and another are not alike is difficult to imagine occurring by a single mutation, it is interesting to note that approaches to similar states to the one described can be attained with simpler choices. A gene could simply determine a preference for a Suppressor cell, or another mutant for Non-suppressor. Such preference genes can be selected because they will tend to establish themselves in linkage disequilibrium with the Suppressor genes such that Non-suppressor preference genes have linkage to cytoplasm Suppressor genes and vice versa. This is because when they are in such association they get an advantage. Subsequent selection favours increasingly tighter linkage.

The above model was originally outlined by Hoekstra,[1] who then commented that 'this scenario appears to work only if the selective differences between the various genotypes are very great, which makes the hypothesis less plausible.' We believe the model to be both plausible and consistent with a sizeable body of data.

The Fundamental Asymmetry of the Sexes: Discussion

The above model shows that if sex involves gametic fusion, a species should evolve nuclear genes to prevent conflict between cytoplasmic genes. In any given pairwise gametic fusion there should be an asymmetry between the cells to mediate the prevention of conflict. We propose that this asymmetry is the fundamental asymmetry of the sexes and is quite different from the system of incompatibility types exhibited by ciliates (see ref. 9) and basidiomycetes (see ref. 10) in which nuclei are exchanged but cytoplasm is not (in basidiomycetes: see refs 11–15; in ciliates: see ref. 16). These incompatibility systems are thought to have evolved to prevent inbreeding and maximize the number of potential reproductive mates, and thus are comparable to the incompatibility alleles of angiosperms (see ref. 17). Significantly, the angiosperms have sexual differentiation as well as incompatibility alleles, supporting the view that sexes and incompatibility types are two separate phenomena. Similarly, in discussing the Foraminiferan *Metarotaliella parva*, Grell argues that the organisms have a system of multiple incompatibility alleles governing choice of mate, but when fusion is initiated the interaction is mediated by a binary mating system. A similar system governs mating in tunicates (see ref. 18).

Table 4.1 presents the incidences of various numbers of sexes and incompatibility types. In the table, having zero sexes is interpreted as not fusing in any way that allows genetically different cytoplasmic genes to become mixed, whereas having '< 1' incompatibility type means being compatible for nuclear DNA exchange or donation with every individual. In the majority where fusion takes place there are just two sexes. When fusion does not occur our argument suggests that sexual compatibility is best thought of as controlled not by 'sexes'.

Analysis of species with gametic fusion suggests that there exist two means for the prevention of cytoplasmic gene conflict. First, and most commonly, there is

Table 4.1. The incidences of sexes and incompatibility types

(Common, convergently attained systems are boxed. Systems without gametic fusion are boxed in thin lines. The table is not exhaustive but conveys the general patterns. Notes on the incidences: (*a*) It is uncertain whether any organism has only one sex but this might be an intermediary stage between the possession of two sexes and their absence. (*b*) Wild-caught samples of *Alexandrium excavatum* have been shown to have more than two sexes. However, it has not been shown that within one population there are more than two sexes, and hence the position of this species in the table is uncertain.)

		Number of incompatibility types	
		<1	>1
Number of sexes	0	acellular slime moulds during plasmodial fusion	(*a*) most ciliates[9] (*b*) hypotrich ciliates if conjugating[26,27] (*c*) basidiomycetes[10]
	1	possibly a few homothallic fungi	
	2	(*a*) most organisms[9,22,23] (*b*) peritrich ciliates[28] (*c*) possibly hypotrich ciliates if fusing[26,27]	(*a*) *Metarotaliella parva* (Foraminifera)[9] (*b*) tunicates[18] (*c*) angiosperms[17]
	>2	*Alexandrium excavatum* (Dinophyta)[29]	acellular slime moulds during gamete fusing, e.g. *Physarum polycephalum*[30,31]

the uniparental inheritance of cytoplasmic genes, and second, there is the rapid segregation of cytoplasmic genes from the two parents after gametic fusion. Uniparental inheritance of organelles can be mediated either by making sperm very small or by the unisexual destruction of cytoplasmic genes just after or before entry into the zygote. The nuclear mating type genes of Chlamydomonads determine the inheritance of organelles such that mitochondria are inherited from the minus type parent but chloroplasts are inherited from the plus type, despite the fact that the zygote inherits both types.[19] Supporting the view that cytoplasmic gene conflict is central to the process, Chiang[20] has shown that the chloroplasts from + and – type parents fuse, and chl DNA from both types attempt to destroy each other by means of restriction endonuclease digestion. Typically, however, the + type DNA is destroyed more slowly than the – type DNA. Other reports claim a unilateral disarmament of the chloroplast genes by nuclear genes in the – type gamete.[21] The same association of isogamy and binary mating types with uniparental inheritance of cytoplasmic genes has been described in a large number of other protists.[22–25]

In yeast, the zygote which is heteroplasmic for mitochondria produces buds which are homoplasmic. Random segregation would be too weak a process to allow this.[32] The above model predicts that a nuclear gene with mating type-specific behaviour controls this process. Diploid *a*/α cells produce buds in different positions from both haploid cells and α/α and *a*/*a* diploid cells.[33] The bipolar positioning of buds in the *a*/α cells is probably contributary to the non-random return to homoplasmy.

Examination of the mating patterns of ciliates reveals more than one phylogentically independent evolution of binary sexes associated with fusion. Peritrich ciliates have lost conjugatory nuclear exchange and have evolved sex by fusion of gametes.[28] Significantly, sexes are binary and fusion is anisogamontic between a macro- and microconjugant.[28] Isogamontic gametic fusion has been described in at least five genera of hypotrich ciliates.[26,27] Normal ciliate conjugation and gametic fusion occur in the same population.[26,27] Significantly, in every locality there can exist multiple mating types with respect to conjugatory mating, but exclusively binary types with respect to gametic fusion. In one instance[26] within a syngen there exist five 'sexes', four inducers of fusion and one superclass of non-inducers. The inducers are mutually compatible, but are not found in the same locality. The viability of between-locality hybrids is typically significantly lower than within-population progeny,[34] suggesting genetic divergence. Alternatively, these hypotrich ciliates might represent a further example of a specialized class of binary neo-mating types which are associated with an asymmetry in the possession and donation of genetic material.[35] The best-known example of this is the bacterial F plasmid which initiates genetic transfer and only those without the plasmid act as recipients.

The outliers in table 4.1 are interpretable such that they are consistent with the theory presented above. Acellular slime moulds such as *Physarum polycephalum* have more than two sexes. Laboratory studies show that any *Physarum* gamete can mate with any other so long as they are genetically non-identical at all of three different polymorphic loci.[36] The inheritance of mitochondria is determined by what appears to be a linear hierarchy of at least thirteen alleles at one of these three loci. The fact that cytoplasmic genes are uniparentally inherited is support for our theory. Meland *et al.*[31] provide further support for our view in arguing that uniparental inheritance involves unilateral digestion of cytoplasmic genes. The problem posed by the *Physarum* case history is why, if multiple sexes can be competent at preventing cytoplasmic gene conflict, do most species have only two sexes? The most obvious explanation is that a system in which the asymmetry of sex is established only after zygote formation is inherently more vulnerable to selfish, zygote-injurious cytoplasmic genes than one in which the asymmetry is established before fusion (as is the case in all other instances). One prediction of this view is that *Physarum* will be vulnerable to selfish cytoplasmic genes which attempt to prevent uniparental inheritance of cytoplasmic genes. It is thus significant that one such unusual gene has already been identified in *Physarum*.[36] A plasmid in the mitochondria of one of the gametes forces fusion of mitochondria at zygote formation. After splitting apart, all of the mitochondria have this plasmid. If this sort of effect is costly for the nuclear genes, then a gamete which preferred to avoid this problem by destroying its own cytoplasmic genes and mating with those gametes which are prepared to let their cytoplasmic genes be transmitted (proto-females) can spread if the advantages outweigh the cost. Hence multiple sexes systems might be expected to collapse to binary types. An alternative view of the collapse of multiple sexes has been proposed by Iwasa & Sasaki[37] who argue that, depending on the mating kinetics, either two or an infinite number of sexes are to be expected. In contrast to the model presented here, these authors do not allow the possibility that gametes might not have any sexual differentiation, and hence they do not explain the origin of binary mating types. However, Iwasa & Sasaki's models are not incompatible with the one presented here and their explanation for the preservation of binary mating types might go some way to explaining why *Physarum*-like 'relative sexuality' is so rare (see also refs 38, 39).

In consistently inbred isogamous organisms a conflict-reducing mutant can spread, regardless of whether the mutant is a nuclear or cytoplasmic gene. This is because in inbred organisms all such genes behave as if they are linked. As a consequence, there is no need for nuclear genes to be asymmetric with respect to the control of the inheritance of cytoplasmic genes. Fusion of plasmodia of the slime mould *Physarum polycephalum* is restricted to individuals which are genetically identical at an array of loci (except when one plasmodia can parasitize the other, as reported by Lane & Carlile.[40]) It is thus significant that, as we predict, there is no evidence of anything other than biparental inheritance of cytoplasmic genes (M. Carlile, personal communication) and hence no indication of any asymmetry between the partners. Inbreeding homothallic fungi (e.g. *Neurospora africana:* ref. 41) and algae would be good candidates for further study. We predict that, if highly inbred, these organisms could have biparental inheritance of cytoplasmic genes. We also predict that microsporidians which lack both mitochondria and chloroplasts have less likelihood of having binary mating types.

Hoekstra[1,42,43] has considered the possibility that the evolution of mating types was forced by problems in finding mates. Hoekstra argues that some cells could evolve to specialize as attractor gametes whereas others specialize to become attracted gametes. This view is not incompatible with the one presented here. Indeed, at the point of invasion it is in the interests of every cell to honestly signal its suppressor type. This could be achieved by amplifying the signal which was used at the outset to discriminate between Suppressor and Nonsuppressors. Such a system does not require, as Hoekstra's model does, that some cells specialize as pheromone producers whereas others specialize to become receivers. Yeast's system in which both cell types produce and respond to mating type-specific pheromones[44] is equally understandable.

An alternative view of the relationship between mating types and uniparental inheritance has been proposed by Charlesworth.[45] Charlesworth argues that if an isogamous species already has sexes, then a selfish cytoplasmic gene which acts to guarantee its transmission by interacting with the + mating type locus to exclude the—type cytoplasmic DNA will spread to fixation. The model does not address the issue of the evolution of sexes.

Acknowledgments

We thank Rolf Hoekstra, David Haig, Alan Grafen, Paul Harvey, Angela Maclean and Andrew Pomiankowski for discussion and review of the manuscript.

References

1. R. F. Hoekstra, The evolution of sexes. In *The evolution of sex and its consequences* (ed.) S. C. Stearns, pp. 59–91. (Birkhauser, Basel, 1987).
2. L. M. Cosmides and J. Tooby, Cytoplasmic inheritance and intragenomic conflict. *J. theor. Biol.* **89**, 83–129 (1981).
3. L. D. Hurst, Parasite diversity and the evolution of diploidy, multicellularity and anisogamy. *J. theor. Biol.* **144**, 429–443 (1990).
4. P. Grun, *Cytoplasmic genetics and evolution.* (Columbia University Press, New York, 1976).
5. I. M. Hastings, Population genetic aspects of deleterious cytoplasmic genomes and their effect on the evolution of sexual reproduction. *Genet. Res.* (In the press, 1992).
6. R. Law and V. Hutson, Intracellular symbionts and the evolution of uniparental inheritance. *Phil. Trans R. Soc. Lond.* B (In the press, 1992).

7. C.-I. Wu, J. R. True, and N. Johnson, Fitness reduction associated with the deletion of a satellite DNA array. *Nature, Lond.* **341**, 248–251 (1989).
8. P. H. Gouyon, F. Vichot, and J. M. M. Van Damme, Nuclear-cytoplasmic male sterility: single-point equilibria versus limit cycles. *Am. Nat.* **137**, 498–514 (1991).
9. K. G. Grell, *Protozoology*. (Springer-Verlag, Berlin, 1973).
10. P. R. Day, Evolution of incompatibility. In *Genetics and morphogenesis in the Basidiomycetes* (ed.) M. N. Schwalb and P. G. Miles, pp. 67–80. (Academic Press, New York, 1978).
11. L. A. Casselton and A. Economou, Dikaryon formation. In *Developmental biology of higher fungi* (ed.) D. Moore, L. A. Casselton, D. A. Wood and J. C. Frankland, pp. 213–229. (British Mycological Soc. Symp. no. 10). (Cambridge University Press, 1985).
12. G. May and J. W. Taylor, Patterns of mating and mitochondrial DNA inheritance in the agaric basidiomycete *Coprinus cinereus*. *Genetics* **118**, 213–220 (1988).
13. W. E. A. Hintz, J. B. Anderson, and P. A. Horgen, Nuclear migration and mitochondrial inheritance in the mushroom *Agaricus bitoquis*. *Genetics* **119**, 35–41 (1988).
14. J. L. C. Baptista-Ferreira, A. Economou, and L. A. Casselton, Mitochondrial genetics of *Coprinus:* recombination of mitochondrial genomes. *Curr. Genet.* **7**, 405–407 (1983).
15. M. L. Smith, L. C. Duchesne, J. N. Bruhn, and J. B. Anderson, Mitochondrial genetics in a natural population of the plant pathogen *Armillaria*. *Genetics* **126**, 575–582 (1990).
16. J. R. Preer, Genetics of the Protozoa. In *Research in protozoology*, vol. 3 (ed.) T.-T. Chen, pp. 129–279. (Pergamon Press, Oxford, 1969).
17. D. De Nettancourt, *Incompatibility in Angiosperms*. (Springer Verlag, Berlin, 1977).
18. R. K. Grosberg, The evolution of allorecognition specificity in clonal invertebrates. *Q. Rev. Biol.* **63**, 377–412 (1988).
19. P. Bennoun, M. Delosme, and U. Kuck, Mitochondrial genetics of *Chlamydomonas reinhardtii:* resistance mutations marking the cytochrome *b* gene. *Genetics* **127**, 335–343 (1991).
20. K. S. Chiang, On the search for a molecular mechanism of cytoplasmic inheritance: past controversy, present progress and future outlook. In *Genetics and biogenesis of chloroplasts and mitochondria* (ed.) L. Bucher, W. Neupert, W. Sebald and S. Werner, pp. 305–312. (North Holland, Amsterdam, 1976).
21. R. Sager, Genetic analysis of chloroplast DNA in *Chlamydomonas*. *Adv. Genet.* **19**, 287–340 (1977).
22. J. M. Whatley, Ultrastructure of plastid inheritance: green algae to angiosperms. *Biol. Rev.* **57**, 527–569 (1982).
23. W. G. Eberhard, Evolutionary consequences of intracellular organelle competition. *Q. Rev. Biol.* **55**, 231–249 (1980).
24. M. Mirfakhrai, Y. Tanaka and K. Yanagisawa, Evidence of mitochondrial DNA polymorphism and uniparental inheritance in the cellular slime mould *Polyshondylium pallidum:* effect of intraspecies mating on mitochondrial DNA transmission. *Genetics* **124**, 607–613 (1990).
25. B. B. Sears, Elimination of plasmids during spermatogenesis and fertilization in the plant kingdom. *Plasmid* **4**, 233–255 (1980).

26. T. Takahashi, New mating types of syngen 1, with reference to stocks that induce total or temporary conjugation in *Pseudourostyla levis* (Ciliophora). *J. Sci. Hiroshima Univ.* B, Div. 1 **27**, 165–173 (1977).
27. J. Yano, Mating types and conjugant fusion with macronuclear union in *Stylonychia pustulata* (Ciliophora). *J. Sci. Hiroshima Univ.* B, Div. 1 **32**, 157–175 (1985).
28. I. B. Raikov, The macronucleus of ciliates. In *Research in protozoology*, vol. 4 (ed.) T.-T. Chen, p. 147. (Pergamon Press, Oxford, 1972).
29. C. Destombe and A. Cembella, Mating-type determination, gametic recognition and reproductive success in *Alexandrium excavatum* (Gonyaulacales, Dinophyta), a toxic red-tide dinoflagellate. *Phycologia* **29**, 316–325 (1990).
30. S. Kawano, T. Kuroiwa, and R. W. Anderson, A third multiallelic mating type locus in *Physarum polycephalum. J. gen. Microbiol.* **133**, 2539–2546 (1987).
31. S. Meland, S. Johansen, T. Johansen, K. Haugli, and F. Haugli, Rapid disappearance of one parental mitochondrial genotype after isogamous mating in the myxomycete *Physarum polycephalum. Curr. Genet.* **19**, 55–60 (1991).
32. B. Dujon, Mitochondrial genetics and functions. In *The molecular biology of the yeast saccharomyces*, vol. 1 (ed.) J. N. Strathern, E. W. Jones and J. R. Broach, pp. 505–635. (Cold Spring Harbor Laboratory, New York, 1981).
33. D. G. Drubin, Development of cell polarity in budding yeast. *Cell* **65**, 1093 (1991).
34. T. Takahashi, Properties of synconjugant clones from total conjugation in *Pseudourostyla levis* (Ciliophora). *J. Sci. Hiroshima Univ.* B, Div. 1 **31**, 17–25 (1983).
35. R. F. Hoekstra, The evolution of male-female dimorphism: older than sex? *J. Genet.* **69**, 11–15 (1990).
36. S. Kawano, H. Takano, K. Mori, and T. Kuroiwa, A mitochondrial plasmid that promotes mitochondrial fusion in *Physarum polycephalum. Protoplasma* **160**, 167–169 (1991).
37. Y. Iwasa and A. Sasaki, Evolution of the number of sexes. *Evolution* **41**, 49–65 (1987).
38. J. J. Bull and C. M. Pease, Combinatorics and variety of mating-type systems. *Evolution* **43**, 667–671 (1989).
39. H. W. Power, On forces of selection in the evolution of mating types. *Am. Nat.* **110**, 937–944 (1976).
40. E. B. Lane and M. Carlile, Post-fusion somatic incompatibility in plasmodia of *Physarum polycephalum. J. Cell Sci.* **35**, 339–354, (1979).
41. R. L. Metzenberg and N. L. Glass, Mating type and mating strategies in Neurospora. *BioEssays* **12**, 53–59 (1990).
42. R. F. Hoekstra, On the asymmetry of sex: evolution of mating types in isogamous populations. *J. theor. Biol.* **98**, 427–451 (1982).
43. R. F. Hoekstra, Evolution of gamete motility differences II. Interaction with the evolution of anisogamy. *J. theor. Biol.* **107**, 71–83 (1984).
44. C. L. Jackson and L. H. Hartwell, Courtship in *S. cerevisiae*: both cell types choose mating partners by responding to the strongest pheromone signal. *Cell* **63**, 1039–1051 (1990).
45. B. Charlesworth, Mating types and uniparental transmission of chloroplast genes. *Nature, Lond.* **304**, 211 (1983).

COVERT SEX†

LAURENCE D. HURST, WILLIAM D. HAMILTON AND
RICHARD J. LADLE

'An Evolutionary Scandal' is how Maynard Smith[1] has described the persistence of asexuality in bdelloid rotifers, a whole taxon in which no male has ever been observed. With so much theory currently predicting such diverse and substantial benefits to having sex, the problem Maynard Smith identified was how anything could survive without sex for as long as the bdelloids probably have.

Rotifers might not be the only taxon with a predominance of asexuality. For instance, Norton and Palmer have suggested that for oribatid mites asexuality is the dominant state and that sexuality is a rare and sometimes derived condition.[2] Aphids have a whole tribe of species (Tramini) that lack males completely (all of them, perhaps significantly, closely controlled by ants).[23] Perhaps, however, by these standards the Protozoa should be considered more scandalous still, for with the exception of a few groups, such as the ciliates, asexuality is considered to be the norm.[3] Indeed, for many groups, such as the amoebae, asexuality is often thought to be a defining characteristic. How is it that when most other organisms require sex, so many of the Protozoa can get away without it? Recent work by Pernin, Ataya and Cariou[3] on the freeliving amoeba *Naegleria lovaniensis* suggests one possible escape from the quandary: the amoebae are not actually asexual, but instead are having sex, albeit in some covert fashion.

The French team collected 71 isolates of the amoeba from the same geographical area (along the course of the Moselle river) and investigated the amount of variation in the population by isoenzyme analysis. If any recombination between individuals was to be found it would almost certainly be found in isolates from the same area. Cariou and Pernin[4] had earlier established that the amoebae were diploid, a condition that is probably unusual for asexual single-celled organisms. Allelic variation at seven polymorphic loci revealed 45 distinct genotype associations. Significantly, analysis of single-locus variation revealed that most of the loci were close to Hardy–Weinberg equilibrium. This the authors present as evidence in favour of the idea that these amoebae have segregation and free recombination between alleles. The conclusion was further

† *Trends in Ecology and Evolution* 7(5), 144–145 (1992).

supported by the finding of a high number of distinct genotypic associations and an absence of linkage disequilibrium between genotypes at the different loci. Most of the multilocus genotypes were unique.

These results were in stark contrast to laboratory-raised populations. The Moselle isolates showed no signs of recombination after having been experimentally reared for three years. Certain reference strains had remained asexual for over seven years. Conversely, the gut-living *Entamoeba histolytica* has shown evidence of genetic exchange under experimental conditions but not in the wild.[5] However, for the *Naegleria* to be at Hardy-Weinberg equilibrium, recombination need not be a regular event. Theory predicts that one sexual generation exhibiting random mating is adequate to restore equilibrium within loci (for experimental verification see Ref. 6), although more such generations are required to equilibrate linkages.[7] In contrast, obligately parthenogenetic species or populations have been revealed to show large deviations from equilibrium.[8–10]

If the amoebae are having sex, how are they achieving it? Pernin *et al.* argue that nothing similar to nuclear fusion has ever been reported in *Naegleria* and that the sexual processes in amoebae remain an enigma. The authors do, however, cite one study of another amoeba, *Sappinia diploidea*, in which amoeba cells forming pairs and containing two or several nuclei have been described.[11] Pernin *et al.* point out that although adherence of one cell to another is a prerequisite to sex, it is by no means justified to conclude that genetic exchange occurs. For instance, an attempt by one protozoan to ingest another of the same species should not be confused with sex (even if such ingestion might be the evolutionary origin of sex[12]).

Could covert sex be an explanation of the survival of other seemingly asexual lineages? Received wisdom on the sex life of trypanosomes was, for many years, that they were completely asexual. In a story with parallels to the *Naegleria* one, isoenzyme analysis of *Trypanosoma brucei* revealed evidence of both diploidy and mating.[13] The suggestion was confirmed by the finding of recombinants in a test cross.[14] The mechanism, however, is still uncertain. Similarly, sexual recombination has been shown in *Plasmodium falciparum*[15] and is suspected in some *Leishmania*.[16] Thus, there exists a precedent for thinking that covert sex might explain many cases of apparent asexuality, and that simply not seeing sex between individuals is not adequate evidence for its absence. (By equal measure, a Martian hovering over planet Earth in a flying saucer might not see much sex in humans!)

Some asexual populations are known to have genotypes highly divergent from Hardy–Weinberg expectations and hence are probably not, or not much, influenced by covert sex.[10,17] If these populations simply represent one species in a genus or family whose other members are sexual, it is often assumed that the asexual species will simply go extinct in the near future. But what about those 'scandalous' groups in which the majority of species are asexual and morphological diversification indicates a long history of asexuality? How are these maintained?

So long as the diversity is not too great, the possibility exists that asexuality within a taxon might be due to the continual production of asexual species from a core sexual species. The asexual species might go extinct within a few hundred years but the species group would be maintained by the regular production of new asexual lineages off the core sexual species. *Naegleria lovaniensis* is only one of six species within the genus and the evidence for recombination in the other species is much weaker than it is for *N. lovaniensis*. Pernin *et al.* comment that at present *N. lovaniensis* is the only 'true species'. If the remaining species are indeed asexual then it could be argued that they might be asexual species branched from the sexual core. The same logic might also be applied to dandelions (*Taraxacum*) and blackberries (*Rubus*): in both instances there are numerous asexual species plus a few well-defined core species which are sexual.[18] If this is the case then the persistence of asexuality within some species groups might not require any special explanation other than one to account for the frequent production of asexual off-shoots. Help from maternally transmitted cytoplasmic genes[19,20] might be appealed to.

The trouble with the bdelloids, on this line of thinking, is that, unlike the *Naegleria* amoebae, they are not merely a genus but an order. Thus we seem to require that not just one but probably several core sexual species are as yet undetected. If they exist and we want to find them, hints suggest that we should look in core stable habitats of the group.[21,22] Just as we look for *Taraxacum palustris* in ancient meadows and for *Rubus ulmifolius* in rich and ancient hedgerows, core bdelloid species might be found, for example, in the softest, moistest of moss cushions where life is most benign and rotifers are abundant.

References

1. J. Maynard Smith, *Nature* **324**, 300–301 (1986).
2. R. A. Norton and S. C. Palmer, In *The Acari* (R. Schuster and P. W. Murphy, eds), pp. 108–136, Chapman and Hall (1991).
3. P. Pernin, A. Ataya and M. L. Cariou, *Heredity* **68**, 173–181 (1992).
4. M. L. Cariou and P. Pernin, *Genetics* **115**, 265–270 (1987).
5. P. G. Sargeaunt, *Trans. R. Soc. Trop. Med. Hyg.* **79**, 86–89 (1985).
6. P. D. N. Hebert, *Hydrobiologia* **145**, 183–193 (1987).
7. S. Wright, *Evolution and the Genetics of Populations, Vol. 2: The Theory of Gene Frequencies*, University of Chicago Press (1969).
8. D. J. Innes, S. S. Schwartz, and P. D. N. Hebert, *Heredity* **57**, 345–355 (1986).
9. J. E. Havel, P. D. N. Hebert, and L. D. Delorme, *J. Evol. Biol.* **3,** 391–440 (1990).
10. P. D. N. Hebert and T. J. Crease, *Heredity* **51**, 353–369 (1983).
11. L. P. Goodfellow, J. H. Belcher, and F. C. Page, *Protistologica* **10**, 207–216 (1974).

12. W. D. Hamilton, R. Axelrod, and R. Tanese, *Proc. Natl Acad. Sci. USA* **87**, 3566–3573 (1990).
13. A. Tait, *Parasitology* **86**, 29–57 (1983).
14. L. Jenni, *et al. Nature* **322**, 173–175 (1986).
15. D. Walliker, *et al. Science* **236**, 1661–1666 (1987).
16. D. A. Evans, *et al. Parasitologia* **29**, 165–173 (1987).
17. J. A. Chaplin and D. J. Ayre, *Freshwater Biol.* **22**, 275–284 (1989).
18. A. J. Richards, *Plant Breeding Systems*, Allen and Unwin (1986).
19. T. Stouthamer, R. F. Luck, and W. D. Hamilton, *Proc. Natl Acad. Sci. USA* **87**, 2424–2427 (1990).
20. L. D. Hurst, H. C. J. Godfray, and P. H. Harvey, *Nature* **346**, 510–511 (1990).
21. G. Bell, *The Masterpiece of Nature*, University of California Press, (1982).
22. Hamilton, W. D., Henderson, P. A. and Moran, N. A., in *Natural Selection and Social Behaviour: Recent Research and Theory* (R. D. Alexander and D. W. Tinkle, eds), pp. 363–381, Chiran Press, (1981).

Reference added in proof

23. N. A. Moran, *Annu. Rev. Entomol.* **37,** 321–348 (1992).

CHAPTER 5

RECURRENT VIRUSES AND THEORIES OF SEX†

W. D. HAMILTON

The review by Lin Chao[1] on sex in RNA viruses presents a balanced view of the merits of what may be called the 'positive eugenical' and 'negative eugenical' theories of sex as applied to these viruses. Judging between these I share his preference for the negative, both on his specific viral arguments and in general. However, his survey steps across an intermediate yet fairly distinct theory which I think has stronger evidence than either extreme.[2] In discussing the influenza virus he comes close to, but doesn't mention, some good evidence for this theory.

Outlining the 'antigenic shifts' of the hemagglutinin (HA) and neuraminidase (NA) surface proteins of the influenza A virus, no mention is made of the now strong evidence that the main rather distinct subtypes of these proteins exist in *limited number* and *recur* in the pandemics. In a recent text of virology[3] it is shown that in the world influenza A epidemics of 1890, 1900, 1918, 1957, 1968 and 1977, hemagglutinins occurred in the order H2, H3, H1, H2, H3, H1—each of the three HAs has returned with periods of 67, 58 and 59 years, returns being variably separated. The sequence for neuraminidase in the same epidemics was N8, N8, N1, N2, N2, N1; here there is less change and recurrence but, notably, one combined serotype (H1N1) returned to humans in 1977 after absence as a worldwide form for 59 years. The evidence for these sequences was based, in part, on what is not very reassuringly referred to as 'seroarcheology' but it is nevertheless cumulatively substantial. The same text points out that the search for new serotypes of either protein in the influenza A viruses of other animals—especially waterfowl, which seem to provide the most important wild reservoir—appears to be nearing exhaustion with a list now comprising 13 main types of HA and nine of NA.

The semiregular cycling of 'flu has many characteristics predicted by the parasite-coevolution theory of sexuality. According to this theory, alleles coding resistant or virulent variants of proteins are at times positively and at times negatively eugenical in populations of hosts and parasites, the benefit depending on the phases of their linked cycles. Effective recombination is most frequent

† *Trends in Ecology and Evolution* **7**(8), 277–278.

when gene frequencies are in the midrange, a state expected commonly under cycling, uncommonly from positive eugenical transitions, and to be absent under the negative or 'purifying' selection scheme. This indicates that selection for recombination has special power under cycling. In addition the 'flu cycles illustrate: (1) recombination creating new successful genotypes, out of (2) limited allele series, which are (3) changing order of abundance, (4) repetitively and/or chaotically, over periods allowing (5) several generations in hosts and many in parasites. Pursuing the last point, 59–68 years allows two generations of humans. Although a model for maintenance of sex on the basis of a two-generation cycle has been given[4] (if contested for realism[5]), human generations are probably not particularly relevant to natural 'flu. In respect of dynamical polymorphism[6] generated between influenza A virus and its natural hosts, it is likely that humans only receive novelty and do not contribute it. The nonhuman species known to be involved with the virus, in rough order of importance, are ducks, pigs, horses, chickens and turkeys. Some of these, notably ducks, could have up to tenfold more generations than humans in the periods in question. A cycle of 20 generations is not unlike mean periods obtained for dynamical polymorphisms in models.[2]

It has to be admitted that most papers concerning the theory so far have tended to downplay the importance of recombination for the parasites, and this could be a cause of Chao overlooking it. One model in its most realistic version had obligately asexual parasites[2]; most other versions have not treated the issue of recombination.[7,8] When a genetically coevolving parasite is given an option of sex in a model it often, though not always, depending on dynamical pattern,[4] quickly fixes asexuality.[2] After noting this, major simulation experiments have tried to simplify the extremely complex dynamics arising with many linked match loci by setting parasites wholly asexual. Often this must be unrealistic. If the host flees its parasite by recombination it is likely to be advantageous for the parasite to hunt by recombination. In other words, parasites may well need not only what the model has left them—mutation, supplying the analogue of 'antigenic drift' mentioned in the article—but also larger changes, which would hopefully parallel similar large changes in the host: these larger changes are the 'antigenic shift'.[1,3]

It has been remarked in the theory[2,9] that where the host possesses a facultative immune system, it deploys, through fast natural selection and reproductive expansion of clones of cells within the body, new sets of hostile environments for microbial invaders and presents these on a time-scale comparable to the invaders' own fast evolution. It is probably the immune-response factor more than any other that ensures that parasites of vertebrates, from schistosomes to viruses, retain sex at least to some degree. Where they lack it or greatly decrease its use, as in the case of trypanosomes, it is probably no coincidence that either the parasite clones have a facultative ability to vary their antigenic properties in a way rather comparable to the working of the immune system of their host, or/and hosts are highly inbred domestic animals which are relatively congenic and unchanging.

Even viruses can have facultative techniques of evasion. However, more relevant to the present theme is that, *potentially*, through segmentation and recombination, viruses escape immune recognition by recombination.[1] Why only potentially? Even if a cell is coinfected by several virus particles, recombination is pointless unless the cell harbours more than one genotype (or the virus is diploid and heterozygous—see below). Heteroinfection probably very rarely happens with influenza A in humans since normally whole populations have only one A serotype in circulation at a time. From the fact that antigenic shifts seem to reach us after being set up in other animal species, it is presumably more common for those animals to carry mixed infections and to have cells colonized by two strains. Animal populations in which this can occur must have a pattern more like that of the mild influenza C in humans: various very different and nonmutually exclusive strains exist at high though inconspicuous levels in the population. Just as C is probably more a native, long-established disease in humans, so may A be for birds. This general scenario fits well with the known diversity of subtypes of HA and NA present in populations of waterfowl.[10]

Failure of the published parasite theory to take seriously the possibility of ancient polymorphisms in pathogens, in contrast to the emphasis of such a possibility for hosts,[2, 6] a failure that I sketchily redress here, may also have tended to exclude the theory from the attention in Chao's article. I have to admit that until recently I have thought of a parasite's problems as more adequately solved, given the faster generation turnover, by mutation. But hemagglutinin and neuraminidase are proteins of comparable complexity to, say, histocompatibility antigens. Thought of crudely as keys and locks, if one technique—recombination—works for a factotum host, then a burglar virus may do well to use it also.

Chao's article does not mention that the highest rates of viral recombination are in the retroviruses, this seemingly connected with their diploidy.[11] A crux in the new and growing controversy over the possible role of polio mass vaccination in starting the AIDS epidemic[12–18] is a conflict of timing of forks of the tree of SIVs and HIVs if brought together with the known human and simian provenances.[19–21] A resolution may be that numerous long-diverged lineages coexist in natural simian populations. This would be similar to influenza A in waterfowl and influenza C in humans. If this view is correct the situation in monkeys is set up for evolution in a complex dynamical polymorphism and the way prepared for the vaccine problem now facing humans.

References

1. L. Chao, *Trends Ecol. Evol.* **7**, 147–151 (1992).
2. W. D. Hamilton, R. Axelrod, and R. Tanese, *Proc. Natl Acad. Sci. USA* **87**, 3566–3573 (1990).

3. B. R. Murphy and R. G. Webster, in *Virology* (2nd edn) B. N. Fields and D. M. Knipe (ed.), pp. 1091–1152, Raven Press (1990).
4. W. D. Hamilton, *Oikos* **35**, 282–290 (1980).
5. R. M. May and R. M. Anderson, *Proc. R. Soc. London Ser. B* **219**, 281–313 (1983).
6. W. D. Hamilton, *J. Genet.* **69**, 17–32 (1990).
7. S. A. Frank, *Heredity* **67**, 73–83 (1991).
8. S. A. Frank, *Evol. Ecol.* **5**, 193–217 (1991).
9. W. D. Hamilton, in *Population Biology of Infectious Diseases: Dahlem Konferenzen* 1982, R. M. Anderson and R. M. May (ed.), pp. 269–296, Springer-Verlag (1982).
10. V. S. Hinshaw, R. G. Webster, W. J. Bean, and G. Sriram, *Comp. Immunol. Microbiol. Infect. Dis.* **3**, 155–164 (1980).
11. J. M. Coffin, in *Virology* (2nd edn) B. N. Fields and D. M. Knipe (ed.), pp. 1437–1500, Raven Press (1990).
12. G. Lecatsas and J. J. Alexander, *S. Afr. Med. J.* **76**, 451 (1989).
13. G. Lecatsas and J. J. Alexander, *S. Afr. Med. J.* **77**, 52 (1990).
14. G. Lecatsas, *Nature* **351,** 179 (1991).
15. L. Pascal, *Univ. Wollongong Sci. Tech. Anal. Res. Program, Working Papers* **9**, 5–46 (1991).
16. T. Curtis, *Rolling Stone* March 19th, 54–108 (1992).
17. W. S. Kyle, *Lancet* **339**, 600–601 (1992).
18. B. F. Elswood and R. B. Stricker, *Res. Virol.* **144**, 175–177 (1993).
19. P. M. Sharp and W. H. Li. *Nature* **336**, 315 (1988).
20. R. M. Anderson, R. M. May, M. C. Boily, G. P. Garnett, and J. T. Rowley, *Nature* **352**, 581–589 (1991).
21. M. Eigen and K. Nieselt-Struwe, *AIDS* **4**, S85–S93 (1990).

CHAPTER 6

FURTHER HOMAGE TO SANTA ROSALIA

Discovery at last of the elusive females of a species of Myrmecolacidae (Strepsiptera: Insecta)

JEYARANEY KATHIRITHAMBY

I first met Bill at Imperial College, London, and he struck me as a quiet but very hard-working scientist. I was researching leafhoppers then, which are not strange and unusual insects, so other than polite greetings we did not engage in long conversations. However, when Bill came to Oxford in 1984 we met often to talk about Strepsiptera, as I had begun working seriously on this group in 1979. Bill was curious about all things strange—hence his interest in these bizarre entomophagous parasites. He regularly stopped by my room to chat about the latest find in Strepsiptera. It was just after Christmas 1999 that I last saw him, before his trip to the Congo in January 2000. He wanted to photocopy something to send Naomie Pierce, and had forgotten his copying key, so I lent him mine.

The undiscovered females in the family Myrmecolacidae (an apparent 'evolutionary scandal' *sensu* Maynard Smith[1]) gave rise to our article in *TREE* entitled 'More covert sex, the elusive female of Myrmecolacidae',[2] which was a follow-up to Hurst *et al.*[3] (also published in *TREE*) on 'Covert Sex'. The latter had observed that many Protozoa are asexual and proliferate without males. Bill suggested to me that we bring Myrmecolacidae into the discussion, pointing out, however, that in this taxa the 'scandal' was not the non-existence of the females, but rather that they had not been found.

He thought such an article would alert biologists to keep a look out for myrmecolacid females while in the field.

In the family Myrmecolacidae, the males parasitize ants and the females parasitize orthopterans.[4,5] The extreme sexual dimorphism and the dual hosts of the sexes fascinated Bill. Heterotrophic heteronomy is found only in two lineages in insects, both of which are parasites: the hymenopteran subfamily, Aphelinidae,[6,7] and the strepsipteran family Myrmecolacidae.[2,4,5,8] Of the 108 described myrmecolacids, only 8 females (without their males) are known; and in only 2 species have the males and females been recorded: matches of them might be conjunctional, as the sexes are based on neither morphology (due to their extreme dimorphism: males are free-living and females totally endoparasitic, except in one family), nor genetics. In the light of recent molecular studies, the sexes of conspecifics in Myrmecolacidae can only be matched by molecular data,[9,10] Kathirithamby *et al.*, unpublished. No males are known for the myrmecolacid *Stichotrema dallatorreanum* Hofeneder, a parasite of a pest of the *Sexava* group of long-horned grasshoppers in Papua New Guinea.[11] This species, however, is most probably a parthenogen (Kathirithamby, unpublished). At the time of writing the 1992 article, 87 species of male myrmecolacid had been described (there are now 108), and none of these had been matched to their conspecific females. An even stranger phenomenon was that hundreds of male *Caenocholax fenyesi* Pierce sensu lato[12] that are morphologically identical are found in traps throughout the Americas,[13] and Vietnam (Kathirithamby unpublished). But not a single female had been located. Later, two females parasitic in *Hapithus agitator* Uhler (Orthoptera: Gryllidae) were found in Texas and presumed to be female C. *fenyesi*,[14] but molecular data showed otherwise.[9] One of the hosts of C. *fenyesi* (now known as C. *fenyesi texensis* Kathirithamby and Johnston[10]) is the red imported fire ant *Solenopsis invicta* Buren in southern USA,[15] but the females have not been found. This has prevented any study on the use of male C. *fenyesi* as a biocontrol agent for this debilitating pest, which has now spread not only in the southern states of the USA but to Australia.

Bill did allow himself two outlandish remarks during the time of writing the article: that perhaps, the males reproduced parthenogenetically, or that the males from Texas flew over to New Guinea to mate with the females there!

Orr *et al.*[16] discussed the use of phorids as biocontrol agents for the fire ant. Bill and I replied,[17] arguing that the red imported fire ant had become aggressive and abundant in an alien environment where its natural parasites were absent. We added that, although the red imported fire ant lost its coevolved parasite when it was introduced into the southern states of United States in the 1940s, it gained another, the male strepsipteran C. *fenyesi*. We suggested that the females of C. *fenyesi* be hunted and that this species be used as augmented control for the red imported fire ant along with other microbial and introduced phorids to maintain a balance.

We have now not only solved some of the many doubts about the life cycle and geography of this species of Strepsiptera, but also found the female (or one of them) and its host.[10] It was Bill's interest in this group and particularly the questions he posed (as always when faced with a new and interesting organism) that motivated us to study some of the interesting aspects of Myrmecolacidae.

What intrigued Bill was the extreme sexual dimorphism of Strepsiptera, in particular the females, which are grub-like (except in one family, the Mengenillidae, where both the males and females emerge to pupate externally from the host). However, the neotenic female, in spite of its simple appearance, is a highly modified organism. Only the head and thoracic portion of the body (cephalothorax) extrude externally through the abdomen of the host. Female Strepsiptera are viviparous, and remain endoparasitic in a live host. The brood canal opening in the cephalothorax (not homologous to the oviduct in other animals) is used both for the transfer of sperm by the male, and also as an opening for the passageway in the apron for the emergence of the free-living 1st instar larvae. The brood canal opens into the genital ducts, which lead into the body cavity, where the oocytes/developing embryos lie free in the haemocoel.[18]

Another aspect of Strepsiptera that Bill was intrigued by was whether they changed the behaviour of their hosts. As early as 1939, Ogloblin[4] stated that stylopised ants 'change their nocturnal habits, acquiring positive phototropism, but evidently lose their social instincts, abandoning their nests and rambling singly, often climbing high on grass and bushes'. Ogloblin reasoned that this change in behaviour was why stylopised ants were never found, since myrmecologists often collected whole nests in order to obtain the various castes within the colony.

Bill and I discussed this at length, and he pointed out that, although ants form the largest number of invertebrates in many habitats, single stylopised ants have seldom been encountered/collected, unlike stylopised bees, wasps, and leafhoppers. Only two records exist of stylopised ants found in the open: Westwood,[19] who described an ant collected by John Nietner in Ramboddo, Ceylon (now Sri Lanka); and Ogloblin[4], who found a few stylopised ants in Argentina and Brazil. To date, the hosts of only 8 male myrmecolacids are known.[10] Although Ogloblin[4] stated that ants wander away from the nest, Bill and I speculated that stylopised ants remain in the nest until the emergence of male strepsipterans in order to avoid predation, and that this might be adaptive for the strepsipteran. In our recent studies of the collection and dissection of whole nests of ants, we found a large proportion of ants with extruded cephalotheca [the anterior head and thoracic regions of the male pupa] within the nest,[8,10,13,15,20] Kathirithamby *et al.*, unpublished. Although myrmecologists do collect whole nests the stylopised ants went unnoticed because the extruded male cephalothecae is cryptic (it lies hidden between the tergites or sternites of the abdomen, and is often the same colour as the host cuticle), and difficult to recognize.

At the end of the last endoparasitic instar the male extrudes the cephalotheca between the segments of the abdomen, and the last instar cuticle tans to form a puparium. The metamorphosis of the male takes place in two stages within the puparium (the pupal stage and pre-adult stage). During the pre-adult stage (the cuticle of the pupa is shed and is therefore not a pharate adult) the cuticle hardens (tans), the sperm matures, the wings expand and the flight muscles are developed (Kathirithamby unpublished). The only task a free-living adult male performs on emergence from the puparium is to excrete its meconium [waste products of pupal metabolism]. It then takes flight immediately, in order to seek and fertilize a female: the adult male has a very short life span (c. 5–6 h) and there is no teneral period [when the cuticle is incompletely hardened] after emergence from the puparium. Unlike in most other insects the metamorphosis to a free-living adult male in Strepsiptera therefore occurs within the puparium. (Kathirithamby unpublished). If a stylopised ant deserts a nest soon after the extrusion of the cephalothecae, it has to wander around for a long while until the male myrmecolacid completes development within the puparium. During this period the host ant will be prone to predation. Stylopised ants therefore remain in the nest until just before the emergence of the male, at which time they leave the nest.

One disappointment that Bill expressed about our article in *TREE* was that a figure of *Caenocholax dominicensis*, Kathirithamby and Grimaldi[21] (Fig. 6.1), described from the Oligocene-Miocene amber, which was

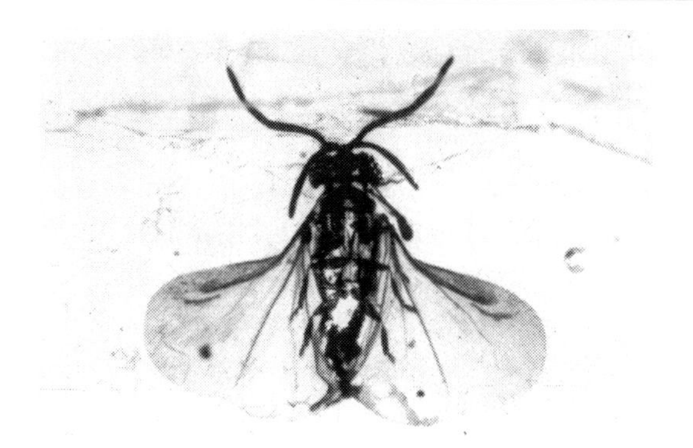

Figure 6.1. Male *Caenocholax dominicensis*
Kathirithamby and Grimaldi occurs in bright yellow piece of amber from the Dominican Republic (25 myo). (with kind permission of D. Grimaldi)

morphologically very similar to extant *Caenocholax fenyesi* Pierce sensu lato, was left out in the final version. We wanted to show that C. *fenyesi* has had a long evolutionary history. Instead, only a figure of extant C. *fenyesi* from Texas was published. The article in *TREE* was reviewed by the *Guardian*[22] and the *New Straits Times* (Malaysia).[23] Bill was rather amused by the publicity and remarked that sex was always a draw.

I have been looking for the female myrmecolacid ever since, and in fact was paying homage to Santa Rosalia (dubbed by Hutchinson[24] the patron saint of evolutionary biology). Hutchinson was looking for water bugs in Sicily that were described a hundred years before from the same location, but were never reinvestigated. Similarly, I was looking for the female C. *fenyesi*, the male of which was described 96 years ago from Veracruz, Mexico. In June 2002 I struck lucky while on a field trip to the Tropical Biological Research Station, Los Tuxtlas, Veracruz, Mexico. I travelled from Catemaco to Los Tuxtlas on the back of a farmer's van, called a taxi truck, driven by a woman. She very kindly wrung a wet towel and placed it on the wet wooden seat (it was raining very heavily) for me to sit on. We travelled for about two hours in this manner from Catemaco to the station, along a bumpy track. The flora around compensated for the discomfort during the journey, as it was reminiscent of that in Malaysia: Veracruz being the northernmost region in the Neotropics of the Amazonian rainforest. At about 8 pm on the first day of our trip, I dissected a batch of crickets. As I pulled the abdomen apart from the thorax of the first specimen, a succulent 4th instar female larva wriggled out. I stared down the microscope, stunned but fairly certain that

I had found the female C. *fenyesi*. One of my first thoughts was: 'If only Bill were here'. The next few days in Mexico were spent collecting, dissecting and fixing: due to sexual dimorphism the females can only be identified and matched to the males by molecular characterization. Using molecular tools, we found that the females parasitic in the cricket were 100% match to males dissected from ants in the same area, but the males from Texas and Los Tuxtlas were 15% divergent (Kathirithamby *et al.*, unpublished). So there is perhaps host-derived speciation in the Fenyesi group. This possibility has now opened up a new study on host speciation in Strepsiptera.

My husband Malcolm and I often invited Bill home for dinner when we had overseas visitors. One such visit in March 1998 when Spencer Johnston was visiting Oxford is worth mentioning. It was past 8.30 pm and Bill had still not showed up, so Malcolm persuaded me to ring him. I was reluctant but finally I did. Bill apologized profusely and said he had totally forgotten about the invitation and had just had a tin of soup but could he still come? He appeared within 15 minutes, and spent several hours talking, over an Indian meal (as with most of us, curries were a favourite with Bill too). Bill and Louisa likewise invited us to Wytham when they had visitors. One such guest was Raghavendra Gadagkar, and during this visit Bill showed us his prized insect collection. It was characteristic that this was of exotic members of most insect groups, unlike the personal insect collections of other biologists, which are usually confined to a few specific groups of interest only to the collector.

In August 1998, while holidaying at Impruneta in Tuscany at the home of Laura Beani, Bill came across aggregations of *Polistes* wasps, which were far from their nests, and sheltered under dead Bramble sprays or in the interstices of walls. He found this rather curious, since these aggregations were too early for hibernation. He brought the specimens back for me (he had previously given me two specimens of *Polistes* parasitized by male Strepsiptera which he had collected in Basel in 1987), and suggested that perhaps they were parasitized by Strepsiptera: similar to a parasitized *Polistes* "fussing around a very similar hibernaculum"[25] [site for overwintering; a group of hibernating individuals][26] "in a very similar month", which he had encountered "in the other hemisphere" [Brazil] "35 years ago'. I dissected the wasps and found 60% parasitization by *Xenos vesparum* Rossi. Bill, inquisitively posed questions about several aspects of the activity of the stylopised wasps in two e-mails to Laura Beani:[27] firstly, he wrote, "those aestivation/hibernation 'gatherings' that we saw around your house, with all the comings and goings, would be good places for exchange and infections of hitherto uninfected wasps to occur. . . . all the activity around the hibernaculum I saw might be transferring triungulin larvae" [minute 1st instar free-living, infective stage larvae with legs of some parasitic insects] "from one to another and perhaps to the uninfected adult who were also undoubtedly visiting the sites. Thus they would be getting chances to get into colonies . . . Perhaps they create somehow an aroma of 'very good hibernaculum' . . . to induce healthy wasps to come to them, or conceal their genetic identity so that they will mix maximally with members of other colonies . . . " But Bill in his characteristically humble way admitted that "I don't know enough about the biology. . . . you overestimate how much I know about stylops . . . ".

An energetic David Hughes studied the behavioural ecology of *Xenos*[28] and initial studies showed that the level of parasitism is high among *Polistes*.[29, 30, 31] Perhaps they do have an "aroma", since the aggregations were composed of females from different colonies. The aggregations occurred from July to early October and were formed of stylopised female *Polistes* that have deserted the nests early, i.e. even before the extrusion of the strepsipteran female cephalothorax and male cephalothecae. In these aggregations, the free-living male *X. vesparum* would emerge from the puparium (formed in the host wasp) by breaking the cap of the extruded cephalotheca and would seek and fertilize the female (endoparasitic in the host). It was hypothesized that nest desertion and aggregation by the stylopised wasps was in reality adaptive parasite manipulation of host behaviour to facilitate parasite mating: a parasite selfish herd.[30] Infections of Strepsiptera are at the larval or egg stage of the wasp, and triungulins are transferred to the nest by uninfected wasps as speculated by Bill.

Bill pondered further that spring aggregations "would seem to give the most immediate prospect of the triungulins reaching the larvae". We found that there are no spring aggregations, and that strepsipteran larvae reach the wasp nest either by being deposited on flowers by their mother (which overwintered inside the host), and are then carried by foraging workers back to the nest (phoretic) [a means whereby small insects use large insects for transport]. The high proportion of parasitized larvae in some nests also suggests a second possible mechanism: a direct release of 1st instar larvae close to combs by a wasp parasitized by a female strepsipteran.[29]

Bill wondered whether it is "possible to imagine these minute delicate animals surviving until nests are well started next spring?" and "how long and where the already-emerged triungulins can survive the exterior of an adult host on or in other wasp-associated materials, such as on the nest, while waiting for hosts, at the right stage. If triungulins can survive for a long time, the idea of parasitized females in spring 'saturating' likely hibernacula with young parasites would have good precedent in the behaviour of *Bombus* queens parasitized by the aberrant nematode *Sphaerularia*... in spring these queens show a very peculiar 'rehibernation' behaviour completely inappropriate to the time of the year and yet very appropriate to getting the likely hibernation sites filled with young worms ready for the autumn".

After being fertilized by the male in late summer/early autumn (which might occur in the aggregations), the neotenic female *Xenos* overwinters in the *Polistes* wasp[30, 31]. The female *Xenos*, during the overwintering period, has developing embryos that would emerge as free-living 1st instar larvae in the following spring (Kathirithamby, unpublished). In the spring, the overwintered stylopised *Polistes* do not build nests, but would wait around while the overwintered, unstylopised gynes [overwintered female] build their nests and lay their eggs. When the eggs hatch into larvae, the parasitized *Polistes* hang around the new nests and "plant/saturate" the "baby" strepsipterans near the new nests. The strepsipteran 1st instar larvae would emerge for the female, and find and parasitize the young *Polistes* larvae in the cells (egg parasitism has also been found).[29] *Xenos* enters a lag phase during the critical developmental stages of *Polistes*.[32] After the emergence of all the 1st instar larvae, the host (along with the now almost empty *Xenos* female) will die. "After all, if they can so completely 'castrate' the wasps without killing them, an ability obviously evolved in an association through an immense period of time, it will not be surprising to find that they also manipulate the behaviour". This is a parallel discussed by Bill involving "another long-evolved nematode of *Bombus*...... also able to pervert the behaviour of queens".

In contrast to the stylopised ants discussed earlier, stylopised *Polistes* desert the nest before the extrusion of the male cephalotheca and female cephalothorax. Hence the behaviour of stylopised *Polistes* is aberrant: early colony desertion and summer aggregates of castrated wasps at leks [sexual display and nuptial arenas] are formed in order to facilitate mating. This is the first example in *Polistes* wasps of behavioural change induced by a parasite.[30]

It was due to Bill's interest in the parasitization of ants and wasps by Strepsiptera that we are making some interesting and unusual discoveries on the host–parasite relationships of these organisms. For a man who was one of the greatest biologists of our time, I found Bill humble beyond measure. Whenever he talked to me about Strepsiptera he would draw parallels from a wide range of animals (with appropriate references), but would always conclude by saying '. . . I do not know enough about Strepsiptera'.

References

1. J. Maynard Smith, Contemplating life without sex. *Nature* **324,** 300–1 (1986).
2. J. Kathirithamby and W. D. Hamilton, More covert sex: the elusive females of Myrmecolacidae (Strepsiptera). *Trends in Ecology and Evolution* **7,** 349–51 (1992.).
3. L. D. Hurst, W. D. Hamilton and R. L. Ladle, Covert sex. *Trends in Ecology and Evolution* **7,** 144–5 (1992).
4. A. A. Ogloblin, The Strepsiptera parasitic of ants. In, *International Congress of Entomology* Berlin (1938) **2,** 1277–84 (1939).
5. J. Kathirithamby, A review of the order Strepsiptera. *Systematic Entomology* **14,** 41–92 (1998).
6. G. H. Walter, 'Divergent male ontogenies' in Aphelinidae (Hymenoptera: Chalcidoidea): a simplified classification and a suggested evolutionary sequence. *Biological Journal of the Linnean Society* **19,** 63–82 (1983).
7. M. Hayat, The Chalcidoidea of India and the adjacent countries. Part 2: family Aphelinidae. *Oriental Insects* **20,** 226–32 (1986).
8. J. Kathirithamby, *Stichotrema robertsoni* spec. n. (Strepsiptera: Myrmecolacidae): the first report of stylopisation in minor workers of an ant (*Pheidole* sp. Hymenoptera: Formicidae). *Journal of the Entomological Society of South Africa* **54,** 9–15 (1991).
9. N. R. Halbert, L. D. Ross, J. Kathirithamby, J. B. Woolley, R. R. Saff and J. S. Johnston, Phylogenetic analysis as a means of species identification within Myrmecolacidae (Insecta: Strepsiptera). *Tijdschrift voor Entomologie* **144,** 179–86 (2001).
10. J. Kathirithamby, and J. S. Johnston, The discovery after 94 years of the elusive female of a myrmecolacid (Strepsiptera), and the cryptic species of *Caenocholax fenyesi* Pierce sensu lato. *Proceedings of the Royal Society, London.* B (Suppl.3), Biology Letters, **271**, S5-S8. (DOI 10.1098/rsbl.2003.0078).
11. T. M. Solulu, Influence of *Stichotrema dallatorreanum* Hofeneder (Strepsiptera: Myrmecolacidae) on the performance of *Segestidae novaeguineae* (Brancsik) (Orthoptera: Tettigoniidae) in Papua New Guinea. MSc. Thesis Univ. of Oxford (1996).

12. W. D. Pierce, A monographic revision of the twisted winged insects comprising the order Strepsiptera Kirby. *Bulletin of the United States National Museum* **66,** 232pp. (1909).

13. J. Kathirithamby, and D. P. Hughes, *Caenocholax fenyesi* (Strepsiptera: Myrmecolacidae) parasitic in *Camponotus planatus* (Hymenoptera: Formicidae) in Mexico: is this the original host? *Annals of the Entomological Society of America* **95,** 558–63 (2002).

14. J. L. Cook, A *study of the relationship between Caenocholax fenyesi* Pierce (Strepsiptera, Myrmecolacidae) and the red imported fire ant *Solenopsis invicta* Buren (Hymenoptera, Formicidae). PhD. Dissertation, Texas A&M University, College Station (1996).

15. J. Kathirithamby, and J. S. Johnston, Stylopization of *Solenopsis invicta* (Hymenoptera: Formicidae) by *Caenocholax fenyesi* (Strepsiptera: (Myrmecolacidae) in Texas. *Annals of the Entomological Society of America* **85,** 293–7 (1992).

16. M. R. Orr, S. H. Selke, W. W. Benson, and L. E. Gilbert, Flies suppress fire ants. *Nature* **373,** 292–3 (1995).

17. J. Kathirithamby, and W. D. Hamilton, Exotic pests and parasites. *Nature* **374,** 769–70 (1995).

18. J. Kathirithamby, Morphology of the female Myrmecolacidae (Strepsiptera) including the *apron*, and an associated structure analogous to the peritrophic matrix. *Biological Journal of the Linnean Society* **128,** 269–87 (2000).

19. J. O. Westwood, Notice of the occurrance of a strepsipterous insect parasitic in ants, discovered in Ceylon by Herr Nietner. *Transactions of the Entomological Society of London* **5,** ser. 2: 418–420 (1861).

20. D. P. Hughes, G. Moya-Raygoza and J. Kathirithamby, The first record among Dolichoderinae (Formicidae) of parasitism by Strepsiptera. *Insectes Sociaux* **50,** 148–150 (2003).

21. J. Kathirithamby and D. Grimaldi, Remarkable stasis in some lower Tertiary parasitoids: descriptions, new records, and review of Strepsiptera in the Oligo-Miocene amber of the Dominican Republic. *Entomologica Scandinavica* **24,** 31–41 (1993).

22. *Guardian*, in G2., Dec. 3, 1992. Parasite lost, p. 11.

23. *New Straits Times* (*Malaysia*), Dec. 15, 1992. Mystery of the missing females. p. 30.

24. G. E. Hutchinson, Homage to Santa Rosalia or why are there so many kinds of animals? *The American Naturalist* **XCII,** 145–59 (1959).

25. Double quotes are from the e-mails of Bill Hamilton and single quotes are quotes of Bill's within his e-mails.

26. Explanations of entomological or other terms are in square brackets.

27. J. L. John, D. Hamilton Archive, e-mails to Laura Beani, September 3, 11:35:59, 1998; March 21, 11:20:00, 1999. British Library (in preparation.).

28. D. P. Hughes, The behavioural ecology of strepsipteran parasites of polistes wasps. D. Phil.Thesis University of Oxford (2003).

29. D. P. Hughes, L. Beani, S. Turillazzi, and J. Kathirithamby, Prevalence of the parasite Strepsiptera in *Polistes* as detected by dissection by immatures. *Insectes Sociaux* **50,** 62–8 (2003).

30. D. P. Hughes, S. Turillazzi, J. Kathirithamby and L. Beani, Social wasps desert the colony and aggregate outside if parasitized: parasite manipulation? *Behavioural Ecology* 15, 1037–1043 (DOI: 10.1093/beheco/arhlll).

31. D. P. Hughes, J. Kathirithamby and L. Beani. Prevalence of the parasite Strepsiptera in adult *Polistes* wasps: field collections and literature overview. *Ethology Ecology and Evolution 16*: 363–375.

32. D. P. Hughes and J. Kathirithamby. Cost of strepsiteran macroparasitism for immature wasps: Does sociality modulate virulence? O,Kos (in press).

MORE COVERT SEX: THE ELUSIVE FEMALES OF MYRMECOLACIDAE†

JEYARANEY KATHIRITHAMBY AND WILLIAM D. HAMILTON

Abstract

Among evolutionary biologists, considerable interest currently surrounds organisms that seem to be represented entirely by females. Here we discuss a less familiar problem—that of organisms in which females appear to be lacking. Our main example is the Myrmecolacidae (Strepsiptera), an unusual group of entomophagous parasites with dual hosts. Males parasitize ants and females parasitize orthopterans. Although the free-living adult males are frequently collected, the permanently endoparasitic neotenic females are elusive and are seldom encountered.

Strepsiptera are a small cosmopolitan order of insects. Their members, known as stylops, are exclusively parasitic and have extreme sexual dimorphism combined with an unusual life cycle. They parasitize 34 families of insects within seven orders.

Stimulated by the recent *TREE* article on taxa that seem to proliferate, even sometimes to ordinal level, entirely without males,[1] we focus here on the converse problem of a group within the Strepsiptera that seems to lack, or almost to lack, females. Although this is almost certainly a problem of the females not having been found rather than not existing, their elusiveness has continued for so long and over such large areas that the situation might well be called another 'evolutionary scandal.' The taxon in question is the family Myrmecolacidae.

Biology and life history of Strepsiptera

Adult male Strepsiptera are small active animals with prominent compound eyes and elegantly branched antennae. Their forewings are mere stumps and their hindwings are fan shaped; males fly in search of the flightless females, which can

† *Trends in Ecology and Evolution* 7(10), 349–351 (1992).

Figure 6.2. Male *Caenocholax fenyesi*
Head to 10th abdominal segment length: 0.93 mm.

only disperse passively through the movements of their hosts. Since males do not feed, this flight is their sole mission. It is also a once-only mission, at least for myrmecolacids. Writing of one of the rare cases where a female myrmecolacid is known, Luna de Carvalho[2] reported aedeagi (copulatory organs) embedded in the cephalothorax of *Stichotrema dallatorreanum*. Young[3] often found aedeagi in the cephalothorax of the same species. From these two accounts it seems that male myrmecolacids (and probably some other Strepsiptera) mate once and die. Such male sacrificial mating is rare: it occurs in honey bees, *Atta* leaf-cutter ants and coccids, and also in species of spiders, mantids and ceratopogonid midges,[4] the females of which sometimes nuptially consume their mates.

Strepsipteran males emerge from a puparium tucked between the segments of an adult insect host, most commonly a bee, wasp, a plant/leafhopper or a plant bug (less commonly thysanurans, cockroaches, flies). In contrast, female Strepsiptera (except of the Mengenillidae) are permanently endoparasitic. They are little more than cigar-shaped sacs bloated with eggs, with a small chitinized cephalothorax area exposed to the exterior of the host. The paired ovaries break up at the second instar of development and the oocytes then lie free in the haemocoel. Once a male has found a female he pierces the brood canal opening (which is not homologous to the common oviduct in other insects). Sperm enters the body of the female and fertilization occurs in the haemocoel.

Females are viviparous, and distribute their 1000 to 750 000 small larvae[5–7] as the adult host moves around. From the point of liberation the larvae reach the young of the host in various ways.

One order of Apterygota (Thysanura) and six orders of Pterygota (Blattodea, Mantodea, Orthoptera, Hemiptera, Diptera and Hymenoptera) are parasitized. After the exit of, say, a male parasite (or even of two of them), host adults remain alive, although only for a day or two. Nevertheless, the effect of the presence of stylops on host fitness is always drastic. The host is ‘parasitically castrated': though mobile, it is unable to reproduce.[8] Thus, while technically a parasite (by the host survival) a stylops is effectively a 'parasitoid'.

Family Myrmecolacidae

As already implied, several examples have shown that the female normally occurs in the same host species as the male. However, this has still to be taken on trust in most of the 532 described species as the host is often not known at all and most Strepsiptera are described only from the flying males.

For the intriguing family Myrmecolacidae, insect hosts of yet unproven associations may be involved. Out of the 87 described species of Myrmecolacidae, five are of females and only two of these have been associated with their males. However, both these females are found in completely different hosts to their males—not only different species but different orders. Where the source of males has been identified (eight species out of 87) their hosts have been ants of both the major ant subfamilies (Myrmicinae and Formicinae). The five females known are from mantids, long-horned grasshoppers and crickets.[9–13]

It is puzzling that more of the females are not known. They cannot always be rare and they are certainly not expected to be tiny. Unlike the usual stylops of a bee or wasp, where the female may not be much bigger than the male, the known myrmecolacid females are giants. In the case of *S. barrosmachadoi*, the female is large enough to distort the abdomen of her mantid host, being 20–30 mm long and some 1000 times more voluminous than the male. O'Connor estimated the number of eggs produced by the mature female *S. dallatorreanum* to be 750 000 (Ref. 7).

Although males of Myrmecolacidae are actually both more numerous and more speciose in the New World than the Old, only one female (*Myrmecolax ogloblini*) has been described in the former.[10] All other known females are from the Old World, examples being from New Guinea, Japan, Thailand, and Angola.

Myrmecolacidae males are not uncommon in *Solenopsis* fire ants from Georgia, USA, to Argentina. Possibly they originated in some other (native) fire ant species but now they occur most commonly in the abundant and aggressively spreading fire ant *Solenopsis invicta*.[14] A host switch (if such has occurred) only exacerbates the puzzle of where the females may be. Since there are not many insects of any kind with distributions ranging from Georgia to Argentina, what can be the mysterious hosts in which *Caenocholax fenyesi* females are hiding? In some areas of the US where the fire ants are a problem, extensive sampling of Orthoptera and other likely and common insects has so far failed to reveal the female.

Whatever strange divergent life cycle *Caenocholax* may have in fire ants and in the unknown host of the females, it is likely to have a long evolutionary history. Recent examination (J. Kathirithamby and D. Grimaldi, unpublished) of male stylops in the Oligocene-Miocene amber has revealed species indistinguishable from this genus and indeed apparently closely related to *C. fenyesi*.

The behaviour of parasitized animals is often abnormal, and the change is often to the advantage of the parasite rather than the host.[15] For example, queen bumble bees (*Bombus*) parasitized by the nematode *Sphaerularia bombi* wander restlessly in spring and perform inappropriate rehibernating behaviour, dispersing nematodes in places where they will be likely to encounter young queens in autumn. Rather similar restless behaviour has been claimed for the bees and wasps carrying adult stylops. In this case it can be imagined that their restlessness might maximize the placement on flowers. However, the same change does not seem to apply to ants stylopized by Myrmecolacidae since stylopized ants are rarely collected unless whole nests are taken.[16] Perhaps here the behaviour is the reverse: they do not wander away from the nest but remain in it until the male strepsipterans emerge. Since the worker ant is much safer from predators in the nest than outside, this could be adaptive for the stylops. The winged male, unlike the female, does not require the movement of a host to attain dispersal of progeny.

Finding the right host: other heteronomy cycles

In the case of Myrmecolacidae progeny, does a newly dispersed young male know that he needs an ant, and a female know that she needs a . . . whatever it may be? Probably not. Extensive breeding experiments[3] on *S. dallatorreanum* in New Guinea showed that when the host ant larvae and tettigonid were infected by the male and female first instar larvae, both sexes entered both the two hosts. In the ant host only the male larvae moulted, and in the tettigonid only the female larvae moulted to the second instar.[3] This indicates both that sex is probably determined by genetic differences and that instinct as to what to enter is not differentiated.

Myrmecolacidae are not the only insects that parasitize different insects to produce their males and their females. A subgroup of a family of tiny parasitic wasps, Aphelinidae, has many members with the same habit.[17,18] The normal and, doubtless, original hosts are various groups of scale insects and mealy bugs (Coccoidea), or else white flies (Aleyrodidea). No females are known to be raised other than as primary parasitoids of insects of these groups. From this a bizarre range of divergence of hosts has evolved between the two sexes. Least strange is the group where males are simply ectoparasitoids of a primary host, while females are endoparasitoids of the same. Next, and more unusual, come cases where male eggs are destined to become hyperparasitoids in other species of hymenopteran parasitoids of the primary host. Stranger still—perhaps strangest of all—there is a group where the male is a hyperparasitoid on larvae of females of his own species. A great difference between the two types of heteronomy is that in the Strepsiptera it is the first instar larvae that seek out and infect the new host, their actions leading to the seeming inefficiency in discrimination that we have already noted. In the Aphelinidae the mother makes the selection of the host; there is no inefficiency because the male-haploid system of the Hymenoptera enables the female to determine the sex of the egg.

The highly male-biased sex ratio in Myrmecolacidae may depend on the ratio of male- and female-producing hosts in the environment. The separation of host types can lead to skewed sex ratios, as in the obligate development of male and female aphelinid wasps, which under host limitation dissolve the trade-off between the production of male and female eggs.[19,20]

Thanks to the energy of past entomologists, such as S.E. Flanders of California,[21] aphelinid distributions, hosts and life histories are well known. Moreover, the group has inspired hypotheses as to how their 'heteronomy' may have evolved. The Myrmecolacidae present an even greater challenge.

Acknowledgements

J. K. wishes to thank the Leverhulme Trust for the Research Fellowship.

References

1. D. H., Hurst, W. D. Hamilton and R. L. Ladle, *Trends Ecol. Evol.* **7,** 144–145 (1992).
2. E. Luna de Carvalho, *Publ. Cult., Cia Diamante Angola* pp. 125–154 (1959).
3. G. R. Young, *Gen. Appl. Entomol.* **19,** 57–64 (1987).
4. J. A. Downes, *Mem. Entomol. Soc. Canada* **104,** 1–62 (1978).
5. J. Kathirithamby, *Syst. Entomol.* **14,** 41–92 (1989).
6. T. Kifune and Y. Maeta, *Insectarium* **27,** 170–286 (1990).
7. B. A. O'Connor, *Papua New Guinea Agric. J.* **2,** 121–125 (1959).
8. J. Kathirithamby, *Biol. J. Linn. Soc.* **10,** 163–179 (1978).
9. K. Hofeneder, *Zool. Anz.* **36,** 47–49 (1910).
10. A. A. Ogloblin, *Proc. Int. Cong. Entomol.* **2,** 1277–1284 (1939).
11. E. Luna de Carvalho, *Publ. Cult., Cia Diamante Angola* pp. 109–150 (1972).
12. Y. Hirashima and T. Kifune, *Mushi* **47,** 75–79 (1974).
13. T. Kifune, (1983) *Kontyû* **51,** 83–89
14. J. Kathirithamby and J. S. Johnston, *Ann. Entomol. Soc. Am.* **85,** 293–297 (1992).
15. J. Moore and N. J. Gotelli, in *Parasitism and Host Behaviour* (C. J. Barnard, and J. M. Behnek, eds), pp. 193–233, Taylor and Francis (1990).
16. J. Kathirithamby, *J. Entomol. Soc. South. Afr.* **54,** 9–15 (1991).
17. G. H. Walter, *Biol. J. Linn. Soc.* **19,** 63–82 (1983).
18. G. M. Walter, *J. Entomol. South. Afr.* **46,** 261–282 (1983).
19. H. C. J. Godfray and J. Waage, *Am. Nat.* 136, 175 (1990).
20. H. C. J. Godfray and M. S. Hunter, *Ecol. Entomol.* **17,** 89–90 (1992).
21. S. E. Flanders, *Entomophaga* **12,** 415–427 (1969).

EXOTIC PESTS AND PARASITES[†]

J. KATHIRITHAMBY AND W. D. HAMILTON

Orr *et al.*[1] in Scientific Correspondence describe the competitive dominance of the red imported fire ant *Solenopsis invicta* at food resources in Brazil, which is diminished when under attack by phorid flies, and suggest that this parasite could be used to control the ant in the southern United States. We can add an interesting note to the American story of *S. invicta*, and comment on the role of parasitism in structuring natural communities.

S. invicta might have escaped regulation by its coevolved parasitic organism in an alien ecosystem when it was introduced into the southern United States in the 1940s, but at some time after this it acquired another exotic parasite species, the strepsipteran *Caenocholax fenyesi*,[2] first described in Mexico, 1909. This parasite is common in South America, but its host in endemic areas is still unknown. We speculate that the strepsipteran was introduced into the southern United States with another host ant species, but now commonly occurs in the red imported fire ant. This is a good example of an introduced species (the red imported fire ant) which has become aggressive and abundant in an alien ecosystem where its natural parasites (including the phorid fly) are absent. Now the niche left by the natural parasite seems to have been taken over by a new parasite (the strepsipteran). It is to be hoped that this parasite may be followed by others (for example, microbial, introduced phorids) and a balance will begin to be restored.

C. fenyesi belongs to the curious strepsipteran family Myrmecolacidae, the males of which parasitize hosts belonging to a different order from the hosts of females. In the New World the male and female and their hosts of only one species are known;[3] of the 87 species of Myrmecolacidae described worldwide, five have been females and only two of these have been associated with their males. Myrmecolacidae are more numerous and speciose in the New World than the Old. The only species of Myrmecolacidae found in the southern United States is *C. fenyesi*, and its range is confined within that of its seemingly new host, *S. invicta*. Based on the five females described so far, we believe that the host of female *C. fenyesi* will turn out to be an orthopteran. But the female of *C. fenyesi* in the southern United States is likely to have switched hosts, just as the male has done.

[†] *Nature* **374**, 769–770 (1995).

A safer course than the suggestion by Orr *et al.* of the introduction of *S. invicta*-specific phorid flies to the southern United States (which has the risk that these too might switch hosts and damage other species in an alien ecosystem) would be to hunt for the females of *C. fenyesi* and use them for the dispersal of the parasites to control the red imported fire ant. Adult hosts parasitized by Strepsiptera are unable to reproduce and are killed only after the emergence of the strepsipteran;[4] hence while technically a parasite (by host survival), a strepsipteran is effectively a 'parasitoid'.[5] Males and young queens are among the individuals affected. But as it most commonly affects the major workers, which are sterile anyway, *C. fenyesi* is perhaps best considered as a macroparasite of a superorganism, the ant colony, which it debilitates in all the above mentioned ways.

In a curious inversion on the other side of the globe, a hunt is now on for the males of another Myrmecolacidae, *Stichotrema dallatorreanum* (presumably in an ant), the females of which parasitize and significantly reduce the long-horned grasshopper *Sexava* sp., a severe pest of oil palm in Papua New Guinea.

References

1. M. R. Orr, *et al. Nature* **373,** 292–293 (1995).
2. J. Kathirithamby and J. S. Johnston, *Ann. ent. Soc. Am.* **85,** 293–297 (1992).
3. A. A. Ogloblin, *Proc. Int. Cong. Ent.* **2,** 1277–1284 (1939).
4. J. Kathirithamby, *Syst. Ent.* **14,** 41–92 (1989).
5. J. Kathirithamby and W. D. Hamilton, *Trends ecol. Evol.* **7,** 349–351 (1992).

CHAPTER 7

HAPLOID DYNAMIC POLYMORPHISM IN A HOST WITH MATCHING PARASITES: EFFECTS OF MUTATION/SUBDIVISION, LINKAGE, AND PATTERNS OF SELECTION†

W. D. HAMILTON

Abstract

A deterministic genetic simulation of the coevolution of a sexual haploid host species with an asexual parasite is described in which parasites gain by matching and hosts by unmatching alleles at three linked and corresponding loci. Realistic parameter states without mutation or population subdivision can generate permanent dynamic polymorphism; however, polymorphism is preserved under wider conditions if mutation and/or migration in a metapopulation is present. Other factors favorable to polymorphism include weak selection, 'zero-sum' patterns of fitness interaction, fast breeding of parasites, and low rather than high or zero recombination. Hard selection was most protective to polymorphism in large populations, but multiple-infection soft (truncation) selection was more protective in small populations. A protective effect of close linkage may strengthen a polygene storage concept of Mather. It may also help account for cases of ancient transpecific polymorphism involving closely linked complexes of loci, as with the major histocompatibility complex and some linkages in plant resistance to pathogens. Typical 'cyclical' dynamics is extremely complex even when selection is moderate in strength. Genotype and gene frequency changes frequently simulate stochasticity/chaos even through wholly determined. Such states seem favorable to permanent polymorphism. This study demonstrates that, via a

† *J. Hered.* **84**, 328–338 (1993).

linkage-control locus, the dynamic states described may provide slow (frequently reversing) selection for high or low recombination. An instance of intermediate confinement of linkage under the selection is illustrated.

When a parasite tries to achieve its goals in a host and the host resists, it is of course not the genes of the hosts and the parasite that interact directly. However, genes on both sides lead to the molecules and processes that, eventually, interact. Peptides are good ligands or bad or none at all, proteases cut or fail to cut other proteins, and so on. As a simplification for a model, therefore, it is not unreasonable to treat the genes as interacting directly or to imagine them participating in a 'gene-matching' phase during or after infection that proves favorable to the host, harmful to the parasite, or vice versa. Ideally we should look at the relation of every gene of the host to every gene of the parasite, cataloguing all the effects; but again it is not unreasonable to suppose that defense and attack have various rather separate facets, so that genes of host and parasite having strong interactions can be paired off. Many highly specific gene-for-gene pairings are in fact known. This is especially so for higher plants interacting with their parasitic fungi and viruses,[1,2] but some similar relationships are known for animals with their pathogens (e.g., ref. 3).

This article presents a simulation of the dynamic, genetic system of a host species with its parasite based on the interaction of three genes of a host and three genes of a parasite with the attack and defense genes paired off and opposed. Without loss of generality, it is supposed that consequences of matching are such that the parasite gains from matching and the host from unmatching alleles at the three loci. Given allelic states in the model marked 0 and 1, it is obvious that we can always choose the allele names in the host (say) so that 0 host meeting 1 parasite is the bad combination for the host.

When fitnesses are derived in the model from the summed matches of loci after particular interactions of host and parasite, it turns out that, even for merely two match loci, permanent dynamic polymorphisms easily emerge. As introduction I outline now unpublished results for a two-locus haploid sexual model in which parasites are not explicitly present but may be supposed in the background to cause the changes in environmental states. This model reveals motivation for the three-locus model to follow. Let the alleles be A,a and B,b giving the four genotypes, AB, Ab, aB, and ab. Suppose over time four different environmental states disadvantage one genotype each while the other three keep standard equal fitness. If the foci have substantial recombination and the four states occur in equiprobable random sequence, such a population will develop, at least for a time, a dynamic polymorphism. The trajectory of the population is centered at the midpoint of gene frequencies but spreads widely. Eventually, in any finite population, distant excursions due to chance repeats of environment lead to fixation. The process resembles a random walk, but a difference is the tendency, in the presence of recombination, to move the population centroid more rapidly centerward than outward. This produces the fairly well-centered distribution.

The addition of any of three reasonable factors in a finite population can prevent permanent allele extinction:

1. Recurrent mutation, or, in formal equivalence,[4] migration in a metapopulation in which the local environmental states are uncorrelated.
2. Balanced repetitive cyclicity of environments. Here 'balanced' (stronger than 'equiprobable') means that within cycles all four environments are equally frequent.
3. Negative frequency dependence of the probability of environmental states—for example, a dependence that increases the likelihood of the most common genotype becoming the next target of elimination.

If strong enough, number 3 produces balanced cyclicity, as does number 2. Anticipating problems of multiple attractors (see below), it may be mentioned that even in the present simple situation of only four genotypes at least two forms of cycling arise. In the more easily imagined form the disfavored genotype follows the sequence *AB, Ab, ab, aB, AB*... or else the reverse: here genotypes take turns in being most common and are decreased or favored in a cyclical fashion and the limit cycle consists of four points of a tilted, centered square in the gene-frequency projection of the phase plot. An example of the second type of sequence is *AB, ab, aB, Ab, AB*.... For this form there are four equiprobable variants from a random start since in addition to the reverse of the example there is also the *AB, ab, Ab, aB, AB* sequence and its reverse. Instead of the first form's tilted squares, cycles now resemble symmetrical long-winged butterflies perched near to the central equal gene-frequency point and looking out along its axes.

Similar results are obtained if other forms of frequency-dependent selection through the environmental regime are imposed. Thus in a 'soft' truncation selection variant, a fixed quota is taken from the total population. In all forms, culling must concentrate on common types if alleles are to be protected, but unexpected outcomes are many and the dynamic performances brought about may be extremely complex. Even in this sketchy account, two points seem worth making about such two-locus match dynamics (see also ref. 5). First, if the 'three-equal-and-better' assumption is dropped, it remains very important for permanent polymorphism to have some kind of nonmultiplicative interaction of the alleles of different loci on fitness. Indeed, if cycles are not completely regular and frequency dependence is absent, this is essential. Second, under anything approaching the 'three-equal-and-better' assumption, neither repeats nor reversals of selection are so dangerous for allele permanence as what may be called 'zigzag' selection—that is, culling sequences like *AB, Ab, AB, Ab*.... Especially when recombination is low, this works unremittingly against a particular allele (*A* in the example), and quickly carries it to extinction.

Among the factors mentioned as protecting dynamic polymorphism, potentiality of mutation/migration (number 1) to prevent allele extinction and to stabilize cycling is well known.[6] Likewise number 3 is far from novel: frequency dependence has been an often invoked agent for polymorphism,[7,8] usually in a context of stable expected outcomes (see also ref 9,10), but the possibility of cycling under strong selection has not been overlooked.[11–15] In contrast to all this, the first effect mentioned above, the centered distribution that can follow under a random sequence of attacks on single genotypes, and the potentially complete protection arising in finite populations from frequency-dependent cyclicity such as parasites are likely to produce, [(2) → (3)], seems to have received surprisingly little attention. Similar protection is probably even easier to achieve in diploid models.[16] However, nonobvious forms of overdominance within loci can play an important part in the permanence in diploid models.[13,17] Hence, I prefer for the present to work with purely haploid selection, looking for facets of permanence that depend only on locus interactions.

Are the models that seek permanent dynamic polymorphism at all realistic? To me it seems that the assumptions, broadly interpreted, are almost inescapable in many biotic interactions and that therefore their potential to illumine genetic and sexual phenomena is vast. This view is still hazarded in the absence of direct evidence of cycling, mainly on the ground that frequency-dependent disfavor to common genotypes arises abundantly and naturally through the coadaptation of parasites: busily adapting parasites are everywhere in the environment, and all species have them. Moreover, cycling is likely to be slow and therefore difficult to observe. There are two particularly important phenomena, as yet only partly explained, that frequency-dependence models based on parasites might clarify. First, there is the general abundance of genetic variability including that involved in polygenic, metric variation. A large part of this puzzle still persists.[18–23] Second, there is the advent and maintenance of sexuality—the very process that makes possible the recombination the model is assuming.[24]

Recently a number of models have tried to explain sexuality or/and variability based on host-parasite coadaptation.[12,13,25–31] Those written by or in conjunction with the present author have concentrated on conditions in which frequently arising parthenogenesis cannot spread even when it has full efficiency—that is, when females suffer no fall in fecundity and males and male functions are eliminated completely. The most successful model with realistic parameter is that of Hamilton *et al.*[24] In the study presented here I describe preliminary results coming from a similar model. In use, so far the new model's objective is more restricted, however, than its predecessor's, being to elucidate only the first of the two target phenomena mentioned above—the maintenance of variability. Obviously a permanent polymorphism is not a sufficient condition for short-term resistance to parthenogenesis. In view of the robust previous findings,[24] however, it will be surprising if asexuality will be able to invade many cases of the present model where dynamic polymorphism is being preserved. Further, if any initial inefficiency of parthenogenesis is assumed,[32,33] slowing its early surge so that it has no time to rise far in frequency within some mean cycle period of the various resistance loci, it is virtually certain that parthenogenesis will be unable to replace sex.

A second focus of this article is finding conditions in which permanent dynamic polymorphism is possible without extra protection from recurrent mutation/migration. In the context of host-parasite coadaptation, this factor is most realistic when interpreted in its migration aspect, supposedly occurring between the demes of an infinite metapopulation. Provided such demes do not synchronize in oscillation [and for low to medium migration rates, there is evidence of a quite strong tendency for them to desynchronize,[27] as also for similar desynchrony in pure ecological models],[34,35] acts of migration return alleles where they are locally diminishing. This has a powerful effect in preventing both local and global extinction,[27,34–38] and this is more realistic than invoking recurrent mutation because it is likely that the alleles involved in the parasite battle in real populations are specifying complex proteins that have evolved as tools, either weapons for the battle itself or for other necessities, and it is unlikely that these can mutate to something even temporarily better except very rarely, while successful 'reverse' mutations preserving effective tool structure are likely to be equally rare. Thus it seems that if mutation/migration has to be invoked to prevent loss of polymorphism, we are almost forced to concede that spatial storing of alleles is occurring at a metapopulation level. This is not quite concession to the 'tangled bank' model for sex,[39] since according to that model different alleles are supposed to have permanent local adaptive values, but it certainly admits the importance of spatial differentiation. My analysis will therefore investigate how far a model for sex can manage without mutation/migration. How far can a haploid model preserve polymorphism on a basis only of panmixia and spontaneous biotic temporal variation, no concessions to spatial effects being made? The seemingly justified pessimism in a review as to single-locus dynamic (or other) polymorphism in haploid models[16] perhaps need not be extended when variation and selection involve several loci.

A third objective of this article is to compare the relative effectiveness of hard selection and two forms of soft selection in preserving dynamic polymorphism when their long-term average strengths of selections are equal. The conjecture that soft (truncation) selection is more conducive to permanent dynamic polymorphism [just as soft selection more generally is known to favor static polymorphism][9] was mooted in previous articles on rather slender evidence.[24,37,38] Since it is easy to set up soft selection runs that mimic exactly the mean intensity of selection experienced by a host in a hard run, the experiment now to be described includes such runs and contrasts them.

A General Model

The model is similar to that described in Hamilton *et al.*,[34] but is deterministic. It tracks all genotype frequencies in supposed infinite panmictic populations of a host and its parasite. A point in the model's state space—that is, specification of a full set of genotype frequencies for both species—fully determines all succeeding points.

Unlike the earlier model (HAMAX), which allowed multiple species of parasites each 'applied' under the gene-for-gene assumption to a different sector of a host chromosome, the new one (named DETSEX), in its present form, allows only a single parasite. In effect, it deterministically develops a reduced version of a rather unrealistic case of the previous model in which a single 12-match-locus parasite was aligned gene by gene to a 12-match-locus host. It is possible, of course, in the present model to cover multiple asexual parasite strains or species by setting parasite recombination to zero, but then these have to be assumed to contend only with the single defense system.

The general model receives and builds its parameters as follows:

1. A number of loci to be handled, L. This includes a sex locus and L-1 match loci.
2. A number of alleles to be handled at each, assumed constant, A.
3. A recombination applying to all match intervals, r_j, dependent on allele j carried at the sex locus of one of the two uniting parents, this parent being chosen at random and considered the female. As an exception to the implied constant map distance per interval, recombination distance to the sex/recombination-affecting locus itself, which if linked is at the low (subscript 0) end of the chromosome, can be different. When different, this recombination fraction is independent of the sex locus and is constant for the whole population at a value T_{sex}.
4. A mutation/migration rate for each locus, assumed constant and bidirectional, m. As an exception, the rate for the sex locus may be different, m_s.
5. An integer, p, giving the number of parasite generations occurring within one host generation.
6. Various switches and parameters concerned with how fitnesses are to be evaluated. These will be described under 'Version Used.'
7. Using (1) and (2) the program creates numbers to base A for all the possible genotypes. These converted to base 10 become serial numbers, A^L in all, for each species, applied to genotypes.
8. Using (6) along with (3) the program creates a Parent–Offspring Table using genetic rules. The table with A^{3L} cells covers the frequencies of all possible offspring from all possible mating combinations.
9. Using (6) with (4), an $A^L \times \mathrm{A}^{\mathrm{L}}$ matrix is set up showing effects of mutation/migration on all genotype frequencies. By matrix multiplication this

matrix transforms any vector of genotype frequencies as if by mutation/migration.

10. Using (6) with itself, a symmetrical $A^L \times A^L$ 'match score' table is created, setting out counts of loci with matched alleles for every combination of host and parasite. The count maximum is L-1 (matching at the sex locus is not considered), and the minimum is zero.
11. Using (10) a function and/or procedure is programmed, which maps match scores into fitness (reproduction) using controls and parameters from (6). Further details are given below.
12. The order in which the above tables and procedures are used once the generation cycle is entered is as follows:

    a. Normalization of genotype frequencies. [Note: although the model handles only genotype frequencies or, temporarily, genotype productions from which the frequencies are obtainable by normalization, relative changes in population can be kept in view if desired in each run by cumulating logs of the normalizing divisors found for each generation. Then at the end, the antilog of the mean of the cumulated sum is the long-term geometric mean fitness (LGMF).]
    b. Host-parasite pairing at random, followed by evaluation and expression of fitnesses.
    c. Mutation/migration. [Note: m is to be interpreted as the mutation rate or else as half an interdeme migration rate.]
    d. Pairing of sexual individuals and formation of their offspring.
    e. Cumulation of offspring to form new genotype productions.
    f. Return to (a).
    g. Intercurrent parasite generations. Hitherto the described sequence implies equal generation times for hosts and parasites, but for most parasites this is unrealistic. Therefore the model arranges for the parasites to undergo p-1 generations, referred to as 'intercurrent' [see also (5) above], in the presence of unchanging host frequencies. On the first generation of each run and on each pth generation subsequently, host and parasite frequencies are both changed.

The Version Used

The model has been run with four loci, two alleles at each ($L=4$, $A=2$). Generally the sex/recombination locus in hosts is made null and consequently will be ignored in discussing the results. Such states follow setting both allelic recombination rates to a single value, r. However, in some runs the 0 allele was set to cause a slightly lower recombination rate than the 1 allele, the difference being as in $r_0 = 0.95r_1$. The locus is then functional, and by observing frequency changes at the s locus the direction of selection on recombination may be detected. In all runs where the s locus is engaged like this, it has been set to be itself unlinked to the match loci ($r_{sex} = 0.5$).

In the parasite species the sex locus was always disabled in a similar way with both alleles specifying zero recombination, so making the parasite asexual. It follows that the eight possible types of parasite can be considered, in view of their interconvertability by mutation, as related biotypes of single parthenogenetic 'species.' Alternatively, when the 'migration' view is taken, they can also be considered unrelated parasite species. However, as species they then must be admitted subject to unusual fixity of type.

(Regarding m under the 'migration' view, the only nonzero rate covered is 0.0001. This is reasonable for a mutation but is unnecessarily low for migration assuming it occurs. However, time constraints precluded including a third level, say $m_h = 0.1$ and $m_p = 0.1$. This would more than double runs in the factorial experiment.)

PROCEDURES AND PARAMETERS

For the determination of fitness from match scores for pairings, the following procedures and parameters were used.

Hard selection. Strength of selection: Separately for host and parasite (subscripts h and p), a parameter s was specified such that maximum and minimum fitnesses would be $1+s$ and $1-s$. The first was obtained by fully matched parasites and zero-matched hosts; the second was obtained by zero-matched parasites and fully matched hosts. Obviously it is arbitrary that parasites benefit from matching and hosts by being unmatched: an opposite decision would give an identically behaving model. In all the runs to be considered the parasite was always assumed to have $s_p = 0.5$.

Form of selection: Fitnesses for the intermediate two match scores, 1 and 2, were interpolated between those for 0 and 3 already discussed ($1+s$ and $1-s$, or vice versa) by fitting a symmetrical arc extracted from a quadrant of a supercircle of degree 10. For this supercircle, a second parameter, b, varied in the runs, gives the angle that the arc, ascending or descending, makes with the match score axis at match score 0 and with a vertical axis at 1. The choice of power 10 determines a high degree of bowing that approaches a step function. This means that as bowing is caused, through b, to be close to either 0° or 90°, the fitnesses corresponding to match scores 1 and 2 will either be very close to the fitness for 0 or very close to fitness for 3. When, however, $b = 45°$, the interpolation is always linear and is independent of the power (the supercircle of infinite size). Explicitly the hard fitnesses are found directly by using the 'bowing' function, $b(b, p, x)$ defined over (0, 1) and with values in the same range. In this function, b is the angle, p the real positive power, and x the fractional match score.

In full:

1. For $b < 45°$ the bowing function used was

$$b(b, p, x) = \frac{1}{1-a} - \left[\frac{1+a^p}{(1-a)^p} - \left(x + \frac{a}{1-a}\right)^p\right]^{1/p}$$

where $a = (\tan b)^{1/(p-1)}$.

2. For $b > 45°$, x in the formula is to be replaced by $1 - x$ and b by $90° - b$.
3. For $b = 45°$ or/and $p = 1$, $b = x$.

The fitnesses are then $1 - s_h + 2bs_h$ for hosts and $1 + s_p - 2bs_p$ for parasites.

Such a function may seem unnecessarily complex in a context where there are only two intermediate fitnesses to be found, but its ready application to more than three loci, as intended for future work with DETSEX, and its ability to bow symmetrically about the 45° line are advantages. Such symmetry is unobtainable, for example, from a generalized power parabola.

Soft sequel runs. Under the truncation soft selection to be applied, it is assumed fitness is always determined as a step function, the position of the step depending on the cull fraction and the composition of the population. Culled individuals have zero fitness so that a parameter like s can no longer be used to determine selection strength. However, Crow's Index of Selection, defined as the coefficient of variation of fitness, variance over mean,[6] is easily determined if a cull fraction is known. It is simply the mortality (or cull) fraction divided by the survival fraction: $I = c/(1-c)$. A corresponding 'Crow' coefficient can be found for any particular generation under hard selection as the variance of fitness divided by the mean fitness: $I = V_w/W$. Thus if the average of such coefficients has been evaluated over a representative series of generations in a hard selection run, this average can be used to specify a soft selection run to follow that will have the same mean Index of Selection by setting $c = I/(1 + I)$ where I is now taken as the known mean.

But having decided in this way how much it is necessary to cull in a soft run in order to mimic the selection strength in a preceding hard selection run, how is it decided which genotypes to cull? Two procedures were followed corresponding to two different ways in which individuals might be supposed to experience their parasites.

1. A single lifetime encounter with a parasite (SIS). All entries in the match score table are ranked, and culling is begun with the most-matched host-parasite elements of frequency that have been determined corresponding to the table. When the cull fraction is found to be due for completion within a set of equal-ranked elements (or a single element), classes are reduced proportional to their existing frequencies until the specified total cull fraction has been taken. All nonculled elements of frequency are then made to 'reproduce' with equal fitness.

2. Multiple encounters with parasites (MIS). Here it is assumed that each host genotype samples a large number of the parasite types present in the

population. Every individual of a given genotype can then be expected to experience approximately (and in the simulation, exactly) the same fitness. The algorithm for this case is therefore simpler. Columns of the match score table are multiplied by the frequencies of the parasite genotypes to which they correspond and then rows are summed to find the 'mean healths' of host genotypes. These means are ranked, and then the culling process is applied as previously to the host genotype frequency vector.

This MIS procedure is rather like assessing the lifetime health impact of the common cold on humans, whereas SIS is concerned with diseases like whooping cough that strike individuals once and cause more or less harm (or even kill), but are always followed by lasting immunity against all other attacks.

OUTLINE OF BEHAVIOR IN RUNS

A reason for using just three match loci is that it is possible to observe three gene frequencies in a 2D projection of a cube. If it is visualized as a wire frame with trajectories contained and if the cube is easily 'rotatable' as seen on a computer's monitor, a good intuitive impression of even very complex dynamics can be obtained. It is of course still far from possible to visualize the full trajectory of the system, which would need 16 dimensions (64, if both sex loci are also brought in). Notably omitted by the cube image are all linkage disequilibria and all the interactions between the host and the parasite. Nevertheless, with different colors for a host and for a parasite (and possibly for different stages of the trajectories), a diagram can be more intuitively revealing than are sets of 2D graphs and still more so than numerical output. The program therefore, at the same time that its summary statistics are evaluated for printing, prepares files of gene frequencies and linkage disequilibria for subsequent visualization.

Experimenting with the system revealed that

1. For a constant parameter set, different starting points in genotype frequency space would sometimes proceed to different attractors.
2. Stable interior point attractors seemed to occur rarely, if ever. Although stabilization of some gene frequencies without loss of alleles can occur, such stability for all loci (as in ref. 12) appeared not to. Some attractors were very compact in the cube (usually but not necessarily central) but still detectibly dynamic. [It is not clear to me how far results of Eshel and Akin[40] apply to this model.]
3. Polymorphisms most characteristically followed complex orbits that seemed not to repeat. By more careful search, a wide variety—one might almost say a zoo—of strange-shaped but truly cyclical attractors was also found.

Confining description to some of the simplest forms and their developments, sometimes a seemingly simple cyclical attractor would lie closely flattened around, say, a 0.5 gene frequency plane of the cube, or be approximated to some diagonal plane, or lie elongated close to some orthogonal or slanted straight line. Yet it was also often found that no truly stable orbit was being approached in this performance; eventually the orbit would move away and, perhaps, after some wide erratic turns, settle into some other equally approximate and temporary form. In this it would be frequency at another locus that was taking a turn to be 'odd one out' and to be held almost constant, or another triplet of alleles might be rising and falling in near synchrony; or the like; and then all these new modes would again change. In short, cases were observed of types of supercyclical or chaotic behavior that must involve components with extremely long periods. All the above possibilities combined to make the results from particular parameter sets very difficult to categorize and interpret. Another aspect of the more complex trajectories is that on shorter time scales what is happening to any single gene frequency often seems like genetic drift in a small population, the resemblance being reinforced by the general form of the spectrogram if time series analysis is applied. On a log plot the characteristic descending line of drift ('pink noise') is seen over most of the high-frequency range to the right while various prominences, corresponding to vaguely favored 'periods,' manifest over the low frequencies to the left. When the effect of selection on recombination in a run is being watched directly in a time plot, the impression of drift (in what must be remembered to be a perfectly determined infinite population) is even more striking. Even after many thousands of generations, the frequency of the recombination allele may still wander irregularly and occasionally across the midline, and it may remain impossible to say whether there is any overall trend. A further impression of pseudostochasticity of frequencies at the sex locus in runs from various starts under a particular parameter set [in which only the location of a close pair of recombination values (using $r_0 = 0.95r_1$) is being varied] can be gained from Fig. 7.1.

In view of the behaviors described, it was decided that the best preliminary survey of such generally pseudostochastic behavior occurring both within and between runs would be to treat a series of runs as if really subject to stochastic variation—that is, to perform a factorial experiment that would include replicate runs using different starting points. Thus, in a way unforeseen at the outset, the plan converged with that used by Hamilton *et al.*[24] in which a truly Monte Carlo simulation addressed a similar problem. If desired, results of the new model could again be analyzed statistically. This might reveal the relative importance of differences found, but what meaning could be attached to probability levels in such tests is hard to say.

The following plan for the runs of the models and its results was made:

1. Runs are to be started from random points constrained to be neither very central nor very peripheral in the genotype frequency space.

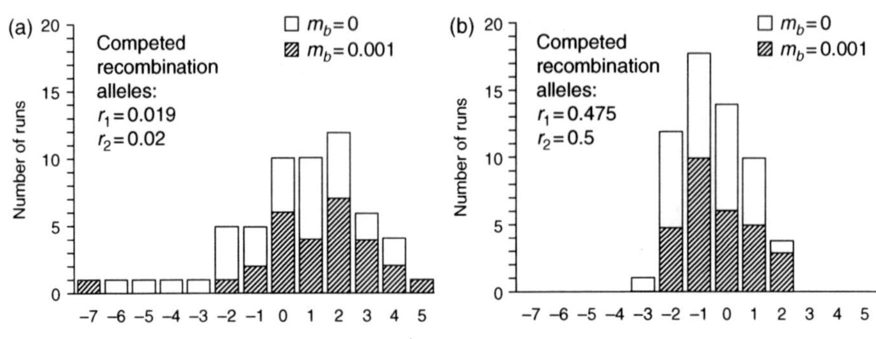

Figure 7.1. Distributions of gene frequency changes at a locus controling recombination of match loci over 2,000 generations. The controling locus is unlinked to the match loci. (a) shows competition of two alleles for low recombination (60 runs), and (b) shows competition of two alleles for high recombination (58 runs). All runs had parasite mutation at $m_p = 0.0001$; host mutation was $m_h = 0.0001$ in half of runs (white area) and zero in others (hatched area). Other parameters of the experiment are given in the text.

2. Unless early abortion is forced through extreme fixation (manifested in floating point overflow), runs are to be of 4,000 generations. In view of the evidence in some cases of prolonged transient behavior, only the last 2,000 generations are to be used to assess results.

3. a. During the last 2,000 generations, a record is to be kept of the lowest frequencies attained by all six match alleles. b. Based on the minimum of six such minima, a run is to be classified as having preserved its polymorphism if no gene frequency sank below 10^{-4}. The attainment of this condition is abbreviated 'P' (polymorphic). It implies approximately, for finite populations, that the polymorphism could be considered to have preserved all its alleles if population size was greater than 10,000.

4. A run is to be considered successfully polymorphic through having a reasonably centered trajectory, if at the end of the run the centroid of all the host population points for the last 2,000 generations was within 0.35 of the center of the cube. The attainment of this condition is abbreviated 'P&C' (polymorphic and centered). When this did not hold, it was deemed that the polymorphism must, in spite of the symmetry of its match loci and alleles, have spent much of its time in some peculiar unbalanced state which would probably, in nature, imply additional danger of losing alleles. Such danger would arise either from a very slow trend of the coevolution or from outside causes. As will be seen from the tables, a considerable number of such peculiar asymmetrical trajectories were observed. The cube visualization has shown that some of these are truly stable limit cycles describing often quite complex courses far out from the center.

FACTORIAL DESIGN

All combinations of all the following factors were made.

1. *Mutation/migration.* Equal-rate reversible mutation of match alleles, in both hosts and parasites (therefore four combinations, *aa, ab, ba, bb*) at $\mu =$ (a) 0 and $\mu =$ (b) 0.0001.
2. *Match locus linkage intervals.* At (a) $r = 0.5$ and (b) $r = 0.02$. The linkage-control locus is always at $r = 0.5$ - that is, in free recombination with match loci.
3. *Strength of selection on hosts.* At (a) $s = 0.2$ and (b) $s = 0.8$. The strength of selection on parasites is constant at $s = 0.5$; the fitness maximum is $1 + s$ and the fitness minimum is $1 - s$.
4. *Parasite selection profile (always hard).* Match score mapping to fitness based on ascending segment of supercircle of power 10. Initial angles of ascent were 0°, 45°, and 90°. Corresponding patterns are referred to as 'peaky,' 'direct' (or 'straight line'), and 'pitty.'
5. *Host selection profile (when hard).* Match score mapping to fitness based on descending segment of supercircle of power 10. Initial angles of descent were 0°, 45°, and 90°. Corresponding patterns are referred to as 'pitty,' 'direct' (or 'straight line'), and 'peaky' (contrast 4).
6. *Parasite intercurrent generations.* The number of parasite cycles within the host life cycle, set at 1, 2, or 8.
7. *Replications.* Two, from differing random, nonextreme genotype starting points. Note that for the two-thirds of runs that are 'soft' in hosts (see below), because factors 2 and 4 were inapplicable, replications are based only indirectly on the fitness parameters apart from match score. Their connection lies through 'inheritance' of a mean index of selection from the preceding hard run.
8. *Host selection form.*
   a. Hard selection: fitness was completely determined by match scores. For selection parameters, see 3 and 5 above.
   b. Soft selection mimetic sequels: phenotypes with lowest match scores in the population were truncated. For full rationale, see 'Version Used' above.
      (i) (MIS): Multiple infection—individual met all parasites and were assigned mean match scores.
      (ii) (SIS): Single infection—individuals met just one random parasite and were assigned a match score for that alone.

Combining these factors gives the total number of runs as $4 \times 2 \times 2 \times 3 \times 3 \times 3 \times 2 \times 3 = 2{,}592$.

Results: Main Effects and Selected Interactions

The model confirms that dynamic polymorphism can persist widely in the parameter range surveyed. Some parts of the range where this happens are far from realistic; others are realistic within the simplicity of limitations imposed. Altogether, a little over half the runs (1,351 of 2,592) preserved polymorphism (P) through the monitored 2,000 generations. A little under one-half (of them 1,167 of 2,592) additionally kept the host trajectory of the monitored 2,000 generations reasonably well centered (P&C) in the gene frequency space. In the range of parameters and patterns of protection considered realistic, both the P and P&C proportions would be much higher.

The most potent factor promoting persistence was mutation/migration (M/M) in parasites. This effect was of such strength that 88% of runs that ended polymorphic and centered (P&C) had M/M working in the parasites (at the level of $m_p = 10^{-4}$) in contrast to only 12% ending P&C without it (Table 7.1). When noncentered polymorphisms were also counted, parasite M/M proved a little less essential (81%), the difference confirming intuition that the effect of parasite migration must be to superimpose a continual nudging of host trajectories centerward from all sides. Because as has been stated, noncentered polymorphisms are suspected of fragility, most of the further results will be discussed only in terms of the P&C successes.

Host M/M at the same level of difference as for parasites also gave a strong (effect (56%)), in comparison with other effects discussed below, although it was not nearly as strong as the effect of parasite M/M (Table 7.1). The difference is probably due to the fact that in most cases where the parasite has one or more intercurrent generations the parasite tends to have greater amplitude of change, so putting alleles in more danger. Hence, if parasite alleles can be saved from fixation by migration, and if parasites by their speed of selection can continually 'head off' extremity in hosts, the polymorphism as a whole is generally protected. Host M/M may come into effective play in those cases where for some reason the host trajectories tend to swing wider.

When the centered polymorphisms were classified according to both host and parasite M/M factors (Table 7.2) it was seen that in fact very few polymorphisms are permanent if they lack M/M in both species. Nevertheless, it is also noticeable that a substantial fraction of these few concentrate in the corner of the parameter space that is considered most realistic, as discussed further below.

About second equal among main effects to host M/M is strength of selection, with weak selection better and accounting for 58% of the P&C runs (Table 7.1). However, noting that under weak selection the worst genotype has a fitness two-thirds that of the best, while under strong it has only one-ninth, it is surprising to find so great a difference in levels making so little difference to the results: in other words strong and weak selection evidently both readily support dynamic polymorphism.

Table 7.1. Simulation of effect of various factors acting in parasites or hosts on persistence of polymorphism in 2,592 runs, each of 4,000 generations; entries are percentages of the *n* runs that maintained polymorphism according to two criteria[a]

Factor	P ($n = 1.351$)	P&C ($n = 1.167$)
Two-class factors		
Mutation/migration (M/M)		
Parasite		
$\mu = 0$	18.7	11.5
$\mu = 0.0001$	81.3	88.5
Host		
$\mu = 0$	40.8	44.2
$\mu = 0.0001$	59.2	55.8
Match locus linkage interval in hosts		
$r = 0.5$	43.9	47.5
$r = 0.02$	56.1	52.5
Strength of selection on hosts		
$s = 0.2$	56.1	58.4
$s = 0.8$	43.9	41.6
Three-class factors		
Parasite fitness profile[b]		
0 (peaky)	33.7	38.1
45 (direct)	28.6	32.8
90 (pitty)	37.7	29.0
Host fitness profile	($n = 477$)[c]	($n = 415$)
0 (pitty)	36.9	32.5
45 (direct)	35.2	36.2
90 (peaky)	27.9	31.3
Parasite intercurrent generation[d]		
1	29.2	29.6
2	34.5	34.4
8	36.3	35.9
Style of selection[e]		
Hard (H)	33.1	35.6
Single-infection soft (SIS)	29.1	31.2
Multiple-infection soft (MIS)	35.6	33.2

[a] P = polymorphic. During the last 2,000 generations of 4,000, no allele had a frequency less than 0.0001. P&C = polymorphic and centered. Same as for P, but also requiring that the centroid of all host population points in the last 2,000 generations had Euclidean distance < 0.35 from the central gene frequency point (0.5, 0.5, 0.5).

[b] Runs by angles to the match score axis of the segment of a supercircular fitness function (see text). Always hard.

[c] Fewer host runs, 864 in all, because soft host runs are not classifiable for profile.

[d] No. of parasite generations within each host generation.

[e] Under multiple-infection soft selection (MIS) each host meets all the parasite genotypes and gains an average match score based on these encounters. Under single-infection soft selection (SIS) each host acquires a match score from only one parasite met at random. In both cases, ranking and then culling is based on simple or average match scores. Hard selection (H) match score is acquired as for SIS but is transferred directly to fitness via the bowing fitness function described in the text.

Table 7.2. Distribution of P&C runs by host and parasite mutation rates

	Host mutation		
Parasite mutation	$\mu=0$	$\mu=0.0001$	P&C totals
$\mu=0$	52	82	134
$\mu=0.0001$	464	569	1033
P&C totals	516	651	1167

Last in magnitude among two-class effects, more P&C success (52.5%) occurred under close linkage (0.02) than under free recombination (0.5) (Table 7.1). Obviously this difference is also small: but if non-centered polymorphisms are included, the effect is somewhat larger (56%). Since totally linked chromosomes must almost certainly fare worse than freely combining sets of loci,[24,41] slight as the effect is, it will be interesting in the future to explore it further. The results hint at existence of an optimum linkage for polymorphism preservation, although, of course, this optimum will not necessarily be the target of the trend in the selection on recombination. Several of the ancient 'transspecies' polymorphisms that are now gaining increasing attention[42] include variation within closely linked regions or superloci (e.g., refs 43, 44). As discussed further below, the finding reported here may bear on the seeming connection between close linkage and ancient polymorphism.

Turning to the factors chosen to have three levels, these have on the whole less dramatic main effects. Leaving aside for the moment the one most marked (parasite fitness profile), consider parasite intercurrent generations and remember that with three-class factors, 33.3% is now the null expectation. A rising trend is found from 30%, when host and parasite generations are equal, to 36%, when parasites have seven intercurrent generations. However, the class remaining, of one intercurrent generation claiming 34.5% of successes, shows that most of the benefit of intercurrency, modest in general, has come already. Tentatively, the pattern as a whole may be understood as arising from the capability that intercurrency gives parasites to 'overtake' and to 'head off' trends of the hosts to approach fixation planes. However, it has to be remembered that while the host is static, the parasite centroid is also heading for boundary planes and not for the host centroid itself: it too runs the risk of going very extreme. Thus such an interpretation is suggestive only.

Now, taking the effects of selection profile (Table 7.1), we see that with the parasites, and as regards the centered dynamic polymorphisms, success increases monotonically as parasites transfer from being on a 'pitty' profile (29% of successes) to being on a 'peaky' profile (38%). This is good for realism since a peaky parasite landscape is more believable (see below and Hamilton).[38] However, the picture of monotonic increase is mildly upset if the noncentered polymorphisms are included. (This suggests that the unrealistic pitty parasite profile may be involved in many of the persistent asymmetric polymorphisms.)

The picture for the main effect of host profile (Table 7.1) is unrevealing, although it should be noticed that it is based on fewer runs due to soft runs not having variable profiles. If anything, direct (linear) dependence on match score of hosts seems best for polymorphism (36%), while pitty and peaky are slightly and equally inferior (32.5% and 31%). However, a much more interesting and clear-cut pattern overshadows all the main effects when the distribution of successes is cross-classified by the parasite fitness profile (Table 7.3). The resulting distribution of P&C successes can be seen to be saddle-like; listed with some categories combined, from best to worst it runs as follows:

- Peaky host with pitty parasite
- Pitty host with peaky parasite

- Direct host with any parasite
- Pitty host with direct parasite
- Peaky host with direct parasite

- Host and parasite both peaky or both pitty.

The gaps in the list indicate the larger differences. The extremely poor performance in the last category is doubtless due to selection under its two regimes being such that one coevolutionary partner can be under very little selection (and therefore possibly almost static in its position) at times when the other partner is intensely selected and rapidly changing. The latter therefore may become extreme in its genotype and gene frequencies while experiencing little frequency-dependent response to the near monomorphism arising. Viewed in another way, however, this result for the pitty–pitty interaction was unexpected. The interaction appears submutualistic with most combinations giving both partners high fitnesses, and mutualism normally produces coexistence and static polymorphism. Perhaps the key here is that 'doing little harm' is not at all the same as 'doing some good,' as happens in a true mutualism.

Table 7.3. Interaction of host and parasite profile type (b_h and b_p) as creating P&C polymorphism in hard selection runs

	Parasite profile			
Host profile	Peaky	Direct	Pitty	Total
Peaky	15	30	85	130
Direct	52	48	50	150
Pitty	74	46	15	135
Total	141	124	150	415

The peaky–peaky interaction on its side might be described as a 'subspiteful' since in most interactions each partner is doing the other a lot of harm for little self gain. That this situation turned out hostile to polymorphism was thus less surprising. Again, the more fundamental explanation, however, is doubtless in terms of the low degree of linkage between rates of selection of the two partners. But, as shown by only moderate performance of the direct–direct interaction, such linkage cannot be the whole story, either. It seems that persistence requires both the 'zero sum' type interaction between species and a degree of non-linearity of the profiles. Clearly, more needs to be done to understand these interaction effects.

The best of all the combinations, peaky host with pitty parasite, is probably not realistic. It would imply something like an 'r-selected' host species kept at low levels by a chronic long-evolved parasite or set of parasites, but occasionally gaining eruptive escape from it or them. The interactions of some weeds, say docks and thistles, coevolving, again for example, with their respective rust fungi, come to mind as possibilities; but even these do not fit well, and the scenario is intrinsically unlikely.

Much more likely to occur in nature is the situation of the second-best interaction, pitty hosts with peaky parasites. Most large long-lived organisms—the organisms most persistently sexual and outbreeding—have a relatively K-selected character. This implies not only a validation of one of the most basic assumptions of the model (constant host population) but also justifies both the potential truncation of worst genotypes and the plateauing of reproductive success for the most successful. These last host effects combine to give the pitty host the selection profile suggested. At the same time, most parasites in nature are relatively r-selected opportunists and probably experience, considering host habitats they can 'key' into, a peaky profile of fitness. One outcome of such a profile for parasites might be, for example, the often discussed and almost ubiquitous clumped distribution of parasite numbers in hosts, while the fact that most hosts are superficially unaffected by their chronic parasite loads until these are very high may be an expression of a pitty profile for hosts.

The last factor to discuss in the experiment is the mode of selection. At first appearance, it is surprisingly uninteresting (Table 7.1): all three modes favor persistent dynamic polymorphism to approximately the same degree. Single-infection soft (SIS) is worst (31%) and hard is best (36%); but if noncentered polymorphisms are allowed, multiple-infection soft (MIS) proves to be best (again 36%).

As with the selection-profile main effects, such a small response to radical changes in mode of selection seemed unlikely to be the whole story. Seeking a more controlled comparison, I therefore took the 449 cases in which all members of each triplet of runs (H, SIS, and MIS) completed the 4,000 generations without any early abortion. Cross-classifying hard and soft sequences according to 'fixed' or 'protected' (P&C) outcomes, I obtained the 2×2 frequency

Table 7.4. Polymorphism protection in hard and *nonaborting* soft sequel runs[a]

Hard	Soft		
	Fixed	Protected	Totals
Soft, single infection (SIS)			
Fixed ($n=25$)	24	1	
Protected ($n=424$)	99	325	
Total	123	326	449
Soft, multiple infection (MIS)			
Fixed ($n=25$)	14	11	
Protected ($n=424$)	58	366	
Total	72	377	449

[a] Distribution according to fixation vs. protection of 449 no-abortion run triplets in which both multiple-infection soft (MIS) and single-infection soft (SIS) runs mimicked selection intensity of a preceding hard run. 'No abortion' means no frequency value in any run of the triplet became so extreme as to cause floating-point overflow or underflow in the 4,000 generation runs. Apart from style of selection, all other applicable parameters are constant within each triplet.

classifications presented in Table 7.4. These reveal that it is common for hard selection to 'rescue' polymorphisms which, under equivalent soft selection, had failed to persist (99/123 = 80.4% rescued against SIS and 58/72 = 80.6% rescued against MIS), while on the contrary it was rare for either soft selection mode to rescue a run that failed to persist under hard selection (1/326 = 0.3% for SIS and 11/377 = 2.9% for MIS). Hence, there are certainly conditions of the model in which hard selection is the most protective of polymorphism, although how realistic such conditions are remains questionable.

Such a result appears intuitive through the known fact that hard selection at a given level diminished its rate of change of gene frequencies as genes become rare, whereas under soft truncation selection against a genotype, gene frequency change increases. Disconcertingly for intuition, however, the effect seems quite different from what was found during early experimentation with the previous model, HAMAX: in those runs, soft selection was most protective, and this fact became the basis of what it seems now may have been an over-generalized claim in my recent work.[24,37,38] Due to its overlapping host generations with low mortality and its single parasite encounters renewed each year, HAMAX can probably be described as having a scheme of infection somewhere between the SIS and MIS conceptions of DETSEX but closer on the whole to MIS. Thus, that MIS proved most preservative for polymorphisms overall in the present model may reduce the puzzle a little, but hardly enough since the effect was small.

A more likely reason for the discrepancy may lie in the small population size (200) used in HAMAX contrasted with the infinite size in used in DETSEX.

Table 7.5. Means of least frequency of any mate allele[a]

	$\mu_p = 0$ ($n = 25$)	$\mu_p = 0.000$ ($n = 299$)
Hard	0.036	0.028
Soft, single infection	0.009	0.055
Soft, multiple infection	0.056	0.121

[a] Based on 449 run triplets with no abortion as for Table 7.4, means over runs of least frequencies of any mate allele are shown for combinations of style of selection and parasite mutation. There are 25 no-abortion triplets for zero parasite mutation and 299 for $m_p = 0.0001$.

Consider again the 449 triplets of DETSEX runs that had no aborting fixations, and extract from these all that remain polymorphic (P). There are 25 when $m_p = 0$ and 297 when $m_p = 0.0001$. Using the mean of minimum frequencies attained by either host or parasite, compare their tendencies to keep well clear from fixation of any allele (Table 7.5). Whether we consider runs with no assistance from parasite mutation/migration or those with assistance, MIS is now the best selection mode for keeping nonextreme gene frequencies. Comparing this with the results in Table 7.4, it seems probable that MIS must be especially inclined to induce wide variation in amplitude over runs so that those runs that do not end categorized as fixed have many that are actually very far from being so. It is likely that the success of HAMAX, which was conducted with more restricted and reasonable parameter sets than used in the present DETSEX experiment, stemmed from the tendency of MIS-like schemes to produce relatively compact trajectories. The figures in Table 7.5 suggest that with small populations with minimum reasonable gene frequencies of about 0.005 as applied in the HAMAX experiment, hard runs would be in greater danger or extinction.

INDETERMINATE LINKAGE AND CHAOTIC DYNAMICS: A DEMONSTRATION

Separate from the main experiment a smaller experiment is now reported which has two aims: first, to substantiate earlier claims of pseudostochastic and drift-like behavior; second, to show that, although spread, the trend of runs gives evidence of an intermediate ESS recombination rate ($0.02 < r_{\mathrm{ESS}} < 0.5$) insofar as the ESS concept still applies under erratic selection.

For this test, two series of 30 runs were started from the following parasite selection parameters: $s_p = 0.5$; $b_s = 45°$; $p = 10$; $m_p = 10^{-4}$. The hosts were supposed under soft, single-infection selection (SIS) with 15% truncated each generation. Host mutation was in half of the runs in each series set at zero and in the other half at $m_h = 10^{-4}$.

At the sex locus two recombination rates were put into competition at 0.5 starting frequency. In a first series the sex alleles specified inter-match-locus recombination rates of 0.019 and 0.02, and in the second series they specified 0.475 and 0.5. The change in frequency of 'higher' alleles at the sex locus in each case was then found over the interval between generations 2,000 and 4,000. The distribution of such movements for the two series is shown in Figure 7.1. It can be seen that from both starting points variants are most commonly favored that give a more intermediate recombination and that presence or absence of mutation in hosts does not affect the issue. Thus neither extreme of recombination is an ESS. At the same time, the spread of the distributions (which as might be expected is greater at low recombination than at high) means that a unique interior ESS, if it exists at all, is extremely weakly stabilized. The differences between the alleles tested are small. Combining this with the always indirect influence on the recombination locus, the weakness of selection observed is not unexpected; but the spread of movements, including so many finishing opposite to the overall trend, illustrates the complexity of dynamics and pseudostochasticity mentioned earlier. The figure suggests that determining an ESS recombination rate at all precisely for this sort of model is not easy. When reliable results are available, however, it will be interesting to compare them with the predictions of Sasaki and Iwasa,[45] and to make the comparison both for regular cycling and, using power modes extracted from time series analysis, for the complex and possibly chaotic cycling.

POLYMORPHISM WITHOUT MUTATION/MIGRATION

Only 52 of the total 1,352 persistent polymorphisms had no mutation/migration in either partner. All were centered. Hence, contrary to the indication of Table 7.1, it appears as if M/M of either host or parasite is necessary for the noncentered polymorphisms to occur. Focusing within the set of 52 on the subset of 24 that occurs under the fitness profiles considered most realistic—pitty hosts with peaky parasites—a core condition of support for persistence from the other parameters can be identified.

This condition has low intensity selection, maximal intercurrent parasite generations, and low recombination. Given this combination, persistent polymorphism arose not only in the hard runs to which the parameters apply directly but also in all their soft sequels. However, it did not seem that the soft runs in the set could have supported appreciable recombination or sexuality since the movements of frequency caused at the sex locus were tiny. Indeed, in the case of the SIS runs all but one of the movements were downward! In contrast, the movements caused at the sex locus in all the hard runs of the set were all substantial (ranging approximately from 0.01 to 0.1 of frequency) and were upward.

The core of protection extended with moderation in various directions. Thus, its intensity of selection was put to its upper level ($s_h = 0.8$), the hard and the MIS run persisted but the SIS polymorphisms did not. If intercurrency of parasites was lowered, the hard runs persisted but soft sequels of both kinds variously dropped out.

Considering hard selection alone, an overlapping core condition existed for all cases in which recombination was low ($r = 0.02$). Persistence here was independent of intensity of selection and intercurrency but otherwise was rather sharply limited. Thus only one run of this kind remained persistent when the recombination was raised to 0.5.

Of the 52 zero M/M persistent runs, only 16 had recombination of 0.5. Twelve of these were hard runs and four were soft sequels, but originating hard runs of the 16 had the peaky host–pitty parasite fitness interaction. This suggests that hosts may need high recombination between relevant defense loci if they are to enter dynamic polymorphism with pitty parasites.

It must be emphasized that apart from the 16 just mentioned, and although few in the overall experiment, almost all of the remaining 36 cases of persistence without migration are in a range of parameters as realistic as any set the model covers. The seemingly rather strong necessity for low recombination under a realistic fitness pattern is surprising and interesting; it may seem to contradict the cases of high recombination aiding polymorphism mentioned in the introduction. However, those were two-locus models, the implications about linkage were variable, and the models involved a frequency dependence acting directly, not a frequency effect emerging through a simulation of a parasite population. A probable though not yet well-tested explanation for the difference may be related to an idea mooted long ago in the context of biometric genetics concerning adaptiveness of polygenes and of polygenic systems.[46] A deficiency of Mather's argument, which possibly the present model can make up, was a lack of a rationale for the long-term maintenance of the polygenic variation he was considering. The idea is also related to another main finding of the work presented here—that a metapopulation with interdeme migration is an effective storage system. The central idea is that by keeping some currently bad alleles closely attached to currently good ones, low recombination, like subdivision and like Mather's 'balanced' pairs and combinations, preserves such alleles from extinction and also from time to time releases them out of 'stored' combinations into others in which, once favorable, they can increase. The difference from Mather is that the '+' or '−' significance of alleles relates directly to fitness and is from time to time changing as the cycles turn.

Old and New in Genetics

The parallel to Mather's idea just mentioned reminds us of other 'unfinished business' of genetics of the past, which themes in this article approach (see also Hamilton).[37] Among such speculative topics are:

- variety in real linkage values (genomes do not 'congeal'),
- evidence for episodic selection,[21]

- positive heritability of fitness,[47,48] contra equilibrium theories,
- the abundance of genic variation including metrical, and
- extreme age and constancy of polymorphic complex alleles.[43]

All these phenomena might become part a view of variation which sees it as dynamic due to parasitism. Given the ubiquity of parasites, such polymorphism would be likely to exist in all species, would involve loci spread throughout the genome, and, at least in some systems, is expected to include codes for anti-parasite technology[49] that are older than the species.

The possibility of extreme antiquity of alleles has been realized since the early days of blood grouping[50] and graft rejection,[51] and recently the topic has gained a firmer footing through new techniques of molecular genetics (e.g., ref. 52). The extremely long permanence of the elements of these polymorphisms is an increasing puzzle to theorists.[10,42,44,53] Until now possible preservation through dynamic polymorphism has not been considered. That some alleles are indeed far older than the species they inhabit, and in some cases even older than the genera, can be considered proved;[54,55] other references in Hamilton *et al.*[24] Discoveries are being spurred on in medical and biotechnology laboratories by precisely the hypothesis that also inspired this article—that the molecules many of these ancient alleles create are well-fashioned and/or well-chosen tools that function against parasites. They may do so either through parasite recognition (e.g., Mhc or lectins) or else as direct weapons (e.g., complement or toxins).

Meanwhile, watching currents in the new human fields begin to divert the flow of discovered organic chemical specification away from university libraries and into patent offices, we perhaps see hints of how costly it must be in the ordinary living world for an individual to 'forget,' as by bad mutation, items of defense technique that its species knows. If this is true, fortiori it must be even more costly for a species to forget and lose not only whole stacks of its blueprints at once but its main faculty for re-creating them or drafting new ones. Perhaps this is the most important setback that follows from a loss of sex, making such loss initiate usually brief episodes of parthenogenesis only, like 'witches' broom' galls sparsely dotting, not overwhelming, the twigs of a great tree.

References

1. R. S. Fritz and E. L. Simms (ed.), Plant resistance to herbivores and pathogens (University of Chicago Press, Chicago, 1992).
2. J. N. Thompson and J. J. Burdon, Gene-for-gene coevolution between plants and parasites. *Nature (Lond)* **360**, 121–125 (1992).
3. I. Edfors-Lilja, *Escherichia coli* resistance in pigs. In: *Breeding for disease resistance in farm animals* J. B. Owen and R. F. E. Axford, (ed.) Wallingford: C.A.B. International (1991).

4. S. Wright, *Evolution and the genetics of populations: Vol. 2. The theory of gene frequencies*. (Chicago University Press, Chicago, 1968).
5. S. Nee, Antagonistic coevolution and the evolution of genotypic randomisation. *J. Theor Biol* **140**, 499–518 (1989).
6. J. Crow and M. Kimura, *An introduction to population genetics theory*. (Harper and Row, New York, 1970).
7. B. C. Clarke, The evolution of genetic diversity. *Proc R Soc Lond B* **205**, 453–474 (1979).
8. B. C. Clarke and P. O'Donald, Frequency dependent selection. *Heredity* **19**, 201–206 (1964).
9. S. Karlin and R. B. Cambell, The existence of a protected polymorphism under conditions of soft as opposed to hard selection in a multi-deme population system. *Am Nat* **117**, 262–275 (1981).
10. N. Takahata and M. Nei, Allelic genealogy under over-dominance and frequency-dependant selection and polymorphism of major histocompatibility complex loci. *Genetic* **124**, 967–978 (1990).
11. B. C. Clarke, The ecological genetics of host-parasite relationships. In: *The ecological genetics of host parasite relationships* A. E. R. Taylor and R. Muller (ed.) Blackwell Scientific Oxford: 87–103 (1976).
12. W. D. Hamilton, Sex versus non-sex versus parasite. *Oikos* **35**, 282–290 (1980).
13. W. D. Hamilton, P.A. Henderson, and N. A. Moran, Fluctuation of environment and coevolved antagonist polymorphism as factors in the maintenance of sex. In: *Natural selection and social behavior: recent research and theory* R. D. Alexander and D. W. Tinkle (ed.) Chiron, New York, 363–381 (1981).
14. R. M. May, Non-linear problems in ecology and resource management. In: *Chaotic behaviour of deterministic systems* G. Looss, R. H. G. Helleman, and R. Stora (ed.) North Holland, Amsterdam, 389–439 (1983).
15. R. M. May and R. M. Anderson, Epidemiology and genetics in the correlation of parasites and hosts. *Proc R Soc Lond B* **219**, 281–313 (1983).
16. J. Felsenstein, The theoretical population genetics of variable selection and migration. *Annu Rev Genet* **10**, 253–280 (1976).
17. J. H. Gillespie, Polymorphism in random environments. *Theor Popul Biol* **4**, 193–195 (1973).
18. N. H. Barton, Pleiotropic models of quantitative variation. *Genetics* **124**, 773–782 (1990).
19. M. G. Bulmer, Maintenance of genetic variability of mutation-selection balance: a child's guide through the jungle. *Genome* **31**, 761–767 (1989).
20. R. Bürger, G. P. Wagner, and F. Stettinger, How much heritable variation can be maintained in finite populations by mutation-selection balance? *Evolution* **43**, 1748–1766 (1989).
21. J. H. Gillespie, *The Causes of Molecular Evolution*. (Oxford University Press, Oxford, 1991).
22. M. Kimura, *The Neutral Allele Theory of Molecular Evolution*. (Cambridge University Press, Cambridge, 1983).
23. R. C. Lewontin, *The Genetic Basis of Evolutionary Change*. (Columbia University Press, New York, 1974).

24. W. D. Hamilton, R. Axelrod, and R. Tanese, Sex as an adaptation to resist parasites. *Proc Natl Acad Sci USA* **87**, 3566–3573 (1990).
25. G. Bell and J. Maynard Smith, Short-term selection for recombination among mutually antagonistic species. *Nature (Lond)* **328**, 66–68 (1987).
26. S. A. Frank, Ecological and genetic models of host-pathogen coevolution. *Heredity* **67**, 73–83 (1991a).
27. S. A. Frank, Spatial variation in coevolutionary dynamics. *Evol Ecol* **5**, 193–217 (1991b).
28. S. Levin, Some approaches to the modelling of coevolutionary interactions. In: *Coevolution* M. Nitecki (ed.) Chicago University Press, Chicago 21–65, (1983).
29. S. A. Levin, L. A. Segel, and F. R. Adler, Diffuse coevolution in plant-herbivore communities. *Theor Popul Biol* **37**, 171–191 (1990).
30. J. Seger, Dynamics of some simple host-parasite models with more than two genotypes in each species. *Phil Trans R Soc Lond B* **319**, 541–554 (1988).
31. J. Seger and W. D. Hamilton, Parasites and sex. In: *The evolution of sex* R. E. Michod and B. R. Levin (ed.) Sinauer, Sunderland, Massachusetts: 176–193 (1988).
32. R. Y. Lamb and R. B. Willey, Are parthenogenetic and related bisexual insects equal in fertility? *Evolution* **313**, 774–775 (1979).
33. A. R. Templeton, The prophecies of parthenogenesis. In: *Evolution and genetics* of life histories H. Dingle and J. P. Hegman (ed.) Springer; New York: 76–85 (1982).
34. H. N. Comins and D. W. E. Blatt, Prey-predator models in spatially heterogeneous environments. *J Theor Biol* **48**, 75–83 (1974).
35. B. P. Ziegler. Persistence and patchiness of predator-prey systems induced by discrete event population exchange. *J Theor Biol* **67**, 687–713 (1977).
36. W. D. Hamilton, Instability and cycling of two competing hosts with two parasites. In: *Evolutionary processes and theory* S. Karlin and E. Nevo (ed.) Academic Press, New York, 645–668, (1986).
37. W. D. Hamilton, Memes of Haldane and Jayakar in a theory of sex. *J Gent* **69**, 17–32 (1990a).
38. W. D. Hamilton, Seething genetics of health and the evolution of sex. In: *Evolution of life* S. Osawa and T. Honju (ed.) Springer, Tokyo, 229–251 (1990b).
39. G. Bell, *The Masterpiece of Nature: The Evolution and Genetics of Sexuality* (University of California Press, Berkeley, 1982).
40. I. Eshel and E. Akin, Evolutionary instability of mixed Nash solutions. *J Math Biol* **18**, 123–133 (1983).
41. M. Treisman, The evolution of sexual reproduction: a model which assumes individual selection. *Theor Biol* **60**, 421–431 (1976).
42. B. Golding, The prospects for polymorphisms shared between species. *Heredity* **68**, 263–276 (1991).
43. J. Klein, Origin of major histocompatibility complex polymorphism: the trans-species hypothesis. *Hum Immunol* **19**, 155 (1987).
44. J. Klein and N. Takahata, The major histocompatibility complex and the quest for origins. *Immunol Rev* **113**, 5–25 (1990).
45. A. Sasaki and Y. Iwasa, Optimal recombination rate in fluctuating environments. *Genetics* **115**, 377–388 (1987).

46. K. Mather. The genetical structure of populations. *Symp Soc Exp Biol* **7**, 66–95 (1953).
47. J. S. Jones, The heritability of fitness: bad news for 'good genes?' *Trends Ecol Evol* **2**, 35–38 (1987).
48. L. W. Simmons, Heritability of a male character chosen by females of the field cricket. *Gryllus bimoculatus. Behav Ecol Sociobiol* **21**, 129–133 (1987).
49. W. D. Hamilton, Pathogens as causes of genetic diversity in their host populations. In: *Population Biology of Infectious Diseases.* Dahlem Konferenzen 1982 R. M. Anderson and R. M. May (ed.) Springer, Berlin, 269–296 (1982).
50. E. von Dungern and L. Hirschfeld, Über gruppen spezifischer Strukturen des Blutes: III. *Z Immun Forsch Exp Ther* **8**, 526–562 (1911).
51. J. Forssman (ed.), Die heterogenetischen Antigene, besonders die sog. Forssman-Antigene und ihr Antikörper. (Gustav Fischer and Urban Schwarzenberg, Berlin, 1931).
52. U. Brandle, O. Hideki, V. Vincek, D. Klein, M. Golubic, B. Grahovac, and J. Klein, Trans-species evolution of Mhc-DRB haplotype polymorphism in primates: organisation of DRB genes in the chimpanzee. *Immunogenetics* **36**, 39–48 (1992).
53. F. Figueroa, J. Gulknecht, H. Tichy and J. Klein, Class II Mhc genes in rodent evolution. *Immunol Rev* **113**, 27–46 (1990).
54. E. Ivaskova, K. Dausset and P. Ivanyi, Cytotoxic reactions of anti-H-2 sera with human lymphocytes. *Folia Biol* (Prague) **18**, 194–197 (1972).
55. G. J. Lawrence and J. J. Burdon, Flax rust from *Linum marginale*: variation in a natural host-pathogen interaction. *Can J Bot* **67**, 3192–3198 (1988).

CHAPTER 8

INBREEDING IN EGYPT AND IN THIS BOOK: A CHILDISH PERSPECTIVE[†]

WILLIAM D. HAMILTON

Sphinx has no more than two kinds of riddle, one relating to the nature of things and the other to the nature of man.

Francis Bacon

The amount of inbreeding in the living world is great if we count species that inbreed[1] much less great if we estimate total inbred biomass. This is because most of the world's species are small-bodied and these are the ones most apt to inbreed. Most of the world's biomass is in big organisms like trees and grass. The relation of inbreeding to size (and concomitant longevity) is little discussed in other chapters of this book. I will later devote part of mine to saying why I think such a relation exists but first will illustrate my claim about size and numbers with some examples chosen from among my own earliest experiences of entomology. Although the examples are anecdotes and as such perhaps inappropriate to a serious book about inbreeding, they nevertheless give entry to an ever-surprising world that is strongly linked to our theme.

Inbreeders are Small

My mother says that shortly after I was born, ants attacked me in my cradle. The place was Cairo, so I like to guess the ant was 'Pharaoh's,' *Monomorium pharaonis*, although there is no such particular probability because the species is a worldwide tramp equally at home in any warm town or even a warm London kitchen. There are plenty of other ants in Cairo that could have been interested

[†]In N. W. Thornhill and W. M. Shields (eds), *The Natural History of Inbreeding and Outbreeding: Theoretical and empirical perspectives*, pp. 429–450 (The University of Chicago Press, Chicago, 1993).

in a baby. However, I like the idea because, besides being a notorious lover of sweet things, this tiny yellow ant is noted for indiscriminate inbreeding within the nest, including brother with sister. Could this be why it is called 'Pharaoh's'—or is it simply that first specimens came from Egypt? The latter is more likely.

To move to my next example, measures that shortly stopped the ants did not, according to my mother, stop another insect which came to eat the buttons on my clothes. Again she can't describe it exactly, but it is very likely to have been the button beetle, *Coccotrypes dactyliperda*, which would have arrived as flying females. The button beetle is even more of an inbreeder. It not only rests its system on a firm foundation of brother-sister incest, like the ant, but continues to other crimes, at least in the case in which a female happens *not* to have been fertilized by her brother. On reaching a button of vegetable ivory (made from seeds of the palm *Phytelephas*) or, probably more commonly in nature, a date stone, such a virgin first excavates a chamber and then, inside, uses parthenogenesis to produce a brood of four or five males. After a wait for the first male to mature she mates with him, eats him except for the hard thorax and head, eats the other sons similarly, and finally proceeds to further enlargement of the chamber and the laying of eggs for her main brood. This consists of about seventy females sired by the first son and about three more males.[2] We see at once that small virgin colonizing arthropods such as Pharaoh's ant and the button beetle are much more keen to start breeding somehow, come what may, than are horned scarabs or female academics at the opposite extreme of *K*-selection, and we see that they act towards this end as if with forethought, paralleling seemingly by ingenious 'ROMs' of DNA the subtle but flawed memes of the god-kings and priests of the same land. In reality, the inbreeding achievements of the pharaohs may have been more in the minds of their subjects—how these looked at royal inheritance, accepted social inequality, although of course these remain memes of the imposers—than in substantive homozygous human tissue. An exciting discovery reported by Rowley, et al.[3] concerning the reputedly incestuous blue wrens of Australia begins to expose a paternity deception, probably initiated by breeding females in this case, that is surprisingly similar. The birds are small as birds go but nearer to humans than to button beetles or mites on the log scale. The point remains that the pharaohs cannot really have gotten far emulating what some of the tiniest minions of their granaries were doing easily. Their usual distaste for their attempt probably was not much different from what the man in the oasis street, or his camel, or his date palm would feel if asked to breed in the same way—and here we can imagine the palm throwing on the other two its mournful shadows of whole towns of happy button-beetle homes as all three ponder the same theme. This is the contrast: the big and the small. I admit I do not know much about the tastes and distastes of date palms or what makes the shadow of one mournful, but the evolution of dioecy by palms in general (and the date palm in particular) plus inescapable natural selection to resist such pests as beetles that destroy their seeds suggests a general direction, which I shall return to below. On whether the pharaohs did ever *successfully* inbreed in terms of descendants, a point that seems not quite decided, I will also come back to some rather lightweight thoughts of my own at the end.

I am not sure whether *Locusta migratoria* ever visited me in my cradle but, being a common insect even when not a plague, it is quite likely to have done so. If it did it probably brought me new examples of inbreeding, for it would have had a good chance to carry a certain parasitic mite, *Podapolipus diander*. This mite is named for having two types of male. One occurs in a brood only as a single precocious firstborn,[4] a son with only one function. As soon as he is born, he shoulders back into his mother's cloaca while she is still feeding and slowly swelling on the haemolymph of the locust. He fertilizes her whether or not she has been fertilized already. After this event she can certainly lay female eggs, for like ants and the button beetle, the mite reproduces by haplodiploidy, the system of reproduction in which haploid males—identical in genotype to their mother's gamete—come from unfertilized ova but normal diploid females appear when ova are fertilized. Later in the brood, different newborn males, uncommon and still very precocious, fertilize their sisters.

But what of the breeding of the locust itself? This animal is admittedly more 'weedy' than a camel or date palm, but a locust certainly cannot tolerate inbreeding at the level of the mite. A connected later anecdote may illustrate this. When I worked at Imperial College Field Station we had in the constant-temperature rooms in the basement of Silwood House a colony of *Schistocerca gregaria*, the other main locust of the Middle East, and after a time it became badly infested with mites. The locusts were depressed and sat sluggishly on the floor of the cage. This time for certain the ants that came to swarm over the locusts and cut them up while so immobilized were Pharaoh's, and the name now suited them in a new, rather gruesome way, for they dismantled their huge prey rather like teams quarrying stone for a pyramid. The remembered unpleasant sight reminds me to be less skeptical of my mother's saucers of water she says she placed under the feet of my cradle, for those ants too were probably after not just a baby's spills but meat and serum, as would certainly have been the case had the ant been another incestuous cosmopolitan possibility, the equally tiny *Hypoponera punctatissima*, which likes flesh only and has no interest in sugar. To conclude a digression, however, the moral here is that the locusts in the inevitably small colony we kept *appeared to suffer from inbreeding*. This was suggested also by their imperfect wings, poor melanization, etc., and it may have been why, in particular, they were so susceptible to parasitism as discussed above (and again below), and why in general they were so physically incompetent as not to escape the ants, who suffered from inbreeding not at all. *Coccotrypes, Podapolipus*, or an inbreeding hymenopteran parasitoid (but see Crozier)[5] would also certainly not so suffer,[6] nor did in fact another middle-of-the-road inbreeder that we had caged in the Silwood basement where it too was doubtless, to me, a returned babyhood companion. This was the small tropical diurnal mosquito, *Aedes aegypti*, the sometime vector of yellow fever.

It is definitely not incestuous by habit but is probably nevertheless well accustomed to small-colony bottlenecks brought about by its normal breeding habitat of rain pools in old tires and similar places. Our stock in the basement was as inbred as the locusts, but escaped females often proved to me in my office that they lacked nothing in agile flight or zest for blood.

My fascination for such small animals, begun in my cradle and later accentuated by admiration as I realized their regal contempt for the taboos of humanity, has since turned up a seemingly endless list of other examples.[7–9] Among these, for compounded incestuous sin it is hard to beat a parasitoid of beetles, *Scleroderma immigrans*, whose antlike female readily remates a grandson born of a daughter got from the mother's previous union with a son.[10] At an equal but different extreme—sin aforethought as one might put it, as judged by structure—consider a case that once came to me on a piece of white-rotten wood pulled from a branch. Pearly droplets adhered to the velvety broken surface of mycelium. Under a microscope the drops proved to have heads and legs, and in water on a slide I could see what was happening inside. Each was the enormous body of a female mite distended with eggs and young. Tiny fully formed adults were bursting from eggs as I watched, joining a throng floating or slowly swimming in the maternal fluid. I soon made out there were three kinds, two of them female (both numerous overall, one kind a chelate dispersal morph and the other more plain) and one kind of male. The males were uncommon, one or two to a brood but they were always first to hatch and very active. They spent all their time searching for new sisters in the crowd and copulating with them. All this was happening *within* the yet unruptured mother (Hamilton;[9] for an illustration see Trivers).[11] This mite is in the genus *Pygmephorus*.

As these examples and others mentioned by Werren[6] make plain, whether incest is or is not exceptional for large higher organisms—for example, for humans, birds, fishes, trees, and palms—it is certainly far from being so for a huge array of small arthropods. The very size of the array in question may be partly a consequence of the incest, for new species formation is facilitated by inbreeding, as explained by Howard.[12] Correspondingly, Acarina, which are perhaps the world's greatest practitioners, as already hinted in my examples, happen also to be the world's second most species-rich class. It has been suggested they may even surpass the insects if they are ever equally studied. Among the insects another huge contribution to the faunal list comes from the Hymenoptera Parasitica, whose exceedingly numerous smaller members are likewise very incestuous. Turning to other groups, within lifestyles as social insects and dead tree inhabitants the smaller ants and the male-haploid 'bark beetles' of tribe Xyleborini also show up more speciose than noninbred sister groups. Whether this reflects *important* (macro-) evolution as well as species splitting is another question, but the inbreeders are unquestionably highly innovative at times.[8]

Groups and examples like those mentioned should not give the impression that all small arthropods mate close relatives by preference or with impunity.

A proposition that really large insects don't practice incest seems firm: no insect the size of the locust mates like the locust's mite *Podapolipus* does, although there are a few fairly large orthopterans and phasmids that are parthenogenetic.[13] But at the other extreme of size even very small arthropods quite often have adaptations that prevent incest. Examples are the unisexual broods of some cynipid gall wasps, gall midges, sciarid flies, and others. For Acarina, opposite tendencies in animals of rather equal size are well illustrated by the mites of our skin. *Demodex folliculorum*, a wormlike mite that typically lives harmlessly around the roots of our eyelashes, is incestuous and has a very biased sex ratio, whereas our other specifically human associate, which burrows and breeds like a mole, but with more damage, beneath the more open pastures of our skin, *Sarcoptes scabiei*, seems well designed as an outbreeder since it has a 1 : 1 sex ratio. Later I will show a plausible rationale for this type of contrast based on the degree of damage to the host; but I have none even in prospect yet for another similarly different pair that I find in British woods. There the scolytids *Xyleborus dispar* and *Trypodendron domesticum* make brood tunnels mixed together in the same newly fallen logs of oak or beech. Both are mutualists with 'ambrosia' fungi that they carry with them and inoculate into the wood, where later special pseudofruit of the mycelium, presented in smooth lawns spread on the walls of their galleries, feed their larvae. The sizes, lifestyles and even habitat of the two species appear nearly identical, and yet the first insect is incestuous, with its males displaying the modifications that usually accompany such a habit, being few, small, blind, and flightless;[7] while the second is an outbreeder with its males common, large, and similar to the females.

Inbreeding and Reproductive Efficiency

One would think that in a species as thoroughly incestuous as the droplike *Pygmephorus*, having sex at all would have lost all its point, whatever the point might once have been. They must surely, one would think, be so inbred that recombination is ineffective. However, extreme inbreeding does not seem to pass easily to pure female parthenogenesis (thelytoky). Thus although there are certainly thousands of species of incestuous 'bark beetles' (for the most part *not* found under the bark but deeper in the wood living on ambrosia, like *X. dispar* above, or else in other obscure plant niches such as the twigs of tea bushes, developing coffee beans, and the date stones and even buttons already mentioned) there is no case yet known to me where such an incestuous species has become parthenogenetic. On the contrary, bark beetles that have adopted parthenogenesis are former outbreeders whose males have well-developed secondary sexual characters.[14] It is not clear that the same is true of cases of thelytoky in the Hymenoptera Parasitica; here there are groups where both this full type of parthenogenesis (female offspring always) and arrhenotoky (males only from unfertilized eggs following from haplodiploidy, females produced normally through fertilization) seem to be common. Elsewhere in animals, however, a rule that inbreeding and parthenogenesis are not found close together in phylogeny seems to hold fairly well. It certainly holds for plants, in which parthenogens typically appear abruptly in outbred and often self-incompatible species and groups, and arise only rarely among selfers. Knowlton and Jackson[15] show that the same is true of marine invertebrate animals, including such cases as the corals and other 'colonials' that have plantlike growth forms. Small animals are the selfers. It seems that when an environment no longer favors outbred sex, species that are adapted to inbreed intensify their practice and reap its efficiency through bias of the sex ratio—that is, they produce fewer of the 'wasteful' males. Inbreeding thus provides an adjustable control of their degree of effective sex. Outbreeders, in contrast, are checked in an increasingly unstable position, barred from inbreeding both by habit and by their load of deleterious recessive alleles. Eventually they must take the whole step to parthenogenesis through finding the right mutation, or possibly, as recent evidence suggests, the right executive microbe.[16] Since species hybridization is so often antecedent to parthenogenesis it may be that inherited microbes gain (or regain) special power over sex when associated with non-coadapted genomes.

How do inbreeders achieve bias of the sex ratio? In some cases the group already has male haploidy and then a mother can produce females at will by control of her stored sperm: a fertilized egg makes a daughter egg and an unfertilized egg makes a son.[6] But how to arrive at male haploidy in the first place? The button beetle is probably of fairly recent entry into the mode, representing one among at least three separate entries in Scolytidae, and it hints at an answer using microbial symbionts and inbreeding together. To see how the idea works, consider first the ordinary XY sex determination mechanism. A good reason why in this system the male is usually heterogametic is that having control rest in the female, with her maternally transmitted organelles or symbionts, would too often lead to disastrous female biases in the sex ratio. As explained elsewhere[9] organelles or symbionts with solely maternal transmission are under strong selection pressure to make a female produce as many daughters as possible, and this can only be done by producing fewer sons—ultimately, if possible, none at all. When females are heterogametic, symbionts are presumably in a strong position to get their way. They can interfere in oogenesis where, for example, they might steer the Y chromosome towards a polar body. In males, on the other hand, almost all kinds of symbionts are doomed and consequently have no interest in the sex ratio: the segregation of the sex chromosomes in a heterogametic male can always proceed unbiased... or can it? If the species is an inbreeder, it comes to be in the *kin-selected interest* of the symbionts in the male to bias the sex ratio towards females for the benefit of 'cousins,' that is, for the benefit of those *related* symbionts that are in the female their bearer is going to mate with. While there is no polar body to which to steer the Y in spermatogenesis, they can attempt to identify the Y-bearing spermatids and kill or disadvantage them. Postfertilization strategies like this by symbionts working against male zygotes in early stages are well known in *Drosophila*. From such rationale comes a possible coevolutionary sequence in which the symbionts in males begin discriminating the Y, perhaps tagging (methylating?) it early in development when it is active and open to recognition, while the host responds by seeking to hide the Y characteristic that is being detected. The host possibly gains a more radical escape from microbial intrigues whenever the real male-determining locus or segment is shifted to another chromosome. Along these lines one may rationalize either of the following courses:

(1) Inactivation of the whole paternally derived set of chromosomes through part of the lifetime of a male, this always being accompanied by rejection of the set during gamete formation. Sometimes staggering onto the printed page under the name parahaplodiploidy, such a condition is surprisingly frequent in small insects.

(2) A sequence in which the sex locus is transferred to an autosome and a previous inactive Y deleted, giving successive constitutions like:

$$\begin{aligned}\text{Males}: \; & 2n\,\mathrm{A} + \mathrm{XY} \rightarrow 2(n-1)\,\mathrm{A} + \mathrm{X'Y'} + \mathrm{X} \\ & \rightarrow 2(n-2)\,\mathrm{A} + \mathrm{X''Y''} + \mathrm{X'} + \mathrm{X} \rightarrow \ldots \\ \text{Females}: \; & 2n\,\mathrm{A} + \mathrm{XX} \rightarrow 2(n-1)\,\mathrm{A} + \mathrm{X'X'} + \mathrm{XX} \\ & \rightarrow 2(n-2)\,\mathrm{A} + \mathrm{X''X''} + \mathrm{X'X'} + \mathrm{XX} \rightarrow \ldots\end{aligned}$$

The latest X^t is the currently effective one determining sex. The sequence ends when all autosomes have been converted to Xs, and male haploidy is attained. The sex of a zygote must now rest entirely on chromosome dosage, and a new group has joined the others to be a touchstone and headache for evolutionary theorists, while, as is most interesting from the point of view of the problem of sex, the group's nonheterozygous males are now ready to contribute their evidence that heterozygote advantage cannot be of great importance for diploid efficiency[6] or for sex, a point to which I return below.

It would take too much space to present the as yet only plausible evidence that exists for such schemes of entry to male haploidy, but two points can be made. Firstly, insects are known whose cytogenetics could illustrate stages on either of the above roads. Multiple sex chromosomes and/or parahaplodiploidy are unusually common in insects known to carry intimate microbial symbionts, as may easily be seen by combining information from the major texts of White[17] and Buchner.[18] (In the case of the button beetle, indeed, symbionts may have been the focus that led Buchner to the strange life history and sex ratio situation I outlined earlier, although that the tiny companions might in some degree guide the evolution of the life history itself seems not to have occurred to him.) Secondly, it is very interesting, and can also be seen from Buchner's general account, that symbionts in species with ovarial transmission by no means fail in their role or disappear, as we might expect to happen from the hopelessness of their situation, when they find themselves in a male. Sometimes they become housed in well-formed mycetomes near the testes and even more often they are found in close attendance on the events of spermatogenesis. Do they crowd here just because they are useful slaves like mitochondria, providers perhaps of an obscure vitamin, their service entirely at the beck and call of their hosts? Or are they there rather as teats are there on the chest of a male mammal, following blind routines and physiological gradients appropriate to another situation, pointlessly aping the motions of cousins in the female gonad? But, *cousins*—might that be the point? I would guess that neither of the previous explanations is the best; instead I suspect that symbionts are evolved to be present and to act positive roles on behalf of related entities residing in the females their bearer is to meet and mate with, along the lines suggested above.

Ecological Correlates of Effective Asex

After sex is effectively abandoned by inbreeding or by the transition to parthenogenesis, it does seem that both types of resulting asex persist best under a certain kind of ecology. Both inbreeders and asexuals tend to be 'weedy' inhabitants of extreme and biotically simplified habitats.[15,19] Or sometimes they are extreme specialists, incumbents of what might be called crannies rather than niches, outside of the mainstreams of life; or finally, they may be organisms specialized in the sense that they live in the shelter of larger protectors.[20] For example, on the last theme, ambrosia beetles mate their brothers while living with fungi that they have planted deep in the wood, far from confamilials in the dizzy arthropodan, fungal, and microbial whirlpool of life in the early decay stages just under the bark. In this more typical under-bark habitat, at least in the temperate zone, the scolytids' approach has been to detect the death of the tree early and get in and out as fast as possible, a policy which however still does not completely save them from the interspecies mayhem of colonizers that follow. Fig wasps (*Blastophaga*) have a 'cranny' provided where they are cultivated to inbreed as well as to become, later, outbreeding agents for the trees that they serve, just as the ambrosia beetles become agents for the dispersal and outbreeding of the large fungal invasions which have fed them and which they help to disperse and implant. We note that in both wasps and beetles, rather similar integumental pockets in the adult females have evolved to transport pollen and spores. With these examples in mind, the inbreeding of *Demodex* mites in our eyelash follicles now prompts a thought that perhaps not only do the mites do no harm, they may even do us good, perhaps, for example, protecting our eyes against other infections. The same could be true of our seemingly harmless pinworms, which, as I learned for the first time in this book[6] are also haplodiploid. As for the button beetle, I still remember my surprise in finding that in

Britain this scolytid species is much more closely related to outbreeding *Dryocoetes villosus*, a common and rather ordinary inhabitant of the familiar typical diverse community of decaying oak phloem, than it is related to *Xyleborus*; but then what cranny could be quieter than a button to support such a parallel trend? Real weeds—the plants—self-pollinate out in the plowed fields. There they may eventually enter the human following, becoming crops in their own right but remaining inbreeders; but all, both weeds and ex-weeds, grow scattered and short-lived, far apart both from each other and from the dense perennial stands in which their close and fully sexual relatives are found wrestling for the space of the meadow. Cnemidophorus lizards, prickly pears, teddy-bear chollas (perhaps also, in Sinai, burning and unburned brambles of Horeb) perform *their* degenerate sexualities, apomixis and vegetative propagation, out in the Sonoran Desert; *Poeciliopsis* fish theirs (gynogenesis) in the pools of the same desert; *Artemia* shrimp theirs in salt pans often in the same desert again; *Hadenoecus* crickets theirs in caves. Viviparous grasses grow seedlings from their flowers on the moraines of glaciers; oppositely and yet the same, viviparous onions sprout bulbs and plantlets out in the sunny fields of Egypt. Asexual mayflies and chironomids live at the insect limits of depth and temperature in lakes and pools, while asexual ostracods, once more oppositely and yet the same, are in the most shallow and *ephemeral* pools... The list goes on. Now it is time to address the question of what makes such extremely various organisms able to evolve an efficiency through reduction of sex that their relatives in denser, more species-rich habitats can't. What is the secret influence connected with size and biotic complexity?

Until we know the answer to this question, as it seems to me, most of the important questions about inbreeding cannot be properly framed, let alone answered. This fact seems to have had too little attention in the rest of this book. I want from here down to give my own view, but as preparation I want first to look at another issue, heterosis, that has been raised in at least three chapters.

Outbreeding: Het Advantage or Hom Combination

As is made clear by Mitton,[21] ever since the earliest years of genetics there has been a question about the nature of inbreeding depression. The alternatives most discussed are that it is due to deleterious recessives in mutation-selection balance, or else that it is due to an intrinsic benefit of heterozygosis itself. Mitton presents evidence mostly favoring the latter, but is cautious and admits contrary evidence at some points. Uyenoyama[22] assumes intrinsic heterosis as the basis for a theme of eugenic behavior in potential parents. Waser,[23] attending only to plants, is also cautious but is more on the side of recessive mutations. In the literature in general the problem has been like painting the Forth Bridge, no sooner finished at one end than begun at the other. While not doubting that both of the ideas above are indeed in their way rust-preventive and play a part, I would like now to stir into the paint of this key bridge of biology a third possibility. Not surprisingly, this new idea accords better than either previous one with my own prejudice on the evolution of sex; whether ultimately it will help any better with the issue of inbreeding depression remains to be seen.

It seems possible that some apparent heterozygote advantage, at least in the wild, is actually epistatic homozygote advantage expressed at unseen loci.

Imagine a viscous population for which a study has shown that heterozygosity at various marker loci correlates with components of fitness (as with the evidence for *Pinus ponderosa*, well summarized by Mitton). My point is that the finding does not really show that loci currently heterozygous are on average giving benefit or loci homozygous giving harm. When an occasional distant outcross arises—a pollen grain has blown a long way to claim an ovule and following this a seed from that ovule has grown to be a seedling and then a tree—the zygote so created is indeed highly heterozygous. If the seedling or tree is unusually fit, it must indeed be an indication of heterosis. However we should note that the *next-generation* offspring of such a cross *remain* more heterozygous than the average. Once descendants become inbred again, as the viscosity makes likely, something new has been added: new homozygous combinations are now present, combinations that may never have existed in the locality before. In such inbred descendants, along with the diminishing heterozygosity from the original event, there will be new *multiply homozygous genotypes*. Some will doubtless be bad and likely to die early. Others may be very good; indeed, the good new multiple homozygotes may be the real reward sought in the striving for the distant outcross. The idea has much in common with Uyenoyama's theme, although the selection for incompatibility is based on a less immediate advantage. The main point is that residual heterozygosity may be found correlating with fitness in a way that only creates an illusion of causation: *there might be no over-dominance in fitness at any locus at any time and still such correlation could appear.*

My approach may seem to betray distrust of heterozygote advantage. It does, and the distrust arises from two sources. One is familiarity with species where perfectly fit and ordinary males cannot possibly be heterotic because they are haploid. It seems to me, for example, that there is hardly any difference either in general evolutionary style or in immediate health and activity between male Hymenoptera and male Diptera. Similarly, there seems to be nothing unhealthy about being a haploid moss[24] or seaweed.[25] My other source of distrust is that when I introduce such advantage into a model for evolutionary maintenance of sex the model is always less likely to succeed. The reason for the failure is easy to understand. Heterozygote advantage makes polymorphism very stable but does nothing to make it efficient: in contrast to sexual heterozygotes, which can't breed true, parthenogenetic heterozygotes can, and this immediately favors the switch. Gene duplication followed by selection to homozygosity in opposite ways at the two loci offers a third and perhaps ideal route to the goal of no wasteful segregation. But this is a more difficult step to take, and moreover, it opens up a new kind of inefficiency, that of carrying too much code. I feel that a tenet basic to a whole line of reasoning on the role of heterozygote advantage, that greater diversity of synthesis within an organism is better, needs to be questioned. Again mosses and the competent haploid males of Hymenoptera and several other insect groups seem to say something different. More generally, an admirable belief critique is that of Clarke.[26]

The essence of 'Red Queen' models that stabilize sex even against fully efficient parthenogenesis (and that therefore would almost certainly support it against inbreeding, whose efficiency is intermediate) is simply eternal change. This is more likely to occur when selection coefficients that are monotonic across the genotypes switch their slant back and forth without heterozygotes ever becoming the most fit. A role for heterozygote advantage is not completely lost in this kind of alternation, but it is changed in kind. When fitnesses throughout such a sequence are considered together, *geometric* mean fitnesses over the many generations still have to show heterozygote advantage, otherwise one or another allele eventually fixes[27,28]; this, however, is a looser advantage than is normally implied.

Now returning in a more experimental mind to the viscous population already discussed, imagine that distant individuals are crossed, and that their progeny are reared at both the parental sites and eventually are scored for their success in life. A greater average success of the F_1 hybrids compared with average success of local controls in this experiment confirms the usual claims: the advantage may be intrinsic heterozygote advantage, local fixation of mildly deleterious mutations, or both together. However, suppose that we inbreed the progeny and continue for one more generation. Now the other effect I have suggested may become perceptible—in addition or instead: Inbred F_2s from the F_1s, similarly reared in two places, may show some individuals *further* to increase their fitness, or else may show fitness increased in some degree even when their parents (F_1s) showed no increase. Heterozygosity would have to be maximal in the F_1, so such further increase in F_2s could not come from that. It could, however, very well be explained by new homozygous combinations. Tests of this point would seem to me well worth trying. As I see it, they must be done in the wild rather than in protected greenhouse, garden, or vivarium conditions so as to expose the broods to the full natural biotic adversity of their habitats. Moreover in such an experiment I would like to see the impact of parasites monitored with particular care, perhaps even experimentally augmented, for reasons that I now discuss.

Outbreeding: Descendant Health

My emphasis on parasites rather than other ecological agencies of stress and selection comes from my belief that parasites are fundamental to the maintenance of sex. If the idea is right then obviously the comparative toll they impose on inbred and outbred individuals is of great interest: the theory says that eventually inbred lines should suffer from the same inflexibility in the face of parasites as parthenogens do and that this will make them die out or else force them back to outbreeding. The reasons to expect parasites to be more effective promoters of sex in hosts than other temporally varying factors of the environment have been given in detail elsewhere[29–31] (and references cited therein) but may here be summarized again as (1) parasite ubiquity, (2) frequency dependence, (3) intimacy, and finally, (4) the combination of (*a*) small size, leading to (*b*) fast turnover, leading to (*c*) fast evolution. I will now raise a few points from such factors that are particularly relevant to inbreeding.

If, instead of heterosis in F_1s, it is the chance to create new combinations for descendants that is the main object of sex, it may be expected that the offspring of wide outcrosses will have adaptations to make best use of the opportunity endowed by their hybridity. For the rapid production of new homozygous combinations, the closer the inbreeding that follows an event of outcrossing the better; and the higher the chance of swift, nonwasteful elimination of bad homozygous combinations, as for example by sibling replacement,[32] the better also. There are few firm data bearing on this, but three suggestive topics may be mentioned. Firstly, we may dimly discern here a possibility for what is happening in some cases of automictic parthenogenesis which, for genes far out on the arms of chromosomes, acts like very close inbreeding after an initial cross. In effect automixis acts like selfing, and parthenogens that have this kind of gametogenesis might be reconsidered from this point of view. The occasional males, if such occur, might in the same spirit be tested for acceptability and competence to inseminate in biotically stressed conditions of the population. (In passing, in fully sexual species such stressed conditions might also, on the same reasoning, be the key to the erratic repeatability in the other kind of 'rare male' study). Unfortunately for the idea about automixis, the best-known exemplars of cyclical parthenogenesis, such as aphids and cladocerans, seem to use apomictic parthenogenesis. Although a process of 'endomeiosis' has been claimed as a variant in the clonal reproduction of aphids, its reality is dubious.[33] The second possibly relevant set of facts concerns some annual plants that do in effect alternate their generations between emphasis on chasmogamous and on cleistogamous flowers. The phenomenon does not at present favor the idea that the alternation has evolved with the adaptive genetic objectives suggested—other factors seem sufficient explanation.[34] Moreover, the seed numbers are rather too low to provide the level of sib competition I would guess to be ideal.[34] Nevertheless, the existence of such cases is intriguing, and it would seem worth looking for weaker trends in plants that are variable but not dimorphic in their flower type and pollination (e.g., small crucifers). A third possibly relevant set of facts concerns termites, which likewise commonly follow their outbreeding as alates with one or more generations of inbreeding.[35] The alary polymorphism that is deeply embedded in every termite colony's full cycle guides us to think of the phylogenetically diverse set of colonial insects that are usually wingless in their endogamous groups but still produce winged morphs to disperse.[8,36] Winged females in the bursts of dispersers are usually fertilized before they leave; however, it has to be that either circumstances are such as to encourage cofounding of colonies by such females, which is indeed often observed, or that winged males occur along with them, showing that at least some outbreeding is intended, as also observed (e.g. Crespi).[37] Visible polymorphism marks such cases as extremes in the array of tendencies to outbreed or inbreed in colonial insects. Similar phenomena can be expected and are found, though without the visible polymorphism, in the breeding systems of many small mammals.[38]

The parasite theory of sex predicts that most social insects will not inbreed, and especially will not if they have large colonies which are very apparent and exposed, which means findable by parasites. How species like Pharaoh's ant, *Hypoponera punctatissima*, and *Cardiocondyla*[39,40] get away with close inbreeding is not clear, but the answers may be in first part that they are the opposite of 'apparent' species and therefore little bothered by parasites, in second part that they are anthropophilic, moving into new man-made habitats where parasites have difficulty in following, and in third and last and perhaps most important part, that they are tramps, and thus like groundsels and dandelions among plants, continually on the move. More specifically in the case of 'Pharaoh's,' it may well be that in the heated buildings in Scotland where their mating system was first studied and they were found to inbreed they had left even more of their parasites behind than in most other places: if studied in native (?) Egypt, for other example, it might turn out that they do *not* inbreed. Likewise that rather parallel tramp *H. punctatissima*, which today often occupies the same buildings, came at least once to the frontier of Scotland, where in an otherwise inhospitable waste of bracken, heather, and bog, it made its home, as archaeological evidence reveals, in the heated living quarters of the soldiers of Emperor Hadrian, built beside his famous Wall. What the Roman guardians of the Wall thought about their unusual tiny visitor and how much incest it practiced in the recesses of their messy floor are matters unknown. However, I have checked for myself with a colony I obtained from a hospital in London that the workerlike, belligerent, yet utterly lazy males readily mate with sisters when these are available. Such matings, however, are only possible for the dominant male, since fighting excludes all others from the chamber of the royal brood, and in fact all non-dominant and defeated males soon die. My observed fighters were brothers or very close, but their adaptations for combat suggest to me that colonies are not always so inbred nor the males so related;[9,40] but see also Murray and Gerrard;[41] Murray:[42] perhaps cofounding of nests is not uncommon in big centers for the ant like London even if one has to doubt the event can possibly have been common at Hadrian's Wall. Did the ants come there in some crate of dried, beetle-infested food from Italy? The incident of colonization in such an inhospitable outer environment seems likely to have been unique.

Like a large multicellular body, a large and homeothermic social insect nest containing a closely packed brood offers a good home for parasitic microbes. Providing spread is possible from one larva to another, the nest is almost like the continuous tissues of a large mammalian or avian body, and nurse adults monitoring for disease in this situation[43] are like lymphocytes. Likewise it is a good living place for larger well-armored predators and for subtle inquilines. I believe that it is the apparency and vulnerability of social insect nests that leads many of the commonest species to be strong outbreeders. They seem to go to great lengths to create diversity of genotypes in their nests.[8,44] The many matings of the honeybee queen and the huge simultaneous mating flights of ants and termites are examples, as are the trends to unisexual broods.[45,46] Pointing the same way is the enormous investment in flying males relative to flightless fertile females shown by true legionary ants. Such ants are particularly abundant and influential in their ecosystems, and doubtless because of this they also support an extraordinary number and variety of inquiline arthropods,[47] not to mention vertebrates like ant birds that follow them to exploit them or their disturbances. One might put it that by daring to become major biotic oppressors in the forest they have ended being also very biotically oppressed.

I hope I have by now shown a reasonable case why large long-lived beings, whether trees, people, or supercolonies, usually retain sex, and for the act itself prefer relative strangers—strangers who should come from as far away, ideally, as their parasites.

Out of the Land of Egypt . . . Genes and Memes

Which has been more than enough on that subject, as Herodotus was wont to say when writing his own endlessly digressive first tourist's account of Egypt in the fifth century B.C.. Legionary ants have made me think of Africa, where my life and this essay started. Fortunately in Africa, *Dorylus*, harsh biter of explorers and the most formidable genus of the group, does not occur above ground north of the desert, and so definitely was not my attacker. We may leave its swarming armies, populous as the town of London, to the south; thence follow the Nile north and time backward to ancient Egypt and so enter my closing topic. I will now add more detail and new speculation to the discussion of a particular famous historical spate of human inbreeding in that country, and I will suggest what may have spread to the world from the event. And in reality this road to Egypt can start not only south of the Sahara but from almost anywhere—from Texas or indeed from wherever science is practiced, as I now explain.

A pioneer paper already cited on the connection of parasites and sex and outbreeding is Levin.[29] I focus now, unusually, not the content of this paper but on its author's name: the name has an odd connection with Egypt. Perhaps

I should apologize at this point for imputing something to a particular name whose origin I don't know about for certain, but Levin is clearly akin to Levins, Levine, Levene, Lewis, etc., and this group as a whole is well known to have Jewish roots. This means, or should mean, that its bearers are more connected in ancestry than most of us are with Egypt. At the very least the root tells that they have *one* more certain connection there, even if a connection across a gap of three thousand years. I thought at first that the names would probably derive from Levi and therefore might claim not just general ancestry through the Israelite period in Egypt but a *male-line connection* to the original Levi, the son of Jacob, who definitely was at least briefly in Egypt according to the Bible story. But surprisingly this root is not considered the most probable,[48] although there are others, such as Levi itself and Levy, for which it is. Such a direct connection to the original Levi is, however, very definitely claimed for another name swarm typified in Cohen. The point I want to make concerns all Jewish ancestry but specially bears on these particular male lines. It is that even for 'Levi' and 'Cohen' there is a strong doubt about the claimed pure male line, and it is one which throws an interesting sideways illumination on the whole theme of this book and on other human matters hitherto outside it. The doubt is over the identity of Levi's supposed great-grandsons, Moses and Aaron.[49]

Inbreeding was certainly by no means lacking in the Jews of Egypt according to the Bible. Levi himself came of a cousin marriage as well as having consanguinity from links in his parents back through the generation of Terah. Further, although Levi himself outbred, his grandson Amran married an aunt (that is, married Levi's daughter). Amran is the stated father of Moses and Aaron. After Amran, Aaron, but not Moses, again inbred distantly, and it is from Aaron that Cohens and others with related names are supposed to descend. If all this were true, the spread of descendants down from Levi would seem to indicate that the mild level of inbreeding practiced among the patriarchs did the tribes little harm. However, we may ask, how believable are the pedigrees? Dubious as may be the genealogy of the kings of Egypt, with their much stronger inclination to inbreed, or to seem to do so, that of the Jewish patriarchs is more enigmatic still, its claims being merely orally transmitted until some six hundred years later than the events described. In the case of the Jews no mummies were carefully laid down for our assistance.

Scholars agree that the names Moses and Aaron are Egyptian. The tale around them has several confusing and improbable features.[49] Difficulties and possible alternative interpretations began to be pointed out in the last century, but a particularly coherent critique and reconstruction is due to Sigmund Freud,[50] himself Jewish and writing his theory with great courage, in old age, under the looming shadow of the Nazis. Freud suggested that Moses was an Egyptian prince or priest: as such he would have *joined* the Israelites, and, as the major point of Freud's argument, would have outbred to them not only his genes (in grandchildren, his wife being foreign even in the story) but his crucial memes as well. Moses' achievement, Freud argued, besides the well-known rescue of the Jews from bondage, was the grafting to their religion of *ideas of a recently suppressed heresy of Egypt which he had kept aflame*. His accepted role in promoting monotheism and iconoclasm in the Hebrew tribes fits very well in time and place with the possibility of his being a missionary or refugee emerging from the counterreformation that convulsed Egypt near to his time. The religion overthrown, of which Moses is suggested to be a surviving adherent, was the sun-disc cult of Pharaoh Akhenaton. It does indeed seem that the presence of the Hebrews in Egypt through the astonishing 17-year revolutionary reign of the heretic king, plus Moses' own name, nature, and described origin, plus the teaching he gave to the Jews as he led them to freedom, is too much to be taken as coincidence. If the thesis is accepted it is likely that his priestly companion and so-called brother, Aaron, was also 'grafted' to the Jewish story at the same time. It then follows of course that Aaron was unrelated to the 'cousin' Elisheba, and more widely, that the inbreeding of the patriarchs described by the writers of the Pentateuch is more to be seen as a record of what was acceptable and likely than as a record of fact. But that is hardly the point; the other implications of Freud's theory, connecting as it does the history of Jewish thought with that of a unique epoch in the history of Egypt, are by far more dramatic. Its particular relevance to the present chapter is that in the transfer, Akhenaton's undoubtedly revolutionary ideas seem to have become separated into two themes. Or rather three; but the third theme though it came from him was not of his will: this, the idea of overwhelming tragedy, will be mentioned later. Of the other two, the first was his extreme personal indulgence of incest, perhaps the most varied combination ever recorded for one man. His relationships seem to have reflected an extraordinarily close and loving family life, as shown in numerous paintings and reliefs of court events. The lovingness of this life in turn appears to integrate into a unique style and belief that was indeed due to the king himself.[51]

The result of the family side of his innovation was disaster. The inability of Akhenaton himself or his sons to produce through their *inbred* unions anything but daughters appeared to the people a confirmation of the withdrawal of the favor of all Egypt's neglected gods, just as the same seemed independently evidenced by the plagues and military setbacks that were affecting the empire from without, coinciding with the monarch's neglect. It was the incestuous, in-turned, family-in-Eden meme of Akhenaton that Moses separated and left behind. Under the laws he gave the Israelites it was still just possible for uncle to marry niece, but other equal and closer unions were prohibited: his tone on sexual behavior generally was stern. Other parts of Akhenaton's beliefs—his iconoclasm, monotheism, and his seemingly almost scientific appreciation that the universe might be interpreted in terms of a single abstract yet life-supporting cause, an attempt, it might be said, at a 'unified theory' of his time—Moses forcefully continued. For Akhenaton, the life-supporting Cause was the sun, and perhaps it was partly so also for Moses, a view which may be faintly transmitted to us in the image of the sun-burned bramble by Mount Horeb, withered, punished, and yet ever renewed, like his people, by the light of a Sun. For the Jews after Moses, the Cause was soon to become (perhaps after another priestly fusion soon after the Sinai period, as Freud suggested) the more human-oriented and moralistic creator Jehovah.

Akhenaton bred in succession with his cousin Nefertiti, at least two unrelated wives, his mother Tiye, a daughter by Nefertiti, and perhaps other daughters. Neither of the three close unions produced surviving offspring, and Nefertiti gave six daughters. The two unrelated wives gave at least each a son. These were healthy but died quite young after brief turns as Pharaoh. Akhenaton's legacy to Egypt was an empire diminished, disheartened, and headed for anarchy. His 15-year-old new capital soon lay deserted and defaced, its temples unbuilt to the ground in counterreformatory zeal that began even in the lifetimes of his sons. Apart from these disasters, however, we can hardly doubt that he left another legacy, an outcome that was almost equally guaranteed by the extent of his failure. The annihilation of his religion almost has to have created emigrants. Moses and Aaron, it has been suggested, were two; for hints of others I turn, following Velikovsky,[52] from the Book of Exodus to another myth, that of King Oedipus in Greece.

Velikovsky's first theory, like Freud's last, seems to me more likely to weather time than most other ideas for which these writers are well known. He claimed to give detailed evidence that the story of the heretic king of Egypt reappeared in the legends of Grecian Thebes. The name of Akhenaton's other seemingly more irreverent Paul who brought the story (he called his hero, for example, 'King Fat Leg,' after the real king's strange physique) has not survived. The story became garbled, barbarized, and mixed with other myths, but still carried along convincing detail from the events of Egypt that are historically known. That genes as well as memes were again implanted is suggested by the extent of the parallel, the very least element of which is the name the Greeks later gave to the Egyptian city, Thebes, linking it to their own. Egyptologists don't seem to like Velikovsky's idea, to judge by his complete noncitation in their books, and it is easy to see how his strong advocacy of a particular version could both cause offense and make him vulnerable. However I cannot see that even the most recent expert accounts of Akhenaton's reign render his general thesis improbable.[51,53]

I have to admit, however, to a bias in my reading of this reinterpretation based on my childhood experiences, as I had better first explain.

Firstly, Velikovsky's theme immediately made sense of something my mother had never properly explained. It was like the 'insects' that ate my buttons: impossible!—how could insects have lived on mother of pearl or plastic? I came to understand that by learning of *Phytelephas* and *Coccotrypes*: what Velikovsky unravelled was how the sphinx of the story that my mother read to me had been able to do things in Greece, even in a myth, when *the sphinx itself* had been there so huge and immovable, as she had also often told me, close to my birthplace in Egypt. That confusion soon combined with another, concerning a kind of 'sphinx' of Britain itself which likewise was based on an early memory. We once took a picnic to Beacon Hill in Berkshire, and ate it inside the ancient ramparts of the hilltop fort. There in the grass was an iron fence enclosing a mound that was called Tutankhamen's Tomb, and something about this fence and the mound and blue sky made my mother and my great-aunt speak in tones of awe, while inside the fence a sheep was stuck and tried to get out. It was in the time when Mr. Hitler did not let the ships go past, and we were evacuated, staying with my aunt. She kept 'chicken,' so we had hard-boiled eggs, and there were soft cheese segments wrapped in foil, a great treat. I rolled down a grassy rampart. Only a little later did Pharaoh Tutankhamen, son of Akhenaton, being buried in Berkshire begin to seem odd to me, but then it remained so for a long time. Of course in the end I realized it was not Tutankhamen but *Lord Carnarvon*, the discoverer of the great tomb, who was buried there. I knew by then that sphinxes were many and could fly and that there was no good reason why the Thebes of the north should not have its own, but even so, granted the mistakes had been childish, Velikovsky still has plenty to be marveled at in his book. Long before reading it, I had thought that although genes and memes don't have to travel together, the bet is that they do whenever transferred detail of culture is great. Moreover, settlement from the civilized to the barbarous is for the most part the more probable, judging by many parallels of recent times. In short, since reading Velikovsky's interpretation it has remained with me strongly that another person or group from Akhenaton's towns did come to Greece. There would be included a person or leader connected nearer to the events than Moses but perhaps more an outsider in other ways—probably not noble or priestly, nor with much interest in the philosophy, perhaps more like a sculptor or a mortician, but above all a storyteller obsessed by a real-life tale. For a parallel of modern times, I think of an ex-hippie of the '60s, perhaps an artist, going native out of the Peace Corps into a tribe of Amazonian Indians and there recounting the Charles Manson affair which he or she happened to have mixed with personally, investing the tale with an atmosphere of Woodstock and disillusion and tragedy. 'It could all have been so beautiful.' Manson was not leader of an empire nor highly original; how much stronger such an implanted legend might be if he had been!

It is somehow appropriate to the barbarity of the recipients that it was the tragedy of the king's life, not the moral messages and still less the philosophy, that became implanted in Greece; and it is also appropriate to the particular genius of Greece that later it would be the art of Akhenaton that sprung back into life there and flourished. For the other, eastward, migration of the ideas, it seems somehow also appropriate to the spirit of Judaism (at least, as it became molded by the event) that it was more the philosophy and the theology that were accepted.

Much later, we find the artistic naturalism and the philosophy of the strange loved-hated king becoming mixed again in the milieu of still more barbarous peoples and making the foundations of the Science we know. Following such chains, not only do I find 'Levis' and 'Cohens' of scientific literature now strongly reminding me of their remote and remarkable connection to Egypt, but, when I argue with owners of such names, it even seems to me that I encounter the slants that I am led to expect: they the priestly descendants expecting the universe to be consistent, organized, subject to kingly power, its logic inviolable; for me, the barbarian, no such deep faith existing, logic ready at all times to fail—and if mine, soon doing so. For me it seems the universe only needs to be beautiful, my 'science' no more consistent or less tragic than Antigone's story or her sculpted head.

Conclusion

Well, to come partly to earth from these stories and to summarize; whether deeply or not, we have to believe in consistency for the purpose of this book and here have a long way to go. From blebs on the side of a locust, and whispering out of a cavity like a caries in an old date stone, listening to the story of their king retold, I have seemed to hear a faint stridulant laughter: 'What a fuss!' Well then, for how long can *they* laugh, either? Where, for example, is the button biotype of *Coccotrypes* now that its host is extinct? Do *we* laugh—or did that biotype change back to the diet of date stones? Is inbreeding a long-term failure even for the likes of this beetle, even for those generic empires, each now more than five hundred species strong, of *Xyleborus* of the ambrosia and *Blastophaga* of the figs? Is it disaster for *our* symbionts—say for our lettuce, which is said to be 99% selfed and has been so, perhaps, ever since toddler princess Beketaton was pictured carrying its early ancestor while her father-brother and mother-grandma (as believed by Velikovsky; their identity as Akhenaton and his mother Tiye in the picture is undoubted) walked hand in hand in front? I would think not disaster for lettuce; but worse than for these, what about the three-thousand-year viviparous (and therefore clonal) Egyptian onion? What is the lifetime of such a cultivar—or of an apple? And what of those still smaller helpful enemies, the microbial symbionts within insects—do they escape, not all die, infect outsiders, if their host goes down? How much do they outbreed, if ever? These yet again apart, what other plagues infect button beetles or their symbionts—or Pharaoh's ant—and what plagues may such plagues themselves have? Finally for the Sphinx's second riddle. What was the plague of 'Thebes'? It cannot be by inbreeding that the Egyptian king was deformed—he was outbred. Then how came he deformed with traits, such as his strange head, that he passed on? Was he a mutant, was he a would-be starter in a new path of male human physique, as strange as that myrmicine ant that took first stride to be a male *Hypoponera*? Was it hereditary infection? How did his father Amenophis rear his boy, if at all? And how Queen Tiye? Tiye, an empire's correspondent to foreign kings, no cipher consort, strangely and extremely loved, what secret did she have? How does a woman make such a son as she had? Was Akhenaton in infancy, like Oedipus in the legend, like Moses, left out to die?—because he was deformed? It seems to me that notwithstanding all the facts and theories in this book (including mine), we hardly begin to know answers to any of these questions.

References

1. W. M. Shields, *Philopatry, inbreeding, and the Evolution of Sex.* State University of New York Press, Albany, (1982).
2. P. Buchner, Endosymbiosestudien an Ipiden. I. Die gattung *Coccotrypes. Z. Morph. Okol. Tiere* **50**, 1–80 (1961).
3. I. Rowely, E. M. Russell, and M. G. Brooker, in *The Natural History of Inbreeding and Outbreeding: Theoretical and Emperical Perspectives*, N. W. Thornhill (ed.) (The University of Chicago Press, Chicago, Chapter 13 (1993).
4. M. Volkonsky, *Podapolipus diander*, n. sp. acarien heterostygmate parasite de criquet migrateur (*Locusta migratoria* L.). *Archiv Inst. Pasteur Algérie* **18**(3), 321–34 (1940).
5. R. H. Crozier, Heterozygosity and sex determination in haplo-diploidy. *Am. Nat.* **105**, 399–412 (1971).
6. J. H. Werren, in *The Natural History of Inbreeding and Outbreeding: Theoretical and Empirical Perspectives*, N. W. Thornhill (ed.) (The University of Chicago Press, Chicago) Chapter 3 (1993).
7. W. D. Hamilton, Extraordinary sex ratios. *Science* **156**, 477–88 (1967).
8. W. D. Hamilton, Evolution and diversity under bark. *Roy. Entomol. Soc. Symp.* 9, 154–75 (1978).
9. W. D. Hamilton, Wingless and fighting males in fig wasps and other insects, in *Sexual Selection and Reproductive Competition in Insects*, M. S. Blum and N. A. Blum (ed.) 167–220. Academic Press, New York, (1979).
10. W. M. Wheeler, *The Social Insects: Their Origin and Evolution.* Kegan Paul, Trench, Trubner, London, (1928).
11. R. L. Trivers, *Social Evolution*, Benjamin/Cummings, Menlo Park, California (1985).
12. D. J. Howard, *The Natural History of Inbreeding and Outbreeding: Theoretical and Empirical Perspectives*, N. W. Thornhill, (ed.) (The University of Chicago Press, Chicago) Chapter 7 (1993).
13. T. H. Hubbell and R. M. Norton, The systematics of the cave-crickets of the North American tribe Hadenoecini. *Misc. Pub. Mus. Zool. Univ. Mich.* **156**, 1–124 (1978).
14. L. R. Kirkendall, The evolution of mating systems in bark and ambrosia beetles (Coleoptera: Scolytidae and Platypodidae). *Zool. J. Linn. Soc.* **77**, 293–352 (1983).
15. N. Knowlton and J. B. C. Jackson, in *The Natural History of Inbreeding and Outbreeding: Theoretical and Empirical Perspectives*, N. W. Thornhill, (ed.) (The University of Chicago Press, Chicago) Chapter 10 (1993).

16. R. Stouthamer, R. F. Luck, and W. D. Hamilton, Antibiotics cause parthenogenetic *Trichogramma* (Hymenoptera/Trichogrammatidae) to revert to sex. *Proc. Natl. Acad. Sci. USA* **87**, 2424–27 (1990).
17. M. J. D. White, *Animal Cytology and Evolution.* (Cambridge University Press, Cambridge, 1973).
18. P. Buchner, *Endosymbiosis of Animals with Plant-Like Micro-organisms.* New York: Wiley Interscience (1965).
19. R. R. Glesener and D. Tilman, Sexuality and the components of environmental uncertainty: Clues from geographical parthenogenesis in terrestrial animals. *Am. Nat.* **112**, 659–73 (1978).
20. R. Law and D. H. Lewis, Biotic environments and the maintenance of sex: Some evidence from mutualistic symbiosis. *Biol. J. Linn. Soc.* **20**, 249–76 (1983).
21. J. B. Mitton, in *The Natural History of Inbreeding and Outbreeding: Theoretical and Empirical Perspectives,* N. W. Thornhill, (ed.) (The University of Chicago Press, Chicago) Chapter 2 (1993).
22. M. K. Uyenoyama, in *The Natural History of Inbreeding and Outbreeding: Theoretical and Empirical Perspectives,* N. W. Thornhill, (ed.) (The University of Chicago Press, Chicago) Chapter 4 (1993).
23. N. M. Waser, in *The Natural History of Inbreeding and Outbreeding: Theoretical and Empirical Perspectives,* N W. Thornhill, (ed.) (The University of Chicago Press, Chicago) Chapter 9 (1993).
24. B. D. Mishler, Reproductive ecology in Bryophytes, in *Plant Reproductive Ecology,* J. Lovett Doust and L. Lovett Doust (ed.) 285–328. (Oxford University Press, Oxford, 1988).
25. R. E. de Wreede and T. Klinger, Reproductive strategies in Algae, in *Plant Reproductive Ecology: Patterns and Strategies,* J. Lovett Doust and L. Lovett Doust (ed.) 267–84. (Oxford University Press, New York, 1988).
26. B. C. Clarke, The evolution of genetic diversity. *Proc. Roy. Soc.* (Lond.), ser. B, **205**, 453–74 (1979).
27. E. R. Dempster, Maintenance of genetic heterogeneity. *Cold Spring Harb. Symp. Quant. Biol.* **20**, 25–32 (1955).
28. S. A. Frank and M. Slatkin, Evolution in a variable environment. *Am. Nat.* **136**, 244–60 (1990).
29. D. A. Levin, Pest pressure and recombination systems in plants. *Am. Nat.* **109**, 437–51 (1975).
30. W. D. Hamilton, Pathogens as causes of genetic diversity in their host populations, in *Population Biology of Infectious Diseases,* R. M. Anderson and R. M. May. (Springer, New York, 1982).
31. W. D. Hamilton, Memes of Haldane and Jayakar in a theory of sex. *J. Genetics* **69**, 17–32 (1990).
32. W. D. Hamilton, The moulding of senescence by natural selection. *J. Theor. Biol.* **12**, 12–45 (1966).
33. R. L. Blackman, Chromosomes and parthenogenesis in aphids. *Symp. Roy. Entomol. Soc. Lond.* **10**, 133–48 (1980).
34. B. K. Schnee, and D. M. Waller, Reproductive behavior of *Amphicarpaea bracteata* (Leguminosae), an amphicarpic annual. *Am. J. Bot.* **73**, 376–86 (1986).

35. T. G. Myles and W. L. Nutting, Termite eusocial evolution: A re-examination of Bartz's hypothesis and assumptions. *Q. Rev. Biol.* **63**, 1–23 (1988).
36. V. A. Taylor, The adaptive and evolutionary significance of wing polymorphism and parthenogenesis in *Ptinella* Motschulsky (Coleoptera: Ptiliidae). *Ecol. Entomol.* **6**, 89–98 (1981).
37. B. J. Crespi, Adaptation, compromise and constraint: The development, morphometrics, and behavioral basis of a fighter-flier polymorphism in male *Hoplothrips karnyii* (Insecta: Thysanoptera). *Behav. Ecol. Sociobiol.* **23**, 93–104 (1988).
38. J. J. Christian, Phenomenon associated with population density. *Proc. Natl. Acad. Sci. USA* **47**, 428–49 (1961).
39. R. J. Stewart, A. Francoeur, and R. Loiselle, Fighting males in the ant genus *Cardiocondyla*, in *Abstracts of the 10th International Congress of the Union for Study of Social Insects*, J. Eder and H. Rembold (ed.) 174 (Verlag, Munich, 1986).
40. K. Kinomura and K. Yamauchi, Fighting and mating behaviors of dimorphic males in the ant *Cardiocondyla wroughtoni. J. Ethol.* **5**, 75–81 (1987).
41. M. G. Murray, and R. Gerrard, Conflict in the neighbourhood: Models where close relatives are in direct competition. *J. Theor. Biol.* **111**, 237–46 (1984).
42. M. G. Murray. The closed environment of the fig receptacle and its influence on male conflict in the Old World fig wasp, *Philotrypesis pilosa. Anim. Behav.* **35**, 488–506 (1987).
43. N. Rothenbuhler, Behaviour genetics of nest cleaning in honeybees. I. *Anim. Behav.* **12**, 578–83 (1964).
44. P. W. Sherman, T. D. Seeley, and H. K. Reeve, Parasites, pathogens and polyandry in social Hymenoptera. *Am. Nat.* **131**, 602–10 (1988).
45. P. Nonacs, Ant reproductive strategies and sex allocation theory. *Q. Rev. Biol.* **61**, 1–21 (1986).
46. J. J. Boomsma, Empirical analysis of sex allocation in ants: From descriptive surveys to population genetics, in *Population Genetics and Evolution*, G. de Jong (ed.) 42–51. (Springer-Verlag, Berlin, 1988).
47. C. W. Rettenmeyer, The diversity of arthropods found with Neotropical army ants and observations on the behavior of representative species. *Proc. North Central Branch, Am. Assoc. Economic Entomol.* **17**, 14–15 (1962).
48. D. Rottenberg, *Finding Our Fathers: A Guidebook to Jewish Genealogy*. (Random House, New York, 1977).
49. M. I. Dimont, *Jews, God and History*. (Signet Books, New American Library, Bergenfield, NJ, 1962).
50. Freire-Maia and T. Elisbao, *Moses and Monotheism*. (Alfred A. Knopf, New York, 1939).
51. C. Aldred, *Akhenaten, King of Egypt*. (Thames and Hudson, London, 1988).
52. I. Velikovsky, *Oepidus and Akhenaten*. (Sedgwick, London, 1960).
53. D. B. Redford, *Akhenaten—The Heretic King*. (Princeton University Press, Princeton, NJ, 1984).

CHAPTER 9

ON FIRST LOOKING INTO A BRITISH TREASURE[†]

W. D. HAMILTON

When I was a boy, Britain was way ahead of the rest of the world in the publishing of books on natural history—indeed until quite recently colleagues from America, Eastern Europe and Japan would gasp at the wealth of our productions and wish they had something similar: the pages and pages of coloured plates of set butterflies and moths, of beetles, bugs, painted toadstools, and the colour-illustrated floras that soon covered not only Britain but also Northern Europe.

The challenge to our supremacy started in America with the Peterson's Guides. That was a shock—we had nothing like it: the compact species-descriptions, the many birds crowded onto each plate, making them easy to compare. By the time the fashion for using colour photographs arrived, it was all over. Nowadays in a book shop in Hay-on-Wye you will more easily find a colourful photographic guide, four inches square, to the fishes of the Chokweneejee National Forest Reserve, Montana (a lake-strewn area once visited by at least one tourist who bought the copy) than any guide to the fishes of the North Sea or any part of the British Isles. That photographic wave was a product of technological development—Japanese cameras and flash lights making it all so easy, no doubt (and British field biologists were too poor to buy the equipment). Now the Japanese are way ahead. Who in Britain, for example, keeps on his bookshelf a colour photographic guide to identify the *weeds* in his garden, like the one I bought in a very ordinary bookstore in Nagoya? But—and here I reach the point of all this—we in Britain actually invented this new publishing format, too. That is, we were the first to use numerous colour illustrations of living animals and plants in books and certainly we were the very first to integrate such photographs into a publishing venture destined to detail every aspect of the natural history of our islands. There is, in fact, a book of fine colour photographs of British weeds and with them a text that ought to teach all of us to pull such plants, when we have to, with more respect than we usually do.

[†] *Times Lit. Suppl.* August 12, 13–14 (1994).

It is written by Sir Edward Salisbury, a former director of the Royal Botanic Gardens, Kew. Who published it? Collins, of course, in the New Naturalist series.

These books, still continuing to appear and now numbering around eighty volumes, have never attempted to be photographic guides, but they were the first series in any country to use numerous colour plates based on photography to sell natural history to the general public. Numerous topics were evidently planned for the series right from the start, and the best naturalists in the land were persuaded to write them. I have always wondered how Collins knew where to look for their writers. Perhaps it is not so strange, given our native eccentricity, that the director of The Royal Botanic Gardens was approached for a book on, say, British botanical collections or British gardens, and said he would rather write one on weeds. But how did Collins find out that a high administrator in Imperial Chemical Industries was Britain's leading authority on spiders? (W. S. Bristowe, *The World of Spiders*, 1962). One imagines invitations cadged to meetings at 49 Queen's Gate, where greying shock-haired enthusiasts could be button-holed, exhibits in their hands, at tea-time in the Royal Entomological Society library; one thinks of publisher's clerks, ordered to find writers on flowers, infiltrating themselves into the Botanical Society of the British Isles. However it came about, beautiful, fascinating monographs began to line up on the bookshop shelves.

The very first volume of this series, E. B. Ford on *Butterflies* (1945)—of which group I was at the time an ardent collector—was given to me as a birthday present by my parents. For a long time I bought none of the series new because, while reviewers rhapsodized on their low prices, these were still beyond my childish resources. I think the first of the series that I bought new was J. E. Lousley's book on limestone plants (1950); I am sure I later bought E. B. Ford's second in the series on *Moths* (No 30, 1955). My own passion for butterfly and moth collecting had almost burned out by then but I was still interested. New Naturalist books are generally reticent about localities for rarities, but who could fail to note here, for example, Cothill marsh, where the rare *bimacula* variant of the Scarlet tiger flies? If one hadn't time to bicycle to see it, there at least was the photograph giving the hot, wet reedy feel.

How I envied the people I knew later who had been able to collect all of the series! Yet how different were those often proudly displayed collections from my own meagre line of books. Whenever I saw a perfect array of the bright and smudgy hieroglyphs of the dust-jackets I would curl my lip just a little: those books had clearly not all been read, or not like mine had. For me the best in the series meant the dirtiest: most of mine had long ago lost not only the dust-jackets but also the gold-impressed titles on the spine. Gradations of dullness and shine of the green cloth born of usage (plus the varying thickness of the different monographs) have for long had to be my main way to pick out the volume I am looking for: Harrison Matthews on mammals (1951/1982) is bulky; C. G. Butler, *The World of the Honeybee* (1945) and S. W. Wooldridge and F. Goldring on *The Weald* (1953) are thin; A. D. Imms on *Insect Natural History*, 1947 (the second one I read and second most important in my life) is unfortunately of rather average thickness—yet it is still detectably thinner than Ford. There are rumoured to be people who unwrap and lay to one side the dust-jacket when they read a book, especially if it is part of a collection, as the New Naturalist books easily become. So perhaps I should not be too caustic about my friends' neat rows—but of course I still am. A loving archivist to my own library in my own way, I pride myself on an independent convergence to the ways of Darwin who, as I learned with pleasure, scribbled notes in all his books, and even once threatened Lyell that if the expanded new edition of his *Principles of Geology* was not printed in two volumes, his first act on receiving it would be to cut it in two down the spine.

Of course I was sorry to see my dust-jackets go and I kept them as long as I could. As to their art, the work of Clifford and Rosemary Ellis, I was very ambivalent about it at first but became a convert as I came to recognize that, for an impression of all the diverse aspects of British nature which they came to include, the designs for them were very good. Now my problem is rather to accept the arrival in the recent series of Robert Gilmour as the dust-jacket artist. He does not use smudges. However, I think that as a boy I probably would have liked him better. But how did Clifford and Rosemary work? It is like finding an obvious Van Gogh painting signed: 'Vincent and Estelle' and thinking about who and how. Does Clifford sketch drawings and Rosemary daub paint or vice versa? Similar problems have occurred to me about the various photographic geniuses of the early series, principally John Markham, Robert Atkinson, Brian Perkins and Eric Hosking, all of whom have photographs in many volumes. It used to seem to me that for people clearly passionate about scenery and natural history they must have had one of the most wonderful jobs imaginable—provided, that is, you could stand cold British wind and rain. The photographic editor of the series no doubt had to economize on the travelling he was commissioning so I imagine him packing his team off to Wales in a small estate car (the kind of those days that had real wooden framing) and giving them a set of directives from the dozen or so busy authors about what photographs were needed. Arrived on a small road up a valley leading to the pass, they would park and get out to survey the terrain. In a few clumsy and laconic words these artists would divide up their tasks. Markham and Atkinson, the hill walkers, would go up opposite sides of the valley to the heights and hope that the drizzle would clear. Atkinson would then photograph the Welsh coal-mining settlement down the valley and at the same time, drizzle or no, go for close ups of *Sorbus cambroendemicus*, the asexual whitebeam growing only, in the whole world, on the screes of this valley. Markham is to do the podsol soil profiles at the top of the landslip on his side, to get the Ebbw black-rumped sheep if he can find them in their characteristic haunt under the bracken, and then go over the top and start for the summits of the Beacons where the Lower Old Red Sandstone shales are well exposed. Meanwhile Perkins will go to photograph the Turquoise Heleborine orchids supposed to be in flower in the wood above the church, its only and new-found station west of the Urals, for V. S. Summerhayes's *Wild Orchids of Britain*, 1951, and also record wild Welsh leeks being demolished by sawfly. Hosking is to set up his hide in the churchyard and do the Pied fly-catchers at the nest box which the vicar has reported. He needs time because he must get his shot with the Flycatcher lousefly crawling on a bird's back for Miriam Rothschild (*Fleas, Flukes and Cuckoos*, 1952), but that of course shouldn't prevent him doing the arrowslits of the Welsh church for H. J. Fleure and M. Davies (*Natural History of Man in Britain*, 1951) while he is waiting, nor the clustered caps of *Coprinus necrosanctus*, the toadstool that grows only on buried miners—an illustration for one of Dr Ramsbottom's more gothic mycological themes (*Mushrooms and Toadstools*, 1953). So, one likes to think, those wonderful wanderlust pictures all came rolling in.

There is almost no end to what was first introduced to me as an intending biologist through the New Naturalist books. They were and remain a treasure of British publishing. How did I first understand why our whole land browns as we travel towards the north and west? That was in W. H. Persall (*Mountains and Moorlands*, 1950); it is not just that there are more mountains but that the rainfall exceeds evaporation and all that follows—the leaching, the acidity, the bog and moorland plants. How did I first learn about the life-cycles of aphids and plantgall formers? From A. D. Imms (*Insect Natural History*, 1947), who taught me also about *Ips*, the harem-forming bark beetles. Sexlessness, inbreeding, dull flowers on high mountains: these are topics on which M. Walters and J. Raven commented in *Mountain Flowers*, 1956, and which, like the parthenogenesis of aphids and gall-wasps and the harems of bark-beetles have become important in my theoretical biological research—how but for the last book would I have known where to look to find certain brilliant beauties in my life, Britain's sea-level alpine flowers, *Primula scotica* and *Oxytropis halleri*, on broken rocky dunes of the north coast of Scotland? How but for Sir Alister Hardy and his *The Open Sea* (1956, 1959) would I know that lone, lost Eskimos from Greenland had kayaked, fishing as they came, to those same remote coasts where, while kindly treated, they died of the porridge diet or a common English cold? The titles in the series go on seemingly for ever; they currently run at about two new books a year. Max Walters as author or part-author now spans three volumes and a period of thirty-seven years. It is not now my pocket money or the supply of books to buy, but my time to read that is sadly fading away.

Yet I am still managing to collect tidbits. When a dog happily paddling back to shore with its master's stick suddenly disappears in the murky water of some south-east midland lake, I will be among the unsurprised *cognoscenti*. This is because I have read P. Maitland and R. N. Campbell (*Freshwater Fishes*, No 75) and will know that the cause is the Giant Danube catfish brought to these waters late in the last century. The dog will have been swallowed to celebrate the fish's

300th kilogram; soon this fish will reach its 300th centimetre, and then it will celebrate with its first fisherman: already in the Danube they 'are reported to have taken dogs—and even children'. Probably no one will believe me, but naturalists are used to being disbelieved. Nature is so fantastic that it is true that we could, if we wanted to, get away with almost anything, and most people suppose that we do. Meanwhile, as the New Naturalist series grows, Collins deserves to be congratulated on this irresistibly increasing brood and to mark this year, the half-century after the first volume was commissioned, with some equally mighty if less drastic celebration.

CHAPTER 10

HOW TO CATCH THE RED QUEEN?

DIETER EBERT

In 1980 Bill Hamilton published a paper with the attention-getting title 'Sex versus non-sex versus parasite',[1] in which he proposed that parasitic diseases are an important factor for the maintenance of sexual reproduction. This hypothesis came to be known as the Red Queen hypothesis because it is based on the idea that hosts and parasites are in a never-ending race, just as the Red Queen in Lewis Carroll's fairy tale cannot stop running, or else she would lose her position. Although the Red Queen hypothesis attracted a lot of attention, more than ten years after it was proposed very little empirical work had yet been conducted on it. Surprisingly, the widespread support that the hypothesis had earned by the early 1990s among many evolutionary ecologists was not based on hard data, but on its plausibility and the increasing recognition that parasites are indeed everywhere. What was actually known about the interactions among hosts and parasites boiled down to only a few studies. Curt Lively, working on snails, and Paul Schmid-Hempel, working on bumble bees, were among the pioneers in the field.[2] However, one of the main reasons for the paucity of data was that there were not many host–parasite systems suitable for efficient experimental work to be conducted. In particular, systems with eukaryote hosts and rapidly evolving parasites had not been used much, although these were the systems where it was suspected that the hypothesis could most easily be tested.

Inspired by the elegance of the ideas involved in the evolution of host–parasite interactions, I decided to switch from life-history studies to evolutionary parasitology for my post-doc. As Bill Hamilton was the key proponent of the Red Queen hypothesis, I visited him in Oxford in the

Summer of 1990 to enquire about a post-doc in his group. Initially, Bill was rather sceptical, as he was not sure whether my background in *Daphnia* life history studies was suitable for a project on the Red Queen hypothesis. However, he liked *Daphnia* as an experimental system, and in the end we agreed that I would try to develop an experimental host–parasite system with *Daphnia*. I had experience with *Daphnia*, so we expected that this part of the work would not pose a problem.

When I joined Bill in Oxford for my post-doc, my first aim was to develop the *Daphnia* system and to use it for empirical work on the Red Queen hypothesis. Around Oxford, there were many suitable places to find *Daphnia*, and I soon had many cultures running, although the laboratory conditions where far from optimal for this type of aquatic work. A bigger problem was the parasites. As I had no experience with parasites, it took me much longer than expected to find them. From time to time, Bill would have a look in the microscope to see what there was to find in *Daphnia*, but for several months we did not see anything resembling what we had hoped for. Finally, a visit to Jim Green's laboratory at St. Mary and Westfield College in London opened the door to the secret world of micro-organisms, showing that there was much to find. Nearly every individual *Daphnia*, in particular *D. magna*, was infected with one or another parasite, mainly bacteria and microsporidian parasites. Within a few weeks, I was able to culture two species of parasites (the bacterium *Pasteuria ramosa* and the microsporidium *Glugoides intestinalis*, formerly called *Pleistophora intestinalis*), and I could concentrate on the second aim of my post-doc, namely, to test the predictions of the Red Queen hypothesis experimentally.

During the course of my explorations in the world of parasites, another post-doc in Bill's group joined the project. Katrina Mangin shared much of my enthusiasm for the Red Queen hypothesis. On one occasion we discovered what we thought might be a new parasite species when we observed individual hosts carrying large amounts of small spore-like structures that we could not identify. After further investigations we found that only males seemed to be infected with this parasite, while females could transmit this infection vertically to her offspring. Apparently, the infection was virulent only in males, which made perfect sense to us, as a vertically transmitted parasite would curtail its own transmission if it harmed reproductive females. When we told Bill about this new parasite, he was eager to see it in the microscope. After he had a good look, he commented in a very polite way that he was not really sure whether this was a parasite, as he vaguely

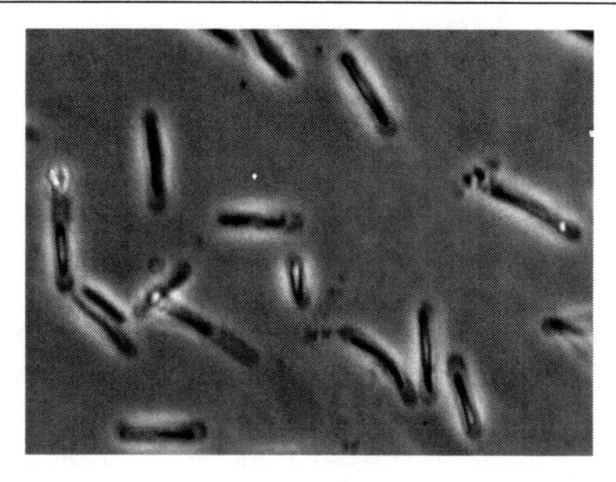

Figure 10.1. Sperm of *Daphnia magna*. (Photo and Copyright by Dieter Ebert).

remembered seeing sperm that looked like this (Fig. 10.1). We had not considered this possibility, although we also had never seen *Daphnia* sperm before: *Daphnia* research focuses largely on the female, due to its asexual mode of reproduction. After some further investigations we had to admit that Bill was indeed right. In our defence, we should add that *Daphnia* sperm looks very different from what sperm usually looks like. Only much later did I discover an old paper with drawings of *Daphnia* sperm that resembled our findings closely. Clearly Bill's wide experience with the biology and natural history of all kinds of unusual creatures had set him quickly on the right track.

After the system was up and running in the laboratory, Bill and I frequently discussed which predictions of the Red Queen hypothesis could be tested experimentally. This turned out to be more difficult than we both had anticipated. The published literature on this topic was of limited help. Mathematical models and computer simulations gave clear insights into certain aspects of host–parasite arms races, but it was often not clear to what extent the predictions of models could be used for real systems. For example, nobody knew the genetics that underlay host–parasite interactions. There were some data from plant–fungal system (the gene-for-gene model), but the proposed mechanism was not verified for animal systems. Most theoretical studies on the Red Queen hypothesis used the matching allele model, for which we have, until now, little empirical support. In contrast to these genetic systems, some verbal predictions were based on quantitative genetics models, i.e. offspring resemble their parents with regard to susceptibility and resistance to infectious diseases. This idea, translated into a coevolutionary scenario, led to the somewhat obscure prediction that heritabilities of resistance should be negative. Other predictions of the Red Queen hypothesis being discussed at this time seemed to depend on contrasting points of view. For example, one could argue that sexual host populations should have fewer parasites than asexual populations, because genetic recombination helps them to fight parasites. However, one could equally well argue that sexual populations are sexual because they have more parasites, and that the asexual populations can only be maintained because they have fewer or no parasites. It may be even more difficult: in dynamic interactions such as coevolving antagonists, the observed relationships between certain variables may change over time, and a clear picture may only become visible over a very long period of observation. The most often cited prediction of the Red Queen hypothesis is that gene frequencies should cycle over time. This prediction, however, is very difficult to test, as it requires that we know the genes under parasite selection and that these genes can be traced in populations. Currently we do not know, for any host–parasite system, enough of the underlying genetics to trace cycling genes, not even talking about cycles in linkage disequilibrium among genes. Furthermore, we need to know on what time scales should we look for cycles. If cycles are slow (e.g. decades) we would hardly be able to see them within a normal research program. Years later, Dybdahl and Lively[3] circumvented the first problem by using clonal snails, in which the entire genome, including the resistance genes, form a linkage group. Surprisingly for many people, but good news for the Red Queen hypothesis, they found that even on short time scales (few years) clones' frequencies can change drastically in response to parasites that attacked common clones proportionately more often than rare clones. Unfortunately, there are only a few systems that allow such studies, excluding the observation of cycles for the majority of systems.

There was clearly some conceptual work to be done before I could start with the experiments. The paper to which this small and certainly rather personal article is the foreword is the result of the discussions Bill Hamilton and I had over numerous teas drunk from styrofoam cups in the coffee area of the Oxford Department of Zoology. Bill's abstract sense for important biological relationships, and my sense for what is a testable prediction, did not always coincide. In numerous incidences, I dismissed Bill's suggestions as being untestable within the normal framework of our research possibilities.

Likewise, Bill dismissed many of my suggestions as already evident, and not in need of further investigation. While his suggestions inspired me to think bigger, my suggestions may have helped him understand that what seemed clear to him was not necessarily acceptable for an experimentalist who views data points as conservative estimates of reality.

It quickly became clear that a key issue was the question of whether hosts or parasites are ahead in the arms race. At this time, it was usually assumed that because parasites evolve quicker than their hosts, they are therefore ahead. If this were true, it would be rather simple to come up with clear-cut predictions regarding host–parasite arms races. However, nobody knew anything about the relative advantages of the two antagonists in the arms race. From bacteriophage experiments we had learned that, rather than searching for patterns created by past evolution, we could allow evolution to happen under controlled conditions and then test the outcome. Applying this form of experimental evolution to evolutionary hypothesis was not yet widespread when we started our discussions and was, until then, a domain of evolutionary microbiologists. However, as Red Queen host–parasite arms races were assumed to happen with considerable speed, they seemed suited for experimental evolution. Thus, we decided to avoid the question of who is ahead in the arms race by making predictions for evolutionary experiments in which one antagonist would evolve, while the population of the other antagonist would be kept genetically constant. In this way, it would be clear that the evolving antagonist would eventually be ahead of its opponent. The predictions made for such experiments form the nucleus for the paper following this article, which we published about 3 years later.

A literature review of studies that allowed parasites to evolve on well-defined host lines provided some support for the predictions derived from the Red Queen hypothesis.[4] In this review, I examined the huge amount of medical, veterinary and agricultural literature on serial passage experiments. Typically, in these experiments, a novel, but related, host is artificially infected and the infection is then transmitted from one host individual to the next (e.g. by syringe transfer of blood). The host strains are usually well defined and of low genetic diversity (inbred lines, full sib families, clonally propagated cell lines). Despite the huge variation in the purpose and methodology of serial passage experiments, several studies consistently reported that parasite virulence increased during serial passage experiments as a result of within-host competition, and that this increase in virulence depends on the host genotype. (Parasites passed through one host-type become 'attenuated', i.e. their pathogenic effects are reduced in hosts different from those in which they were passed. Attenuated parasites are useful vaccines, for they can elicit an immune response without causing harmful effects, e.g. Sabin's polio vaccine, smallpox, rubella, measles, mumps.) Given these results, the question arises, 'Why does virulence not increase under normal, "non-passage" conditions?' The Red Queen hypothesis offers a solution (but not the only solution[4]) to this problem. During serial passage experiments, parasites are exposed to a narrow range of host genotypes. Infection of a novel host, usually a different host species or a different cell line, results in parasite attenuation, indicating that growth and virulence are adaptations to the host-genotype in which it evolved. Ebert and Hamilton[5] proposed that virulence does not usually escalate in natural populations because genetic diversity among hosts prevents the parasite from evolving host genotype-specific virulence. In other words, because the parasite suffers from attenuation whenever host-to-host transmission occurs, it rarely has sufficient time to evolve high virulence on a single genotype. Genetic diversity among hosts hinders the escalation of virulence. Under such conditions, rare host genotypes have a selective advantage.

Serial passage experiments have not been designed to test the Red Queen hypothesis and therefore do not convince in all aspects. However, an increasing number of experimental studies have been published indicating that arms races, such as those envisioned by Bill Hamilton when he proposed the Red Queen hypothesis, are a natural part of host–parasite interactions. These arms races are not only intriguing by their complexity, but also by the strength and speed of their dynamics.

References

1. W. D. Hamilton, *Oikos* **35**, 282–290 (1980).
2. See Chapters 5 and 18 in W. D. Hamilton, *Narrow Roads of Gene Land*, Vol. **2**. *Evolution of Sex* (Oxford University Press, Oxford, 2001).
3. M. F. Dybdahl and C. M. Lively, *Evolution* **52**, 1057–1066 (1998).
4. D. Ebert, *Science* **282**, 1432–1435 (1998).
5. D. Ebert and W. D. Hamilton, *Trends in Ecology and Evolution* **11**, 79–81 (1996).

SEX AGAINST VIRULENCE: THE COEVOLUTION OF PARASITIC DISEASES†

DIETER EBERT AND WILLIAM D. HAMILTON

Abstract

Reciprocal selection is the underlying mechanism for host–parasite coevolutionary arms races. Its driving force is the reduction of host lifespan or fecundity that is caused by a parasite. Parasites evolve to optimize host exploitation, while hosts evolve to minimize the 'parasite-induced' loss of fitness (virulence). Research on the evolution of virulence has mostly emphasized the role of parasite evolution in determining virulence. However, host evolution, accelerated by sexual recombination, contributes to the evolution and expression of virulence as well. The Red Queen hypothesis predicts that genetic variation among host offspring facilitates selection for reduced virulence. Here, we outline a synthesis between current thinking about the evolution of virulence and the evolution of sex.

Parasites—here broadly defined as damage-producing organisms, including microbial pathogens, traditional parasites and small herbivores—are ubiquitous and influence either directly or indirectly almost every conceivable level of biological organization. The impact parasites have on the evolution and ecology of their hosts depends on their virulence, the driving force in host–parasite coevolution. Virulence, *per se* beneficial for neither parasite nor host, cannot be a property of a parasite alone; rather, it is a product of the host–parasite interaction. Different host genotypes from the same population do not suffer equally when infected with the same parasite strain, and different parasite strains cause variable levels of virulence in the same host genotype.[1–3]

Most studies on the evolution of virulence have concentrated on parasite evolution, assuming that virulence is maintained by genetic trade-offs between

† *Trends in Ecology and Evolution* **11**(2), 79–82.

virulence and other fitness components of the parasite. For example, parasite-induced host mortality was shown to be negatively correlated with host recovery rate (which contributes to parasite mortality) in Australian rabbits infected with the myxoma virus[4,5] and positively correlated with the multiplication rate of a microsporidian parasite in *Daphnia* hosts.[3] Therefore, it has been suggested that to maximize fitness a parasite should optimize the trade-off between virulence and other fitness components.[5] This optimality concept for the evolution of virulence, however, largely neglects genetic variation among hosts in their interaction with parasites. Such variation results in differential reproductive success among hosts and would, in the absence of parasite evolution, lead to reduced virulence. Given the high evolutionary rate of parasites,[4,6] host evolution can often be ignored in a first approximation, but for a better understanding of the evolution of virulence it is essential to understand the host's evolutionary response and in particular the role of genetic recombination in host evolution.

It has been suggested that sexual reproduction of hosts is a means to overcome the disadvantage of the low evolutionary rate that an asexual host would have in comparison with its rapidly evolving parasites.[7–9] Combining current theory of the advantage of genetic recombination and outbreeding with the theory on the evolution of virulence, one would predict that hosts continuously evolve to reduce virulence, while their parasites evolve to keep virulence as close as possible to an optimal level for their own life histories. In this arms race, a high evolutionary rate would benefit both opponents. Since parasites already have a very high evolutionary rate intrinsic to their short life cycle, hosts would be selected for increased evolutionary rates, even if this has costs. Sexual recombination could provide such an increase.[7,8,10,11] The principal underlying assumption for this hypothesis is that genetic variation for host–parasite interaction exists within populations and gives differential fitness to both the host and the parasite. Such genetic variation has been shown for different host–parasite systems [e.g. refs 2, 12, 13, including variation at the major histocompatibility complex (MHC) in relation to infectious diseases in humans (severe cerebral malaria[1] and chronic lyme arthritis[14])]. Given continuance of such variation, genetic recombination creates novel gene combinations. In a sexual population, every host constitutes a genetically unique environment for the parasite. Therefore, parasite adaptation to one host genotype is only of temporary benefit. Host diversity hinders evolution towards an optimal level of virulence, and we expect a level reflecting not only the evolution of the parasite to optimize host damage, but also the evolution of the host to minimize damage.[15,16]

Observing the Evolution of Virulence

The hypothesis that virulence in naturally coevolving populations is on average sub-optimal for the parasite allows us to make testable predictions.

If host evolution is experimentally restricted by reducing host genetic variability, parasites would be expected to adapt to the predominant host genotype by shifting virulence upwards, towards their optimum. There is some empirical evidence that virulence increases during adaptation to a new host genotype. Influenza virus increases in virulence and multiplication rate when propagated experimentally within chickens with the same genetic background.[17,18] Virulence of the measles virus increases when transmission occurs between siblings, compared with transmission between non-related members of the same host population.[19,20] The primate malaria agent *Plasmodium knowlesi* initially causes mild infections in humans; however, after 170 artificial human to human transfers, its virulence had risen sharply, indicating its adaptation to the new host environment (discussed in Ref. 21). The well-known case of initial decrease of virulence of the myxoma virus in Australian rabbits does not contradict this. The above prediction assumes that host and parasites are in a dynamic balance and it is when this holds that slower evolutionary change of the host will allow the parasite to increase virulence. The myxoma virus, however, was released to control rabbits and was therefore chosen to be as deadly as possible.[4] Since it was in a non-natural host the optimal level of parasite virulence was far off the optimum—in this case, far above the optimum. The strong decline of myxoma virus virulence over the first few years after its release indicates the strong potential of parasites to respond to changes in their genetic environment. Although ecological factors might have contributed to all these described changes in virulence, adaptation to the novel host genotype appears to us to be the more likely explanation.

The adaptation of a parasite to a new host genotype that was formerly rare or absent often results in the loss or a decline of virulence in the host of origin.[17] This finding was used in the development of vaccines, by employing strains in immunization programs that became non-virulent for their normal host after being kept for some time in foreign hosts (e.g. vaccines against yellow fever virus and poliovirus.)[22,23] This process emphasizes the dynamic, 'Red Queen' nature of host–parasite interactions.[7,10]

Local Adaptation

Support for the coevolutionary hypothesis of virulence also comes from studies using the reverse of the above argument. Namely, a parasite that infects a novel host (a host of a different genotype or population from that to which the parasite is adapted) should initially have, on average, a lower virulence in this new host, compared with that in its original host. In nearly all experiments where parasites were brought into contact with novel hosts, virulence and transmissibility decreased; this has been shown for viruses,[24] fungi,[25] helminths,[26,27] protozoans[3] and herbivores.[28–30] The reduction in virulence and transmissibility was stronger the more the novel hosts differed genetically from the host with which the parasite was associated before the experiment.[3] Experiments that did not find significant advantages for the parasite in its original host showed very high levels of genetic interactions between hosts and parasites[31–33] and this could contribute to the masking of local adaptation. Failure to detect local adaptation in cases where it is present can have various causes. Statistical power is weak when the within 'genetic unit' (e.g. host population, geographic area) variation is much larger than the variation explained by genetic isolation. Misjudgment of the scale of local adaptation might also lead to problems; for example, if parasites adapt to individual hosts rather than to host populations, detection of local adaptation across host populations might be difficult. Detection of local adaptation may also be hindered by acquired immunity of hosts, maternal effects on resistance, asymmetric gene flow between populations (source and sink populations) and insufficient time for adaptation. To our knowledge, no evidence has been presented against local adaptation.

Attention-attracting as they may be, cases where parasites showed devastating effects after accidental introduction into new host populations (e.g. rinderpest in Africa, Dutch elm disease, chestnut blight, HIV) appear to be exceptions.[3,13,16] There are likely to have been numerous failed introductions that have passed unnoticed. Most studies conducted under controlled conditions (see above) clearly show that parasites cause, on average, most harm in the host populations to which they are adapted.

Testing the Red Queen Hypothesis

In summary, genetic diversity of host populations appears to be crucial in hindering the parasite to evolve an optimal level of virulence. Sexual recombination benefits an outcrossing host through the production of variable offspring. In the context of host–parasite relationships, novel and rare genotypes have intrinsic advantages and may be selected. Asexual offspring, in contrast, cannot escape the antagonistic advances made during the previous generations by their parent's parasites. Genetic variability among hosts forces a parasite to adapt anew whenever it encounters a new host genotype.[10,34] The more different genotypes a host population consists of, the lower the frequency of each and the smaller the chance that a parasite will encounter the same genotype in successive hosts.

So far, direct evidence for the benefits of genetic diversity for host populations is weak, although numerous studies suggest such benefits. The vulnerability of monocultures to pathogen attack is notorious. Cereal monocultures have been shown to be more prone to attack by rapidly evolving, clone-specific, fungal diseases than genetically mixed cultures are.[35] Also, virulence of viruses was suggested to be higher in human populations with low genetic diversity at their MHC than in MHC-diverse populations.[20] Studies in natural populations are difficult, because frequency dependence of the host–parasite arms race is time-lagged, and therefore one cannot expect a positive correlation between the parasite prevalence and frequency of host clones.[36]

The theory that sex is advantageous in the presence of rapidly evolving parasites[7,8] became known as the Red Queen hypothesis. Support for the 'sex against parasites' hypothesis has come mainly from comparative studies (e.g. Refs 9, 37–39). Experimental studies are so far limited, with some support coming from the effects of herbivores on long-lived plants.[40,41] There is a strong need for experimental tests to clarify our understanding of arms races and the advantage of outbreeding. Boxes 1–4 summarize testable predictions derived from this review. Our predictions are stated for clonal host populations, but they are valid for all host organisms that can be brought to a lower level of genetic diversity than that found in normal wild populations (for example, by cloning, by using offspring from crossed inbred lines, or by using full-sib families). It is also helpful if strains of parasites and/or host can be 'genetically frozen' over time, to be used as a control in later experiments.[42] We hope that these predictions will stimulate research on the coevolution of sex and virulence.

Box 1. Testing the Red Queen: do parasites express higher virulence in host genotypes to which they are adapted?

(1) Variation in virulence reactions is expected in both hosts and parasites. 'Wild-sampled' hosts will vary in fitness when exposed to a single parasite strain and 'wild-sampled' parasite strains will vary in virulence and reproductive success when applied to a single host line.

(2) Wild-sampled parasite strains will initially evolve an increase in virulence and reproductive success when kept in monoclonal host populations.

(3) Parasites taken from one host population should be, on average, less virulent in hosts from other populations. Virulence in these novel hosts should, on average, decrease with decreasing genetic similarity between the host of origin and the novel host. 'Average' should be stressed here, since occasional highly virulent poorly reproducing parasites can be expected.

Box 2. Testing the Red Queen: does outbreeding and host genetic diversity hinder parasite adaptation?

(1) Parasites kept in monoclonal host cultures should evolve higher levels of virulence than parasites in multiclonal host populations. To avoid inevitable selection on host genotypes, the genetic composition and gene frequencies of the mixed host populations must be kept constant throughout the selection period. This can be done by continually reconstituting the host population from the same mixture of the same stocks. Without such replacement, selection for the least susceptible host genotype will occur and confound the result. The same replacement procedure must be done in the monoclonal host populations, to avoid selection on mutants and provide comparable handling and interference.
(2) Parasites kept in monoclonal host cultures should lose (reduce) their virulence when combined with other host clones, even if they were previously adapted to these clones.
(3) Parasites adapted to a monoclonal host population should express, on average, lower virulence in sexually outbred descendants of their hosts compared with inbred or asexually produced offspring. Variance of virulence and parasite success, however, should be high in the outbred offspring generation, and particularly high variance due to new homozygous combinations may become apparent in grand offspring and in later descendants from the outcrosses.

Box 3. Testing the Red Queen: do parasites mediate selection and polymorphism in natural host populations?

(1) Natural host populations that suffer intense parasite pressure should maintain higher levels of genetic variability than host populations without parasites. In clonal or cyclic parthenogenetic host populations, in which the whole host genome represents one linkage group, this effect can be investigated by correlating diversity of genetic markers with some measure of parasite selection pressure (e.g. mean parasite richness or abundance) across host populations. For sexually reproducing hosts, the allelic diversity of the defence loci can be studied (e.g. MHC haplotypes in vertebrates).
(2) In clonal and cyclic parthenogenetic host populations, linkage and selection by parasites are likely to produce genotype frequencies that differ between parasitized and uninfected hosts. In host–parasite systems where genes involved in defence are known (e.g. the MHC in vertebrates), the frequency of these genes can be studied directly.

Box 4. Testing the Red Queen: do parasites induce temporal changes in their host populations?

(1) Following prediction (2) in Box 3, if associations between parasites and host genotypes are detected and host generation time is short, parasite-mediated selection may be tested by re-sampling the same host population and monitoring changes in the frequency of the particular genotype (clonal markers or MHC genes, respectively).
(2) When parasite strains can be stored unchanged (e.g. by freezing of microbial pathogens), their re-introduction into the same host population many host generations later should reveal a change in their average virulence—most often a reduction in virulence. Due to ongoing host evolution, some originally highly virulent strains are likely to have become much less virulent.
(3) A similarity should be found between patterns of virulence in hosts over space and genetic distance [hosts vary spatially: see prediction (3) in Box 1] and changes in virulence when a genetically 'frozen' parasite is applied to the same local population over time [hosts vary temporally: see prediction (2) in Box 4].

Acknowledgements

We thank E. A. Herre, J. Lawton, C. Lively, K. Mangin, S. C. Stearns and J. Wearing-Wilde for helpful comments on earlier versions of the manuscript.

References

1. A. V. S. Hill, *et al.*, Common West African HLA antigens are associated with protection from severe malaria, *Nature* **352**, 595–600 (1991).
2. J. J. Burdon, Variation in disease-resistance within a population of *Trifolium repens*, *J. Ecol.* **68**, 737–744 (1980).
3. D. Ebert, Virulence and local adaptation of a horizontally transmitted parasite, *Science* **265**, 1084–1086 (1994).
4. F. Fenner and K. Myers, Myxoma virus and myxomatosis in retrospect: the first quarter century of a new disease, in *Viruses and Environment*, E. Kurstak and K. Maramorosch (ed.), pp. 539–570 (Academic Press, 1978).
5. R. M. Anderson and R. M. May, Coevolution of hosts and parasites, *Parasitology* **85**, 411–426 (1982).
6. M. S. Hafner *et al.*, Disparate rates of molecular evolution in cospeciating hosts and parasites, *Science* **265**, 1087–1090 (1994).
7. J. Jaenike, A hypothesis to account for the maintenance of sex within populations, *Evol. Theory* **3**, 191–194 (1978).
8. W. D. Hamilton, Sex versus non-sex versus parasite, *Oikos* **35**, 282–290 (1980).
9. C. M. Lively, Evidence from a New Zealand snail for the maintenance of sex by parasitism, *Nature* **328**, 519–521 (1987).

10. W. D. Hamilton, Haploid dynamic polymorphism in a host with matching parasites: effects of mutations/subdivision, linkage, and patterns of selection, *J. Hered.* **84**, 328–338 (1993).
11. R. Ladle, Parasites and sex: catching the Red Queen, *Trends Ecol. Evol.* **7**, 405–408 (1992).
12. M. P. de Nooij and J. M. M. van Damme, Variation in host susceptibility among and within populations of *Plantago lanceolata* L. infected by the fungus *Phomopsis subordinaria* (Desm.) Trav., *Oecologia* **75**, 535–538 (1988).
13. J. J. Burdon, Fungal pathogens as selective forces in plant populations and communities, *Austr. J. Ecol.* **16**, 423–432 (1991).
14. A. C. Steere, E. Dwyer, and R. Winchester, Associations of chronic lyme arthritis with HLA-DR4 and HLA-DR2 alleles, *New Engl. J. Med.* **323**, 219–223 (1990).
15. S. A. Frank, Evolution of host–parasite diversity, *Evolution* **47**, 1721–1732 (1993).
16. A. F. Read, The evolution of virulence, *Trends Microbiol.* **2**, 73–76 (1994).
17. E. D. Kilbourne, Host determination of viral evolution, in *The Evolutionary Biology of Viruses*, S. S. Morse (ed.), pp. 253–271 (Raven Press, 1994).
18. J. Schulmann, Effects of immunity on transmission of influenza: experimental studies, *Prog. Med. Virol.* **12**, 128–160 (1970).
19. M. Garenne and P. Aaby, Pattern of exposure and measles mortality in Senegal, *J. Infect. Dis.* **161**, 1088–1094 (1990).
20. F. L. Black, Why did they die? *Science* **258**, 1739–1741 (1992).
21. P. W. Ewald, *The Evolution of Infectious Disease* (Oxford University Press, 1994).
22. M. Theiler and H. H. Smith, The use of yellow fever virus modified by *in vitro* cultivation for human immunization. *J. Exp. Med.* **65**, 782–800 (1937).
23. A. Sabin, W. A. Hennessen, and J. Winser, Studies of variants of poliomyelitis virus. I. Experimental segregation and properties of avirulent variants of three immunological types, *J. Exp. Med.* **99**, 551–576 (1954).
24. J. R. O. Dawson, The adaptation of tomato mosaic virus to resistant tomato plants, *Ann. Appl. Biol.* **60**, 209–214 (1967).
25. M. A. Parker, Local population differentiation for compatibility in an annual legume and its host–specific fungal pathogen, *Evolution* **39**, 713–723 (1985).
26. V. S. Files and E. B. Cram, A study on the comparative susceptibility of snail vectors to strains of *Schistosoma mansoni, J. Parasitol.* **35**, 555–560 (1949).
27. C. M. Lively, Adaptation by a parasitic trematode to local populations of its snail host, *Evolution* **43**, 1663–1671 (1989).
28. S. Mopper *et al.*, Local adaptation and agents of mortality in a mobile insect, *Evolution* **49**, 810–815 (1995).
29. G. F. Edmunds and D. N. Alstad, Coevolution in insect herbivores and conifers, *Science* **199**, 941–945 (1978).
30. R. Karban, Fine-scale adaptation of herbivorous thrips to individual host plants, *Nature* **340**, 60–61 (1989).
31. M. A. Parker, Disease impact and local genetic diversity in the clonal plant *Podophyllum peltatum, Evolution* **43**, 540–547 (1989).
32. J. J. Burdon, A. H. D. Brown, and A. M. Jarosz, The spatial scale of genetic interactions in host–pathogen coevolved systems, in *Pests. Pathogens and Plant Communities*, J. J. Burdon and S. R. Leather. (ed.), pp. 233–247 (Blackwell, 1990).

33. A. M. Jarosz and J. J. Burdon, Host–pathogen interaction in natural populations of *Linum marginale* and *Melampsora lini*: II. Local and regional variation in patterns of resistance and racial structure, *Evolution* **45**, 1618–1627 (1991).
34. O. P. Judson, Preserving genes: a model of the maintenance of genetic variation in a metapopulation under frequency-dependent selection, *Genet. Res. Camb.* **65**, 175–192 (1995).
35. J. K. M. Brown, Chance and selection in the evolution of barley mildew, *Trends Microbiol.* **2**, 470–475 (1994).
36. M. F. Dybdahl and C. M. Lively, Host–parasite interactions: infection of common clones in natural populations of a freshwater snail (*Potamopyrgus antipodarum*), *Proc. R. Soc. London Ser. B* **260**, 99–103 (1995).
37. J. S. Schrag *et al.*, Ecological correlates of male outcrossing ability in a simultaneous hermaphrodite snail, *Am. Nat.* **143**, 636–655 (1994).
38. C. Moritz *et al.*, Parasite loads in parthenogenetic and sexual lizards (*Heteronotia binoei*): support for the Red Queen hypothesis, *Proc. R. Soc. London Ser. B* **244**, 145–149 (1991).
39. A. Burt and G. Bell, Mammalian chiasma frequencies as a test of two theories of recombination, *Nature* **326**, 803–805 (1987).
40. W. R. Rice, Sexual reproduction: an adaptation reducing parent-offspring contagion, *Evolution* **37**, 1317–1320 (1982).
41. S. Y. Strauss and R. Karban, The significance of outcrossing in an intimate plant-herbivore relationship. I. Does outcrossing provide an escape from herbivores adapted to the parent plant? *Evolution* **48**, 454–464 (1994).
42. M. Travisano *et al.*, Experimental tests of the roles of adaptation, chance and history in evolution, *Science* **267**, 87–90 (1995).

CHAPTER 11

BETWEEN SHOREHAM AND DOWNE: SEEKING THE KEY TO NATURAL BEAUTY†

WILLIAM DONALD HAMILTON

People divide roughly, it seems to me, into two kinds, or rather a continuum is stretched between two extremes. There are people people and things people. I undoubtedly fall on the side of the things. From the earliest childhood I can remember I was content to be making objects, looking at things, playing in silence with no one near. Yet I don't think I was an asocial child. I certainly enjoyed games with others, slap stick, practical jokes, and the making up and enacting of stories which, of course, always involved people or else the objects or animals that were people surrogates. But in the games with others somehow it turned out that I was usually the one out of step and the slowest to pick up the rules, I the most apt to miss my move because day dreaming of something else. Likewise in the stories I made up I laid great emphasis on what I considered an exciting landscape setting, less on the human relations and complexities. My story line was often just a long march of exploration in which my characters—often represented physically by inch-long twigs—would scale the cliff-like mossy roots of an overthrown tree or penetrate the bamboo jungles of the wild grass of a corner of the garden, heroically surviving dangers which were sometimes of a human nature but more often animal of inanimate. Miniaturized landscapes that I could imagine myself to be in and admire from a 'tiny' point of view were a passion. (I remember excitement discovering an article on Japanese *bonsai* landscapes in my father's encyclopaedia: at once, of course, I tried *bonsai* cultivation myself but, with neither skill nor patience, all my treelings soon died.)

Corresponding to all this I suspect that fewer of my earliest memories than with most children are of people—of social incidents, friends, adults kindly or hateful, and the like—and more are of dramatic physical and especially visual experiences. As one example, I remember as a startling gift of nature my first

† Inamori Foundation Kyoto Prize, Commemorative Lecture (1996).

sight of how oil spreads on a pool and makes it become alive with colours; how easy it was to cause this magnificent display again with a drop from the household can! I remember as a favourite environment the bare ploughland at the foot of our garden where amid screaming lapwings I would wander hunting for colourful flint stones and fossils, and how in that same field in summer I learned, for example, that potatoes came from plants with beautiful flowers and later bearing fruits mysteriously resembling small green tomatoes—yet unlike tomatoes poisonous, my mother said. In the same field again after rain I remember how at the lower end, water coursing in the furrows laid out flat deltaic fans of pale silt beautifully marked. Firm underfoot to a first step, if trampled a little, these fans, losing their braided patterns and turning to mud, would suck my bare feet down as if with a living appetite. The shape or colour of a new flower, even a known one seen again, could make me weak with joy. I would stare, long drinking the colours, longing to visit it as a bee, to be somehow joined with it. I realized that such intense feelings were incommunicable to others and generally unshared, even by my siblings. Hence most of the time I kept them to myself, ignoring or making light of them when with company. Of course I didn't think of them then as love, not knowing the adult emotion, but what I am describing are in fact almost exactly the feeling of romantic love and I think it is probably true in us things people that there occurs some aderration of a natural sequence that has been evolved for a purpose of bonding person to person. In us this sequence has grown awry somehow and gained untypical intensity directed towards inhuman objects. Yet the same misdirection, which is so often disastrous socially, leanding to whole fields of skills unlearned and therefore later badly applied, can be very helpful in the making of scientist, an engineer, or the like. Thus it is probably not wholly maladaptive. I believe it is in essence an aderration of this kind that makes me a successful scientist.

Love leads to endeavour (often unsuccessful) to understand, predict, and control a loved object. Successful prediction of course is what a scientist always aims for, but unlike with the typical lover he has another objective too which is to generalize and not to be too concerned about any single instance. The people people for their part learn others, both as individuals and in the mass, with intensity equal to the way we learn things. They accomplish their general understanding via an averaging that is not unlike that of the scientist but differs from it in stopping short of principled understanding. People people tend to be satisfied by rules-of-thumb about their fellows, and have little interest to find, indeed distrust as rather inhuman, any system of clear and simple, interlocking principles that might predict human action. Perhaps when the object of interest is humanity, to hope for a simple and principled understanding will always be illusory. If anyone claims to predict me, I at once try to devise behaviour that will confound him: I dislike the idea that I am predictable.

Nevertheless, that some generalization about 'normal' human behaviour is possible is probably admitted even by the most fanatical anti-hereditarian and, oddly enough, it may be things people who are more able to find the laws.

Standing back we may be better placed to notice and understand patterns that others just accept. In this way, at least indirectly, my work may have contributed to generalization even about humans. I am told that politicians say that they have little interest in the new discipline of human sociobiology; whether it is right or wrong they say they don't need it since they learned all that it can tell about people long ago—in the market place, in the lawyer's office, or wherever. This is probably true. Through constant interaction they know already how to appeal to people. Probably they have learned much better the manners, the words, and even the lies that sway the mood of a crowd—or indeed sway the warmth of any particular person to judge by the scandals that seem so often to surround politicians. But at the same time they can probably give no explanation for their skill or not any deep one. Politicians are probably the least likely among all the professions to be things people, expecially today.

In contrast to this, leaving aside for a moment the scientist, consider necessary attributes for an engineer—a profession which, encouraged by my father who was one, I was also once inclined to. If an engineer is to be creative he must as far as possible concentrate an inward vision on what he means to make and must anticipate in his mind all the difficulties that his design is likely to encounter. In short he must test his construction mentally so far as possible before he even starts to put it together. It is hard to imagine that the intense introspection needed for all this can go on in a busy place with people continually interrupting him. Spatial thought is probably relatively very demanding of neuronal space and activity in the brain just as our 3-D graphics programmes are demanding of memory and CPU time in a microcomputer. For scientific modellers the case is doubtless much the same and I am certainly one of these. For such a person it is hard to imagine that even a passive flow of music or poetry into thinker's ears can fail to disturb his concentration. On the other hand, as with the naturalist in presence of the insect, plant or stone, the concentration the engineer brings to bear is also proportional to his involvement in the thing he envisions, his love of it. In some minds this may be so intense that outside stimuli can be shut away. The adaptiveness in the evolutionary scheme of the engineer/inventor's concentration is obvious enough—such people can build houses, make clothes and live in places that would otherwise be uninhabitable to an erstwhile tropical hairless ape—but even the concentration of the naturalist if moderate in degree probably also becomes understandable when we look back to the importance which hunting, gathering and cultivation have had in our human past. It is probably the things person who eventually, for all the time he seems to waste on 'useless' sides of the observation of nature, will read the tapestry of ecology correctly, know how to find in its complexity what matters to human life—what plants grow where and when or can be made to, where the wild quarry of the hunt is likely to move, and so on.

On the basis of these thoughts it seems clear to me that things people constitute, when not so extremely oriented as to be pathological (as with autists), an integral and adaptively maintained part of the human pattern of variation.

Human groups which lack things people may be uncreative, and, even on the social side, may be too Macchiavellian for their own good. On the other hand human groups that have things people to the exclusion of all socialites may simply fail to hang together, fail to direct ideas and inventions into cooperative enterprises and useful channels. The weakness of both extremes implies a pressure towards some degree of a mixture—a polymorphism.

So altogether I would like to think. I am saying all this, I guess, to give some justification for a particular lover of things, one lazy dreamer, stone-turner, bird's-nester, flower-picker, butterfly hunter who wandered about my home county, Kent, as a boy.

On the hilltop of the North Downs where I grew up, 200 metres above the level of the not-distant Thames Estuary and English Channel, I experienced, I am sure, a wonderful world for a boy so inclined. My mother and father had bought a house with a small-holding of about five hectares in a part of the first real countryside outside the suburbs to the southeast of London. My father commuted into the capital but did so not quite every day. Following his own bent as an engineer and inventor he had converted a large shed that came with the house, designed for battery chickens, into a workshop. There he built models, tried his engineering experiments, designed tools, and mended absolutely everything that ever broke down in our set of increasingly antique possessions. Hoping that his boys would follow his footsteps as an engineer, he strongly encouraged us to use all the tools and materials of the workshop for our own purposes. As I will explain in a minute, as regards a career my mother's inclination to natural history eventually won over my father's engineering; nevertheless I much enjoyed making things and I became quite proficient with his tools. Besides cages for rearing butterflies for which I was becoming rapidly impassioned, I built many elaborate and battery-powered mechanical models—cranes, cable cars and the like, setting these to perform in the miniaturised outdoor landscape settings I have already mentioned. But the outdoors itself was the trouble. Just outside the shed where I worked landscapes of all kinds called to me whether I had models to put in them or not. There lay my mother's marvelous garden with its edges trailing into the unfenced fields and woods. How easy to throw down one's tools! What child addicted to living patterns could fail to be drawn out on any fine day by all the gliding and humming and crawling marvels that came to my mother's flowers, drawn by the flowers themselves, by the birds in her apple trees, the miniature orchards of her current bushes? Who could not wish to stroll beyond eventually into the even more endless marvels of the woods where, as an added incentive, no sudden call to dig potatoes or wash the car could follow, and where a book might be carried almost unnoticed in one's shirt, to be read in peace in some quiet sunlit place? The outside world was an irresistable competitor with the engineer's workshop, and in the long run it lured me away.

Beyond simply drinking the wonder of a new natural pattern or activity that I discovered I always wanted to understand better what I was seeing. How did the patterns come about? Why should the green fruits in the potato fields be similar to tomatoes? Why did rain water in the furrows leave the delicate braided patterns which I saw on the silt? It was just water, wasn't it, if so why not go evenly over the flat soil like the water went from my bath? Why, in short, should there be patterns upon patterns everywhere I looked—what made them?

I soon found that books had keys to many of the mysteries. A picture in a small book on geology, showing fossils akin to those I had already found in our field, and the stated fact that these fossils proved that our two hundred-metre hilltop had once been under the sea—a marvel whose near incredibility has hardly left me to this day—made me for a time want to be a geologist. I think this ambition was just after my social wish to grow up to be a knight in armour, whilst it was a wish to be an astronomer that was shortly to follow (our field within in its dark surround of woods was a good, though very cold, observatory for the stars and my father possessed a small telescope that he had won as an engineering prize). But after the astronomer a still more profound impulse was soon to push me even farther away from the knight in armour which had been perhaps the only trace of an inclination have a 'people' ideal in my life.

When training as a doctor in New Zealand, my mother, unusually both as intending doctor and as intending mother, seems to have paid close attention in her courses to a brief outline of evolution and natural selection. Much sooner than it comes to most children, if it ever comes, she had passed on to me the general idea of Darwinism. I remember what she told me even now as a revelation. Suddenly I could see why the potato fruit and the tomato were so alike, and alike too to the fruits of the bittersweet, the native weed of our hedges. These plants were in the same group, they were a real family! Darwin had proved that a certain surname, Solanum, which Linnaeus had given them, jokingly and for convenience presumably, really meant something: they were indeed cousins—and had been cousins even when as far apart as the snowy Andes and our humble field, as with bittersweet versus the others! Clear to me at a stroke now was why parents loved their children, why people were afraid to die, even such small matters as why it so hurt to stub one's toe and why such a painful and useless appendage as a toe nail might still exist—too recently it had been our useful claw. All this and much else I slowly put together. I had not developed the idea for more than a few such cases before I knew what I wanted to do most in my life, and since making that decision, I have hardly wavered. Whatever my profession was to be, most of all I wanted to understand all that the idea of evolution could tell me. It was to be a torch with which to peer into the mechanism of a gigantic machine more ingenious than any of my father's; a key to unlock to pedigree of life's single enormous family.

I realized that I had made as yet only the faintest beginning: given time, the idea might illuminate everything—myself, for example and even the things

I thought. At the very least it should serve to interpret all of those visual patterns that fascinated me that were not inanimate. At that time I had hardly heard of a university. My parents referred to 'geologist' and 'astronomer' as professions one might possibly aspire to but had never once mentioned 'evolutionist.' Why should they indeed, evolution being to this day a pursuit so useless as to have no existence in the world of professional expertise. As to the geologists and astronomers, I imagined they worked in offices in towns just as my father did. Having no idea that to become an evolutionist as such could be an option, I continued to think of other professions, but thought always of how I might side-track the work involved towards evolution. I hoped to find ever more marvellous correspondences in unexpected organs of animals and plants. Above all, I wanted to understand for what purposes the transformations had been brought about. Professions of school teaching, bee keeping, carpentering, and novel writing (I was good at stories at school) passed through my mind. I rejected engineering rather definitely. Perhaps my father, being so good at it, was also a factor, and perhaps there was a bit of Freudian perversity and competitiveness towards him combined. Mainly, however, I think I sensed that engineering would be too absorbing: one might daydream of evolution in the intervals of handling bees or planning planks, but one could not when working on any best design of a thing. I knew well how even at night a design would never leave me until it was completed—and then immediately, if I was in the profession, there would be another. Similar thoughts applied to geology and astronomy: they were at once too close to my dream and too absorbing and creative in themselves. They would hence distract from my main aim.

I have said that I was not asocial, and as a matter of fact, ideas about human life and questions of right and wrong did interest me extremely in a somewhat abstract way—this is an interest I now vaguely trace back to my knight-in-armour phase. I was an avid reader of novels and always eager to judge and compare characters. Usually I was persuaded to what seemed the author's own view of his creations, but sometimes not. I came to realize eventually that the best authors just gave you characters as such—humans—hardly even pretending to have decisions of their own about them. Wanting to understand the moral issues affecting human life does not, of course, make me a people person: people people just need people to interact with, not necessarily the understanding of them; They tend to be conformist and are seldom more than superficially critical of any ethos of their time. The best authors and artists, however, while virtually having to be of the people orientation, give us not only life as it is but also hold up before us, if only by an implication, a vision of a utopia in which human life could be better. Sometimes this vision is quite radical.

A British painter, engraver and poet whose work was unusually utopian and who affected me greatly during my childhood was William Blake. I was strongly drawn by the mysticism of his paintings and his poetry, as also by the mysterious polytheistic world that he seemed to inhabit. In later life this admiration dwindled. Doubts came about his panglossian moral philosophy and with this a realization that his initially convincing robust nudes actually ripple with muscles that are almost as non-anatomical and impossible as angel's wings. Surely an artist that could be satisfied with such slap-dash and unreal anatomy would be content with a slap-dash and poorly thought out utopia too: I came to contrast Blake with my growing admiration for some of his own masters whose style he had followed but, in this sense, debased. Michelangelo's nudes too could be exaggerated but at least in essence they would 'work' and at its best Michelangelo's naturalism could be utterly convincing. In short, he could really draw and sculpt and Blake couldn't. However, concerning Blake's sincerity and originality, as well as concerning his eccentricity, likewise appealing to me, there could be no doubt.

Then later in my mid-twenties, when already engaged on research, I learned that Blake had had a disciple painter named Samuel Palmer, and that this man had worked on a famous series of paintings and drawings in a village that was a mere mile or so from my childhood home. Discovery of the peculiar art that Samuel Palmer had executed during a brief seven-year period when he had lived in Shoreham, our neighbour village, and almost only during that period, came as a revelation and a shock to me. Here was an artist who seemed to have redeemed, to an extraordinary extent, what Blake had lost—and yet kept also much of what he had gained! More than this, there in his paintings were our Kentish churches between the hills and among their trees, there my own ancient hollow oaks of Lullingstone Park where I had been accustomed to bird-nest for jackdaws and owls, there the steep grey fields scabbed with ant hills and the bare chalk showing under fence and stile; and all of this grossly heightened, steepened, twisted into a kind of religious phantasy, a vision of an ideal rural peace in England of the early 1800s. Enormous harvest moons rose behind the woods and hills that I knew; ears of wheat sized like pine-apples swayed behind the head a man reading his Bible in a field; near at hand a tree held out its nest and eggs and bird, and hawthoms and apple trees were bowed under their weight of flowers or fruit. All this was Palmer's world and it was also my own. My own ecstasy in that beautiful countryside could be felt in his penstrokes.

In these works the intense vision of Blake had been hybridised with the reality that I knew intimately, where I too had felt myself to be seeing Eden remade before my eyes. But just as the bliss of the irresponsibility of childhood had by this time ended for me, so Samuel Palmer's vision of those seven years had had to end for him. It was as if it took all seven years for the fact to dawn on him that despite its appearances, Shoreham of 1830 was very far from being a paradise for those who lived in it. Its people, it seems, he knew only from a distance. Even more than his scenery, his rustic figures were highly idealised, their poses often in fact borrowed from other known paintings. In spite of his strong utopian leaning I have ended thinking that Palmer was not really a people person, in this again being more like me. Out there in the real Shoreham, peasants were starving for want of even of bread and how could there be starvation in the famous 'Valley Thick with Corn' that he painted? The countryside in the reality included riots and hopeless drunkenness. The glorious fat wheat ricks that he celebrated were being set aflame. It was a time when countrymen were hanged for stealing one sheep to feed their family and others exiled in chains to Australia for snaring one pheasant. As suddenly as it began, Palmer's visionary period ended. He married respectably and transformed throughout a long remaining career into a competent but somewhat uninspired English landscape painter, not much to be distinguished in style from all the rest who were contributing to the decoration of the walls of the drawing rooms of respectable Victorian England. It was in one of these in London, a drear bed-living room where normally I laboured on the theory of kin selection, that, turning on my knees the pages of the newly discovered book illustrating his paintings and about his life, I was to cry myself for his disillusion. His reaction to landscape and his social vision had been mine too exactly: I too had loved Blake and had come in the end to see his social vision to be hopeless, as far from feasible as his false muscles are from composing limbs that would work, or for that matter, far as true justice is from the wistful chivalry of knights and *samurai*.

I am not a superstitious person, and do not believe in ghosts in any way that allows them an independent existence. It may make some sense to talk of the 'spirit' of a place, but as to a spirit that of itself chooses to talk to or influence a place's inhabitants or visitors, all that is nonsense. I cannot have been influenced by Samuel Palmer in my childhood because I had never heard of him, nor had my parents. Yet my reaction to his pictures shows that I certainly was influenced by the scenes he saw and painted, and in the same way, I have come to think of this influence as formative of, or at least somehow representing, the utopian side of my nascent interest in evolution.

In this light it is certainly an odd thing in my background that if you stand on Badgers Mount, or better still, if you stand on the adjacent even higher ground of Well Hill, of which Badgers Mount is really just a shoulder, and if you then turn your back to Lullingstone Park and the Shoreham Valley, with their rolling farmlands and sharp hills and woods, and thus facing exactly the opposite way towards the northwest, you find yourself looking across a further grooved tableland of the North Downs hill towards where, slightly farther of than Shoreham, there is a village called Downe. Here Charles Darwin the evolutionist lived and wrote down all his ideas. He too, as it happens, was all his life an intense lover of the scenery around him, whether in farthest Chile where he traveled or here in Kent. And he wrote and died there within my view from that hill. He chose to end the famous book which launched his theory with a description of a roadside 'tangled bank,' describing how in his imagination its ecology might have been shaped through the processes he had deduced. In this case, by the time I first read this passage in about my fourteenth year (having chosen the book for a school prize) I already knew that Darwin had been our fairly close neighbour. I had been on a walk with my mother and the other children to his house. Reading the passage, I realized how I was privileged to know exactly the banks that he was describing. They were those of the deep incised roads crossing the chalk ridges towards his home, worn out of the hill by the passage of hooves and wheels and feet over centuries. They were there on the Shoreham side too. Darwin had obviously been moved by them like Palmer but in a differently creative way. Darwin and Downe came to stand in my own mind for the less emotional, less humane side of my nature, the pure science in which the problems of man and the possibilities of utopia need figure relatively little, a view of life where I would just wish to understand and not at all to improve. In contrast, Palmer and Shoreham stood for my social and utopian side.

Would I have grown up with the same interests as I did if I had lived as a child in, say, Chicago, far away from haunts of Darwin and Palmer? Actually I do expect so. Chicago and Illinois have in fact reared excellent evolutionists. Closer to home, though, if raised simply farther from Shoreham, on the other side of Downe, say in Croydon or Epsom, where there is nothing much except a race-course, very much a people people's thing, to enthuse about, would I still have avoided to be a more pragmatic, more worldly, less emotional scientist? Again I expect so: had misfortune made me a Croydonian I believe my genes would still have made me much the same and I would now be hanging a basically similar life upon other hooks—on Jane Austen, it might have been, and the beauties of Box Hill. To believe otherwise, indeed, would make my receiving this prize as unfair as if it had been awarded by astrology or drawing a number. Nevertheless, I think these historical and geographical associations of my boyhood can help, if you wish, to explain how I am made.

As a young man, like most, I was certainly interested both in utopias and in ideas of social reform. But from starting with somewhat naive ideas about eugenics that soon arose out of my evolutionary interest, I slowly came to realize that there were major unsolved problems about the organization of life that, until solved, must almost preclude eugenic prescriptions. I saw this not just in terms of the usual reasons given in those times for decrying eugenics—such as that we didn't yet know enough in detail about genetic disease or about what would be really genetically progressive, and anyway it was all Nazi-Germanic and horrible. It was a much more sweeping difficulty that I saw: despite Darwin and the revelations of his greatest book we didn't yet understand how the whole scheme of life had got itself involved, firstly, in creating these species all around us—the cats versus dogs—and secondly, in storing within them huge amounts of recombinable variation. Until we understood what variation, sex and genetics were all about, how could we begin to say whether one variety was more desirable than another, or even say that a seemingly gross abnormality was bad? There was a warning on this that I already knew. One might think that the gross distortion and inadequate function caused to human blood cells caused by the presence of the 'sickling' gene of West Africa was a trait that just had to be bad: however, given the presence of deadly malaria in any area, to which disease only the sicklers were resistant, this was not the case. Might this prove to be only the tip of an iceberg of similar examples? More important for a pure evolutionist, would a more general understanding of sex and genetic variability prove possible along the line this example suggested?

Another thing I wanted to understand, as a preliminary to thinking about what might be 'best' for humanity as seen by my limited people-regarding side, was the source of the passions that seem inevitably involved in any mere discussion of such issues as eugenics and population control. Was the reason for these passions that the issues were indeed among the deepest, most important humans ever needed to think about, as my evolutionary paradigm would suggest, and was it that precisely because of that huge importance, their discussion could not easily be made a part of normal human intercourse, perhaps not even admitted to normal consciousness? Perhaps the vast new assets of *Homo sapiens*, language and rationality, so magnificently useful for all contingent issues, prove themselves inappropriate in a problem that is so fundamental. The pressure of our genes to proliferate may be like the motion of a heaving ship or of a tennis ball which, if made the object of conscious thought, only becomes the harder to allow for. But perhaps it is not that rationality is inappropriate, rather that only with pain and under social censure that rationality on a subject like this can be expressed. It is of course hard to imagine any more offensive and arrogant purpose language could be put to than saying to another person what the evolutionary rationale suggests one should sometimes think: 'You are having too many children: stop, permit me and my family to produce them instead. We will make a better world than you will.' Ghosts of implications of this kind inevitably haunt even the most general discussions of population control, leave alone do they they haunt discussions of eugenics. It remains, then, that it ought to be possible, just as Galton suggested, both to take the sting out from natural selection and to bring to birth out of humanity a happier and less wastefully selfish and crime-prone species by eugenics; and yet it remains very difficult even to see any justifiable and acceptable policy, leave alone to begin to implement one in a humane way. Natural selection in fact seems extremely likely to continue its sway into the forseeable future.

As the niche expansion that has been allowed by technology reaches its limits, we already begin to see both in small, in the bitterly embattled local populations that were formerly peaceful, and in large, in the resurgence of infectious disease that an increasingly dense, mobile and world-dominant population brings to itself, some of the forms the full resumption of natural selection is going to take. Nowadays, every way one looks from Well Hill, one cannot escape the lights and the rooves of the new suburbs and towns that did not exist in Palmer's and Darwin's time. Their presence, of course, calls up from afar yet another great British savant, that gloomy prophet who stands behind Darwin as Blake behind

Palmer (and he too behind Palmer at least as at the time of his disillusion). This prophet is Malthus.

In spite of the widening shadow of overpopulation now hanging close over all the habitable lands of the earth—and one has only to go to slums in anciently civilized habitable lands like India to see what life under the final shadow is like—in spite of this darkness, I still have hopes that Samuel Palmer's initial vision was not so entirely hopeless as he himself seems to have come to think. Indeed, I retain even some hope that still gleams from his mentor Blake's most perfect poem, where he wrote he would not rest from mental fight 'Till we have built Jerusalem/In England's green & pleasant land.' I said I did not believe in ghosts in a real sense, but as to the continued existence of such artists as Palmer and Blake through their work, there can be no argument. I feel as if I myself still strive with both their ghosts as I try to see a way that part of their vision can be redeemed without the crippling illogic and false interpretation of life that they originally had. For my own peace of mind, I have to try to hybridize such artists, reconcile them some way, with ghosts of my other side, with Darwin and Malthus. I have never considered building utopias to be my strength: I realized long ago that one needs to be more of a people person to be good at that. However, rather surprisingly, I do feel I have made some progress towards effecting a reconciliation among my old heroes as well as between them and myself.

The first steps in my research career, and what I was actually struggling to achieve when I happened upon the book that revealed Palmer, would, I believe, either have dismayed or angered the young man—dismayed in so far as he understood, angered where he did not. In the great sociobiology debate which E. O. Wilson opened through his brave book, I feel fairly sure that the ghost of Blake lines up with the opposition: I easily imagine him raging and lamenting with Science for the People. Palmer probably would have joined that side at least at first. It is indeed a gloomy prognosis for human society and for altruism if nothing can be found to underpin our cooperation except for nepotism. Only the worst of Nazi utopians could wish to believe nepotism is all the hope there is for humanity. Fortunately there also now exist other well-defined and workable concepts in this field of the study of social evolution. One began under the term 'reciprocal altruism' and is now better called simply 'reciprocation.' I have had only a very small part in this concept, far less than the other pioneers such as Trivers and Axelrod, or Richerson and Boyd: nevertheless, I did a little and I think Blake and Palmer will appreciate that. The topic is more 'humane' but unfortunately it doesn't involve anything that can be called real altruism: the concept comes down to no more than farsighted self-interest. However, there is another new field bearing indirectly on social behaviour to which I feel I have been contributing more, and this for me at least has turned out more dramatic and unexpected. It comes from an approach to the problem of species and sex. Together with other authors I have shown a way the sickling disease case that

I mentioned earlier may indeed lead into a much wider set of cases, a set which may touch on almost every aspect of human life. The theme gives a reason both for the existence of sexuality itself and for why we are as almost infinitely varied as we are, (as well as being as prone to gamble with mates and with genetics as we are, and as resistant to the evolutionary seductions of nepotism as we are). The theme is simply, again, disease.

The idea comes in two main versions. One sees the relevant disease problem as only due to genes that have somehow mutated and gone wrong—genetics here provides both the disease and then, by sex, its cure. The other version accepts the existence the same constant load of bad mutation as the first, but at the same time implicates an additional, even more important source of ill health, against which variability and the operation of sex must try to protect. This comes to a species actively from the outside—by pathogen infection. The second is the view I support. According to it, the variability of sexual species, including presence of forms like those anaemic sicklers (which, when malaria is absent, might seem merely mutations of the worst sort, straightforward agents of the first theory), is intrinsic, a group character that cannot be eliminated without risk of later population extinction. The notion this version has of a stored disease—control technology applies to species even across the boundaries of their local races. More specifically, the idea tells that if you want your offspring to be innately immune to malaria, say, then your best chance is to marry a West African: he is the type most likely to have evolved anti-malaria genes because the disease's prevalence in West Africa has been recently greatest. Indeed we now know that there, such genes are there for the taking that are nothing like as drastic or as makeshift as the gene of the sickling trait is. If on the other hand, you put more emphasis on your offspring's need to be immune to tuberculosis, then you must go to marry an Eastern European: this group has been through the world's most recent severe TB epidemic and as result, a mate from that region will give raised chance that your offspring has a gene to resist it. Based on such thoughts, a two-generation program of arranged marriages might get your grandchild both such genes together and a kind of freedom of both continents. But perhaps you should hurry and I advise to find first the now identifiable malaria-resisting emir into whose harem you must enter yourself or persuade your daughter. Maybe as his genetic prize gets to be known there will be a rush for him! The main message, however, is don't just lazily breed with your cousin. There you are almost sure to get neither new nor resurgent lucky genes, and in fact, nothing much but what you already have. In this only partly joking picture I epitomize a situation that exists within all outbreeding animal and plant populations, and exists not just for pairs but for many more elaborate combinations of resistances. What I write about the pair in the human case will doubtless seem wild and exaggerate to many, perhaps repugnant: but it is based on facts, not speculation.

Clearly such a view of what can be attained through our sexuality and gene recombination looks to a level of necessary diversity in the genome that was not contemplated in the early days of evolution theory and genetics. And obviously too while it by no means says that eugenics is impossible, it warns that the subject needs much circumspection and that in general it is not a good idea to reduce any overall span of variation that humans have. This is so as long as it is desired to retain a naturally healthy physique as the basis of the human species.

Most interestingly in the context of eugenics, however, both the externally driven infectious-disease version sex theory that I support, and that more internally driven (and at present, better accepted) pure mutation-elimination version, lead to a similar conclusion, that a high level of selective death of zygotes has been a normal and necessary part of the maintenance of the health of species. The only escape from this for our own is either a level of genetic engineering and cellular intervention that is at present not remotely in sight, or a series of technological fixes after, or before birth, for both, all the old diseases of humanity and the new ones that will increasingly appear and accumulate. The problem is not only with the major new infectious diseases or the major gene defects. There will also be needed physiological fixes for all the small bad mutations are constantly being added to the human gene pool. The natural system of life was to arrange deaths after some sort of testing through competition. Generally in a species with parental care these deaths will evolve to occur as early in life as their effects can be made to appear. Such deaths eliminate multiply bad and/or currently inappropriate genotypes. The multiply disadvantaged genotypes are constantly being created by recombination along with other 'clean' genotypes that are likely to survive in their place. The idea that the elimination of the former class is natural and even eugenically necessary, of course, runs much against our humane instincts and it is doubtless partly for this reason that genetics is sometimes referred to as 'the gloomy science.' In the face of such a bleak outlook of constant deterioration, our instincts are almost guaranteed to be pre-set to tell us: 'Even if that may be true in general, of course it doesn't always apply—and surely anyone can see it doesn't apply in my wonderful family.' But according to the old system, which the new one of medical tinkering is very far as yet from being able to replace and perhaps, even in principle, never will replace (and certainly won't before the Malthusian crunch begins to make medical progress much more difficult), death must cull from almost every family. No family is so intrinsically healthy against all infections or so shielded from mutations that it is not being carried steadily down hill, in need not at all of the 'Rassenhygiene' of our mistakes of the past but, as the least, of just a natural wild culling of badly endowed foetuses and neonates. To my mind in its complete opposition to the natural system of death that has undoubtedly served to raise us up to what we are, the Roman Catholic Church and like-minded organizations could hardly be more wrong or more cruel to the species they profess to believe the most special on earth than by trying to prevent the death of every fertilized zygote. As can be seen I am here more on the side of Christian Science and Jehovah's Witnesses with their doctrines of no medical intervention at all, although for the more self-aware stages of human life I still have to part from them. Even in the stories of Christianity, Jesus, it may be noticed, was never said to cure or bring back from the dead a sick infant, while the Old Testament has examples of exposure of infants described without censure. No one can look on such matters without distress but to me the incomprehension, the lack of fear of the human neonate, seem like a divine providence for humane infanticide or at least for a letting of 'nature to take its course.'

It is over issues like these, then, that my ghostly fight goes on. Here, not particularly hopeful or even wanting to convince people generally in a world that normally reacts with horror to the very mention of such thoughts as I am raising above, I wrestle with dead neighbours of my youth instead, with Blake and Palmer and even with Darwin. Darwin I think is already largely on my side, firstly because of both his trained understanding and intrepid truthfulness, both stronger than my own, and secondly because he knows the intense anguish of losing his daughter Annie to disease when she was far into childhood. He contrasts this with the sorrow of deaths of some three others of his offspring as infants. One, incidentally, he knew to be handicapped, and Darwin certainly also knew dread, both of crippling ill health and of hereditary disease. I fancy I have further to go with the other two but Samuel Palmer has grown slightly sympathetic to my view. A stubborn eccentric in many ways, even after the conversion of his art into saleable respectability, I like to read, for example, how at all times, even the most bourgeois and proper, he was inclined to dive a hand into the huge pockets of a workman's jacket and produce nuts which he cracked with powerful teeth. That is me teasing my dentist wife exactly—and as it happens sometimes with nuts I may have pulled down from branches of the very hazels that trained Palmer's teeth a century ago! Crunching nuts in the woods of the High Hill above Shoreham that my childhood house looked out on, his ghost regards me quizzically and he listens: he too lost children at ages of three and nineteen to disease, and perhaps he believes less strongly than he once did in the religious dogma. I know the second death, that of his eldest son, was for him, like Annie's to Darwin, a terrible sorrow. Could this have been foreseen and preempted in utero, or even after birth, what would he have chosen, I ask him. As for Blake, I still only see one faint point of hope. One poem that is for me the second most perfect of all that he wrote, or even perhaps the very best, 'Oh rose, thou art sick! The invisible worm . . .,' shows that he too may have glimpsed a problem of this kind and seen how it can be in the gift of art to transform its sadness.

It may be gathered from what I have written that the straightforward visual beauty of natural patterns became somewhat lost sight of in the research I ultimately entered on, and for which I have become best known, a laying bare of the roots of sociality. I have not in fact made any discovery concerning how the patterns of animal shape and colouration or even the patterns of ecology come about. Though I followed some embryo ideas enunciated by Darwin even in this, my excursions, so to speak, have turned out to be more on the Shoreham side of my life than I intended, less on the side of Downe, so that at times my work has almost entered a people person's domain—even though by such persons themselves the ideas tend to be much resented, as I have imagined them to be by Blake. But what lures one into the garden of Science doesn't necessarily define where one goes when one is there. The problem that I felt inspired to attempt first was in fact more or less being solved at the time I was thinking of it in my last year at school, although it required some thirty years before I was to know this. This was the definition of what are now called the fractal patterns of nature and it was done more than by any other by Benoit Mandelbrot. Through him I think I understand now much better that beauty of order-in-confusion that still so delights and torments me, whether purely visually in the fern leaf or tree shape or braided sand, or in part temporally as in the eddies of the river, or in boiling clouds. I understand now that natural selection guides and often reduces these 'easily' attained fractal patterns into the more 'difficult' actually less natural ones, as Mandelbrot saw. Contrast the fern-leaf pattern with that of the fruiting Lunaria plant, where a non-fractal pattern (an almost perfect circular flat disk) is produced for each fruit. It is the circle and the stiffly arranged seeds within it makes the Lunaria 'higher' plant in contrast to the more fractal fern alongside with its myriad sori. However, that is hardly the point: I have always liked finding answers without having to go through the long agony of working them out for myself. But whether I understood the patterns or not, I still believe that the pattern addictions of my childhood that drew me in, addictions richly supplied out of what might be described as a vast psychedelic drug enterprise of nature that surrounded my home—drugs supplied out of every flower, each changing leaf—was not wasted on me as a scientist. For beauty is beauty always, and in my opinion, nothing is so likely to determine the depth of a scientist's contribution as the aesthetic standards that somehow are set to work in him. Human arts, like human science, can never be more than a pale reflection of beauty that is taken from nature, from outside of the human workshop. At their worst, both activities come to present admired uglinesses and run-away fashions in which all objects are second hand, and their appreciation more concerned with human style than with the reality behind them. While I certainly exaggerate above for explanatory purposes by putting such emphasis on certain illustrious neighbours of my boyhood, I do not think I exaggerate at all in saying that the richness of the world I had as a boy and the varied beauty in it I was helped to appreciate has determined my course and my success all through my life.

CHAPTER 12

BORN SLAVE TO THE QUEEN OF LIFE[†]

W.D. HAMILTON

What use is evolution? Can it make money and does it raise your social standing? Starting with my own case and answering the last two questions, I would say, on the whole, no. As result of my infatuation with it, I am a half man—a person who can't name to you one film-star, two ministers of his own government, three models of car... in short, in the terms of the man in the street, I'm a disaster, a 'Professor Cosinus'. All these shortcomings undoubtedly follow from my having spent my life in a 95% effort to understand evolution better.

But what rewards have come? And am I equally a disaster for humanity? One reward to me, of course, has been honour—being regarded as a significant contributor to this science, being chosen by this Foundation, and listened to by such a gathering of French scientists as are around me. But really I think I can honestly say that, for me, the honour is a by-product of an addiction I just can't shake off; for me, the satisfaction of increasing my understanding has been far more important than anything else.

I can assure you I am happy about this prize and have no difficulty about thinking how to spend the money. On the contrary, such a prize is something I can show to my family with pride and with a feeling, for once, of being a good father. I can say, look, I haven't been so mad in never knowing about those tax concessions, those investment schemes, or in never deciding whether or when or where I ought to be owning a house. All of such mundane but important things I generally fail in just because it takes time away from study. I shall try to give, in a minute, a sketch of what I believe started me out to be such a fanatic. I think my outline of this will be like Pierre Jaisson's generous description of my life a few minutes ago, but retold, this time, with an accent on passion—a bringing forward, as it were, of the *femme fatale* in the background who is always, for me, Evolution. As with all love affairs, you may imagine the misery of this kind of passion as well as its occasional triumph. And yes, this image of a love affair is a kinder, juster image

---

[†] Allocution du Professeur William D. Hamilton, Laureat du Prix International Fyssen 1995.

of my state of mind than to call it a drug addiction. In addiction, there is just the drug and all its in-turned and short-term effects; in a love affair there really is something beautiful, free-living, outside oneself, that one is chasing.

But before coming in a few moments to how evolution affects me, should I not also speak about how evolution—whether my favoured branches or any other—may add wealth or well being to others, and help the world as a whole? Well, in fact evolution can so affect the world and begins to prove it. At the present time I think we are seeing a proliferation of ways in which the understanding of natural selection is helping medicine, psychiatry, agriculture, and even industry. But all this I will return to shortly.

First, the fanatic.

Most important is my mother. Educated in New Zealand beyond the expectations of most women of her generation and so that she become a doctor, my mother somehow in her courses picked up a smattering of Darwinian evolution. Not long after she had begun to practise, she decided to leave medicine entirely and devote herself to raising a family. I was her second beneficiary—actually her third, since I suppose my father was the first. By the age of about 12 I had a working acquaintance with the idea of natural selection; at 14, I won a school form prize and bought a cheap copy of Darwins' 'Origin of Species'. From that age onward I hardly wavered: I wanted really to understand all this about natural change; it was my key to unlock the world.

Another important factor was that I grew up in countryside and, again with my mother's encouragement, became a naturalist. Just possibly a still further factor was that I grew up only about five miles from Downe House, the country mansion near to London in which Darwin worked and wrote for most of his life. It is true that my mother took me on a walk there when I was very young. That was a memorable day, but as much memorable for the long walk as for the museum-like house at the end. I think that if anything, the visit estranged me from Darwin rather than drew me to him. The house was so large and the amenities he worked with so great, and all this by inheritance and the expectations of a gentleman/capitalist. It made me feel that the best I could hope was to be a dabbler by Darwin' standards, an amateur with an interest. The world of his secluded garden where he walked each day, supervised his gardener, oversaw his experiments, was alien. I had no idea at that time that there even existed niches in universities that might support me in study or that might give me amenities not so far different from those which Darwin himself had. The world outside the garden, however, was a different matter, and later it came to seem to me a great privilege that I had learned by my independent parallel experiences as a naturalist exactly the chalky fields Darwin worked in for his pollination experiments with wild orchids, that I knew exactly the 'tangled banks' of the sunken Kentish roads around Downe that had inspired the wonderful closing passages of his book. But it was not Darwin who taught me to see that world: it was my mother, who had encouraged her children to make flower collections, and me in particular to collect butterflies. It was she who had explained to me, long before that visit, that if flowers of cabbage and turnip and wild cardamine looked alike, shining their little crosses yellow, white or pink all set to the same arrangement, it is because they were indeed related, were cousins; and that if the 'comma' butterfly that I pursued so perfectly disappeared to become a dead leaf, it was because millions before it had gained advantage from just such predators as me by just such an act.

Thanks to these experiences, I think I soon realised that Darwin's work had treated two main and rather separate aspects of evolution. *Firstly*, there was *his evidence*, the fact of evolution having occurred. Here I realised he was far from being the first. Jean-Baptiste Lamarck, Darwin's own grandfather, and several others had preceded him (Buffon and Cuvier, also from France incidentally, but others were included, even from as far back as Lucretius, the Roman Epicurean). Historians since that time have made it even more clear to us that *the fact* of evolution had been forcing itself to the notice of naturalists and savants during at least a whole century prior to Darwin. *Secondly* there was the question of the *mechanism* by which evolution took place. Even in this case the right idea was obviously floating to the front, rocked forward on a king of ocean swell, and soon this wave must break in a way that all the world would see. To my mind this inevitable movement is well shown in the poem by Tennyson containing the phrase 'Nature, red in tooth and claw' which came to be so often quoted, stating besides of Nature how 'so careful of the type she seemed, / so careless of the single life'. This was written *ten years before* Darwin or Wallace published anything on the subject; in fact both Wallace and Darwin were doing little more than putting rational and factual flesh to current ideas and exploring their consequences. To my mind, Lamarck's notion of a mechanism supporting his very clear and explicit recognition of the *fact* of evolution could be likened to one of those small false breakings of a wave crest as a great roller nears the coast. Lamarck's idea of a mechanism was plausible, so plausible indeed that Darwin, as he grew older, under pressure from physicists telling him his time was too long and of engineers telling him his heredity was unworkable, became more of a Lamarckist, less of a selectionist. Even so, Darwin's early stance was right and Lamarck's idea was wrong. In fact, a little thought should have shown anyone Lamarck's idea never had a hope to provide a complete and satisfying explanation of all evolution. I think I can convince you of this by means of one simple example, and it's one I have already mentioned; the comma butterfly and me chasing it. This butterfly lives all its juvenile life feeding on leaves of hop or currant; during this time as a caterpillar it always seeks the greenest of leaves. After its metamorphosis, each butterfly emerges from his chrysalis bearing perfectly drawn on its underside, and further perfected in the shape of its wings, the features of a dead and twisted leaf. From then on it flies and acts as though it had been brought up intimately among dead leaves; it seeks them whenever it wants to sleep, or hibernate—or just to hide from a pursuing boy. Even if characters acquired by effort within a lifetime *could* be inherited, as Lamarck supposed, how could any such 'effort to resemble' have been begun in a larva or pupa towards resembling something it hasn't seen? And the theory here, you will note, cannot even be saved by a supposition that the adult butterfly earnestly compares itself to dead leaves and then transmits the information into its egg in order to enable the next generation to improve the resemblance. This is because a butterfly's head is so placed that it can't see the undersides of its own wings—to be Lamarckian a butterfly would have to carry a mirror! The reality of the present case is that only the predator can be judging the resemblance. The mechanism of natural selection, as I am sure you realise, has no difficulty whatsoever with this case of marvellous adaptation: those butterflies whose genes happen to code for best dead-leaf resemblance survive best, they leave the most offspring; and so it goes on.

It will be fair to add to this, however, that recently, rather like Darwin, I have come to make some concessions toward Lamarck's position. Surprisingly there exists, even today, an intermediate idea. You will find it indexed in textbooks under 'The Baldwin Effect'. However, I don't believe that in most textbooks you will find a very clear explanation of how the Baldwin Effect works: if there were such an explanation, I like to think I would have believed in it much sooner than I did. I have to say almost with shame that it is a philosopher, Daniel Dennett, who has at last made it clear to me. Certainly for a long time I have been a disbeliever in the Baldwin Effect and regarded the supposed argument as a covert attempt to bring Lamarckism back to favour through a back door. But during the past year I have become an equally strong believer and now see the Baldwin Effect as likely to be very important.

It would take too long to explain it to you in detail and must suffice to say that the idea carries the prediction that a species able to be plastic in its within-lifetime adaptation has much more chance of acquiring radical genetically endowed characters of the right kind in its descendants—these through natural selection and eventually replacing the plasticity. Thus, species in which individuals have plastic, effort-driven responses, whether these are automatic or by trial and error, are able to evolve more rapidly. *Homo sapiens* is a very plastic unspecialised animal and doubtless so were his ape-like ancestors. The plastic tendency being present, the Baldwin Effect may have helped human progress in many ways. In particular, for example, it may have helped the rapid evolution of our brain and our linguistic ability, and more generally, may account for our almost unprecedented rate of physical evolution of the past 5 million years. In its essence, such an idea could be described as semi-Lamarckian: striving has reentered the picture. If striving in a strange environment allows an organism to survive there at all, it may be giving to natural selection almost *its only chance* to find the genes that can eventually complete the conquest of a new way of life.

Returning to my passion for natural history, as also complexed with my special addiction to evolution, I would now like to speak of another influence almost equal to that of Darwin in my boyhood. I had an unmarried great aunt who, with eccentricity perhaps even exceeding her niece's later neglect of her New-Zealand medical degree, had collected beetles and had amassed a large cabinet of them. She also had, in her home which I often visited, several books by a French naturalist, Jean-Henri Fabre. Seeing that I was interested, my aunt gave them to me. As I read, I remember being shocked and disappointed at first by Fabre's obvious antagonism to evolutionary thought. This antagonism recalled to me Linnaeus's indifference to ideas that seemed equally demanded by his work, and the indifference of other great naturalists. But what astounding discoveries Fabre has made! I was completely enthralled by my aunt's collection of the translated stories from the *Souvenirs Entomologiques*. The world of insect behaviour that Fabre revealed, offered to us like and a reworked edition of the Fables of La Fontaine, was amazing. It was robotic, totally without sentiment, strange in its acts and achievements almost beyond belief. And above all there was no difficulty, no heresy (as came to be very important to me later) in my trying to imagine genes for all the behaviour he reported!

I have never ceased to be amazed at the production of this schoolmaster from the South of France, and how little, even now, a lot of his work is known. How many biologists know, for example, who first gave evidence of individual recognition in a non-mammal and what animal was it? It was Fabre and it was a scarab. Who discovered which group of animals have the lowest fecundity and the highest level of parental care? It's not us, we are nowhere near; again it's scarabs and Fabre discovered this. In visiting his home in Sérignan I have felt there the same sort of reverence as when I visited Downe. Even in boyhood, because he had so stressed in his writing his background and even his poverty, I came to draw encouragement from him that I was unable to draw from Darwin. It has always seemed to me that French biologists give Fabre less honour than is his due, treating him as a quaint naturalist story teller with an 'incorrect' style, a fringe figure in biology. Well, so he was and so he had a strange style—but what stories and what implications! We don't belittle Newton because, besides leading a grand new movement in physics, he remained lifelong a dabbler in alchemy.

Before finishing on the origins of the evolution addict you see before you, I need to add a little about my father, an engineer, and how it was probably from him that I received my feeling that mathematics has a place in, and provides the right language for, the ultimate truths of evolutionary thinking. It is not as yet true (as it may be already with physics) that you can't say what you need to about evolution without mathematics. But it is the case that mathematics provides the tools and models to test your ideas. If you can't or won't clarify with a model and idea you are telling me and which I don't understand, then either what you are saying is obvious and you're being pointlessly obscure, or you're wrong—or as a third faint possibility, you might be right but not yet have seen the way to put it all together. What has been important for me, I think, has been an instinct similar to my father's to make models when I was not sure whether something will work. Thinking through a model always helps towards truth; usually and fortunately, this truth ends as something that I can explain in words. In all this activity I like to think that I inherited my father's capability attitude by which he built bridges in wild rocky ravines and across grand rivers, as well as inheriting a certain distant admiration that he had for the real, ongoing mathematics—whence descend those minor tricks that engineers and we evolutionists both use.

Now I return to my first question, what use is evolution?—and how is evolutionary philosophy going to change us? Early on, evolution gained a bad name among humane people. The earliest brands of eugenics and social Darwinism seemed to show that if evolution was useful at all it was going to be in painful ways. However, those movements had taken up very poorly understood themes; the versions which gained momentum I am sure would not have had the approval of either Darwin or Galton. Moreover, it is fair to state that Social Darwinism and Fascism were not the only political movements that were stultifying evolutionary thinking at the time, forcing it to fit with their prejudices. Like Fascism, Marxist Communism grew upon, and finally collapsed, due to a wrong understanding of evolution. Marx believed Darwin had been writing about what was his own (Marx's) preoccupation—group conflicts. Group selection, which would be the natural selection mode corresponding, is actually a very weak force (and when not weak, it may be noted, it is actually a more frightful mode from the 'humane' point of view even than individual selection); Marx also believed in a sort of Lamarckian infinite malleability of human nature. It was hard to see the terrible consequences that lay latent in such seemingly harmless and even likable ideas. But we note that before long, Marx's devotee Stalin was killing more people even than Hitler killed. When 'wishful' social ideas are wrong, I think it is common that their product is eventually both inhumane and very destructive. In using evolution to further social needs, we have to be tough-minded in order to be kind, and it is hard to get over this hump. Nevertheless, I think we have reached a stage where we have much more reason to be optimistic that we now have a working understanding such that any one who wants predictions of biological social consequences of our actions can have them. This understanding once gained, it is just a matter of adjusting each desired course of action to each Utopia—and, presumably, mass courses, those that politicians adopt, can be determined as usual through democracy. Unfortunately, the state of acceptance of evolution in the world is very far from helping people to see this, and innumerable weird and potentially destructive myths are still afloat. I think just because it is so obviously rational and right, most of these myths see evolution as a dreadful opponent.

Meanwhile, the fact that the evolutionist paradigm plainly does work and is providing an ever more solid basis for the high technology of modern medicine and agriculture, must be causing some doubts even to the most tenacious adherents of supposed infallible revealed truth—I mean here, of course, ideas such as the creation of the universe in a literal seven days and the like. This applies whether the adherents are Christians, Muslims, Jews, or any other.

Amongst the modern evidence that should give these people pause for thought are the phylogenetic trees for AIDS viruses, monkeys being predictably better models for human diseases than are mice, and fulfilled expectations of related drugs coming from evolutionarily related plants. As it happens, I don't always 'believe' in a Utopian sense in the high-tech advances in medicine, say, that are made in these ways; I am often a sceptic about their supposed benefit; but one cannot but regard with awe the precision of what is coming to be achieved under the evolution rubric. Validation of both *the fact* and the *mechanism*—natural selection—are involved in this progress at every stage.

As to psychiatry, here there are applications that are closer to what I personally am supposed to have achieved—the importance of relatedness, of reciprocation, fundamental conflicts leading to deception, and so on. I have already talked about them in the large, as it were, in discussing the failures of some mass social movements of the recent past; but it is certainly also true that we can expect a more incisive, effective psychiatry as we attain a more realistic understanding of our evolved psyche and its past. I believe sociobiology is rapidly giving this to us. It is too much to hope that cures for all psychiatric problems will be forthcoming since unfortunately a lot of psychopathic and neurotic problems arise out of dilemmas for which there really is no resolution: it's not much help telling the patient: 'Gee, that rape strategy certainly would have made you a big hit in Paleolithic Europe!' However, it is virtually certain that the doctor with real understanding of his patient's condition can do better than the doctor without; and actually even to make the above remark to the rapist might cheer him up a little and so help him towards seeing his problem and how he must fight it. There would be no need to stress an accompanying remark that ought, logically, to have followed: 'A hit in the Paleolithic; yes but here in today's London such a strategy (and you, sad to admit) have fundamentally no hope.' Of course, ideas of psychiatry are not very easy to measure and I cannot give you any examples of any *quantitative* predictions made by my social theorising that have been fulfilled. But I can give a quantitative idea of how well the underlying paradigm is supported for an allied matter surprisingly strongly connected to the above claims, the matter of predicting sex ratio.

Finally, I mention a way in which evolution by natural selection, in symbiosis with the computer, is invading even engineering and industry. This is the topic of the 'genetic algorithm' (GA), the idea that when confronted with very difficult design problems whose nature is to admit many false local optima of design that may stand 'a valley or two away' from something that is much better, then the best course is to set up an evolutionary model of natural selection in a computer and to hope to evolve to the something better. Such a model will specify genes arrayed on artificial chromosomes, with the genes themselves varied in the many individuals so as to make available all the elements that the designer believes may be important. Finally, the method simply puts mutation, recombination and natural selection into motion, with the best chromosomes always rewarded with more descendants and the progress watched. Any one who starts work with this seemingly 'blind' and rather stupid technique is usually astonished at its rapid achievements. Already the method is evolving better designs in many branches of industry and science. I'm not sure that there is any well-known item on the market yet where one can say, look at this marvellous thing, who would have thought of the way it works?—and then add that actually nobody ever did think of the way, the gadget was created by natural selection in a GA. But that time is surely coming. I have experimented with the technique enough to see the potential. True to my character as I earlier described it, I haven't yet applied my GA experimentation to anything very practical. Instead with my ever-continuing cloud-cuckoo perversity, I am turning this evolutionary technique in upon itself, using it to understand further one of my own other evolutionary puzzles. This is the evolution of sex itself: and sex, of course, is already basic to be Mendelian approach which the GA employs!.... However, when I've retired and all my money from my Fyssen Prize is used up, I'm quite looking forward to the time when I will start applying my knowledge to becoming rich. I'll put up a brass plate announcing *'Consulting Evolutionary Analyst'*. Barons of industry from all over the world will beat the path of my door, and soon you may hear me telling one of them:

'I think what you need for your problem, sir, (the optimal power distribution net that the man wants, let us say, for continental China) will be a toroidally wrapped stepping-stone host metapopulation with a lek breeding structure based 80% on the genetic quality of its males. A 10% of choice will go on idiosyncratic disassortative preference based on the locus matching, and then perhaps 5% on imitation while a last 5% will be random. On the 'Red Queen' side to keep things moving, we need what we're now calling 'running dog parasites' (a concept you'll find in the manual) and a generation ratio of one to six against the hosts. You need a mean selection index adequate to start marginal chaos in the **GA** but then of course will want to control away from the gene-space boundaries by a high reversible mutation—a rate of ten to the minus four for parasites usually suffices, don't make it much lower even if you can, or you'll have some parasites evolving as hyperparasites—and then a non-optimising ecosystem. This may tell you to send power to Tibet or somewhere, ha, ha ! (Cough) Well, check with me again when you've started—and that advice will be twenty thousand dollars, please.'

He gives me twenty thousand dollars. Later, by my method, he actually obtains a very efficient grid, saving his country many billions.

After this, who will doubt Darwinism?

CHAPTER 13

FOREWORD TO S. TURILLAZZI & M. J. WEST EBERHARD (EDS) *NATURAL HISTORY AND EVOLUTION OF PAPER WASPS*[†]

WILLIAM D. HAMILTON

Social wasps are among the least loved insects. Even dogs dislike them. Sharing an Englishman's jam sandwich half-way to his mouth, massing in hostility to the Amazonian's gliding and bumping canoe, raiding and even predating their angelic and useful social cousins the honeybees, and perhaps above all simply being born with that bold, swaying and ever-returning flight, that remorseless buzz—all these things estrange us. Joylessly, most of us read texts stating them to be important predators, that without them flies and caterpillars would be even more abundant. Yet, where statistics will not alter a general impression, another approach might. Every school-child, perhaps as a part of religious education, ought to sit watching a *Polistes* wasp nest for just one hour. While probably not more than one in a thousand will be converted to become the kind of wasp addict that contributes to this book, I think that few will be unaffected by what they see. It is a world human in its seeming motivations and activities far beyond all that seems reasonable to expect from an insect: constructive activity, duty, rebellion, mother care, violence, cheating, cowardice, unity in the face of threat—all these are there. The ancient Greeks, moved seemingly by parallel observations, frequently presented *Polistes* wasps (as well as perhaps other wasps disproportionately) in varied paintings and ornaments. Latreille, the French systematist who named the genus (Linnaeus had them in *Vespa*) acknowledged both the Greek attention and the social traits by his selection of the plural of the word for a Greek city state. Between Latreille in 1802 and to the present, however, it took an American shopkeeper, Phil Rau, and an Italian biologist,

---

[†] In S. Turillazzi and M. J. West-Eberhard (ed), *Natural History and Evolution of Paper Wasps*, pp. v–vi (Oxford University Press, Oxford, 1996).

Leo Pardi, to bring these astonishing insects back into the forefront of biological attention—and along with *Polistes*, to some extent, the rest of their papery clan.

Clearly Rau and Pardi watched and were entranced. Many following their writings, including the present writer, have become entranced likewise. From his careful observations, Pardi gave us the concept of an animal dominance hierarchy; this predating the establishment of the idea for any other group. The concept of unconditional instinctive behaviour in animals, undifferentiated by individual propensity and as predictable as the bouncing molecules of a gas had from this time to begin its wane. If mere insects could be so human, what of the rest? Far beyond the humble paper nest of a *Polistes* wasp, the whole idea of prediction in ecology was, or ought to have been, immediately shaken—although perhaps we should admit here that another concept, also largely home-grown within biology, that of mathematical chaos, was needed before the full consequence of the new individualism could be seen.

From *Polistes*, wasp studies branch out to all the rest, as the chapters in this book show. As social insects, paper wasps are probably a monophyletic group. Certainly they have elements of a common style, however much they differ in physique. How different *Vespa mandarinia* is, for example, from a gnat-like *Parischnogaster*, whose slim and tiny nest the mighty hornet, largest of all social insects, might hardly deign to devour, even if she discovers it! Yet when working a ball of wood pulp or plant hairs into a firm papery wall, or in knowing when to consume a comrade's egg, both of these types, the huge and the tiny, turn out to be much the same.

In this book almost the entire sweep of the activities of wasps is represented, although most of the emphasis in the chapters is on the internal, social side, rather than their impact on the rest of ecology, great as this is. Thirty years ago, when I was an unknown admirer of Pardi and Rau, this social side was my interest too, and *Polistes* likewise my favorite. Sure, I had known long before this a *V. germanica* clypeus from a *V. vulgaris* (their owners peering at me, perhaps, equally nervous and displeased, as it might be, from inside their hole in a plum). And soon I would be proud to tell a *Mischocyttarus* from a *Stelopolybia* (now *Agelaia*) even while in flight and even when, in coloration and shape, the gentle *Mischocyttarus* was a near perfect mimic of the ferocious other. But it was to social life that wasps were providing my touchstone puzzles.

Much of that skill, as well as some of the intensity of my curiosity, has now gone. At the Castiglioncello conference that originated this book, I wandered rather like a man in paradise who finds God's memory embarassingly better than his own, listening to a list of loving and conscientious answers, detail by detail given upon all those questions that were once so desperately pursued, saved up, and then . . . well, simply forgotten. This is not to say that there is not still an endless line of questions concerning this ill-starred, under-appreciated group, but this book is a milestone—and not least I hope in giving certain marvelous animals a better image.

CHAPTER 14

BILL HAMILTON'S INVOLVEMENT WITH THE OPV THEORY

'Medical Science's most Hated Hypothesis'

EDWARD HOOPER

I knew Bill Hamilton for the last six years of his life, and our relationship was almost exclusively based around a mutual interest in how the AIDS pandemic began.

It was also, however, an intensely personal relationship–so much so that after his death, I was for some time unable even to mention his name without weeping. The reasons for that are many, but in retrospect, I believe they mainly involve certain qualities of his which I find both exceptional and moving—his lack of hubris, his searing honesty and his intellectual generosity.

By the time Bill and I first met, in September 1993, I had been working on AIDS for seven years, and researching its origins for three. By 1992 I had done enough literature research into the earliest evidence of HIV-1 and AIDS to know that the pandemic epicentre was located in the African countries formerly administered by Belgium—the Democratic Republic of Congo, Burundi and Rwanda, rather than in Uganda, Gabon (and even Haiti), as was then being proposed in scientific journals.

I also knew that almost all of the theories about the origins of AIDS were unsustainable. The only apparent exception was the hypothesis that the virus had entered humankind when African hunters or market-sellers had killed, or butchered, an animal infected with the simian immunodeficiency virus (SIV) that was directly ancestral to HIV-1. By 1990, it was known that the probable host of this immediate ancestor was the common chimpanzee, *Pan troglodytes*. The main problem with this 'cut hunter' or 'bushmeat' hypothesis of origin involved the timing. A second AIDS virus, HIV-2, had been discovered in West Africa, and the absence of these two viruses from North America and the Caribbean before the 1970s strongly suggested that neither virus had existed in Africa during the time of the Slave Trade. So why had two AIDS epidemics evolved since that trade ended in 1865, when Africans had been eating chimps (and sooty mangabeys, the ancestral host of HIV-2) for millennia? It seemed possible that the Hand of Modern Man might have been involved.

In 1992, I first heard about another theory, one initially proposed by Louis Pascal, an armchair philosopher from New York. Pascal had been amazed to learn that polio vaccines had been routinely grown in primate kidney cells, and further research revealed an oral polio vaccine (OPV) called CHAT, developed by the Polish-American scientist, Hilary Koprowski, which had been tested on a million 'volunteers' in the Belgian colonies in the late 1950s.[1] Pascal came to an amazing conclusion—that CHAT vaccine was responsible for the arrival in *Homo sapiens* of the precursor virus of HIV-1, and therefore for the birth of AIDS. He sent carefully written papers to many eminent scientists, most of whom didn't reply, and to several scientific journals, all of which rejected them. His powerful essay, 'What Happens When Science Goes Bad',[2] eventually had to be published as a 'working paper'.

The only considered response to Pascal had come from Bill Hamilton, and I decided to seek an interview with the one major scientist who seemed to take the OPV theory seriously. Bill lived up to my vision of the eccentric genius—a shock of white hair; a house littered with papers; a shy, self-effacing manner; and a gift (when he did speak) of describing important ideas in accessible language. By that stage, I had interviewed several hundred scientists, and immediately recognized that here was someone special, in terms of both breadth of knowledge and clarity of reasoning. I told him about my research, and he made his responses, sometimes simple and sometimes profound, but always based on a bed-rock of sound judgement. That first meeting lasted eight hours, and a bond was forged between us. We became partners in pursuit of the putative iatrogenic event at the source of AIDS.

There were many factors that fired this search, but one was key. Koprowski had been testing CHAT at Camp Lindi, a huge colony of chimpanzees and bonobos sited just outside Stanleyville (now Kisangani). To Bill and me, the coincidence between the world's first mass trials of OPV, its earliest cases of HIV-1 and AIDS, and perhaps its largest chimpanzee colony seemed too significant to ignore.

By the end of our second meeting, on New Year's Eve, 1993, I felt I knew Bill well enough to ask a sizeable favour. Knowing that he had recently won several major scientific prizes, and having finally exhausted both my own savings and the largesse of my parents, I asked him whether he could lend me the money to make the two remaining research trips (to the US and Belgium) that I felt were needed before I could start writing my book. Bill asked for a breakdown of the costs and then, without further hesitation, wrote me two cheques totalling £2,000, adding that he would like me to consider these as a grant, rather than a loan. (Much later, after the deaths of my parents, I tried to repay Bill—but my letter was found in his papers after his death, with the cheque still inside. He had decided to ignore it.)

The next month we visited Stockholm to see Hans Wigzell, the head of the Karolinska Institute, where I had discovered that some unopened vials of 1958 CHAT vaccine were stored in the freezers. Professor Wigzell agreed to our request to have them tested for the presence of HIV and SIV, but declined to release any portion of the samples to us, to test for the mitochondrial DNA of the primate cell substrate.

Later that January, Bill prepared a long letter to *Science*, in which he sought a fairer hearing for the OPV theory. The letter was rejected, so Bill wrote a follow-up letter to the editor, Dan Koshland, further pleading his case. He was told that he was 'superbly qualified to comment' on this issue, but still *Science* declined to publish. Bill was sent copies of sections of the referee's reports, which revealed that one referee had highlighted 'the possibility of local contamination [of OPV] by chimpanzee tissue in Central Africa', an eerily prescient suggestion. Nevertheless, he voted against publication, and against the testing of CHAT samples (on the grounds that even if found positive, they would only prompt a lawsuit—and that scientists were already well aware of the dangers of potential iatrogenic disasters). Details that feature in this referee's report reveal that the author can only have been the eminent British retrovirologist Robin Weiss.[3,4*]

Bill then submitted a similar, but stronger, letter to *Nature*, which was also rejected.[5] I have recently learned that for many years, major AIDS submissions to *Nature* have been routed through Robin Weiss, so it seems that he may have been involved in a two-fold rejection of Hamilton's plea.

* See section 5(b) in reference 4.

In January 1995, the Karolinska faxed Bill their findings: the CHAT vials were negative for immunodeficiency viruses. However, one intriguing detail was highlighted, for both the original 1958 vaccine from the Wistar, and further vaccine that had been prepared therefrom in cynomolgus cells at the Karolinska in 1963, were described as 'CHAT pool 10A-11'. The truth dawned slowly. Pools (or lots) of OPV represent material prepared at a certain level of attenuation, but it is the specific batches prepared from those pools that are homologous—not the pools themselves. Different batches of CHAT pool 10A-11 had been prepared at different times, in different labs and (it seemed) in different substrates. It was the history of the batches, not the pools, that was crucial. It was therefore not legitimate to argue—as some had—that a pool of CHAT fed in Africa must have been uncontaminated, because the same pool had been fed without problems in Europe.

Over the next four years, Bill and I were in contact by phone or letter every few days. In addition, about once a month I would drive up to Oxford, or else (more occasionally) he would visit me in West Sussex, and later Somerset. It did not then strike me as remarkable that whenever I called, he always had time for me. But amidst all the serious talk, we also had the odd bit of fun. When I told him that one of the Belgian doctors recalled vaccinating along the eastern shore of Lake Kivu, and remembered seeing the clouds changing colour to russet when they passed above the volcano of Mount Nyiragongo, Bill spent some hours analysing maps and the curvature of the earth, to try to determine where my witness might have been.

In 1996, we collaborated on a letter to the *Lancet* that attempted to unravel the mystery of the so-called 'Manchester sailor' (an apparent AIDS case from 1959). There seemed to have been lab contamination, but we were still unable to explain how four of six tissues from the case, and none of six from the control, had tested HIV-positive in the original double-blind study.[6]

Bill also helped greatly as I began writing *The River*, providing not only a fine foreword, but also some suggestions for the opening sentences of text, inspired by the book's title. They show something of his love of the natural sciences, and of his clarity of thinking, and I adopted them almost wholesale.

> What is a source? Where does a river begin? In this valley is a spring, but higher up the hillside lies a dripping rock..... That ultimate source on the ground is almost never easy to identify, and some would say the search is meaningless. But the resulting geography—the nick in the hillside, the steep-edged valley, the mature river, the floodplain, the estuary—although it never ceases to evolve, remains firm enough to allow description on maps. These features are the visible consequences of that tiny source, and it is these that make their immense impact on humanity.[7]

What I found most remarkable about the foreword he wrote for *The River* was the extent to which he was prepared to allocate responsibility for the genesis of AIDS. From the opening words ('Every time two people put their heads together, Truth suffers...'), he weighed in against his fellow-scientists, against pharmaceutical houses, and against governments.[8] He spoke with conviction and quiet anger, and went further than I was then prepared to go.

Not all his peers liked the foreword. Shortly after Bill's death, Robin Weiss told me he considered it 'bullshit'. At that stage I too had some misgivings, partly about Bill's range of targets, and partly because of the praise he had lavished on me. Now, however, in 2005, I find his central argument a marvellous piece of reasoning, one that shows the fearlessness and foresight that evolved from his years of lonely study of biological processes.

Shortly before *The River* was published in September 1999, Bill and I made our one and only safari together, spending ten days in the breathless humidity of Kisangani, Democratic Republic of Congo (DRC), where the *Laboratoire Medicale de Stanleyville* (LMS) had coordinated the 1950s CHAT trials, and ten miles from Lindi camp, where some 400 chimps had been utilized as the scientists 'put the finishing touches' to CHAT vaccine.

The journey was a success in terms of research, but a personal disaster, for we had three volcanic arguments. In the end, Bill largely concentrated on collecting faeces from pet chimpanzees to test for SIV, and I on trying to discover more about the history of the LMS and Lindi camp. Yet I have many fond memories of his Congolese exploits. Bill rushing, vortex-like, across the hotel courtyard to greet a fresh arrival of banana leaf-wrapped chimp shit, with passport, notes and money spinning in his wake. Bill astonishing a large crowd beside the Congo ferry crossing, by whipping out a butterfly net and executing a series of startling manoeuvres in pursuit of an especially glorious specimen. Bill, surrounded by children, inventing a drawing game in his notebook which prompted whoops of delight. And the two of us returning in a huge motorized dug-out from the site of Lindi camp, now overgrown by rain forest, but still heavy with significance and collective memories. We sit in facing plastic armchairs, watching the banks of the Congo idle past, as he explains the evolutionary similarities between the strangler figs of the Amazon and Congo basins. The common denominator of all these memories is the sometimes unworldly, but always single-minded, scientist.

Bill's uncompromising approach to travel included an unwillingness to rely on pharmaceutical products, and he refused anti-malarial prophylaxis; not surprisingly, he contracted the disease. (He felt that the best way to fight illness was to experience the worst, and build up natural immunity, but there was also something stubborn and old-fashioned here: a true explorer does not complain. I found his disregard for his personal safety quite at odds with his professionalism as a scientist.) The last image I have of Bill as an active player in life is of an ashen man standing alone at the baggage belt at Heathrow, awaiting the emergence of his battered rucksack.

The trip scarred us both, but the months that followed eased the hurt. Bill was as happy as I about the burgeoning, and largely positive, response to the publication of *The River* in the UK. At the end of November, after three months of silence in the US, the *New York Times* published a lengthy article,[9] and suddenly all the news media were phoning. At my request, Bill did an interview with CNN,[10] in which he once again stated his position on OPV/AIDS . . .

> It's not only the origin of AIDS which is in question here, it's also the conduct of Science towards this hypothesis, which has been one of almost paranoid rejection . . . I think I would not exaggerate to describe it as medical science's [most] hated hypothesis.

Sadly, I was never again to speak with Bill in person. I planned to see him just before he flew back to the DRC in January 2000 (this time to collect faeces and urine from wild chimps), but my car broke down. He called once briefly from Kisangani by satellite phone, but the next news I heard was that he was comatose in a London hospital, having collapsed from a massive intestinal haemorrhage the day after his return. Once again he had contracted malaria, this time the cerebral variety, and although he had apparently recovered by the time he returned home, it may be that the strain which this placed on his system exacerbated a pre-existing gastric condition. Whatever the precise cause of death, those who loved him were incredulous. I spent half a day with him in hospital, holding his hand, and telling that great still body the latest news on the debate. But this time, when I paused, there was no quiet, reflective response, no impish smile.

All February I was racing to complete a new postscript to the book, and it was arranged that once I finished, I would come up to see him one final time. I was working on the penultimate footnote on the morning of March 7th when the phone rang, and his long-time partner, Maria Luisa Bozzi, told me the sad, but not unexpected news.

At the funeral and the remembrance event there were tears, but also power struggles taking place in the wings. Some of Bill's former colleagues, embarrassed by his involvement with the OPV theory, began to propose that he was merely an open-minded scientist seeking to test a rather far-fetched hypothesis. It was largely because of Luisa Bozzi that this position was unable to take hold. She read through his personal and professional letters, and at the Lincei conference in September 2001 gave a moving and powerful speech in which she confirmed that Bill was '95% persuaded' that the OPV theory had merit.[11] In reality, during his final years, Bill was intensely involved with OPV/AIDS research, and he effectively risked his life in order to collect more of the hard data which, he felt, would support the hypothesis.

Shortly before that final safari, Bill had persuaded both the Royal Society and the *Accademia Nazionale dei Lincei* (in Rome) to stage conferences at which the origins of AIDS could be debated. The co-organisers of the London conference, Robin Weiss and Simon Wain-Hobson, took over sole responsibility after Bill's death. They managed to keep the conference on track (not an easy thing, since while Bill was still comatose a campaign had been waged, mainly by American scientists, to declare the debate one that damaged Science, and to persuade others not to attend. Two of the principal supporters of the bushmeat theory, Beatrice Hahn and Bette Korber, simultaneously withdrew, while Koprowski's former deputy at the Wistar, Stanley Plotkin, implied that he and Koprowski might join them.) After Bill's death all these scientists came back on board, but at a price. Two extra speakers were allowed to the anti-OPV camp, while I was refused the chance to nominate a full speaker to replace Bill.

Many attendees felt that the conference was far from the level playing field that had been promised, but instead afforded a prepared stage for an official refutation of the OPV theory, focusing on the half dozen samples of CHAT vaccine that Koprowski's Wistar Institute had belatedly released for independent testing—which were found negative for HIV, SIV and chimpanzee DNA. But there was more. A team led by Stanley Plotkin had approached many of the scientists I had previously interviewed, and obtained signed statements from some that contradicted their previous, tape-recorded statements on key issues. (Later I discovered many instances of improper approaches being made, including one case in which a witness was badgered to sign a prepared statement which was patently untrue.)[4*] Robin Weiss also played an unwelcome role, for his closing speech was frankly biased. He implied that the theory had been fatally wounded, and not unexpectedly, the press followed his lead.[12,4†]

* See section 5(e) of reference 4.
† See section 5(d) of reference 4

Seven months later, the world's two leading scientific journals, *Nature* and *Science*, took the unusual step of reporting simultaneously on what was termed new phylogenetic dating 'evidence', and test results from another sealed vial of CHAT, again originating from the Wistar. Weiss claimed incorrectly that this CHAT material was from the same batch that had been used in Africa (when it was merely from the same pool),[4*] and concluded his *Nature* commentary: 'Some beautiful facts have destroyed an ugly theory'.[13] *Science* headed its commentary 'Disputed AIDS theory dies its final death'.[14] This blanket rejection of the theory had an enormous impact, and most neutral scientists and lay persons now seem to believe that the debate is settled.

But on what grounds have the OPV sceptics reached their conclusions? They have five main arguments: that local chimps are not infected with SIV; that local chimps are not infected with 'the right SIV'; that chimp tissues were never used to make the vaccine; that phylogenetic dating indicates that HIV-1 predated the OPV trials; and that there is anyway no correlation between the CHAT feedings in Africa and the first appearances of HIV-1 and AIDS.

I believe that each of these arguments is flawed, and that recent scientific and historical findings actually offer very strong support to OPV/AIDS.

Simon Wain-Hobson, who had agreed to collaborate by testing the samples of chimp faeces and urine that Bill collected on his two trips to Kisangani, has sadly never reported the details or results of his SIV testing. Furthermore, he has failed to make any sensible response to the five detailed e-mails I have written to him over the last four years, requesting feedback on his findings, or else that he release the samples so that others can do the work. Given how much Bill invested in this research, I think that Wain-Hobson's performance has been disappointing. Beatrice Hahn, another committed opponent of the OPV theory, was given aliquots of these samples in 2001, and she repeatedly found protein bands typical of SIV. However, she did not report these findings (which ran counter to her general hypothesis) until 2004, when they appeared in a brief communication in *Nature* entitled (misleadingly, to my mind) 'Contaminated polio vaccine theory refuted'.[15,16]

Hahn's findings indicate that chimps from one of the very sections of the DRC rain forest where Lindi chimps were collected are SIV carriers, and Paul Sharp has also reported that 13% of a single wild troupe of this same subspecies were SIV-infected.[17] If that percentage applied to the 400 chimps used during the Lindi polio research, then approximately 50 would have

* See section 4(a) of reference 4.

been naturally SIV-infected before arriving at the camp, where co-caging of pairs and groups was routine. However, Hahn and Sharp also argue that the Lindi chimps are from the 'wrong subspecies', pointing out that the very closest HIV-1 relative discovered to date comes from a *Pan troglodytes troglodytes* from Cameroon or Gabon, rather than a *Pan troglodytes schweinfurthii* from DRC.[18] This is true, even if relatively few chimpanzee SIVs have so far been sequenced, and even if it leaves Hahn having to postulate an infected chimp-hunter who failed to spark AIDS in Cameroon or Gabon, but who migrated hundreds of miles southwards to spark infection in the HIV-1 epicentre in the DRC. However, one of the 54 Lindi chimps for which there are surviving records came from Mbandaka territoire in the west, which is as near to the range of *troglodytes* as *schweinfurthii*. This animal spent over two years at Lindi, and clearly could have introduced a *troglodytes* SIV to the camp. But perhaps it didn't need to. Because exactly the same genes are found in HIV-1 and in chimp SIV (whether *troglodytes* or *schweinfurthii*), a recombination event looks to be the most parsimonious explanation for bridging the genetic gap between the chimp and human viruses.

The official reason for Lindi camp was to test the susceptibility of chimps to orally administered poliovirus, and to safety test the Koprowski vaccines by intraspinal inoculation, but Koprowski's group mentioned these tests in conference discussions in 1959, revealing that only 89 chimps has been involved.[19] In reality, the chimps served other purposes too, as I discovered during a second visit to Kisangani in April 2001, when I conducted further interviews with the surviving Lindi 'caretakers'. They confirmed that almost all the chimps had been sacrificed, with blood and organs frequently being obtained from anaesthetised chimps, just before sacrifice. (The significance is that the best method for preparing tissue culture, for instance for vaccine cultivation, involves removing organs from living animals.) I also interviewed several technicians, former workers at the LMS, who reported that tissue culture had been mainly prepared from chimpanzees, and that the head of the virology department had been 'making the polio vaccines' in his lab, namely propagating vaccine in locally prepared tissue culture, to boost both vaccine titre (concentration) and quantity.[4] These African testimonies have since been confirmed and enlarged upon by Belgian sources, including one eminent doctor who stated that the principal purpose of the Lindi chimpanzees was for 'the preparation of the vaccine'.[20]

This unique aspect of the Congo CHAT trials (making fresh batches of vaccine locally in chimpanzee cells) is the key detail that is missing from

The River, albeit largely because of the denials of the Belgian and American vaccinators. Over the last three years, every stage of the local preparation process has been multiply confirmed by different sources.[21] The vaccinators continue to issue strenuous denials, but these are often self-contradictory, and their attempts to explain away the counter-evidence are increasingly implausible.[4*] For example, they stress that the LMS annual reports mention nothing of chimpanzee tissue culture, or of local polio vaccine propagation. This is true, but it merely highlights that the use of chimp cells was a secret, even back in the 1950s.

Further research has revealed that propagating OPV locally (either from a sample of vaccine or from a seed pool) was routine practice in the 1950s; it happened with the vaccines of Sabin and Lepine, as well as Koprowski, and in places as far apart as Switzerland, the USSR and South Africa, as well as the Congo. This demonstrates that the CHAT batches that need to be tested for SIV and chimpanzee DNA are not those produced at the Wistar Institute, but those that, uniquely, were administered (and also prepared) in central Africa. Robin Weiss believes that samples of the vaccines used in Africa no longer exist.[22] I suspect that they do, but doubt that they will be released for testing.

The geneticists have a different 'disproof'. Those who favour the concept of 'phylogenetic dating' for HIV-1 argue that the most recent common ancestor (MRCA) of all the AIDS viruses seen today existed in 1931, plus or minus 15 years—namely, before the OPV trials. But their calculations are based on a constant molecular clock, and they ignore recombination, which, according to the OPV theory, could have occurred in a tissue culture based on chimp cells. Documents prove that primitive chimpanzee tissue cultures prepared at the LMS in 1958 also contained 'isologous serum' (serum from other chimpanzees) as a nutrient medium,[23] suggesting that these chimpanzee cultures were effectively pooled, which further increases the likelihood of *in vitro* recombination.

Immunodeficiency viruses are inherently recombinogenic. Recent studies indicate that the intrinsic recombination rate of HIV-1 is some ten times greater than its mutation rate (which is what phylogenetic dating measures). They also show that ignoring recombination would lead one to place the MRCA too far back in time.[24] At the Lincei meeting, Mikkel Schierup highlighted the evidence not only for substantial recombination, but also (crucially) for substantial *early* recombination, even before the virus diversified

* See sections 4(b), 5(d) and 7 in reference 4.

into subtypes. But whether or not early recombination occurred, the phylogenetic dating of HIV-1 is invalid, being based on a false premise.[25]

Finally, there is the epidemiological argument. I am pleased that Bill Hamilton's friend (and first post-grad student), Peter Henderson, has agreed to co-author a statistical study that compares several different hypotheses of AIDS origin (including the ranges of different chimp subspecies; and proximity to transport routes, major towns, and centres of health delivery). His analysis detects a significant link in only one instance: when the early foci of HIV-1 infection are compared to places where CHAT vaccine was given in Africa in 1957–1960. The correlations are highly significant both on a macrocosmic scale, across central Africa, and on a microcosmic scale, in Burundi alone.[26] This study substantially undermines the one full epidemiological paper that was presented at the Royal Society, which inexplicably ignored the CHAT vaccinations in Rwanda and Burundi (over half of the African total), and then concluded that there was no association between CHAT and AIDS.[27]

This new evidence (especially about local vaccine preparation) is revelatory. I would argue that every one of the alleged 'disproofs' of the OPV theory presented at the Royal Society and in *Nature* and *Science* has been intrinsically flawed.[28]

Bill, who realised long before I that several of his peers were more interested in disposing of an ugly theory (with frightening implications) than in examining that theory in a cool, dispassionate, scientific manner, would have derived great pleasure from these latest developments. I suspect he would also have been pressing for those individuals, institutions and governments that staged and backed the trials to be brought to book, and (if found culpable) to be made to accept some degree of responsibility for the terrible aftermath.

To him, there was only one way to practice Science, and that was with absolute integrity—and his pessimism about our 'human future' was at least partly based on his growing belief that integrity is a vanishing virtue.

Many have commented on the beauty of Bill's 'last testament', which describes his body being laid out in the jungle and consumed by Amazonian beetles, and through them borne aloft beneath the stars.[29] It seems that Bill may have wanted to die in the field, in such a way that his work and his spirit lived on, and I believe that metaphorically, at least, this final wish was granted.

NB Many of the Royal Society and Lincei articles are accessible on http://www.aidsorigins.com.

References

1. G. Courtois, H. Koprowski *et al.*, 'Preliminary report on mass vaccination of man with live poliomyelitis virus in the Belgian Congo and Ruanda-Urundi', *Brit. Med. J.*, **2(i)**, 187–190 (1958).
2. L. Pascal, 'What happens when Science goes bad. The corruption of Science and the origin of AIDS: a study in spontaneous generation'; 1991; Working Paper No. 9; Science and Technology Analysis Research Programme, University of Wollongong, Wollongong, Australia.
3. Part of 'Review #2', enclosure to letter from C. Gilbert [*Science*] to W. D. Hamilton, April 28, 1994; Hamilton papers.
4. E. Hooper, 'Dephlogistication, imperial display, apes, angels, and the return of Monsieur Emile Zola. New developments in the origins of AIDS controversy, including some observations about ways in which the scientific establishment may seek to limit open debate and flow of information on 'difficult' issues'; *Atti dei Convegni Lincei*, **187**, 27–230 (2003).
5. Hamilton's letter to *Nature* is reprinted in Julian Cribb, The White Death, (Angus and Robertson: Sydney, Australia, 1996); 254–257, with further discussion on pages 182–184.
6. E. Hooper and W. D. Hamilton, '1959 Manchester case of syndrome resembling AIDS'; *Lancet*, **348**, 1363–1365. [See Section 14.2] (1996).
7. E. Hooper, *The River—A Journey to the Source of HIV and AIDS* (London: Penguin and Boston: Little, Brown; 2000); 3. [Updated paperback version.]
8. W. D. Hamilton, Foreword, in E. Hooper, *The River*; xxvii–xxxiii. [See Section 14.3]
9. L. K. Altman, 'New book challenges theory of AIDS origins'; *New York Times*, November 30th, 1999, F1 and F6.
10. 'Insight' programme, broadcast by CNN on December 1st, 1999 (presented by Jim Clancy).
11. M. L. Bozzi, 'Truth and Science—Bill Hamilton's Legacy'; *Atti dei Convegni Lincei*, **187**, 21–26 (2003).
12. See papers in published proceedings of the Royal Society conference: R. A. Weiss, S. Wain-Hobson (ed.), 'Origins of HIV and the AIDS Epidemic'; *Phil. Trans. Roy. Soc. Lond.* B; **256**, 771–977 (2001).
13. R. A. Weiss, 'Polio vaccines exonerated'; *Nature*, **410**, 1035–1036 (2001).
14. J. Cohen, 'Disputed AIDS theory dies its final death'; *Science*, **292**, 615 (2001).
15. M. Worobey, B. H. Hahn et al., 'Contaminated polio vaccine theory refuted'; *Nature*, **428**, 820 (2004).

16. E. Hooper, 'Why the Worobey/Hahn 'refutation' of OPV/AIDS theory is wrong'; www.aidsorigins.com (2004).
17. M. L. Santiago, P. M. Sharp, G. M. Shaw, B. H. Hahn *et al.*, 'Foci of endemic simian immunodeficiency virus infection in wild-living eastern chimpanzees (*Pan troglodytes schweinfurthii*)'; *J. Virol.*, **77**, 7545–7562 (2003).
18. M. L. Santiago, P. M. Sharp, B. H. Hahn *et al.*, 'Amplification of a complete simian immunodeficiency virus genome from fecal RNA of a wild chimpanzee'; *J. Virol.*, **77**, 2233–2242 (2003).
19. H. Koprowski and G. Courtois, 'Discussion', in *First international conference on live poliovirus vaccines: papers presented and discussions held*, (Pan American Sanitary Bureau, 1959, Publication No. 44); 201 and 227.
20. M. Kivits, personal communications, 2000–2004.
21. 'The Origins of AIDS', MFP/Galafilm television documentary; January 2004; personal communications to the author, 2001–2005.
22. 'The Origins of AIDS', MFP/Galafilm television documentary; January 2004.
23. W. Henle, G. Henle and F. Deinhardt, 'Studies on viral hepatitis'; Annual report to the commission on viral infections of the Armed Forces Epidemiological Board [US]; 1959, p. 5.
24. S. Wain-Hobson, J-P Vartanian, A. Meyerhans *et al.*, 'Network analysis of human and simian immunodeficiency virus sequence sets reveals massive recombination resulting in shorter pathways'; *J. Gen Virol.*, **84**, 885–895 (2003).
25. M. H. Schierup and R. Forsberg, 'Recombination and phylogenetic analysis of HIV-1'; *Atti dei Convegni Lincei*, **187**, 231–248 (2003).
26. E. Hooper and P. A. Henderson, 'Statistical analysis of competing hypotheses to explain the origin of HIV-1 Group M in central Africa'; in preparation.
27. K. M. De Cock, 'Epidemiology and the emergence of human immunodeficiency virus and acquired immune deficiency syndrome'; *Phil. Trans. Roy. Soc. Lond.* B; **256**, 795–798 (2001).
28. E. Hooper, 'The latest scientific evidence strongly supports the OPV theory'; www.aidsorigins.com (2005).
29. W. D. Hamilton, 'My intended burial and why'; reprinted in: *Ethology, Ecology and Evolution*, **12**, 111–122, and in this volume, Chapter 3 (2000).

1959 MANCHESTER CASE OF SYNDROME RESEMBLING AIDS†

EDWARD HOOPER AND WILLIAM D HAMILTON

Bailey and Corbitt's letter to *The Lancet* about the 25-year-old man who died in Manchester Royal Infirmary, UK, in August, 1959, with a clinical syndrome resembling AIDS[1] is welcome but it leaves several points unresolved, including some raised by a science journalist in March, 1995.[2]

A particular puzzle is that the original polymerase chain reaction (PCR) study[3] was claimed to be of a randomised double-blind design. Properly applied, such a design makes it difficult for an interpretative bias to generate a false positive or negative result, and impossible for random contamination to do so. On application of Fisher's exact test to the results of 1990, the probability that random contamination of the test and control samples would produce four positive results in six test samples and none in six controls is 1 in 33. Occurring in 1990, before the dangers were fully appreciated,[4] accidental contamination in the first PCR study of a potential early case of AIDS would be understandable. However, the subsequent failure to address the statistical anomaly above and the neglect of other anomalies is not. We wish to highlight not only the mystery of how random contamination could have led to the results but also five more questions. For the third and fifth and partly for the fourth we suggest possible answers; the other two remain open. The questions are:

(1) How did the original tissue samples from the patient come to be found HIV-1 positive by PCR when these results cannot now be repeated?

(2) How have archival human tissues, which were apparently well enough preserved in 1990 to allow human and viral genetic analysis after 30 years in storage, apparently ceased to be so in the past five years?

(3) Accepting contamination,[1] what is its likely source?

(4) How have four (and possibly five) different human genotypes been reported for HLA-DQα in tissue samples claimed to be from one cadaver?

(5) What was the patient's fatal disease?

† *The Lancet* **348**, 1363–1365 (1996).

Contrary to speculation mainly, but not wholly, in the non-medical press, investigations by EH have shown no evidence to suggest that 'the Manchester sailor' (MS) was either homosexual or bisexual, or that he ever visited Africa. In early 1957 his ship did dock in Gibraltar for a fortnight. A day trip (well recalled by members of the ship's company) was made by about a dozen sailors to Tangier in Morocco, but a member of that party has no recollection that MS was present. Even if he was, or there were other day trips to Tangier, and even if (as has been hypothesised) he had sex in a brothel during such a visit, this can hardly be characterised as a high-risk episode. HIV prevalence varies widely across Africa and the seroepidemiological evidence suggests that Morocco has always been among the least affected countries. The earliest evidence of HIV infection in the country pertains to 1984–87, when seven of 8161 individuals (0.086%) tested positive, all from Casablanca. Six were in high-risk groups (gay men, male prisoners, and female prostitutes), the seventh was one of 3577 blood donors. None of 283 blood donors and pregnant women tested in Tangier in 1991 proved to be HIV-1 positive.[5]

Questioning MS's fiancée, family, friends, colleagues, and doctors suggests that he was neither sexually adventurous nor very experienced, and that he was not an intravenous drug user and had received no blood transfusions. Clearly one sexual encounter could have been enough, but everyone who knew him rates him as an improbable candidate for HIV infection. Those closest to him were saddened, indignant, and (rightly as it now appears) near to incredulous at the suggestion that he might have died of AIDS.

That incredulity is now borne out by Bailey and Corbitt,[1] who have joined Zhu and Ho[6] in concluding that the posthumous AIDS diagnosis was unsound, and that certain of the archival tissues made available to them may have been or have become contaminated with a modern (subtype B or 'Euro-American') strain of HIV-1. They suggest contamination 'sometime from sectioning onwards', and that the most likely source 'would be from within our own laboratory'.

The following scenario might go some way towards explaining the facts. The positive control used during the PCR work on MS was a CEM cell line infected with CBL-1.[1] In 1991, Weiss reported that CBL-1 had 98.0% identity with LAV-1 BRU (or, as it is now referred to, LAI) and 97.8% identity with HTLV-IIIB in *env, tat*, and *nef*.[7] An accompanying commentary on this 'remarkable similarity' cited laboratory contamination as the possible cause,[4] and reported that Gerry Myers of the HIV Sequence Database in Los Alamos considered that up to 3% divergence in *env* usually indicated different isolates from the same person, whereas, at the other extreme, genuinely unlinked isolates usually diverged by more than 10% in the envelope gene.

The earliest versions of LAI are the French patent application sequences bearing the Genbank/EMBL acquisition numbers A04321 and A07867, and Fergal Hill, of the MRC Laboratory of Molecular Biology in Cambridge, has characterised A04321 as 'apparently the most similar sequence to the Manchester isolate sequence currently known—at approximately 90% identity over large tracts, including the envelope gene'. Hill concludes that 'this high degree of sequence similarity, and the fact that CEM/CBL-1 was grown in Manchester, *strongly* [his emphasis] suggest that the Manchester isolate is...derived from LAI via its derivative CBL-1'. Clearly Hill believes that repeated passaging of CBL-1 (for instance in Corbitt's laboratory) could explain the 10% divergence between this positive control and the MS isolate. Myers is less convinced, considering that 'the contaminant may have been a lab strain, or...another patient sample'.

We have already mentioned that only MS's tissues came to be contaminated in spite of their random interspersion with the controls. Thus conventional significance points either to earlier contamination, before the coding and dispatch of the samples to Corbitt's laboratory (in which case considerations of the last paragraph suggest that the CEM cell line might also have been present in the source laboratory) or to error during the breaking of the codes. Sections were cut 'with separate knives for case and control and with careful cleaning, with alcohol soaked swabs, of knives between blocks'.[3] If we accept that the procedure was as stated, the best scenario at this point would seem to be that a knife cleaned neither before nor between section cutting happened to be contaminated with modern HIV-1-infected tissue and thus passed not only HIV-1 DNA but also appreciable human cell material to the first four sections, which happened to be from MS. By the fifth and subsequent cuttings the knife supposedly had wiped itself clean. As discussed below, however, there are still many problems.

The hypothesis of prior contamination might be clarified by a detailed description of the storage and location of the two sets of tissues, and of how and where sectioning was undertaken. EH learned from one of the doctors involved that for at least a part of the period of the PCR investigation the blocks were being stored in Williams' home, and Williams later confirmed this.

Both Corbitt and Williams told EH that the code had been broken during a telephone call, in which Corbitt read through the list of numbered samples, indicating for each whether or not the presence of HIV had been demonstrated, and Williams then broke the codes, indicating which samples had come from MS and which from the control patient. Corbitt states that nobody else was in the room at the time; Bailey was waiting outside. A more appropriate method might have been an exchange of sealed envelopes and the presence of witnesses when the envelopes were opened.

Further examination of the original MS tissues and of the PCR products from Corbitt's laboratory is needed. In the past, Williams has stressed that there was little tissue available and that he had been keeping a judicious eye on what remained to ensure that not all was used up.[8] But he acknowledges that about 40 blocks were taken at necropsy. These originated from a wide variety of skin lesions, together with bone marrow, heart, lung, and central nervous system, and abdominal viscera (including liver, kidneys, pancreas, and spleen), and even if most of the tissues are not ideal for finding lymphotropic virus, some DNA from an overwhelming virus infection should be detectable. Extraction of human DNA should be feasible from any of the samples. Perhaps the Central Manchester Health Care Trust could reveal exactly what tissue remains and perhaps some of the blocks could be examined by another laboratory. One laboratory, experienced in PCR and in sequencing lentiviruses, made a written offer to test tissues from the patient in March, 1995, in response to Williams' statement[2] that he would 'be quite happy to supply tissue to anyone who would take it on'. This offer was apparently forwarded to the Trust but was neither acknowledged nor accepted.

Five human genotypes for MS have been mentioned.[1,6] Zhu and Ho found that three HLA-DQα genotypes had been sent to them, with traces of a fourth. In material from Corbitt they found type 1.2,4 'with traces of 2,3' in kidney and 1.2,3 in bone marrow. In material from Williams, on the other hand, they found 3,4 in thyroid, liver and kidney. Bailey and Corbitt now report that, working on samples received from Williams in 1989 (those from 1995 having been found unusable), they detected 2,4 in liver and brain. They also found human type 2,4 in the CEM line that was their HIV-positive control in 1990. The frequency of 2,4 in Britain is likely to be well below 5%.[9]

If Zhu and Ho's interpretation of their bands was at all equivocal and '2,4 with a trace of 1.2,3' for kidney and bone marrow is a possible alternative to their stated '1.2,4 with traces of 2,3' the inconsistency of the New York and Manchester accounts would be greatly lessened: 2,4 could then be due to the contaminating CEM cells, and Zhu and Ho's technique, perhaps more sensitive than that of Bailey and Corbitt, could be revealing the underlying tissue type 1.2,3, exactly as found by Zhu and Ho in bone marrow which had seemingly escaped contamination.[3] MS would then have a puzzle of only two genotypes; a third would be due to the CEM cells.

Perhaps both DNA and proteins of the wax block material were so degraded that they provided weaker and sometimes undetectable signals relative to those provided by a recent cell contaminant, when present. This is further suggested by the partial and wholly negative results obtained, respectively, by Bailey and Corbitt and by the UK Forensic Science Service.[1] However, the idea that contaminant CEM cells explain all the genotyping and viral results since 1989 still involves many difficulties, whether that contamination arose in the laboratory where sectioning took place or in Corbitt and Bailey's laboratory.

Turning to the nature of the patient's disease, Bailey and Corbitt express themselves puzzled and reiterate that the symptoms were, retrospectively, very suggestive of AIDS. We believe, however, that the diagnosis has become the least of the problems of the case. It would be flippant to suggest that a patient with five HLA genotypes—more diploid combinations, it may be noted, than are known for any chimera apart from a few Panamanian strangler fig trees[10]—would of necessity be a simmering cauldron of autoimmunity and immunocompromise. Let us propose two plausible alternatives. MS may after all have had Wegener's granulomatosis. This was the working diagnosis for the final two months of his life and for more than seven weeks after his death the gross post-mortem findings were being described as 'consistent with [this] diagnosis'. Only when the microscopic findings revealed cytomegalovirus and *Pneumocystis carinii* was this diagnosis abandoned.

A second possibility is CD4+T-lymphocytopenia (CTL). This condition was christened 'AIDS without HIV' when its existence was first announced at the Eighth International Conference on AIDS in 1992.[11] Other publications quickly followed (e.g. Laurence *et al.* in 1992[12]). Rezza *et al*[13] mention a 39-year-old man without HIV infection who died as a result of a wasting syndrome, *P. carinii* pneumonia, disseminated cytomegalovirus infection, and neurotoxoplasmosis. Apart from the *Toxoplasma* infection, the clinical profile matches that of MS.Dr T B Stretton, one of the MS physicians in 1959, now leans towards this retrospective diagnosis.

If MS did die from AIDS it is vital to our understanding of the early history of primate immunodeficiency viruses that an authentic sample of HIV DNA from such an archival case be made available for sequencing and phylogenetic analysis. Besides the controversial postmortem tissues, biopsy specimens were taken from sternal marrow, scalene region (including a lymph node), and ulcers and skin lesions. Perhaps these are still available at the Manchester Royal Infirmary.

If, however, as we believe, this patient did not have AIDS, and if there was either substantial contamination with modern HIV DNA or tissue samples from other patients came to be included in the PCR investigations, then this man's family and fiancee are owed an apology for the distress which this episode has caused them.

Unsourced information in this article is based on tape-recordings and notes of interviews between EH and the various scientists mentioned, personal letters from some of these scientists, and medical records of the patient, viewed with permission of his next-of-kin.

References

1. A. S. Bailey, G. Corbitt, Was HIV present in 1959? *Lancet* **347**, 189 (1959).
2. S. Connor, World's first AIDS case was false. *The Independent* (March 24, 1995).
3. G. Corbitt, A. Bailey, G. Williams, HIV infection in Manchester, 1959. *Lancet* **336**, 51 (1990).
4. P. Aldhous, Spectre of contamination. *Nature* **349**, 359, (1991).
5. HIV/AIDS Surveillance Database, US Department of Government, Bureau of the Census, June 1994.
6. T. Zhu, D. D. Ho, Was HIV present in 1959? *Nature* **374**, 503–04, (1995).

7. R. A. Weiss. Provenance of HIV strains. *Nature* **349**, 374, (1991).
8. S. Garfield, The end of innocence: Britain in the time of AIDS. (Faber and Faber, London: 1994) 10.
9. K. Tsuji, M. Aizawa, T. Sasazuki, eds. HLA 1991. Proceedings of the 11th International Histocompatibility Workshop and Conference. Vol 1. (Oxford University Press, Oxford: 1992).
10. J. D. Thomson, E.A. Herre, J. L. Hamrick, J. L. Stone, Genetic mosaics in strangler fig trees: implications for tropical conservation. *Science* **254**, 1214–16, (1991).
11. G. Williams, T. B. Stretton, J. C. Leonard, Cytomegalic inclusion disease and *Pneumocystis carinii* infection in an adult. *Lancet* **ii**, 951–55, (1960).
12. Anon. AIDS without HIV. *BMJ* **305**, 271–72, (1992).
13. J. Laurence, F. P. Siegel, E. Schattner, I. H. Gelman, S. Morse, Acquired immunodeficiency without evidence of infection with human immunodeficiency virus types 1 and 2. *Lancet* **340**, 273–74, (1992).
14. R. Rezza, P. Pezzotti, F. Auiti, Acquired immunodeficiency without HIV infection: epidemiology and clinical outcome in Italy. *BMJ* **311**, 785–86, (1995).

FOREWORD TO E. HOOPER, *THE RIVER*†

W. D. HAMILTON

Every time two people put their heads together, Truth suffers; when many put their heads together, she suffers more. A major point of this book is that when the heads are great ones and have owners with much to lose (employed perhaps in giant companies or government departments), Truth can be made so ill that we should all shiver.

Evasion and untruth have long been known to be beneficial at many levels and useful to people in many ways. They can be presented as virtues—the little bads that add to a greater good, with a proviso, of course, that the good is of a kind that the colluders believe only they know how to attain. 'Don't we have faith in ourselves?—let's keep it simple for their—for all our sakes.' Even for God's sake: this version has been abundantly illustrated by religious leaders ever since Christianity became official in the Roman Empire, with disastrous effects upon other faiths—and a fiery impact upon a myriad of free-thinking 'witches,' as well as the occasional literary loner like Giordano Bruno. Once there is acceptance by an 'establishment,' there is often no need to whisper about it anymore: in those who have jointly suffered to win, say, the Queen's Commission in the British armed forces, or the privilege of saying the Hippocratic Oath, a solidarity springs up automatically, and with it a deep conviction that the purpose of the discipline, whatever it be, must be good. And yet, knowing the untruths that emotions arouse, especially in groups, Plato amazingly denied roles even for poetry and music in his ideal Republic.

Most of the daily untruths communicated need not be taken too seriously: we have become accustomed to them and in a sense self-vaccinate. However, when eminent rivals in an ancient profession are seen to be uniting to crush an outside critique, and when the best-funded branch of science, to which the rivals belong, draws almost all its practitioners into line behind them (as Louis Pascal and then Tom Curtis in the case treated in this book had already experienced, even

---

† In E. Hooper (ed), The River: A Journey Back to the Source of HIV and AIDS, pp. xxvii–xxxiii (Harmondsworth: Allen Lane, The Penguin Press. Boston: Little, Brown, 1999).

before Hooper), and when an expectant and immensely wealthy international industry is also seen marching in step with the profession in question, it is time for the rest of us to wake up.

The thesis of *The River* is that the closing of ranks against inquiry may, in this case, be preventing proper discussion of an accident that is bidding to prove itself more expensive in lives than all the human attritions put in motion by Hitler, Stalin, and Pol Pot. Furthermore, essentially unwarned by what we have recently done, we may be moving rapidly toward further and perhaps even worse disasters of the same kind. Some aspects of genetic engineering may indeed be dangerous, but a situation in which the general public has greater concerns about mystical subversion of the chemicals in soy sauce than about the risk of viruses in live animal products that are already administered, almost compulsorily, to our bodies, is near to absurd. In parallel to this, our doctors' Hippocratic Oath warns them of various temptations and dangers, but it says nothing of how they need to guard themselves, and their profession, against the effects of the millions of profit that dangle before the nascent industry proposing to transplant organs into humans from other species.

These are the foreground dangers emphasized by Hooper in this book. Its background has another danger, which is still more insidious. Litigation has been used to suppress the publication of discussions about a hypothesis; litigation is again being used as a threat to Hooper. In the same vein and equally unsettling, we have seen the best known and seemingly most independent science and medical journals join forces on the side of the countercritique, while generally avoiding publishing details of the original issue. Again it is time for us to wake up and consider what is happening to freedom of discussion and to the spirit of science.

It is the foreground, the potential repercussions in the next thirty or so years, which will probably most arouse the reader of this book. Perhaps something is being tardily seen by the establishment. A few months ago, the British Medical Association announced revisions to the Hippocratic Oath British doctors must take; then just a week ago, as I write, the Association's organ, the *British Medical Journal*, published for the first time an admission of a likelihood that Simian Virus 40, established as an infection in millions of humans by the Salk polio vaccine, is causing human cancers. 'Salk,' it may be remembered, is the 'dead' and therefore safer polio vaccine—safe supposedly not only from reversions to virulence but from the possibility of 'extraneous agents.' It is quite different from the type focused upon in this book—the type we now all receive. On another front, committees in recent months have enjoined slowness and caution with xenotransplants, but not before the first baboon liver transplant into a human was attempted—an operation that perhaps fortunately failed. Meanwhile heart valve implants from pigs, a species known to harbor retroviruses that can live in human tissue cultures, are in trial and application.

All this is why the world still very much needs lone researchers like Edward Hooper. They reach truth faster than committees. Shortly after I first knew him, I introduced him to someone as a journalist, knowing he had formerly been one in Africa. Later he asked me, pained, 'Why journalist? Couldn't you call me a *writer?*' I did so from then on but stayed puzzled. Weren't journalists supposed to be the guardians of our free world, the para-predators ranging our savannah and making even the most lordly lions take care of their actions? Weren't they (the best at least) even cousins to us scientists, ferrets setting themselves to bolt the most willfully concealed and elusive truths of history where we scientists deign only to chase the immobile targets, such as atoms and missing links? Why should one not want to be a journalist? After reflection and listening to the talk of 'paparazzi' and the like that came after Princess Diana's death, I think I see better now the perspectives that journalists dread—but just as hyenas do less scavenging and far more primary predation than was once thought, so also do the best journalists.

Whatever, this book, with its almost 2,500 footnotes, demonstrates how Hooper has finished up. Not only is he the kind of predator that all in Big Science should fear, but he is a writer and historian as well. Even that is not all. He has self-taught his way to 'honorary' status in several branches of science—to be almost virologist, almost geneticist, almost evolutionist. To most of us, however, these achievements just provide the reassurance that he is writing sense in his diverse fields; in contrast it is the writing itself and the history—dare I say even the first-class journalism?–that will keep us bent over the pages that follow. What scoops, what personalities, what landscapes, what far places! Above all what enigmas, what awful inexorable tragedy (tragedy at its deepest, gnawing within millions of homes—a scale perhaps grander than any ever before described) stand there behind!

In 1995, in Africa for another purpose, I tried to help Ed by looking for some of the Ugandan friends who had helped, nearly a decade earlier, with the research for his first book, which described the AIDS disaster in that focal area close to the shores of Lake Victoria. There were two men in particular whom he wished to contact and to thank. As I discovered after some questioning, both had died. I was led to the father of one, and he in turn took me to a neat private graveyard in his *matoke* plantation and showed me the newly heaped mounds, six in all. They were for his wife and all his children. One mound, with a stone slab, was for the son Hooper knew, a local government official (who had been, perhaps, a little more important locally than the others). The old man sat on a corner of the slab and read the letter Ed had sent, while two grandchildren, come into his care after the last death, watched from nearby. The children were lively and healthy but very quiet, and I hoped the infection was going to miss them. Such graveyards, I found, were everywhere in the district, though they are not much seen from the roads. Orphans, too, were everywhere: a generation had been scythed out from between those who were too young and too old to be readily infected. I saw children in groups ranging from teens to tots seemingly loose and self-foraging in the countryside, which included as it happened trying to forage from me, the passing foreigner. Presumably these were the children not lucky enough to have grandfathers and grandmothers who were still alive. Both in the robust elderly and in these youthful gangs I felt I was seeing how Africa would survive, if only after a period of great suffering. Yet it may end up less changed, it seemed to me, than will the continents of the First World, in spite of our lower expected mortalities.

After that brief experience in southern Uganda—a few days only—I understood better what had been driving Hooper to follow up on the lighter and more emotional book he had already written about the epidemic in Africa. I suspect he had no idea, at the start, of the magnitude of what he was undertaking, nor of the nine-year odyssey of research and travel it would require. Even before he read Louis Pascal's extraordinary paper 'What Happens When Science Goes Bad...' and had realized the full tragic possibility about the origin that it raised, he had been aroused by personal indignation to far more energy over the epidemic than had most of the rest of us. In the late eighties in Nairobi and Kampala, he had seen friends sicken and die around him. Despite this, in the nineties he was still finding Westerners who claimed it was all untrue, and that there was no epidemic. Instead, false trails and absurdities were glibly promoted; hypotheses were floated that seemed aimed, even from the first, to lead into impenetrable bush. At the same time, as he found later, much better hypotheses about the epidemic were studiously ignored and had needed tortuous paths to achieve any public notice at all. The ideas and research of New York-based Louis Pascal, for example, had to be published in Australia, and the investigations of science journalist Tom Curtis went perforce to an outlet in a popular magazine, *Rolling Stone*. Neither piece was much followed up.

Without question it is science that will shape the human world of the Third Millennium. Even if science can only direct us back to a dark age it will still be our cause and our guide. But it could be made to do better or worse. There is a risk that science is going to lose its fertility and change radically away from that spirit of free inquiry and exchange that first inspired the Greek and then later the Renaissance experimenters and philosophers. Indeed, this process seems to be starting already; patenting and secrecy about gene sequences are perhaps one symptom. Science may bring on us not so much a dark age in the old sense, via some spectacular collapse, but rather a super-technological state whose monstrous futures—if they could be shown to us dearly through the present smoke of excitement about more and ever more technology—would only arouse our dread. While still working its miracles on the outskirts, science may already, at its center, like a great city, be slowly dying of its very success. Dictators and businessmen everywhere want to use all the technical products of science and, if possible, to control the rights and the how-tos for creating more. They would also like to be free to hide the results of their unsuccessful or disastrous experiments.

After reading Pascal's paper, it was a great shock to me that when I passed out copies to others whom I thought would be interested, including a journalist who had written on AIDS for a major popular science magazine, I met with exactly the wall of silence Pascal had described. From being at first impressed mainly by his theme about the origin of AIDS, I thus began to believe his arguments about scientific integrity as well—arguments that at initial reading had seemed to me just overreactions generated in a sensitive, frustrated man. Only one person (from the medical fraternity, surprisingly) replied to my mailing with any sign of taking the paper seriously. Even my old mother, a doctor, told me, 'You are going to be very unpopular if you pursue that one—polio of all things, that one is sacred! Anyway, if it's true, it's all happened and what could you do?' Well, personally I didn't pursue anything very far; after several tries with the editors of both *Science* and *Nature*, I lapsed back again into the general silence. Overall I have left it to Pascal, Curtis, Julian Cribb, and now Hooper. I have simply watched from the sidelines as each in turn has held aloft his blazing but strangely unregarded torch. However, I have become, with each new revelation, and particularly with the discoveries of Hooper, which you can now read about for the first time, more and more a convert to the underlying theme. The new facts in the case still tend to be widely separated and none by itself amounts to a proof; however, taken together the steady trend and accumulation has become very impressive. At the very least the OPV theory of the origin of AIDS now merits our acute attention.

I have pondered very much about what sorts of people should be encouraged to try which sorts of tests: Hooper also in the book gives his list. There are some that could be decisive. However, the factual case was already quite strong after Pascal, and the present situation adds up to reiterating that Pascal was also right in his other theme, and that very major questions need to be asked about why supposedly 'free' science has been so slow to listen to what should have been taken very seriously from the first. If the topic had somehow been far from Big Science and had lacked any implications touching on issues like politics and professional pride, I have little doubt that its questions would have been much more discussed and investigated by now. I very much hope this book will cause the questions to be asked and the tests to be undertaken, and that it will also stimulate a lot more of the kind of sociology and science critique which Brian Martin in Australia promoted during (and supportative to) the building of the present story. How much more useful his effort is than so much that is done under the name of the sociology of science!

Forensic high-tech analysis has been enthusiastically applied to the hair of a historic corpse, Napoleon, in order to try to separate the natural events, accidents, and malfeasance that might have played a part in his death. He was a great man by any standard and also, looked at a bit more sourly, was instrumental in causing hundreds of thousands of deaths. Most would agree that these attributes of Napoleon justify the considerable interest historians have in how he died. But this level of interest makes it all the more remarkable that another historical issue with already far more deaths to its tally, and its Waterloo not even in sight, receives currently only a single historian's effort. Vaccine vials, which are surely much more accessible than samples of Napoleon's hair, stay untested in the Wistar Institute freezers. Through turning a blind eye to the OPV/AIDS hypothesis, our establishment actively avoids testing and hearing about the plentiful though scattered evidence that the AIDS epidemic may have had a medical accident at its origin—an accident possibly compounded, more recently, by a desire by certain protagonists to conceal the evidence.

In getting together the materials for his book, Hooper has worked harder and for much longer than any of his forerunners. Several times he has countered my plea for a start on the writing by saying there just had to be this further trip to Belgium or that one to the United States. His work has amounted to more than six hundred interviews in all, he tells me, and this says nothing of the library research. I believe no one, not even a person 'speaking as a scientist,' is going to call this book 'the wildest of lay speculation'—the criticism that was leveled, even then unfairly, at Tom Curtis's much briefer accounts in *Rolling Stone*. If the OPV theory of AIDS origin comes to be proved, I think the new standards of *evolutionary* caution in medicine that their publications will eventually engender (especially regarding all treatments that use live products from other animals on humans) should merit for Hooper and Pascal jointly a Nobel Prize. As a species we ought to have known somehow in our culture, or even genes, that intimate invasions of live animal products, especially those coming from closely related species, are inherently dangerous. I have conjectured elsewhere that these dangers may be the main reason why separate species exist generally. That notion and what happens next in the present case are all in the lap of the gods. There are as stated, however, tests which can prove convincingly whether or not AIDS was our medical mistake. Meanwhile, Hooper deserves great praise for having so tenaciously carried through his investigation and for bringing to light so many more facts affecting the main question—facts that are almost all further challenges to the null hypothesis of 'coincidence only.' Even if the OPV theory is eventually rejected or remains permanently in limbo, he has done a great service in putting so many details of the early spread of AIDS on record. He has in fact given us the best history of the epidemic.

I have seen the cost the task has had for him manifested in many stages of tiredness, illness, and despair, which however he has always managed to overcome. Truly it has been like watching an explorer—Burton or Livingstone—making his halting progress toward some center of mystery that is far inland from the obvious coastal hills which we have all been seeing. Most strangely, as it may seem at first, his story wends toward exactly the same center of Africa as those Victorian explorers sought. This comes to seem a little less strange, however, once we reflect on our evolutionary origins. What dramas on all scales have been played out in the human population in the same geographic region, around the spine of Africa and in those places where the savannah and the forest meet. Almost all of these things were happening long,

long before there was anyone who could write or even speak about them. Upright we became... trying for new social structures, for tools, for speech, for fire... Finally out of Africa, our home, there came this new disease and on its heels, in this case, a *written* drama of *how* it came. Both themes are gravid with our future, and the written one is like Sherlock Holmes, Professor Challenger, Augustus Caesar, and Mark Antony all rolled into one.

Everyone should read this book, both for its story and in order to think hard on all that it implies—all this before Truth, more white and sick even than with AIDS, quietly rejoins us through another door.

CHAPTER 15

HAMILTON AND GAIA

(A preface to 'Spora and Gaia: How microbes fly with their clouds')

BY TIM LENTON

About, about, in reel and rout
The death-fires danced at night;
The water, like a witch's oils,
Burnt green, and blue and white.

From 'The Rime of the Ancient Mariner' by Samuel Taylor Coleridge

The story of my work with Bill unfolds with the spring of 1997, and I will tell it in his written words, where possible. I was in the third year of my PhD studies at the University of East Anglia in Norwich, and had been absorbed in an effort to reconcile what I was learning about the planetary effects of life and what I knew about the mechanism of natural selection. This was summarized in a review article I had written and (rather optimistically) submitted to *Nature* (eventually to be published as 'Gaia and natural selection'[1]). I was eager to resolve what I saw as some long-held misunderstandings between 'Gaiaists', as Bill called them,[2] and evolutionary biologists. I was aware of Bill's interest in Gaia from things he'd written[2,3] and through my friendship with Jim Lovelock. The two met in January 1997 and 'a bit unhappy about the way the discussion ended, seemingly in deadlock'[i], Bill then wrote Jim a long letter clarifying his views on Gaia.

[i] Quoted from letter, W. D. Hamilton to J. E. Lovelock, 19 January 1997.

I received a copy of this from Jim and an encouragement to send Bill a copy of my review, which I duly did.

Bill's letter to Jim helps set the scene for what came next. Bill accepted that there exists a planet-scale system with some remarkable stabilizing properties, but like most evolutionary biologists, he objected to the use of the words 'organism' or 'superorganism' to describe a system that isn't subject to natural selection: 'That n.s. [natural selection] [is] going [on] within and among the component life forms on the planet doesn't help to justify the term "organism", super or not. However, once we have a *principle* concerning to why system-stabilising outcomes in this n.s. are more likely than system-de-stabilising outcomes it is certainly justified to look for a term that has a fairly equivalent meaning and (if you like) dignity. Gaia seems to me good.'[1] Bill saw the search for principles as most important, and he speculated (with great prescience, I think) about their nature: 'I am hesitant myself as to whether when this set of principles is discovered it is going to involve n.s. in a big way or something else. I suspect one will be able to refer to n.s. but it won't be quite the idea as we normally think of it—vaguely I imagine that "learning" through *repetitions over time alone* in a sufficiently complex system has to be shown able to replace the currently understood (and I am sure much more powerful) "learning" through repetitions over both time and space, which is n.s. as we know it.'[i]

On 4 March 1997 I opened my e-mail box as usual and was delighted to find a message from Bill Hamilton that was both flattering and intriguing. Bill's attention had clearly been grabbed by something I had written about dimethyl sulfide (DMS) production by marine phytoplankton. This is a widespread trait that both of us were having difficulty rationalizing in adaptive terms. As Bill put it then: 'I always regarded Jim's ideas about DMS as one of the most challenging to make sense of in terms [of] EB [Evolutionary Biology]; all the feedbacks seem so diffuse and long range in space and time, very unlike Daisyworld.'[ii] However, Bill saw a ray of hope and had a host of questions about the DMS producers, which, after chatting with a local expert, Gill Malin, I tried to answer, sending him back a stack of papers. Unbeknown to me at the time, Bill had a wonderful capacity for finding needles of interest in haystacks of otherwise unpromising papers.

Clearly there was something exciting going on, but with customary care Bill wasn't revealing his hypothesis until he was sure there was something

[ii] Quoted from e-mail, W. D. Hamilton to T. M. Lenton, 4 March 1997.

in it. Then, on 27 March I received a long and detailed letter in which he revealed that: 'What caught my attention on a second reading of your ms was your saying that DMS was an "osmolyte" and that its original presence in the algae probably had to do with buffering the alga against changes in salinity of the ambient water and perhaps acting as an antifreeze. ... it sparked the following idea. What could antifreeze have to do with algae, many of them probably never going near to seas with freezing temperatures? Then I thought of them being whipped up occasionally off the water in waterspouts and the like and travelling high in to stratosphere. Then they might well need osmolytes to cope with (a) their ambient water droplet drying and becoming completely salt-saturated and crystallising around them, and (b) freezing if they were wafted high enough. They might reduce some of this if they were emitting a gas that could aid the formation of cloud droplets on and around them.'[iii] Furthermore: 'In the course of all this travel through the air, a more subtle benefit would have occurred to the alga managing to survive it: it would almost certainly normally be deposited at quite a distance from the patch of sea where it started'[iii] and 'dispersal is advantageous to organisms *per se*.'[iii] All of which led to the central hypothesis 'that *algae may travel with the clouds they cause to form*. And try to travel with them.'[iii] More generally 'clouds of certain types may be in part biogenic ... and they may be serving as dispersal vehicles, in different situations, for bacteria, blue-greens, micro-algae, and spores and conidia of fungi, some of these over land and some over sea.'[iii]

Bill was clear from the start that his hypothesis didn't offer the kind of Gaia principle he was after, but it did show how adaptations can come to have global effects. In his words: 'I originally regarded this supposed adaptive "feed back" as about the most hopeless of all that Jim Lovelock had proposed because it just wouldn't be possible to argue that the links were *evolved* to be the way they are—the system must be regulatory by luck. As a result of what I think are just passing comments in your paper I now see, to my surprise, much more hope. It all hinges on how we can argue that the *individual* DMS emitting algae, or else a *clonal* patch of such algae (if these exist which will be a very important question), can benefit *itself* by the emission it makes—not just benefit some other individual or patch. Thus I am just trying to make my normal kind of argument but even if I succeed we will just have another seemingly fortuitous case where the distant outcome happens to be

[iii] Quoted from letter, W. D. Hamilton to T. M. Lenton, 27 March 1997.

increased stability for the planet: there will be no *principle* in sight yet (at least to me) to explain why this case turned out to be stabilising rather than destabilising on a planetary scale . . . Things that impress me, however, are that (a) this DMS case seemed an unlikely one for an adaptive explanation such a distant object as a cloud and (b) if my explanation is valid it is obvious that the adaptation of the algae can indeed have an effect on a vast scale, fully capable of influencing world climate. In other words, one may not need many of such stories before one has a substantial part of the world climate explained plus (for me) a mysterious finding that most or all of the biological effects involved are stabilising!'[iii]

It was all wonderfully adventurous blue (and white) skies research to my mind and I eagerly joined in the hypothesis construction, search for evidence and testing of the numerous predictions that Bill was already making. Having been working on what to many colleagues was the rather eccentric topic of Gaia it was a great comfort to see such an eminent scientist display such a creative and bold scientific imagination, and not be afraid to air a radical hypothesis. Over the following months, I found the scientists I spoke to about the ideas falling into two camps: the sceptical and dismissive majority, and the open-minded minority. I was lucky that early on I caught the open-minded support of Peter Liss, one of my PhD advisers and leader of a group researching DMS. Peter pointed me to a body of research on the mechanism of bubble bursting at the sea surface and its potential to inject micro-organisms into the air. I also alerted Bill to the topic of biogenic ice nucleation (IN) as a possible descent mechanism, having read about its potential to trigger precipitation in 'The Ages of Gaia' (pages 86–87 of Lovelock[4]). For a month or so, both Bill and I spent a lot of time ferreting around in our respective libraries. I fed Bill with a steady stream of suggestions, papers, and back of the envelope calculations (of descent rates and the like), and waited to hear something.

Then, to my pleasant surprise, there arrived a letter and a first draft of a paper then entitled 'Spora and Gaia: do microbes fly with their clouds?' (As our confidence increased the question in the title was to become a statement.) There ensued a flurry of e-mail and letter correspondence. I spotted Andy Andreae and colleagues' promising results of ship-borne measurements correlating DMS in the water with condensation nuclei (CN) and cloud condensation nuclei (CCN) in the air above. It struck both Jim Lovelock and me that there could be a positive feedback between condensation-induced convection, increased windiness, and increased sea–air transfer of DMS, driving further convection. My weekend walks on the North Norfolk coast helped bring the subject out of the library and into life. That June, the sea was milky white with a spring bloom, and an abundance of *Phaeocystis* foam could be found blown onto the adjacent beaches, where exposed chalks served as a reminder of the antiquity of some of the organisms involved in our story. It was at this time that Bill mentioned 'The Rime of the Ancient Mariner', noting that: 'The mariner who must have originated the idea of Coleridge's poem was fairly obviously describing, in the episode of the long becalming within a stinking, slimy sea, the conditions of a great algal bloom. However, it wasn't in the Sargasso or the Doldrums but rather, according to the poem, somewhere far to the South. Wherever it was, may be that is where *Phaeocystis* really needs its DMS!'[iv]

We both had moments of doubt on seeing the weaknesses of our hypothesis for adaptive DMS production. The weakest link remains, I think, the time delay between DMS release and latent heat production in condensation. At one point Bill described how 'I had been becoming extremely depressed by that figure of 24 hrs which you say Prof. Liss gave you for the time it would take an emitted DMS molecule to be ready to play a part in forming a cloud droplet and last night I had a crisis of thinking we will have to abandon the paper until we can see a way to make the DMS side of the story make more sense... However, this morning some thoughts have occurred that make things seem a little more hopeful and since the field seems replete with theories that are launched on heuristic grounds long before there is any rationale for them—e.g. that of Gaia as a superorganism itself—perhaps we can go ahead.'[iv] On chasing the latest chemical kinetic studies I found that the chemical oxidation pathway from DMS to sulfate and potential condensation nuclei can be much shorter (e.g. at the time when mad dogs and Englishmen are out, under the midday sun). Jim Lovelock added that the alternative oxidation product methane sulfonic acid (MSA) will form much faster and combined with ammonium in the air will tend to form better CCN (cloud condensation nuclei) because it is a stronger acid.

After a trip to the Alps, during which I spotted some interesting aerobiology, including spiders lofted to the top of mountains, Bill and I finally met in Oxford in late July and finalized the paper. We decided to submit it to *Science* (as I was already embroiled in one long slog with *Nature*) and

[iv] Quoted from e-mail, W. D. Hamilton to T. M. Lenton, 26 June 1997.

needless to say, they rejected it without review. It found a rather unusual home in *Ethology Ecology and Evolution*, due to a long-standing promise Bill had made to the editor Francesco Dessi. The one referee, Steve Frank, was kind and we were thus able to publish without altering what we wanted to say. When the article came out the following March it generated such interest[v] that a high-profile journal outlet hardly seemed necessary. *New Scientist* ran a front-cover feature,[5] which Bill had mixed feelings about.[6] But this in turn led to interest from an independent television producer, Annamaria Talas, with whom both of us became friends. Annamaria helped encourage William ('Chad') Marshall (then at the Marine Biological Association in Plymouth) to test one of the links in our causal chain. Chad was filmed collecting airborne spray samples in Plymouth Sound and later culturing and analysing them in the lab. This demonstrated how readily many members of the plankton get airborne. Jos Wassink, from the Dutch television station VPRO, made a programme with Annamaria, and had the fun of taking cloud samples from a friends' light aircraft, using the same technique as aerobiology pioneer F. C. Meier, of an agar plate held out of the window at arms length. On leaving the plates in his fridge he produced a rich collection dominated by fungi, which (alas) were never properly analysed.

Others were stimulated to extend our DMS hypothesis, and Bill was particularly taken by Dave Welsh's ideas about the dispersal of macro-algal spora, which he described as 'a major unifying extension since I had worried about the macro-algae and have never liked explanations where a seemingly single adaptation actually has one interpretation here and another there.'[vi] We made minor contributions to the resulting paper[7] and Bill took the opportunity to correct a small error we had made concerning film and jet droplets. This we realized through a fascinating correspondence with Duncan Blanchard who pioneered experimentation on the transfer of micro-organisms across the water/air interface by bubble bursting and made some wonderful, serendipitous discoveries.[8]

Progress testing our adaptive DMS hypothesis has been slow, but our related hypothesis of adaptive biogenic ice nucleation in clouds has generated more interest. Bacteria have been found to be metabolically active in super cooled cloud droplets collected high in the Alps.[9] A project is underway in the UK to characterize the diversity of cloud-borne bacteria, to

[v] 'Kleine Götter.' *Der Spiegel* 18: 198 (1998), 'Life in the clouds.' *The Economist* March 28, 112–13 (1998).

[vi] Quoted from e-mail, W. D. Hamilton to T. M. Lenton, 28 August 1998.

detect whether some carry ice-nucleating genes, and to try and find out if they are actively ice nucleating in supercooled cloud droplets. Latest samples are dominated by *Pseudomonas fluorescens*-related bacteria, of which certain strains are known to have ice-nucleating capabilities (K. Walsh and B. Moffett, personal communication).

After finishing our paper, Bill and I continued to debate Gaia and search for principles. Bill saw the need for models but was rather dismissive of Daisyworld; it is a special case in that the daisies alter the environment in the same way at the individual level and the global level.[1] Furthermore, for Bill it would be: 'much stronger evidence if gaia-like properties emerge in a model that was conceived completely independently of the idea it turns out to illuminate.'[i] This fuelled his interest in the results from community assembly models.[2] He also suggested that I 'Make models of assembly of interacting macro systems, for example, some of which are assumed destabilising and some stabilising, and try to show that the former tend to lose out and be replaced by the more stabilising ones before the global system dies.'[iii] When we met in Oxford I took along some of the early papers on homeostasis in complex systems, e.g. ref. 10, which greatly impressed him. Soon after, Bill noted that: '. . . the literature seems to be saying that admission of new species (and perhaps mutant morphs of existing species in some papers?) tends to do a good job of building increasingly homeostatic purely biotic ecosystems, it seems quite likely a system including the physical background also would become more homeostatic.'[vii] We discussed how to extend the existing work: 'I think some modelling is needed because it looks as though none of it yet focuses the feed back of the whole growing community to the abiotic environment, making it more benign, more generous of its materials, etc., or less so.'[vii] Alas, I never had the pleasure of pursuing such work with Bill (having a PhD to finish on a rather different topic), but happily he did start working with Peter Henderson on a model that came to be known as 'Damworld'.

Damworld yielded interesting results and some two years after we first corresponded, I received an excited e-mail message from Bill that '. . . a Genghis Khan species may be less likely to be about to destroy life on the planet than I had previously speculated.'[viii] 'Even with an "unexpandable resource base" to the model we do find some accumulation of resistance to disaster from the next species added; and when the model is endowed from

[vii] Quoted from letter, W. D. Hamilton to T. M. Lenton, 25 July 1997.
[viii] Quoted from e-mail, W. D. Hamilton to T. M. Lenton, 12 March 1999.

the first with physical possibility that its resource base can be expanded if the right species are acquired, very disturbance-resistant communities sometimes appear in our model. There seems to be no "rapid" asymptotic disappearance of the chance that newly offered and accepted species will cause a vast catastrophe but on the whole things do get more unlikely to suffer disaster and also more short-term stable . . .'[8] These were just the kind of principles Bill had been seeking all along.[2] It seems an appropriate, guardedly optimistic note on which to end. As should be apparent, Bill made an important contribution toward the development of a principled Gaia theory. Had he lived longer, I like to think that he would have become the Newton to Lovelock's Copernicus.

References

1. T. M. Lenton, 'Gaia and natural selection.' *Nature* **394**, 439–47 (1998).
2. W. D. Hamilton, 'Ecology in the Large: Gaia and Genghis Khan.' *Journal of Applied Ecology* **32**, 451–53 (1995).
3. W. D. Hamilton, 'Gaia's benefits.' *New Scientist* **151**(2040), 62–3 (1996).
4. J. E. Lovelock, *The Ages of Gaia—A Biography of Our Living Earth, Second Edition*. (Oxford University Press, Oxford, 1995).
5. L. Hunt, 'Send in the clouds'. *New Scientist*, 28–33 (1998).
6. W. D. Hamilton, 'Cloud cover.' *New Scientist* (2140), 52 (1998).
7. D. T. Welsh, *et al.* 'Is DMSP synthesis in Chlorophycean macro-algae linked to aerial dispersal?' *Ethology Ecology and Evolution* **11**(2), 265–78 (1999).
8. D. C. Blanchard, 'Serendipity, Scientific Discovery, and Project Cirrus.' *Bulletin of the American Meteorological Society* **77**(6), 1279–86 (1996).
9. B. Sattler, *et al.* 'Bacterial growth in supercooled cloud droplets.' *Geophysical Research Letters* **28**(2), 239–242 (2001).
10. K. Tregonning, and A. Roberts, 'Complex systems which evolve towards homeostasis.' *Nature* **281**, 563–564 (1979).

ECOLOGY IN THE LARGE: GAIA AND GENGHIS KHAN[†]

W.D. HAMILTON

The existence of a balance of nature is folk wisdom but what is the evidence? How is it all supposed to work? In the history of ecology, predicting origins and extinctions of populations and all their changes in between has been like a holy grail. So such prediction remains including, sadly, the lost faith brought by time. Through Pimm's *The Balance of Nature?*,* we are better able to see what we are to be disillusioned with, what not and, as with the abandoned grail, we find out about many worthwhile quests that have been effected along the way.

In nature a degree of balance is obvious. This year will reveal roughly the same species as last year; and for most, even their numbers will be about the same in spite of attrition and reproduction. Observation also admits that many populations change, some widely and some permanently. Every naturalist knows examples from personal experience. The collared dove has come to Britain to stay; the turtle dove is fading: these, this book would say, are the 'reddened spectra' of change. As a whole its scope is the current state of theoretical ecology, mainly at the level of populations and species, and viewed as broad landscape rather than in detail.

The book is excellent; but for ecology as a predictive science its answers are rather gloomy. If the weather is difficult it was almost a foregone conclusion that the prediction of natural populations would be more so. New understanding under the surprising heading 'chaos' (the idea of exponentially divergent and yet bounded system trajectories of change) explains a part of this pessimism; but there is worse. Other organisms may be less wilfully and thoughtfully perverse than we are, but at the same time they are never like molecules of a gas. Molecular uniformity yields gas laws, rates of chemical reactions, classical physics: all very predictive as we know. Contrast with this a

* **S. L. Pimm** (1992) *The Balance of Nature?* Pp. xiii + 434. University of Chicago Press, Chicago. Price US$71.25 (hardback)/US$30.95 (paperback). ISBN 0-226-66829-0 (hardback); 0-226-66830-4 (paperback).

---

† *J. Appl. Ecol.* **32**(3), 451–453 (1995).

specimen of Pimm's dry humour invoking the osprey's stunning dive and crash into the water, its laboured rise and departure (with fish in his version) to its perch. All this, the ensemble of instances of such a dive, he comments, is coolly boiled down by the theoretician into $a_{ij}X_iY_j$, one term in a recurrence equation. The real situation is obviously vastly more complex than such a term and equation can represent. For a start we ought to be reducing to some much less tractable dynamic equation the cases where the osprey errs from stomach ache caused by a tapeworm or the fish falters in its escape because of a cold in the gills. Even without such k- and l-species added (and needing *simultaneous* consideration), ospreys, fish and parasites still never approach the behaviour of bouncing or joining molecules. The bird may have noted, for example, the territory holder of the lake watching, flashing the realization upon it that the fish may have to be relinquished even if captured; best perhaps for this osprey to wait, try for the same fish tomorrow... How could it have become itself the territory holder, free from such constraint? Again, this may be partly a matter of its innate resistance to various parasites. Theoretical ecology has still fewer analyses capable of coping with quality differences and social effects such as these than it would have for straightforward three-way and four-way simultaneous interactions. In short, whether or not most organisms can be excluded from the vagaries that beset beings that are thinking and planning in our sense, it is certain that they are packed with their own very radical complications which even the best weapons of the theoretical ecologist's armoury still cannot neutralize.

Accepting all this as true, should we not be the more impressed if any light shines through? Pimm argues that, bad as the situation may seem, it gives no reason for despair. Such a message is heartening from a man who is no stranger to the real complexity of nature. Just as with the weather, useful generalities emerge independent of the grail of prediction. None of this book is my field: my review must be taken as an outsider's impression only. Nevertheless, for such an outsider, the book was an illuminating and exciting read, tinged only occasionally with scepticism. It is difficult for me to imagine a better guide either to the theories currently in play or to the current state of the evidence. The section on food-web theory gave particular pleasure, enlightenment and disbelief strangely mixing, as I saw a kind of erector-set technology appear as if from empty air: food webs, it seems, share some construction principles with girder bridges! Triangles are again at a premium.

Experts might find some of the efforts for clarity in the book bordering on the repetitive. R. A. Fisher would have written at less than half the length; but he would also have waited more than twice as long for biologists to understand him. For myself, the summaries of progress and the lack of need to refer back constantly were definitely assets. The text is well balanced between theory and evidence. As stated, it is obviously a strength that the author has experience of both sides as well as possessing a vast command of the literature. His approach to modelling is one I strongly endorse myself. He deliberately avoids time-consuming and sometimes misleading detail in favour of simple guide posts that lead on to further work. These are furnished often by bare-outline quantifications whose aim is mainly to back up or destroy intuitions and so to feel the way ahead. As one example in this style, it is shown how an analysis by Schaffer and Kot[1] turns population ecology's oldest example of 'control', by density-independent and quasi-random physical factors (supposedly mostly the weather), into a very promising example of deterministic chaos. Trajectories of the system fall on an odd-shaped but *thin* attractor; they do not reveal the expected fuzzy clouds of true randomness. The data concern an Australian thrips and are due to Davidson and Andrewartha.[2,3] Thrips chaos may be no better than weather chaos for prediction, but philosophically the change is very important and may eventually lead, if this be needed, to better ideas about control.

Pimm's own models are equally exciting. In addition to some that might be described as mere substantiation and partial quantification for various guesses and common-sense ideas (e.g. communities being easier to invade when simple than when complex, their specialists being in more danger of extinction than their generalists, and so on) there are other more unexpected and challenging results. Thus, I urge to the attention of Gaiaists Pimm's finding that a successful invasion was just as likely to cause a big cascade of extinctions in a late stage of development of a simulated developing community as it was in the early stages. These Post and Pimm[4] Gaias certainly seem to acquire more stability in the sense just mentioned: that of progressively increasing their resistance to invasion; but the protection the system accords to its members is frail and in any case cannot be said to evolve in the normal sense. Gaiaists might profitably search the book and its references, however, in other places, as where other possible emergent properties of real microcosms are tangentially discussed. Indeed, it seemed to me a pity that Gaia is not addressed more directly in the book: right or wrong, Gaia presents the claim for an evolution of a supreme 'balance of nature'. Authors cited in the book seem to be showing the same neglect. In the elegant work of Drake on *in vitro* community assemblies (aquarium microcosms) no drive is evident to determine whether systems became more resistant to disaster, or more homeostatic in other ways, as time and the series of introductions went on.

More to the point, however, one may ask whether Gaia proponents themselves are studying development (evolution?) of microcosms in order to test their concept. If they are, in the theoretical versions are they providing random mutations and properties, rather than particular biased sets that are likely only to promote stability, as was the case with the black-and-white daisies in Lovelock's only model? Invaders that have, contrary to the daisies, the 'ecological habits of Ghengis Khan' in Pimm's phrase, need also to be considered. One curious species, which seems well to fit the image of Mongolian shepherds transforming suddenly to the world-predatory and all-destroying armies came to my notice from a recent news item. The phytoplankter, *Pfiesteria piscimorte*, is at times a primary producer but at others becomes group predator on the topmost normal predators (the large fish) of its estuarine ecosystem. How will such a species integrate, one wonders, into Gaia's benign flock—or integrate into any food-web theory for that matter, which it challenges just as much? Gaia already admits to one Genghis Khan, of course. This is *Homo sapiens;* but I have the impression that Gaia's proponents see human properties as reflecting an out-of-touch and overweening intelligence, in other words that special perversity of *Homo* that I alluded to above, which, presumably, unicellular *Pfiesteria* does not parallel.

It interested me to find Pimm reviewing, with ambivalence similar to my own, the triangle of simple-analytical, complex-analytical, and computer-simulation approaches to modelling. As he does in quantitative ecology, so do I in genetical and evolutionary problems. Both of us dream of simple models able to explain and predict everything; of those inverse square laws of universal ecology that will never be. Far short of this, however, we are both made reasonably happy with guide-posts, hints of tracks in the forest; and hopeful that these will soon lead to other signs pointing the same way. Yet increasingly the landscape ahead is so dense and entangled and, given the time for real-life field studies or experiments, is so seemingly hopeless for analysis, that simulation of fairly complex models seems the only way ahead. Rough chainsaw clearings as such models must be, they can show where it is worthwhile to look for the real paths. The long study by Post and Pimm,[4] already mentioned, is an example of this approach and it is perhaps characteristic that I found its results among the most interesting in the book. As pointed out in a recent review,[5] simple *analytical* principles, expressible by neat formulas on a page and forming parallels to the great predictive principles of physics, may really be unattainable in ecology. Yet this is far from saying that principles do not exist. A statement like 'Invasion becomes more difficult as a community matures...' is still a principle. Even if there should be no formula or even rationale for it, it is worth knowing. It becomes the more so, of course, if simulation or/and the natural evidence can add some quantification like '...with the difficulty increasing approximately as the square root of the elapsed time or the number of attempted invasions'; and, still more so, it is worth knowing if a clear rationale can also be given. This is the ideal. Meanwhile, desperation to know and contentment with the partial enlightenment of simulation represent, in my opinion, a more scientific spirit than the purist's demand that there must be a proper analysis and a fully intelligible theory or else nothing. This is especially so when the purist in practice is forced simply to abandon whole areas of the research forest as impenetrable. Chaos theory, which I rate as perhaps the century's greatest advance in scientific thought, shows that some of those areas from which the classical methods of the exact sciences at present simply avert their attention (e.g. from multilocus models in my own field), may have to be recognized as impenetrable by those methods for ever.

My main criticisms of the book, as of its underlying discipline, concerns its relative neglect of two topics: the genetical and evolutionary change occurring within species; and parasites and microbes as actors.

Some concessions are made by Pimm in both these cases. On the genetical issue, attention comes in the space devoted to inbreeding and drift in small populations. However, this seems to be picked up only because it is becoming increasingly obvious that genetic impoverishment sets points of no return to small populations and is thus part of the prelude to extinction which is a major focus in the book; the topic is not treated, as it could be, as just one facet of pervasive local and short-term genetical adaptation. Moreover, even inbreeding and loss of variability are not covered with quite the care which I hope applies in Pimm's more familiar fields. Thus I comment that, contrary to his account, *increasing* inbreeding is indeed a problem for populations in its exposure of deleterious recessives; but, once made *permanent*, inbreeding soon empties the heterozygote pool of its rubbish and, if the population can survive the crisis of the emptying, it thereafter has no worse consequences than outbreeding: or at least it should not from this purely 'recessive' (and I would add certainly inadequate) point of view. This is a mere detail and the passage as it stands must be counted much better than total neglect: on the other hand, theoretical ecologists should be focusing and understanding these issues more sharply. Populations and species cannot be treated as entities resembling molecules any more than individuals can. As much is sometimes admitted, as where Pimm briefly alludes to the susceptibility which Amerindians showed to the diseases of Europeans on first contact; but this too, surely, ought to be treated as a part of a much wider theme. On these lines an island, or simply a more peripheral population, may have quite different properties from a mainland or central one; and this is especially so with regard to the diseases that populations have evolved to resist. It is understandable, for example, that mainland species will invade offshore islands more than island endemics re-invade mainlands. The Goths and Napoleon are exceptions, not the rule.

A parallel treatment of the problems of gene demography and of classical ecology could have brought mutual enlightenment. In reading I noted many counterparts to ideas that have been forced on me by an interest in gene extinction and in dynamic genetic polymorphism. Thus, Pimm's discussion of *persistence* of species in a community closely parallels the population geneticist's topic of *protection* of alleles within a species or population. In the genetic case, overriding the question of the stability of polymorphisms comes the question of whether or not a system of selection protects all its alleles from extinction. I noted several passages in the book which recognize how a metapopulation of communities, via asynchrony of the component dynamic events and repeated reintroductions, potentially preserves species more securely than could happen if all communities were merged (a key phrase here is 'checkerboard distributions' but there were also others), a theme which is familiar in my own study of the parallel problem far alleles. The question of whether we should expect strict stability, that mainstay of formal (Lyapunov) analysis, or should expect simply something looser but still held within bounds, is addressed several times. In my opinion we probably can't expect much true stability in realistic population genetics or ecology, and can expect still held less when the two are conjoined. The very existence of Mendelizing sex tends to confirm this: if accepted, such a view encourages a recourse to simulation, since there is nothing else. We are everywhere finding ourselves forced away from the comfortable neighbourhoods where linear approximations and the classical methods for determining stability are sufficient.

Finally, my second worry: a subject not covered. 'Disease and parasite introductions, consequences of': that is all there is in the index: one entry and one page number. Words such as 'parasite' and 'pathogen' have no entries. To my mind, parasites and their effect on the health of hosts, in every sense from death to dominance-hierarchy, have great consequences in every aspect of life. I see their influence spilling in force right up to the scale Pimm is considering. Effects of parasites must be as pervasive in such 'pure' species ecology as I believe them to be in ecological population genetics, and no account of quantitative ecology is ever going to be complete without them. Yet in ecology still I see far more talk of predators than of parasites, with only an exception for the increasing role being recognized for small herbivores such as insects in the ecology of perennial plants. Parasites are why there is sexuality and, proceeding from this, they underly also the construction of social behaviour. The reviewer's opinion over this matter was perhaps already sufficiently apparent in his choice of the additional potential species to bear on the osprey–fish example; therefore, let me leave it at that, merely claiming that in this direction will lie the wave of the future. For the present I should confess myself again to be only a naive reviewer and yet add that, in its summarizing and extending current thinking on population ecology, both in its theory and its evidence, this book is far the best I know. It seems to me that every ecologist who cares about quantitative species relations and ecosytem development should read it.

References

1. W. M. Schaffer and M. Kot, Chaos in ecological systems: the coals that Newcastle forgot. *Trends in Ecology and Evolution*, **1**, 58–63 (1986).
2. D. Davison and H. G. Andrewartha, Annual trends in a natural population of *Thrips imaginis* (Thysanoptera). *Journal of Animal Ecology*, **17**, 193–199 (1948a).
3. D. Davidson and H. G. Andrewartha, The influence of rainfall, evaporation and atmospheric temperature on fluctuations in the size of a natural population of *Thrips imaginis* (Thysanoptera). *Journal of Animal Ecology*, **17**, 200–222 (1948b).
4. W. M. Post and S. L. Pimm, Community assembly and food web stability. *Mathematical Biosciences*, **64**, 169–192 (1983).
5. O. P. Judson, The rise of the individual-based model in ecology. *Trends in Ecology and Evolution*, **9**, 9–14 (1994).

SPORA AND GAIA: HOW MICROBES FLY WITH THEIR CLOUDS†

W.D. HAMILTON AND T.M. LENTON

Abstract

We hypothesise that marine algae and various common microbes of the atmosphere (spora) use chemical induction of water condensation to enable or increase their wind dispersal between their aquatic, terrestrial or epiphytic growth sites. Biogenic chemical cloud condensation nuclei (CCN) and ice nuclei (IN), sometimes co-occurring in single species (e.g. *Pseudomonas syringae*), release heat energy of phase change, thus contributing to local air movements that can be used both for lofting and for lateral dispersal of their producers. The phase-change catalysis may occur on the microbial surface (e.g. *P. syringae*) or may happen more distantly through the release of chemical precursors for suitable ions (e.g. plankton-derived dimethylsulfide [DMS] forming atmospheric sulfate). Small phytoplankton and bacteria take off from water through bubble-burst processes especially in 'white-caps', these often themselves caused by convective winds.

Selection for local induction of wind is likely to be most effective at the level of clonal microbial patches. Algal blooms having high DMS emission may represent attempts to create winds for dispersal; if so, algal cell changes occurring in such blooms may have features convergent to terrestrial dispersal cyclomorphoses of aphids, locusts, subcortical insects and others. It is already established that biogenic cloud formation occurs on a scale fully capable of affecting world climate. This fact features prominently in the Gaia Hypothesis. However, in contrast to the evolutionary scenario for microbe dispersal that we present, the claim of an *adaptive* function at the world level still lacks an explanatory mechanism.

† *Ethology Ecology and Evolution* **10**, 1–16 (1998).

Introduction

Herrings falling with rain miles inland in Scotland, frogs and a juvenile turtle being found in American hailstones,[1] and live bacteria and fungal spores collected by rocket more than 50 km from the Earth's surface[2] all demonstrate that both terrestrial and marine organisms are sometimes raised very high by extreme atmospheric events. Allergies and airborne diseases further prove the troposphere to be, in most places, a dilute suspension, or 'spora', of microbes, viruses and pollen.[3] All these organisms owe their presence in the air to atmospheric instabilities. The basic principles are physical but some of the details may be biological. In particular the seeding of condensation either to water or to ice, which greatly adds to the local energy of convection, has for a quarter century been suggested to be, even for the world as a whole, mainly biogenic.[4,5] Whether causative organisms *adaptively evolved*[6] their propensity to seed cloud formation remains controversial,[7] there being grave problems with all the selection scenarios so far proposed.[8–10] Here, based upon the universal imperative of organisms to disperse, which applies even if new habitat is never better than old[11] we suggest a new hypothesis that some microbes have evolved to seed cloud formation to create local dispersal vehicles for themselves, winds and clouds.

Lofting Microbes from Water: Bubble Bursts and Convection

When wind over an expanse of water reaches a velocity of about 20 km/h, wave tops start to break. Rising bubbles in the 'white caps' concentrate planktonic microbes at their lower surface; when the bubbles break these microbes, sometimes aided by cell-surface hydrophobicity,[12] are tossed into the air within minute droplets formed from the burst surface film. Their concentration in such drops is almost always elevated over that in the surface sea water, sometimes by more than two orders of magnitude.[12–15] The bubble-burst mechanism of take off which has been principally investigated for bacteria[12] must also work for microalgae that are significantly smaller than the bubbles. Since the size range of bubbles in breaking sea water is 20 μm to over 1 cm in diameter, with peak concentration at about 100 μm,[12] there is scope for lofting small algae, although they will not be concentrated or ejected to the same degree as bacteria. Direct evidence for algal take-off seems lacking but various studies suggest it is common. Working with inshore and oceanic sea water from latitudes between about 42° and 32° in the Atlantic, including in the Sargasso Sea, Wallace and Duce[16] (1978) found about half the organic matter in surface water was transported to surface froth by induced bubbles. Conditions for white caps on the sea are also conditions for air turbulence above it and the updraughts must catch some of the algae thrown from bubbles and lift them higher. In another study 20% of the particulate organic matter in air 15–20 m above the Sargasso was found to be different from the rest. Firstly, its carbon was sea-derived, and, secondly, its particles were larger, the diameters ranging upwards from about 0.75 μm.[17] The fall rate for a 20-μm spore or pollen grain is only 0.5 cm/s[18] and microplankton cells important for our argument (see below) are mostly smaller (e.g. *Emiliana huxleyi*: 5 μm while eucaryotic picoplankton range down to about 0.5 μm). Since pollen grains of small anemophilous weeds such as *Plantago coronopus*, with a 20-μm diameter, liberated from only a few centimetres above ground level, are found in high spora,[3] it is no surprise that microplankton are also present there[19,20] with potential to be carried for long distances.[21,22] Actual wind transport of marine algae[23] is most striking when toxic dinophytes in marine aerosols blowing onto land cause severe skin and respiratory problems to humans[13,24] and even occasionally cause death to livestock.[25] However, turbulence forced by wind over an uneven surface is not the only lifting mechanism. Surface plankton that intercept sunlight warm their water layer and transfer heat quickly to the air above.[26,27] Hence even in initially still air, surface plankton patches tend to cause thermals to form above them. Rising like huge bubbles[28,29] or columns,[22,30] and seemingly, again, with a propensity to concentrate spora at their lower margins when these exist,[3,31] thermals lift spora high in the troposphere. They initiate clouds[3,18,28,31,32] and in the process release latent heat, so drawing more air from below. If the atmosphere over a warmed sea happens to be metastable on a large scale, the effects of condensation may be very dramatic: tropical storms, waterspouts, even monsoons, may be started.[27] Above the storms, convective cloud tops may go as high as 15 km and inject spora well into the stratosphere.

The crucial effect in our evolutionary interpretation, however, does not require extreme events or high altitudes. It relies on the fact that thermals and cumulus clouds almost never rise vertically and that air generally has lateral motion.[30,33] A tiny organism or group which could cause a thermal, be launched into it, and which could later help to seed its condensation[34] so as not only to release more lofting power for its vehicle (via latent heat) but also to provide itself UV shelter and eventually to begin the aggregative condensation that may return the organism to the planet's surface,[34] would be very effectively dispersed. Whether or not adaptive causation of this kind is believable, the literature in plant epidemiology[35] and entomology[21] abundantly attests that spores are distantly and successfully transported. Sometimes spores even survive transit across oceans,[22] but that is clearly exceptional. From here on we expand only the hypothesis that certain classes of microbes may have evolved to add to the effectiveness of atmospheric transport, most commonly with effects ranging from a few tens of metres to tens of kilometres.

Dimethyl Sulfide and Local Wind

Consider the marine examples again and the controversy about 'adaptive' dimethyl sulfide (DMS). Lovelock and co-workers suggested that emissions of this gas by various marine algae, especially the abundant bloom-forming micro-plankton in the unrelated groups Haptophyta and Dinophyta,[36] might seed cloud formation on a planetary scale and thus affect world climate.[37] Correlations between marine DMS emissions and the concentration of cloud condensation nuclei (CCN) over parts of the South Atlantic and the Southern Ocean support the first part of this proposal.[38,39] However, fundamental gaps remain in our understanding of this biosphere-climate interaction,[40] not least concerning its biological basis. The precursor to DMS in marine algal cells, dimethylsulfonioproprionate (DMSP), is plausibly evolved as an osmolyte buffering plankton cells against salt concentration changes and sometimes ice damage.[41,42] However, the step of claiming that an initial side effect of DMSP synthesis—namely DMS production—was adaptively seized on in some way to regulate the climate of the planet, is controversial. Indeed in the claim's simplest form a *darwinian* mechanism for additional expenditure for such an end can be excluded: a benefit that increases the welfare of an entire group considered in isolation does not increase the frequency of a causative element.[43–45] Worse, if creating the benefit had a cost, the gene causing the group benefit definitely declines.[46,47] We thus need a selection process at a lower level, although not necessarily at that of the individual.

Before DMS emitted by an alga can release the latent heat energy of condensation various processes must occur including sea–air transfer, oxidation to sulfate, nucleation to form aerosol, growth into condensation nuclei (CN) or the larger CCN, and finally advection of these nuclei into supersaturated air.[48] The potential for an alteration of wind speed in the region of the responsible organism seems at first to be small because the oxidation of DMS alone is widely thought to require a period of the order of a day[49] and this must be followed by the other processes, estimated to take a total of over 3 days.[50] During this time an air parcel may typically have travelled hundreds of kilometres. However, these figures refer to means and the earliest formation of CCNs could be much sooner. Moreover, much faster routes have been conjectured with a first and highest peak of H_2SO_4 in only 6 h.[51] The simultaneous measurements of Andreae *et al.*[39] over the tropical South Atlantic show a strong correlation between sea-to-air DMS flux (as also atmospheric DMS concentration) and CN concentration, with over 60% of the variance of CN attributable to the DMS. As mentioned by these authors (and later reiterated by Andreae and Crutzen[40]), this adds weight to various proposals that DMS oxidation, aerosol formation, and particle growth can sometimes occur much faster than had been assumed. Especially on sunny days with their high midday peak of radicals in the air, just when local convection is strongest, there seems a good chance for some situations in which convection due to algal DMS emissions generate a local increase of wind speed in a matter of hours. If white tops are augmented or initiated by this increase, then the take-off process already described can potentially pay back to a DMS-emitting alga at the individual or clonal patch level of selection (or to individuals via inclusive fitness) an extra possibility for causative genes to become airborne,[28,36] and to disperse rapidly away.[13,19]

Dispersal and Cyclomorphosis

Dispersal is extremely important to life, indeed for self or progeny it can be considered an organism's third priority after survival and reproduction. It remains a crucial necessity even if growth conditions at the point of landing are never better than at the point of take off,[11] although if they are sometimes better the incentive is of course all the stronger.[52] With data on algal relatedness in plankton patches and also on the scale of the biophysical effects, the very adverse calculations presented by Caldeira[9] referring to adaptive alteration of climate by DMS might be modified to seem much less hostile, at least as applied to adaptive cloud formation of some kind. The selection at the individual or patch level would not be strong[47,53] but assuming several hundred million repetitions (that is counting algal blooms in both spring and autumn) to have occurred since, say, coccolithophores formed chalk in the Cretaceous period (while a still higher figure could be suggested for the dinophytes whose record extends to the Archaean),[36,54] weak positive selection could easily accomplish what is found.

Under such a scenario other facts concerning algal DMS emission fall into line. Various workers have expressed perplexity at the rather low and variable correlations in blooms between DMS production and chlorophyll concentration;[40,49,55] algae, it seems, are sometimes not forming DMSP or else are storing it. Andreae[49] in particular has noted the puzzlingly high emissions from tropical oligotrophic seas where there are few algae. The *Emiliana huxleyi* strain from the Sargasso Sea proves innately a higher DMS producer than strains from temperate seas.[49,56] Algae in such warm sunlit water with low mean wind velocity may have both special needs and extra chances for seeding local convective disturbances. Air samples collected during a cruise in the South Equatorial Current of the Pacific captured nitrate ions and particulate matter believed to be of planktonic origin in much higher concentrations at 1200 and at 1400 than at 1000 h during the day.[57] This would accord with both a convective dependence on wind and also with possible wind causation. Sulfur compound emissions from the sea were believed (but not tested) to be peaking in parallel. These facts further support the suggestion above of rapid processing of DMS to form CN and CCN. For comparison, similar samples from the same cruise collected in latitudes 40 °S and 60 °S showed only slight midday nitrate ion increases.

Regarding the combination of bloom formation with the toxicity often accompanying in the set of DMS emitting algae, we note that (i) blooms commonly occur in marine frontal systems where nutrient rich and poor waters mix, conditions encouraging immense multiplication and, probably, clonal patch formation,[58] (ii) again given such clonal conditions, which even unicells might be able to recognise,[59] local resources may be further adaptively exploited through group predation[60] or dasmotrophy,[61] thus increasing the chance that a communal emission will be powerful enough to trigger a dispersal event, (iii) differing temperatures of frontal waters may help create local tropospheric quasi-instabilities that will be responsive to seeding, (iv) plankton in blooms concentrate especially in the extreme surface layers, as appropriate for take-off,[54,58,62,63] (v) as plankton exhaust local nutrients they may switch from using nitrogen-based osmolytes to DMSP[55,64–66] and thus, via DMS, to the chance to initiate or augment convective events. Findings that DMS release greatly increases when phytoplankton blooms are ceasing growth[64,67] or are being grazed by zooplankton[56,68] are consistent with a dispersal function. In the case of predation the emission increase may represent not so much a direct effect of ingestion by zooplankton[55] as an adaptive reaction by the phytoplankton to a perception of predation, bacterial attack,[55,56] or virus presence[63,69] in their environment, indicating a time ripe for dispersal. Many fast-breeding organisms of other habitats use cues of adverse factors to initiate production of their defensive, dispersive or resting forms, or to begin other facultative changes (e.g. to induce wings in aphids; see ref. 70). Cyclomorphosis induced by predators and grazers is known[71] in Insecta,[72] Cirrepedia, Cladocera, Rotifera, Bryozoa and in both zoo-[73] and phytoplankton.[74,75] (In phytoplankton successions of morphs or 'phases' are indeed general; see refs 36, 76). Predator-induced change also occurs in higher plants.[77] Tests like those used to prove these cases to be facultative over a series of generations could demonstrate the suggested chemical 'cyclomorphosis' of phytoplankton, or at least show individual changes in DMS production. If aggregations are largely clonal, selection of altruistic DMS-specialising morphs are a further possibility. More generally, chemical morph differences do indeed sometimes affect chances of microbe take off, as shown by Blanchard and Syzdek[78] with the bacterium *Serratia marcescens*. A red strain was several hundred times more concentrated in bubble drops due to its higher hydrophobicity than a white strain while a pink strain was intermediate. This suggested a single compound mediating both effects but instead, in later work, redness and hydrophobicity were proved separable and not to depend on the same molecular pathway.[79,80] Thus, as is actually more likely for the different features of an adaptive morph, they appear to be due to some general switch occurring under conditions hostile to growth.[81] On this view the redness (due to the pyrrole, prodigiosin) may, as with various other organic pigmentations[82,83] be for UV protection for cells destined for exposure to sunlight. The same explanation for colour change might apply to 'red tide' blooms generally.[84]

At least two of the above characters revealed by bloom-forming algae (toxin production and rise to the water surface in suitable weather conditions) apply also to blooms of cyanophytes. Procaryotes of this group seem to be fairly common in spora over land,[23,85] but, rather surprisingly not over the sea. Although on the whole rather few have yet been tested,[86–89] marine cyanophytes seem to show little production of CN-potentiating sulfur compounds.[90,91] However, a set of freshwater species belonging to genera well-known in the aerial spora[23] are indeed producers.[90,92] Generally it further fits to our theme that small ('ultra-') plankton, including seemingly some cyanobacteria, are greater DMS emitters than large plankton. Thus the highest DMSP concentrations recorded for dinophytes in the list of Keller *et al.*[89] come from small species (in *Amphidinium* and *Pro-rocentrum*) and the two most negligible concentrations are from two large species (in *Ceratium* and *Pyrocystis*). With $<50\ \mu m$ and $>100\ \mu m$ given respectively for maximum dimensions in these two pairs,[76] the two former genera provide subjects obviously more suited for lofting in bubble droplets.[12]

Terrestrial Cloud Formation, Ice Nucleation and Descent

Little evidence exists for important natural sulfur emissions over land and fresh water apart from that just mentioned for some cyanobacteria.[90,92] Sulfur on land is generally scarcer and in any case DMSP as an osmolyte must be relatively unneeded except possibly in salty habitats.[93] However, a rapidly expanding theme of terrestrial microbiology, ice-nucleation (IN)[94–96] suggests alternative adaptations for cloud transport and especially for its termination.[34] Although IN and CN properties sometimes coexist in the same organisms,[34] ice nucleation is usually more important as the trigger for precipitation than for forming clouds. This is because the minute water droplets in cloud grow too slowly by aggregation to form rain. However, if seeded by minute ice crystals (or particles capable of initiating them), the disparity of the equilibrium vapour pressures of ice and water causes the ice crystals to grow rapidly at the expense of evaporation of neighbouring cloud droplets. The resulting crystals soon achieve substantial fall rates and then accrete more droplets by collision. If at lower altitudes ice crystals melt, their fall and further enlargement continues as rain.

Spora may often be in danger of remaining airborne too long and consequently being damaged by radiation and freeze drying.[34] The danger will be most acute for very small organisms like bacteria. Although some spora collected at 48–77 km altitude by the Russian rocket were still alive[2] germination percentage generally falls with altitude of collection.[3] Thus the smaller the species, the greater is its likely advantage from CC and IN abilities. Most recent research on biogenic ice nucleation refers to a few species in three genera of bacteria (*Pseudomonas, Xanthomonas*, and *Erwinia*) with the first and last presenting CC ability too.[34] However, a few species (both saprophytic and pathogenic) in the hypomycete fungal form genera *Fusarium*[91,97] and *Penicillium*[32] have revealed similar ability. The IN temperatures for the best strains of these various cases range up to −2 or −1 °C making them amongst the best IN agents of any kind, organic or inorganic. If IN ability is not adaptive it has to be a very unusual chance coincidence that *all five last-mentioned genera are well known in high air spora*.[3] Their IN[32,98,99] and in some cases CN[34] properties, presumed to be maintained during transport, have already been speculated to be important in causing rain or snow; protection during flight and return to earth have been adaptive riders to the suggestion.[34] However, as with DMSP in the sea, there is a more individualistic explanation for IN in these terrestrial examples than through events in the troposphere. All five of the mentioned genera are also known (through at least some members) as blight, fungal-spot and superficial root-rot pathogens on plants. When these species attack living plant surfaces, a means of initiating frost damage at high sub-zero temperatures may aid subsequent exploitation. In the high aerial stages too the case for immediate benefit is stronger and more 'individual' than it is for DMS because the ice-nucleation is by molecules seemingly inseparable from the cell surface, perhaps even facultatively exposed there.[100] (DNA sequence analysis suggests that in the bacteria the proteinaceous IN products derive from a single origin[94] but the compounds used by *Fusarium* appear to be different[101]). Each cell possessing IN molecules could help itself, first (if in a frost-prone situation) to obtain nutrients and grow on its plant; second (especially where cloud seeding may occur direct into the ice phase) to gain a cloud-cover when dispersing; and, third, to be rained, snowed or hailed back to earth,[4,34,99,102] where, fourth, it is likely to encounter suitably damp and rain-cleared leaf habitat on which to grow.[103]

Ice-nucleating bacteria are also present in the sea, often mixed with DMS emitting algae.[20,98,104,105] Algae from an inshore bloom were particularly effective ice-nucleators,[104] perhaps suggesting bacteria originating from land.[105] In summary, putting together terrestrial and marine data, many but not all hints in the literature support a hypothesis that biogenic IN can adaptively aid aerial transport and descent of spora. More speculatively, the highly variable and possibly facultative expressions of CCN and IN abilities of spora[100] (which, in fact, for IN abilities of leaf-surface bacteria is observed changeable over a period a few hours[106]), may be connected with the puzzle that the ascending columns that form cumulus '...do not, always, have a fixed source at the ground, but must often carry their energy supply with them as they move'[33]; that energy supply might be latent heat of water phase change facultatively released by microbes.

So far *Cladosporium,* the terrestrial genus whose spores are usually most common in collections over both land and sea[3,32,102,107,108] have only revealed a rather marginal IN ability[32] while the next most common genus, *Alternaria,* has not yet shown any. However, it is not clear that tests have yet been done on conidiospores rather than on vegetative mycelium. Perhaps significantly, IN ability was noted rising in older cultures of *Fusarium,*[91,97] and also was only found weakly in one replicate of one sample of some tropical isolates tested.[91] Since the mainly saprophytic forms such as *C. herbarum* (anamorph of *Mycosphaerella tassiana,* a common secondary agent of leaf decay, and the most commonly reported species in spora) might gain no advantage from ice-nucleating in already dead leaves, it is important to test spores directly.

Back at earth level and once again considering take off, we note that many observations suggest moulds and fungal pathogens of genera found in the tropospheric spora release spores just when thermals are likely.[28,109,110] Trees appear the most typical sources and even at herb level, sporulation in a 'canopy' is reported best for take off.[111] The situation for bacteria may be similar;[4,103,106,112] some, together with yeasts, seem to favour and perhaps cause foggy conditions. This may give low-level dispersal directly[113] or, in the case of nocturnal and morning radiation mists, it may hold spora in readiness for more distant convective dispersal later in the day.[111,113]

Co-Dispersal

So far we have related facts relevant to both individual and group dispersal strategies within a species. Polymorphism suggesting either individual facultative changes, altruism or parasitic adaptation is indicated by bacterial species having IN ability only in a small minority of the individuals in cultures.[86,87,114] In such diffuse dispersal enterprises as we are suggesting much literal 'free-riding' is unavoidable. Physical linkage to coclonal cells is a dubious way to combat a 'parasite load' in passively wind-dispersed organisms because of the increased settling rate of larger units, but one haptophyte alga, *Phaeocystis pouchetii*, best known of all plankton for its DMS production, has a special bloom phase where it is colonial in a gelatinous and often foamy matrix. This phase seems well adapted to blow low across the often windy oceans where it occurs and it certainly does so.[36,65] In conditions appropriate to our theory many algae form slime.[115] Some eukaryotes like *P. pouchetii* as well as some cyanobacteria form gelatinous colonies and of these some (e.g. diatoms, see ref. 116) are known to increase foam. Suspicion of similar causes extends to oceanic foam generally.[117] Marine wet aerosols blowing on land may contain higher plankton biomass than the surface water from which they derive.[118]

More interesting, however, is to consider whether several species might contribute in different roles to an event of convective dispersal. Thus, in the mixtures of algae and procaryotes in the sea which Schnell and Vali[20] have documented, we might imagine the microalgae providing the winds and the accompanying IN bacteria travelling with them providing the ice nuclei that get the team back to the sea after dispersal. This sets up an unlikely but not impossible scenario, its key necessity being that the mixed-species 'team' normally remains together throughout the journey. The reason is best seen by looking at alternatives. If due to differential settling by size, two species that were emitted together by a patch tend to land separately, then DMS-emitting algae will often have lofted bacteria that fail to assist the return by rain, indeed which abandon them to slowly lose even their own within-species correlation. Likewise good IN genotypes of bacteria will often be rained into the sea along with algae that had not been good lofters and who will not serve them in the next dispersal episode (in the case of our conjecture for DMS, sites a little downwind of an actively emitting patch are presumably best for take off). Because we are proposing team relations in an assemblage each member of which relies on propagule smallness for dispersal, multi-species parcels are unlikely. Thus in spite of the existence of some correlations of dispersing types in spora, as in some algal assemblages[119,120] and also the very common *Cladosporium-Alternaria* combination, the prospect for evolving close interspecies cooperation seems generally poor.

Clear evolutionary achievements of attached dispersal, like that shown in lichen soredia and thallus fragments, which certainly can be distantly airborne,[3,121] are rare. The general problem is similar to that of rationalising adaptive mutualism in the 'team' of *Sphagnum* species that build raised oligotrophic bogs; but whereas in the case of bogs we might invoke accidental carriage and co-planting of compatible spores on feet of moorland birds, a similar carriage of plankton by sea birds simply replaces the need for any 'team' to fly at all. The notion of a reciprocatory team[122–126] becomes vastly more plausible if physical linkage of the mutualists is guaranteed.[126,127] This happens most definitely when the larger party enwraps the smaller within itself. The varied affinities of the 'chloroplasts' of phytoplankton are relevant here. Some indeed seem to show cellular associations still changing at the present time (e.g. in *Peridinium*, see ref. 36) although the relevance of the intracellular combinations here to any issues of wind and cloud is, of course, quite unknown. However, strong ice-nucleating abilities are known in lichens,[91,97,128,129] and it should be investigated whether this extends to their soredia. Following recent work by Queller[123] and Frank,[124–126,130–132] we see the key to behaviour in such potential interspecies teams in regression predictors of allospecific social phenotypes by actor's genotype (or more precisely, by actor's breeding value; see ref. 131), these regressions being evaluated for the associations as they occur in nature. Non-randomness in associations will be normally due to relatedness but there are other possibilities.[133] In effect we need to know how the social goods and ills dispensed by an individual are 'paid back' to causative genes by the system into which they are sent. Hence the first step in making a case for 'teams' will be to show that associations of genotypes both within and between species are non-random in appropriate ways.

Fred Campbell Meier: History and Conclusion

The idea of precipitation *adaptively* aiding return of spora to earth is more than 10 years old,[34] whilst the broader but not specifically evolutionary idea of bubble processes in marine whitecaps lofting living and other particles into the troposphere, where they may affect earth climate, traces back at least to the 1950s.[134] These allusions are brief, however, and neither topic central to the paper in which it appears. This is also true of an enigmatic briefer and earlier comment that may refer to *self-assisted* microbial dispersal by wind and cloud. The comment is due to F.C. Meier,[135] pioneer in the study of high air spora and originator of the term 'aerobiology', who died in his forties when the plane provided for his research was lost in the Pacific Ocean in 1938.[136] Little other than brief summaries of the late aerobiological phase of his work have been published, but in one abstract outlining some collecting flights over the Caribbean, following a comment on clearance of air spora by rain below cloud level, Meier states: '. . . viable spores taken from air currents above the clouds show that dissemination of certain fungi may occur regardless of and *ultimately aided by rainfall at lower levels*' (our italics).[135] Does he imply here the energising of lofting through water condensation or more simply that rain is likely to be the eventual agent of a safe return of spora to the earth? Whichever, the rain, ice or thermals affecting his own aircraft may have been disastrous for Meier himself and those who flew with him. During his last collecting flight on the 23rd of July 1938 the final message of the radio operator defined the dense cumulus and strato-cumulus clouds into which they were flying at 9100 ft, proceeding from San Francisco towards the equator; last words of all expressed the operator's slight concern with a problem of 'rain static'.

It is a long leap from Meier's probably fatal clouds and from ours to planetary stability. While all facts and arguments we have brought forward remain far below that level, the potential for impact on the global system of the adaptations we have conjectured is vast. The mechanisms we describe do not directly bring us any nearer to discovering why life influences that are stabilising to the planet should be more common than destabilising ones[137,138] and actually the slant even of the present case is somewhat equivocal.[40,139] But a proof that large side effects, stabilising or not, can arise from activities that are adaptive, either at patch or individual level, for thoughtless aerial and marine plankton, strengthening the expectation of large influences from similar unpromising systems,[7] can perhaps help clear a path towards a principled theory.

Acknowledgements

We thank J. E. Lovelock for the contact initiating this paper and S. A. Frank, P. S. Liss, G. Malin and A.J. Watson for helpful comments.

References

1. T. Gislén, Aerial plankton and its conditions of life.*Biological Reviews* 23, 109–126 (1948).
2. A. A. Imshenetsky, S. V. Lysenko and G. A. Kazakov, Upper boundary to the biosphere. *Applied and Environmental Microbiology* **35**, 1–5 (1993).
3. P. H. Gregory, Atmospheric microbial cloud systems.*Science Progress, Oxford* **55**, 613–628 (1967).
4. R. C. Schnell and G. Vali, Atmospheric ice nuclei from decomposing vegetation. *Nature, London* **236**, 163–165 (1972).
5. J. Lovelock, A geophysiologists's thoughts on the natural sulphur cycle. *Philosophical Transactions of the Royal Society of London (B)* **352**, 143–147 (1997).
6. J. E. Lovelock, The ages of Gaia: a biography of our living Earth. (W.W. Norton, New York, 1988).
7. T. M. Lenton, Gaia and natural selection: the case for a self-regulating earth. (Submitted) (1997).
8. R. Dawkins, *The extended phenotype.* (Oxford University Press, Oxford, 1982).
9. K. Caldeira, Evolutionary pressures on planktonic production of atmospheric sulphur. *Nature, London* **337**, 732–734 (1989).
10. K. Caldeira, Chapter 18. Evolutionary pressures on planktonic dimethylsulfide production, pp. 153–158. In: S. H. Schneider and P. J., Boston (ed.) *Scientists on Gaia.* (MIT Press, Cambridge, Mass., 1991).
11. W. D. Hamilton and R. M. May, Dispersal in stable habitats. *Nature, London* **269**: 578–581 (1977).
12. D. C. Blanchard, The production, distribution and bacterial enrichment of the sea-salt aerosol, pp. 407–454. In: P. S. Liss and W. G. N. Slinn (ed.). *Air-Sea Exchange of Gases and Particles.* (D. Reidel, Dordrecht, 1983).
13. A. H. Woodcock, Note concerning human respiratory irritation associated with high concentrations of plankton and mass mortality of marine organisms. *Journal of Marine Research* **7**, 56–62 (1948).
14. D. C. Blanchard and L. D. Syzdek, Mechanism for the water-to-air transfer and concentration of bacteria. *Science* **170**, 626–628 (1970).
15. E. R. Baylor, M. B. Baylor, D. C. Blanchard, L. D. Syzdek and C. Appel, Virus transfer from surf to wind. *Science* **198**, 575–580 (1977).
16. G. T. Wallace Jr and R. A. Duce, Transport of particulate organic matter by bubbles in marine water. *Limnology and Oceanography* **23**, 1155–1167 (1978).
17. R. Chesselet, M. Fontugne, P. Buat-Menard, U. Ezat and C. E. Lambert, The origin of particulate organic carbon in the marine atmosphere as indicated by its stable carbon isotopic composition. *Geophysical Research Letters* **8**, 345–348 (1981).

18. J. M. Hirst, O. J. Stedman and W. H. Hogg, Long-distance spore transport: methods of management, vertical spore profiles and the detection of immigrant spores. *Journal of General Microbiology* **48**, 329–355 (1967).
19. R. E. Stevenson and A. Collier, Preliminary observations on the occurrence of airborne marine phytoplankton. *Lloydia* **25**, 89–93 (1962).
20. R. C. Schnell and G. Vali, Freezing nuclei in marine waters. *Tellus* **3**, 321–323 (1975).
21. D. E. Pedgeley, *Windblown Pests and Diseases: A Meteorology of Airborne Organisms*, (Horwood, New York, 1982).
22. D. E. Pedgeley, Aerobiology: the atmosphere as a source and sink for microbes, pp. 43–59. In: J. H. Andrews and S. S. Hirano (ed.) *Microbial Ecology of Leaves*, (Springer, New York, 1991).
23. H. E. Schlichting Jr, The importance of airborne algae and protozoa. *Journal of the Air Pollution Control Association* **19**, 946–951 (1969).
24. T. J. Hart, Some observations on the relative abundance of marine phytoplankton in nature, pp. 375–393. In: H. Barnes (ed.) *Some Contemporary Studies in Marine Science*, (Allen and Unwin, London, 1966).
25. P. A. Machado, Dinoflagellate bloom on the Brazilian South Atlantic Coast, pp. 29–36. In: D. L. Taylor and H. H. Seliger (ed.) *Toxic Dinoflagellate Blooms. Proceedings of the 2nd International Conference On.* (Elsevier/North Holland, New York, 1979).
26. A. Mazumder, W. D. Taylor, D. J. McQueen and D. R. S. Lean, Effects of fish and plankton on lake temperature and mixing depth. *Science* **247**, 312–315 (1990).
27. S. Sathyendranath, A. D. Gouveia, S. R. Shetye, P. Ravindran and T. Platt, Biological control of surface temperature in the Arabian Sea. *Nature, London* **349**, 54–56 (1991).
28. H. E. Schlichting JR, Meteorological conditions affecting the dispersal of airborne algae and protozoa. *Lloydia* **27**, 64–78 (1964).
29. R. R. Rogers and M. K. Yau, *A Short Course in Cloud Physics*, 3rd edn., (Pergamon Press, Oxford, 1989).
30. J. S. Malkus. Trade wind clouds, *Scientific American* **189** (5), 31–35 (1953).
31. H. A. Heise and E. R. Heise, The distribution of ragweed pollen and *Alternaria* spores in the upper atmosphere. *Journal of Allergy* **19**, 403–407 (1948).
32. K. Jayaweera and P. Flanagan, Investigations on biogenic ice nuclei in the arctic atmosphere. *Geophysical Research Letters* **9**, 94–97 (1982).
33. J. S. Malkus, Cumulus, thermals and wind. *Soaring* **13**, 6–8 and 12 (1949).
34. J. R. Snider, R. G. Layton, G. Caple and D. Chapman, Bacteria as condensation nuclei. *Journal de Recherches Atmosphériques* **19**, 139–145 (1985).
35. J. M. Davis and C. E. Main, Applying atmospheric trajectory analysis to problems in Epidemiology. *Plant Disease* **70**, 490–497 (1986).
36. C. Van Den Hoek, D. G. Mann and H. M. Jahns, Algae: *An Introduction to Phycology*, (Cambridge University Press, Cambridge, 1995).
37. R. J. Charlson, J. E. Lovelock, M. O. Andreae and S.G. Warren, Oceanic phytoplankton, atmospheric sulphur, cloud albedo and climate. *Nature, London* **326**, 655–661 (1987).
38. G. P. Ayers and J. L. Gras, Seasonal relationship between cloud condensation nuclei and aerosol methanesulphonate in marine air. *Nature, London* **353**, 834–835 (1991).

39. M. O. Andreae, W. Elbert and S. J. de Mora, Biogenic sulfur emissions and aerosols over the tropical South Atlantic. 3. Atmospheric dimethylsulfide, aerosols and cloud condensation nuclei. *Journal of Geophysical Research* **100** (D6), 11,335–11,356 (1995).
40. M. O. Andreae and P. J. Crutzen, Atmospheric aerosols: biogeochemical sources and role in atmospheric chemistry. *Science* **276**, 1052–1058 (1997).
41. A. Vairavamurthy, M. O. Andreae and R. L. Iverson, Biosynthesis of dimethylsulfide and dimethyl propiothetin by *Hymenomonas carterae* in relation to sulfur source and salinity conditions. *Limnology and Oceanography* **30**, 59–70 (1985).
42. U. Karsten, K. Kuck, C. Vogt and G. O. Kirst, Dimethylsufoniopropionate production in photrophic organisms and its physiological function as a cryoprotectant, pp. 143–153. In: R. P. Kiene *et al.*, (ed.) *Biological and Environmental Chemistry of DMSP and Related Sulfonium Compounds.* (Plenum, New York, 1996).
43. J. B. S. Haldane, *The Causes of Evolution.* (Longmans, London, 1932).
44. S. Wright. (1961 printing). Genetics of populations, pp. 111D–112. In: W. E. Preece (ed.) *Encyclopaedia Britannica.* (Benton, Chicago, 1948).
45. R. Fisher, *The genetical theory of natural selection*, (Dover, New York, 1958).
46. W. D. Hamilton. The evolution of altruistic behavior. *The American Naturalist* **97**, 354–355 (1963).
47. W. D. Hamilton, Innate social aptitudes of Man: an approach from evolutionary genetics, pp. 133–155. In: R. Fox (ed.) ASA Studies 4: *Biosocial Anthropology.* (Malaby Press, London, 1975).
48. R. J. Fevek, R. B. Chatfield and M. O. Andreae, Vertical distribution of dimethylsulphide in the marine atmosphere. *Nature, London* **320**, 514–516 (1986).
49. M. O. Andreae, The ocean as a source of atmospheric sulphur compounds, pp. 331–362. In: Buat-Menard P. (ed.) *The Role of Air-sea Exchange in Geochemical Cycling.* (Reidel, Dordrecht, 1986).
50. F. Raes, Entrainment of free tropospheric aerosols as a regulating mechanism for cloud condensation nuclei in the remote marine boundary layer. *Journal of Geophysical Research* **100**, 2893–2903 (1995).
51. X. Lin and W. L. Chameides, CCN formation from DMS oxidation without SO_2 acting as an intermediate. *Geophysical Research Letters* **20**, 579–582 (1993).
52. H. N. Comins, W. D. Hamilton and R. M. May, Evolutionarily stable dispersal strategies. *Journal of Theoretical Biology* **82**, 205–230 (1980).
53. W. D. Hamilton, The genetical evolution of social behaviour, I. *Journal of Theoretical Biology* **7**, 1–16 (1964).
54. H. W. Paerl, Nuisance phytoplankton blooms in coastal, estuarine, and inland waters. *Limnology and Oceanography* **33**, 823–847 (1988).
55. C. Leck, U. Larsson, L. E. Bagander, S. Johansson and S. Hajdu, Dimethyl sulphide in the Baltic sea: annual variability in relation to biological activity. *Journal of Geophysical Research* **95**, 3353–3363 (1990).
56. G. V. Wolfe, M. Steinke and G. O. Kirst, Grazing-activated chemical defence in a unicelluar marine alga. *Nature, London* **387**, 894–897 (1997).
57. J. Rosinski, P. L. Haagenson, C. T. Nagamoto and F. Parungo, Ice-forming nuclei of maritime origin. *Journal of Aerosol Science* **17**, 23–46 (1986).

58. R. D. Pingree, P. R. Pugh, P. M. Holligan and G. R. Forster, Summer phytoplankton blooms and red tides along tidal fronts in the approaches to the English Channel. *Nature, London* **258**, 572–577 (1975).
59. G. Beale, Self and non-self recognition in the ciliate *Euplotes. Trends in Genetics* **6**, 137–139 (1990).
60. J. M. Burkholder, E. J. Noga, C. H. Hobbs, H. B. Glasgow and S. A. Smith, New 'phantom' dinoflagellate is the causative agent of major estuarine fish kills. *Nature, London* **358**, 407–410 (1992).
61. K. W. Estep and F. Macintyre, Taxonomy, life-cycle, distribution and dasmotrophy of *Chrysochromulina*—a theory accounting for scales, haptonema, muciferous bodies and toxicity. *Marine Ecology-Progress Series* **57**, 11–21 (1989).
62. H. Fudge, The 'red tides' of Malta. *Marine Biology* **39**, 381–386 (1977).
63. G. Malin, P. S. Liss and S. M. Turner, Dimethyl sulfide: production and atmospheric consequences. (Chapter 16), pp. 303–320. In: J. C. Green and B. S. C. Leadbeater (ed.) *The Haptophyte Algae.* (Clarendon Press, Oxford, 1994).
64. S. M. Turner, G. Malin, P. Liss, D. S. Harbour and P. Holligan, The seasonal variation of dimethyl sulphide and dimethylsulfonioproprionate concentrations in nearshore waters. *Limnology and Oceanography* **33**, 364–375 (1988).
65. P. S. Liss, G. Malin, S. M. Turner and P. M. Holligan, Dimethyl sulphide and *Phaeocystis*: a review. *Journal of Marine Systems* **5**, 41–53 (1994).
66. C. J. Macdonald, R. Little, G. R. Moore and G. Malin, NMR spectroscopy as a probe for DMSP and glycine betaine in phytoplankton cells, pp. 45–54. In: Kiene R. P., (ed.) *Biological and Environmental Chemistry of DMSP and Related Sulfonium Compounds.* (Plenum, New York, 1996).
67. B. C. Nguyen, S. Belviso, N. Michalopoulos, J. Gostan and P. Nival, Dimethyl sulphide production during phytoplanktonic blooms. *Marine Chemistry* **24**, 133–141 (1988).
68. J. W. H. Dacey and S. G. Wakeham, Oceanic dimethylsulphide: production during zooplankton grazing on phytoplankton. *Science* **233**, 1314–1333 (1986).
69. G. Bratbak, M. Levasseur, S. Michaud, G. Cantin, E. Fernadez, B. Heimdal and M. Heldal, Viral activity in relation to *Emiliana huxleyi* blooms: a mechanism for DMSP release. *Marine Ecology Progress Series* **128**, 133–142 (1995).
70. A. F. G. Dixon, *Aphid Ecology*. (Blackie, Glasgow, 1985).
71. D. A. Roff, The evolution of threshold traits in animals. *Quarterly Review of Biology* **71**, 3–35 (1996).
72. A. E. Hershey and S. I. Dodson, Predator avoidance by *Cricotopus*: cyclomorphosis and the importance of being big and hairy. *Ecology* **68**, 913–920 (1987).
73. H. W. Kuhlmann and K. Heckmann, Interspecific morphogens regulating prey-predator relationships in Protozoa. *Science* **227**, 1347–1349 (1985).
74. F. R. Trainor, Cyclomorphosis in *Scenedesmus armatus* (Chlorophyta)—an ordered sequence of ecomorph development. *Journal of Phycology* **28**, 553–558 (1992).
75. E. Fialkowska and A. Pajdak-Stos, Inducible defence against a ciliate grazer, *Pseudomicrothorax dubius*, in two strains of *Phormidium* (cyanobacteria). *Proceedings of the Royal Society of London (B)* **264**, 937–941 (1997).
76. R. R. Kudo, *Protozoology* Springfield, Illinois: Charles C. Thomas (1966).
77. D. Gibson, D. R. Bazely and J. S. Shore, Responses of brambles, *Rubus vestitus*, to herbivory. *Oecologia* **95**, 454–457 (1993).

78. D. C. Blanchard and L. D. Syzdek, Seven problems in bubble and jet drop research. *Limnology and Oceanography* **23**, 389–400 (1978).
79. R. Barness and M. Rosenberg, Putative role of a 70 KDa outer-surface protein in promoting cell-surface hydrophobicity of *Serratia marcescens. Journal of General Microbiology* **135**, 2277–2281 (1989).
80. H. C. Vandermei, M. M. Cowan, M. J. Genet, P. G. Rouxhet and H. J. Busscher, Structural and physicochemical surface properties of *Serratia marcescens* strains. *Canadian Journal of Microbiology* **38**, 1033–1041 (1992).
81. I. N. Rjazantseva, I. N. Andreeva and T. I. Ogorodnikova, Effect of various growth conditions on pigmentation of *Serratia marcescens. Microbios* **79**, 155–161 (1994).
82. T. Swain, Plant-animal coevolution: a synoptic view of the Palaeozoic and Mesozoic. (Chapter 1), pp. 3–19. In: J.B. Harborne (ed.) *Biochemical Aspects of Plant and Animal Coevolution.* (Academic Press, London, 1978).
83. J. B. Harborne, Flavonoid pigments. Chapter 11. In: G. A. Rosenthal and M. R. Berenbaum (ed.) *Herbivores: Their interactions With Secondary Plant Metabolites*, 2nd edn (Academic Press, San Diego, 1991).
84. T. Wyatt and J. Horwood, A model which generates red tides. *Nature, London* **244**, 238–240 (1973).
85. R. M. Brown, D. A. Larson and H. C. Bold, Airborne algae: their abundance and heterogeneity. *Science* **143**, 583–585 (1964).
86. L. R. Maki, E. Galyan, M. Chang-Chen and D. R. Caldwell, Ice nucleation induced by *Pseudomonas syringae. Applied Microbiology* **28**, 456–459 (1974).
87. S. E. Lindow, D. C. Arny and C. D. Upper, Bacterial ice nucleation: a factor in frost injury to plants. *Plant Physiology* **70**, 1084–1089 (1982).
88. R. White. Analysis of dimethyl sulfonium compounds in marine algae. *Journal of Marine Research* **40**, 529–536 (1982).
89. M. D. Keller, W. K. Bellows and R. R. L. Guillard, Dimethyl sulfide production in marine phytoplankton, pp. 167–182. In: E. S. Saltzmann and W. J. Cooper (ed.) *Biogenic Sulfur in the Environment*. (American Chemical Society, Washington, D.C., 1989).
90. M. J. Bechard and W. R. Rayburn, Volatile organic sulfonides from freshwater algae. *Journal of Phycology* **15**, 379–383 (1979).
91. C. Richard, J. G. Martin and S. Pouleur, Ice nucleation activity identified in some phytopathogenic *Fusarium* species. *Phytoprotection* **77**, 83–92 (1996).
92. D. Jenkins, L. L. Medsker and J. F. Thomas, Odorous compounds in natural waters. Some sulfur compounds associated with blue-green algae. *Environmental Science and Technology* **1**, 731–735 (1967).
93. J. W. H. Dacey, G. M. King and S. G. Wakeham, Factors controlling emission of dimethylsulphide from salt marshes. *Nature, London* **330**, 643–645 (1987).
94. P. K. Wolber, Bacterial ice formation. *Advances in Microbial Physiology* **34**, 203–237 (1993).
95. C. J. Warren, A bibliography of biological ice nucleation. *Cryo-letters* **15**, 323–331 (1994).
96. W. Szyrmer and I. Zawadzki, Biogenic and anthropogenic sources of ice-forming nuclei: a review. *Bulletin of the American Meteorological Society* **78**, 209–228 (1997).

97. S. Pouleur, C. Richard, J. G. Martin and H. Antoun, Ice nucleation activity in *Fusarium acuminatum* and *F. avenaceum*. *Applied and Environmental Microbiology* **58**, 2960–2964 (1992).
98. R. C. Schnell, Airborne ice nucleus measurement around the Hawaiian islands. *Journal of Geophysical Research* **87** (C11), 8886–8890 (1982).
99. M. B. Baker, Cloud microphysics and climate. *Science* **276**, 1072–1078 (1997).
100. G. Caple, D. C. Sands, R. G. Layton, W. V. Zucker and J. R. Snider, Biogenic ice nucleation—could it be metabolically initiated? *Journal of Theoretical Biology* **119**, 37–45 (1986).
101. Y. Hasegawa, Y. Ishihara and T. Tokuyama, Characteristics of ice-nucleation activity in *Fusarium avenaceum* IFO-7158. *Bioscience, Biotechnology and Biochemistry* **58**, 2273–2274 (1994).
102. P. Mandrioli, G. L. Puppi, N. Bagni and F. Prodi, Distribution of microorganisms in hailstones. *Nature, London* **246**, 416–417 (1973).
103. J. Lindemann and C. D. Upper, Aerial dispersal of epiphytic bacteria over bean plants. *Applied and Environmental Microbiology* **50**, 1229–1232 (1985).
104. R. C. Schnell and G. Vali, Biogenic ice nuclei: Part I. Terrestrial and marine sources. *Journal of Atmospheric Science* **33**, 1554–1564 (1976).
105. R. Fall and R. C. Schnell, Association of an ice-nucleating pseudomonad with cultures of the marine dinoflagellate, *Heterocapsa niei*. *Journal of Marine Research* **43**, 257–265 (1985).
106. S. S. Hirano and C. D. Upper, Diel variation in population size and ice nucleation activity of *Pseudomonas syringae* on snap bean leaflets. *Applied Environmental Microbiology* **46**, 1370–1379 (1989).
107. S. M. Pady and L. Kapica, Fungi over the Atlantic Ocean. *Mycologia* **47**, 34–50 (1955).
108. M. Hjelmroos. Relationship between airborne fungal spore presence and weather variables—*Cladosporium* and *Alternaria*. *Grana* **32**, 40–47 (1993).
109. S. M. Pady, Spore release in some foliar saprophytic and parasitic fungi, pp. 111–115. In: T. F. Preece and C. H. Dickinson (ed.) *Ecology of Leaf Surface Micro-Organisms*. (Academic Press, London, 1971).
110. D. W. Li and B. Kendrick, Functional relationships between airborne fungal spores and environmental factors in Kitchener-Waterloo, Ontario, as detected by canonical correspondence analysis. *Grana* **33**, 166–176 (1994).
111. J. Lindemann, H.A. Constaninidou, W.R. Barchet and C.D. Upper, Plants as sources of airborne bacteria, including ice nucleation-active bacteria. *Applied and Environmental Microbiology* **44**, 1059–1063 (1982).
112. D. C. Gross, Y. S. Cody, E. L. Proebsting Jr, G. K. Radamaker and R. A. Spotts, Distribution, population dynamics and characteristics of ice nucleation-active bacteria in deciduous fruit tree orchards. *Applied Environmental Microbiology* **46**, 1370–1379 (1983).
113. S. Fuzzi, P. Mandrioli and A. Perfetto, Fog droplets—an atmospheric source of secondary biological aerosol particles. *Atmospheric Environment* **31**, 287–290 (1997).
114. S. S. Hirano, E. A. Maher, A. Kelman and C. D. Upper, Ice nucleation activity of fluorescent plant pathogenic pseudomonads. Proceedings of the 4th International

Conference on Plant Pathogenic Bacteria. *Beaucouzé, France: Institute National Recherche Agronomique* (1978).
115. I. R. Jenkinson, Oceanographic implications of non-newtonian properties found in phytoplankton cultures. *Nature, London* **323**, 435–437 (1986).
116. W. B. Wilson, Production of surface-active materials by culture of marine phytoplankton. *Journal of Phycology* **3**, 4 (1967).
117. A. T. Wilson, Surface of the ocean as a source of air-borne nitrogenous material and other plant nutrients. *Nature, London* **184**, 99–101 (1959).
118. G. A. Dean, The iodine content of some New Zealand drinking waters with a note on the contribution from sea spray to the iodine in rain. *Journal of Science, New Zealand*, **6**, 208–214 (1963).
119. I. Rosas, G. Roy-Ocotla and P. Mosiño, Meteorological effects on variation of airborne algae in Mexico. *International Journal of Biometeorology* **33**, 173–179 (1989).
120. G. Roy-Ocotla and J. Carrera, Aeroalgae: responses to some aerobiological questions. *Grana* **32**, 48–56 (1993).
121. K. Hamata and M. Olech, Transect for aerobiological studies from Antarctica to Poland. *Grana* **30**, 458–463 (1991).
122. R. Axelrod and W. D. Hamilton, The evolution of cooperation. *Science* **211**, 1390–1396 (1981).
123. D. C. Queller, A general model of kin selection. *Evolution* **46**, 376–380 (1992).
124. S. A. Frank, Genetics of mutualism: the evolution of altruism between species. *Journal of Theoretical Biology* **170**, 393–400 (1994).
125. S. A. Frank, The origin of synergistic symbiosis. *Journal of Theoretical Biology* **176**, 403–410 (1995).
126. S. A. Frank, Models of symbiosis. *The American Naturalist* **150**, S80–S99 (1997a).
127. W. D. Hamilton, Kinship, recognition, disease and intelligence: constraints of social evolution, pp. 82–102. In: Y. Itô (ed.) *Animal Societies: Theories and Facts.* (Japan Scientific Press, Tokyo, 1987).
128. E. N. Ashworth and T. L. Kieft, Ice nucleation activity associated with plants and fungi, pp. 137–162. In: R. E. Lee *et al.*, (ed.) *Biological Ice Nucleation and its Applications.* (APS Press, St. Paul, Minnesota, 1995).
129. M. R. Worland, W. Block and H. Oldale, Ice nucleation activity in biological materials with examples from antarctic plants. *Cryo-letters* **17**, 31–38 (1996).
130. S. A. Frank, The Price Equation, Fisher's fundamental theorem of natural selection and causal analysis. *Evolution* **51**, 1712–1729 (1997b).
131. S. A. Frank, Multivariate analysis of correlated selection and kin selection, with an ESS maximisation method. *Journal of Theoretical Biology* **189**, 307–316 (1998a).
132. S. A. Frank, *Foundations of Social Evolution.* (University Press, Princeton, 1998b).
133. D. S. Wilson and L. A. Dugatkin. Group selection and assortative interactions. *The American Naturalist* **149**, 336–351 (1997).
134. A. H. Woodcock, Salt and rain. *Scientific American* 197 (4): 42–47 (1957).
135. F. C. Meier, Collecting microorganisms from winds above the Caribbaean Sea. *Phytopathology* **26**, 102 (only) (1936).
136. R. J. Haskell and H. P. Barss, Fred Campbell Meier, 1893–1936. *Phytopathology* **29**, 293–302 (1939).

137. W. D. Hamilton, Ecology in the large: Gaia and Genghis Khan. *Journal of Applied Ecology* **32**, 451–453 (1995).
138. W. D. Hamilton, Gaia's benefits. *New Scientist* **151**, 62–63 (1996).
139. J. E. Lovelock and L. R. Kump, Failure of climate regulation in a geophysical model. *Nature, London* **369**, 732–734 (1994).

IS DMSP SYNTHESIS IN CHLOROPHYCEAN MACRO-ALGAE LINKED TO AERIAL DISPERSAL?[†]

D. T. WELSH, P. VIAROLI, W. D. HAMILTON AND
T. M. LENTON

Abstract

Dimethylsulphoniopropionate (DMSP) has been proposed to be a compatible (osmotic) solute in marine macro-algae, allowing adaptation to changes in the osmotic pressure of the growth medium. However, whilst DMSP undoubtedly does contribute substantially to the overall osmotic pressure of the cytoplasm, several studies have demonstrated that its intracellular concentration is not directly regulated by the osmotic pressure of the growth medium. Thus, DMSP does not behave as a compatible solute sensu stricto and therefore its role may not be strictly osmotic.

Recently, Hamilton and Lenton[1] proposed that DMS emissions associated with blooms of DMSP accumulating marine phytoplankton may be linked to an aerial mode of dispersal via the induction of cloud formation and local convective winds. Only micro-algae were discussed. All macro-algae, however, also have unicellular stages as gametes or spores which could become windborne.

In this paper, we review the literature on the life histories, growth cycles and ecology of marine DMSP synthesising chlorophycean macro-algae, in order to assess whether a similar dispersal mechanism may exist within this group. Whilst only direct experimental evidence can resolve if these macro-algae are dispersed or not by an aerial mechanism, many features of their growth, reproductive and life cycles are consistent with such a mechanism. Such an aerial dispersal mechanism might represent an especial ecological advantage in this group, allowing colonisation of areas separated by land masses or denied to normal waterborne dispersal routes by directional currents.

[†] *Ethology Ecology and Evolution* **11**, 265–278 (1999).

Introduction

Many species of marine and freshwater micro-algae and macro-algae synthesise dimethylsulphoniopropionate (DMSP) as part of their intracellular compatible solutes (osmolytes) pool[2–8] The release of intracellular DMSP to the water column resulting from osmotic down-shock, grazing, viral lysis, cell death or due to passive diffusion across the cell membrane can lead to the production of dimethylsulfide (DMS) due to enzymatic cleavage of DMSP by DMSP-lyases produced by bloom associated bacteria or the micro- and macro-algae themselves.[8–15] The resultant fluxes of DMS to the atmosphere have been proposed to complete the global sulfur cycle by coupling the marine and terrestrial cycles.[16] Additionally, the oxidation of DMS in the atmosphere to sulfate aerosols which can act as condensation nuclei for water vapour, increases cloud formation over the open oceans and may influence climate on a global scale due to increased cloud albedo.[17] The influence of DMS fluxes on climate is supported by data demonstrating a direct correlation between fluxes of DMS and cloud condensation nuclei over the oceans,[18,19] providing an important strand in evidence for the Gaia theory.[20,21] However, it remains a puzzle as to how such a climate regulating trait could arise consistent with natural selection since the proposed benefits of the trait are to the community, not directly to the individuals which bear the energetic cost of DMSP synthesis.[22]

Recently, it has been proposed that DMSP synthesis and lysis of DMSP to DMS and acrylate may be involved in the protection of micro-algae from grazing[23,24] and/or play a role in micro-algal dispersall[1] and these functions may be adaptive at the individual or clonal patch levels. This may be particularly true for dispersal, which ranks as an organism's third priority, following survival and reproduction, even when conditions at the point of arrival are never superior to those at the point of departure.[25] The dispersal hypothesis for marine micro-algae proposes that the rapid oxidation of DMS in the atmosphere and resultant cloud formation above the algal blooms causes local convective winds due to the release of latent heat of condensation. These local winds may stimulate transfer of the micro-algae to the air due to the formation of 'white caps', whose bubbles can both concentrate small cells during their ascent and eject them into the atmosphere when they burst at the water surface. Once in the air, the rising thermals above the blooms, further energised by the proposed incipient clouds, would lift the cells into the atmosphere leading to their dispersal by air currents.[1]

In this paper, we review whether a similar dispersal mechanism could exist for DMSP synthesising macro-algal species. Within this group dispersal may be particularly important as many species are adapted to specific coastal habitat types, e.g. substrate type, water depth, current speeds etc., and these biomes may be physically separated by large areas of unsuitable habitat. Additionally, in the coastal marine environment where most macro-algal species occur, headlands or pebble and sand bars, combined with strong, often highly directional water currents may act as physical barriers and limit macro-algal dispersal by the conventional waterborne mechanisms normal in the sea.

Distribution and Function of DMSP in Marine Macro-Algae

DMSP is known to occur at high intracellular concentrations (10–50 mmol kg wet weight^{-1}) in many species of green algae (Chlorophyceae) and red algae (Rhodophyceae), including species of *Ulva, Enteromorpha, Monostroma, Halidrys, Ulothrix, Acrosiphonia, Blidingia, Pelvetia, Codium* and *Polysiphonia* species.[3–6,26] Although, its distribution may be much more widespread, since only a relatively limited range of species have been tested. In these algae, DMSP has been proposed to function principally as a compatible solute, balancing the osmotic pressure of the growth medium.[2] However, measurements of the intracellular DMSP concentrations as a function of salinity of the growth medium in *Codium fragile, Polysiphonia lanosa, Ulva lactuca* and *Enteromorpha intestinalis* have demonstrated little or no influence of growth medium osmotic pressure on the intracellular DMSP content of these algae.[4,5,26] Thus, whilst the high intracellular concentrations of DMSP in these algae undoubtedly do contribute to the overall osmotic pressure of the cytoplasm, DMSP does not appear to function as an osmoregulatory solute sensu stricto. Since, DMSP is a relatively fixed, constitutive component of the intracellular osmotica and it is other intracellular solutes such as proline and sucrose which regulate the overall osmotic pressure of the cytoplasm in response to variations in the osmolarity of the environment.[5] These facts, plus that DMSP production and its subsequent lysis to DMS is certainly not without cost to the algae, prompt a search for other adaptive functions for DMSP.

In the micro-alga *Emiliania huxleyi*, DMSP has been proposed to act a deter rent against grazing,[23,24] as lysis of the ingested algae mixes DMSP and DMSP-lyase, resulting in the lysis of DMSP to DMS and acrylate which is proposed to be the active agent.[27] This hypothesis is supported by recent data which demonstrates that some protozoan grazers are unable to subsist on strains of *Emiliana huxleyi* which exhibit high DMSP-lyase activities, and all the tested grazers preferentially ingested prey with low enzyme activities when offered prey mixtures.[24] Whilst there is no direct evidence of a similar role for DMSP in marine macro-algae, this function cannot be excluded and these algae are known to synthesize a range of other anti-herbivorous compounds.[28,29]

In order to be dispersed by an aerial mechanism similar to that which has been proposed for marine micro-algae,[1] macro-algal species would require a suite of secondary characteristics other than simply synthesising large quantities of DMSP. A fundamental requirement for aerial dispersal would be the formation of a dispersal phase of a suitably small size, to be lifted and transported by atmospheric currents, and this dispersal phase would need to be capable of surviving under the harsh conditions which prevail in the atmosphere. There would also be a requirement that the algal blooms show a high degree of clonality to reduce or eliminate selection problems,[22] where the benefit of the trait may not be directly to the individual organism bearing the costs. Similarly, there would need to be some degree of co-ordination of the algal growth and reproductive cycles so that entire clones of individuals produce their dispersal phase(s) at the same time and that this period corresponds with release of DMSP and environmental conditions suitable for generating convective winds. Other potentially adaptive but not necessarily essential characteristics could be that the dispersal phase(s) are motile and show some form of phototactic or geotactic response which allows them to concentrate at or near to the air-water interface, and the production of DMSP-lyases to catalyse the rapid conversion DMSP to DMS.

The word 'near' above deserves emphasis since Hamilton and Lenton,[1] misreading Blanchard,[30] believed bubble *film* droplets to be the main vectors for aquatic microbes into the atmosphere. Instead bubble *jet* droplets are the vectors, and these eject hydrophobic microbes scavenged mainly to bubble lower surfaces during the ascent of bubbles. Since the surface water tends to be pushed aside by an upwelling bubble cloud, thus locally clearing objects on the water surface film,[30] the optimal waiting place for ejection by bubble activity is not at the water surface itself but some centimetres below the surface, although not lower than the maximum depth of the bubble plumes created by 'white tops'.

Below, we discuss using data from the literature whether such secondary characteristics consistent with an aerial mode of dispersal do indeed occur in marine macro-algal species. This discussion is confined principally to green algae (Chlorophyceae) of the Ulvales group, as the ecology of these species has been intensively studied due to their propensity to form mass blooms in eutrophied coastal environments. However, the mechanisms discussed may also be relevant to other chlorophycean and rhodophycean DMSP synthesisers.

Life Histories and Dispersal Phases

The life histories of members of the Ulvales group, like those of most marine macro-algae, are generally complex. At the simplest level, there is an alternation between diploid sporophytic and haploid gametophytic phases, which are morphologically indistinguishable (see ref. 31 and references therein for details). Sporophytic thalli produce quadriflagellate zoospores of one of two mating types, which germinate to produce gametophytic thalli, which in turn produce biflagellate gametes completing the life cycle. Additionally, gametes can germinate parthenogenetically to produce parthenosporophytes which can remain haploid or may partially or completely undergo diploidisation[32] and both haploid and diploid forms produce zoospores.

Some species of *Ulva* have also been observed to vegetatively produce diminutive, floating, monostromatic globose plants which released motile 'swarmers'.[33] Thus, in chlorophycean macro-algae several small (μm-10s of μm) single cell dispersal phases are formed, which may be suitable for dispersal by an aerial mechanism. All these dispersal phases are motile and could concentrate at or near to the water-air interface, or in the case of the swarmer cells produced by floating globose plants, may actually be liberated at the water surface.[33] In at least some species, such as *Enteromorpha*, zoospores are positively phototactic and spend relatively long periods in the plankton,[34,35] which may allow them to concentrate at or near to the water surface. Similarly, in *Ulva lactuca* gametes are initially positively phototactic and only become negatively phototactic upon pairing (ref 36 cited in ref. 37) and again could potentially congregate in the surface waters. Whilst these positive phototatic responses could also play a role in water-borne dispersal mechanisms, allowing the exploitation of surface currents and/or maintaining the cells within the photic zone, it is interesting to speculate that positive phototactism may be a specific adaptation to an aerial mode of dispersal, since to germinate, zoospores and gametes must first attach to a suitable substrate.[38] Indeed, without adaptive intent to disperse, a phototaxis which would result in a tendency of zoospores and gametes to swim away from the seabed and thereby potential germination sites and towards the surface, would seem somewhat difficult to rationalise.

In order to be dispersed by air currents, the macro-algal dispersal phases would need to exhibit characteristics which would allow them to survive under the harsh conditions which prevail in the atmosphere. Tolerance of desiccation and freezing or freeze drying in the higher atmospheric strata would be of particular importance. In this respect, the compatible solutes, sucrose, proline and DMSP,[2,5–7] which are commonly accumulated by chlorophycean macro-algae, may also play a role. During osmotic stress, compatible solutes are generally agreed to function by being preferentially excluded from the hydration sphere of proteins, thus the changes in the intracellular concentration of solutes have little effect on the overall hydration status of the proteins or their structure and activity.[39–42] The exclusion of the solute from part of the cellular water creates a thermodynamic disequilibrium. Therefore, the presence of high concentrations of compatible solutes tends to promote protein sub-unit interactions and the folding of proteins into their tertiary, active form, since these effects reduce the volume occupied by the proteins, and thereby the volume of cell water from which the solute is excluded and the degree of the thermodynamic disequilibrium.[39,41,42]

Freezing can be considered as an equivalent to osmotic stress, since damage results from freezing-induced dehydration and/or the resulting high intracellular ionic solutes concentrations when the cell water becomes partially frozen. Although, freezing can be considered as a more severe form of stress due to the greater effective removal of cell water during freezing. However, at least 5% of the cell water is considered to be unfreezable[42] and therefore protection can be accounted for by the same mechanism, preferential solute exclusion that is proposed to function during osmotic stress. Thus, due to the similarities between freezing and osmotic stresses, it is unsurprising that compatible solutes also afford protection for both cells and isolated cold sensitive enzymes against freeze damage[43–45] and DMSP may be a particularly effective cryoprotectant.[46]

Desiccation and freeze drying represent harsher forms of stress, since the residual quantities of water in the dried cells are insufficient to maintain even a single mono-layer of water around intracellular macromolecules and therefore the exclusion principle of solute compatibility is unable to account for protection against these stresses (see ref 42 for review). Disaccharides, such as sucrose, have been demonstrated to promote tolerance to both freeze drying and desiccation in a wide range of organisms and to increase the stability of dried or freeze dried isolated membranes and individual enzymes (see ref. 42, 47, 48 for reviews). Disaccharides are proposed to protect whole cells and isolated membranes through direct hydrogen bonding to the polar head-group of membrane phospholipids and to a lesser extent with the hydrophilic regions of proteins.[42,47–49] Additionally, during desiccation but not freeze drying, sucrose and other disaccharides form amorphous intracellular glasses in which the intracellular components are entrapped. Within these glasses, molecular motion is extremely limited, and thus degradative molecular reactions are kinetically insignificant[42,50] and the entrapped bio-molecules are stable for long periods in the dry state even at elevated temperatures.[51]

In conclusion, whilst there is no direct data concerning the tolerance of macro-algal dispersal phases to freezing, freeze drying or desiccation, the compatible solutes commonly accumulated by these algae would inherently afford some degree of protection against these stresses. This may be particularly true for sucrose, which has been shown to play a vital role in the ability of pollen to survive drying and storage[42,52] and thereby long periods in the atmosphere. Thus, it would seem probable that the dispersal phases of sucrose accumulating macro-algae could survive under the harsh conditions in the atmosphere, at least for the periods of hours or days that would be required for their efficient dispersal by air currents.

Growth, Bloom Formation and Reproduction

Similarly to most of the insect and other groups having cyclomorphoses connected to dispersal,[1] the members of the Ulvales group are classified as opportunistic or ephemeral bloom forming species, characterised by their rapid nutrient assimilation and growth rates, which allow them to efficiently exploit transiently available nutrient sources[53,54] Generally, vegetative growth commences in spring or early summer as the water temperature increases, blooms crash in summer when the nutrient supplies become exhausted and in some cases, secondary blooms may occur in late summer or early autumn when nutrients from the mineralisation of the primary bloom are recycled to the water column.[55–57] These blooms can occur on a massive scale, attaining peak biomasses of significantly greater than 10 kg wet weight $\times$ m^{-2} in some highly eutrophic environments and can cover areas of 10s of km^2,[55–58] although biomasses of 10s–100s g wet weight $\times$ m^{-2} may be more typical of the coastal environment in general.

The dominance of vegetative reproductive mechanisms of kinds giving structure to thalli and whole populations during bloom formation indicates that blooms may be clonal or at least show a high degree of relatedness. Many macroalgal populations appear to be exclusively asexual and reproduce by fragmentation and the formation of neutral biflagellate swarmers or quadriflagellate zoospores or in some cases both.[59–61] Against local clonality necessarily following from this it must be remembered that populations of characteristically asexual species often consist of mixtures of distinct clones[62–64] whose coexistance is probably and sometimes provenly[65] due to frequency-dependent selection. Nevertheless, the presence of asexual phases and also more permanent asexual modes of reproduction must always make clonal patches more likely.

As a result of their high biomasses and tissue DMSP concentrations of between 10–50 mmol kg wet weight^{-1},[2,4–6] macro-algal blooms represent extremely large stocks of DMSP which may be released to the environment during a relatively short time period when the algal blooms crash and decompose. Assuming efficient conversion of this DMSP to DMS, water column DMS concentrations could potentially be 100s of μM, approximately 1–3 orders of magnitude higher than those generally associated with micro-algal blooms.[66–68] In most cases it is not known whether DMSP-lysases which could cleave DMSP to DMS and acrylate are produced or not by members of the Ulvales group as only a relatively few species have been tested. However, in a single study of DMSP-lysase activity in crude extracts of micro- and macroalgal tissues, the highest enzyme activities were detected in extracts of three *Enteromorpha* sp. (*E. intestinalis, E. clathrata*, and *E. compressa*) and a single isolate of the micro-alga *Phaeocystis* sp.[15] Further studies have demonstrated DMSP-lyase production by *Ulva lactuca*[12] *Ulva curvata*[8] and *Enteromorpha clathrata*[14] and *Enteromorpha bulbosa*,[15] indicating that DMSP-lyases may occur commonly in chlorophycean algae. DMSP-lyases have also been identified in the DMSP synthesising red (rhodophycean) algal species *Polysiphonia lanosa*[69] and *Polysiphonia paniculata*.[70] Additionally, even in blooms of macroalgae which may not produce DMSP-lyases or have only low enzyme activities, DMSP-lyases produced by heterotrophic bloom associated bacteria could effectively convert DMSP to DMS.[9,10] In pure culture studies of

DMSP-lyase producing bacterial isolates it has been demonstrated that these bacteria actually grow on the liberated acrylate, and that the DMS moiety of DMSP is not further metabolised and thus would tend to accumulate.[71,72] In contrast, some bacteria are able to degrade DMSP directly via a demethylation pathway, yielding methylmercaptoproprionate or mercaptoproprionate following a second demethylation as intermediate products.[73–75] Similarly, dissolved DMS may be bacterially or photochemically oxidised to dimethyl sulfoxide (DMSO),[76–78] although at least a proportion of this DMSO may be reconverted to DMS by facultative anaerobic bacteria in the sediment or anoxic bottom waters.[79,80] However, despite these potential sinks for released DMSP and DMS, due to the large stock of DMSP represented by the algal biomass, it is highly probable that ample DMS accumulates in order to generate sufficiently high atmospheric fluxes of DMS to induce local convective winds by the mechanism proposed by Hamilton and Lenton.[1]

In order for macro-algal species to exploit the convective winds generated by DMS-induced cloud formation for dispersal, they would need to co-ordinate the formation of one or more of the potential dispersal phases discussed previously, with the summer biomass crash when DMS fluxes to the atmosphere would be greatest and climatic conditions most sensitive to the seeding of thermals. In principle, the bloom collapse would appear to be a period ripe for the production of reproductive, dispersive and resting phases. Since further vegetative growth is restricted by limited nutrient availability and the plants may be expected to invest the reserves accumulated during bloom formation in the production of reproductive and dispersive stages, and in chlorophycean macro-algae these functions overlap as discussed above. There is some evidence in the literature to support this proposal, for example, in *Ulva fasciata* nitrogen depletion has been found to enhance rates of gamete formation, whereas high nitrogen availability favoured vegetative growth and asexual reproduction.[81] Similarly, healthy actively growing thalli of *Ulva mutabilis* have been reported to produce an unidentified heat labile sporulation inhibitor,[82] which may lead to a coordinated reproductive effort at the bloom level at the end of the growth phase. Additionally, an in situ study of *Ulva lactuca* in Mumford Cove, Connecticut also demonstrated an increased production of reproductive tissues as percentage of total biomass throughout the summer months, significant liberation of swarmers (gametes and zoospores) by this tissue and this increased investment in reproduction coincided with a decrease in total biomass.[83]

Other Contributing Factors

The propensity of members of the Ulvales group to form mass blooms covering several km^2 and exhibiting biomasses exceeding 10 kg wet weight $\times$ m^{-2} [55–58] or to be concentrated by winds or currents in sheltered areas and the rapid changes in water-column chemistry which can occur when these blooms collapse and decompose, could also contribute to convective wind and cloud formation. Firstly, the dense floating mats of algae would absorb sunlight more strongly and therefore warm more quickly than surrounding clear water areas, thus columns of rising warm air (thermals) would naturally tend to form above the algal mats on sunny days even if DMS emissions were low or zero. These thermals would interact with atmospheric instabilities which arise due to the differential warming of the land and seawater in the coastal zones where these blooms occur and these inherent instabilities may be particularly sensitive to seeding by DMS. Additionally, provided if algal mats do not cover a high proportion of the surface (which is usually the case), the shallowness of coastal waters may make them particularly prone to the formation of 'white caps' whose bubbles can effectively concentrate small cells and eject them into the air when they burst at the surface. Even where the water surface is largely covered by, macro-algae, oxygen bubbles from photosynthesis or bubbles of fermentatively produced CO_2 and hydrogen during dystrophic crisis, must rise through and between the mats and could serve as launch vehicles for small cells. Similar to the occurrence of 'white tops', such bubbles will rise most when 'shaken out' by water movements caused by winds.

Secondly, at macro-algal biomass concentrations above 2–3 kg wet weight $\times$ m^{-2}, density dependent factors become dominant and drive the system towards dystrophy[54] and in eutrophied areas biomasses may often surpass 10 kg wet weight $\times$ m^{-2} [55–58] When these mass blooms collapse, the decomposition of the algal biomass creates physicochemical conditions in the water column which may also favour the transfer of dispersal phases to the atmosphere or promote the induction of convective winds. The massive release of dissolved organic matter from the algal tissues would tend to reduce the surface tension of seawater, promoting the formation of 'white caps' and/or surface foams and thereby enhance transfer rates of both dispersal phases and DMS to the atmosphere. Similarly, the mineralisation of the algal biomass in the water column also releases substantial quantities of ammonium, stimulating ammonia fluxes to the atmosphere. Alkaline ammonia can dissolve in acidic cloud sulfate droplets and this process is thought to be important in cloud nucleation[84] and would stimulate the formation of convective winds due to the increased release of latent heat of condensation. In sheltered areas where water exchange is limited, the decomposition of the algal biomass causes complete anoxia and bacterial sulfate reduction leads to the formation of free sulfides in the water column at μM to mM concentrations (refs 85, 86 and G. Giordani, R. Azzoni and P. Viaroli unpublished data). The resultant fluxes of highly reactive hydrogen sulfide to the atmosphere and its oxidation to sulfate would augment or even dominate rates of sulfate aerosols formation from DMS. Additionally, the high concentrations of highly basic sulfides in the water column would locally increase pH pushing the ammonium-ammonia equilibrium (pKa = 9.2) towards ammonia, thereby increasing ammonia fluxes to the atmosphere, which may also aid the induction of convective winds as discussed above.

Conclusions

Previous studies have concluded that the role of DMSP accumulation by marine macro-algal species is principally osmotic. However, although the high intracellular DMSP concentrations in macro-algae would contribute substantially to the overall osmotic pressure of the cytoplasm and thereby osmotic balance with the bathing medium, some studies have demonstrated that intracellular DMSP concentrations are not directly regulated by the osmotic pressure of the growth medium.[4–6,26] Thus, DMSP does not behave as a true osmoregulatory solute and its accumulation may be linked to other ecological adaptations such as the aerial dispersal mechanism which has been proposed for DMSP synthesising micro-algae.[1]

Whilst direct experimental evidence would be required to elucidate if DMSP synthesising chlorophycean macro-algae are dispersed or not by an aerial mechanism involving local convective winds induced by DMS fluxes, several aspects of the life, growth and reproductive cycles of these macro-algae are consistent with such a dispersal mechanism. The algae produce small motile gametes and zoospores which may be suitable for aerial dispersal and at least in some cases these are positively phototactic and thus may naturally concentrate at or near to the air-water interface. Chlorophycean algae are ephemeral species, often forming dense blooms which tend to collapse in summer when nutrient supplies become limiting and the bloom collapse may be associated with a heightened reproductive effort. The crash of the blooms could potentially release massive quantities of DMSP which may be rapidly converted to DMS by DMSP-lyases produced by the algae themselves or associated heterotrophic bacteria, and the rapid oxidation of this DMS in the atmosphere would enhance the thermals which naturally occur above the blooms on hot, sunny summer days. Finally, in eutrophied areas where blooms can attain densities of greater than 10 kg wet weight $\times$ m^{-2} the physicochemical conditions induced in the water column by the decomposition of the algal biomass may result in substantial fluxes of ammonia and hydrogen sulfide to the atmosphere which would also enhance convective winds. The combination of these events would favour the formation of 'white caps' and the rising columns of air above the algal mats would lift any cells thrown into the air by bursting bubbles higher into the atmosphere, leading to their dispersal by air currents.

Such an aerial dispersal mechanism would be adaptive in these species, allowing colonisation of areas separated by land barriers or directional water currents, which may act as physical barriers to waterborne dispersal phases in the coastal environment. Thus, aerial dispersal would allow colonisation of isolated areas such as coastal lagoons or the heads of estuaries, which may be difficult to colonise by conventional waterborne means.

Acknowledgements

The authors thank Prof. J.A. Raven, University of Dundee, for helpful comments on drafts of the manuscript. D.T. Welsh is currently supported by the European Environment and Climate programme ROBUST (Grant No. ENV4-CT96-0218). This work is a contribution to ELOISE programme (ELOISE No. 091).

References

1. W. D. Hamilton and T. M. Lenton. Spora and Gaia: how microbes fly with their clouds. *Ethology Ecology and Evolution* **10**, 1–16 (1998).
2. D. M. J. Dickson, R. G. Wyn Jones and J. Davenport. Osmotic adaptation in *Ulva lactuca* under fluctuating salinity regimes. *Planta* **150**, 158–166 (1980).
3. M. A. Ragan. Chemical constituents of seaweeds, pp. 589–626. In: C. S. Lobban and M. J. Wynne, (ed.) *The Biology of Seaweeds.* (University of California Press, Berkeley and Los Angeles, 1981).
4. R. H. Reed. Measurement and osmotic significance of β-dimethylsulphoniopropionate in marine macroalgae. *Marine Biology Letters* **4**, 173–181, (1983a).
5. D. M. Edwards, R. H. Reed, J. A. Chudek, R. Foster and W. D. P. Stewart. Organic solute accumulation in osmotically-stressed *Enteromorpha intestinalis. Marine Biology* **95**, 583–592 (1987).
6. U. Karsten, C. Wiencke and G. O. Kirst. Growth pattern and β-dimethylsulphoniopropionate (DMSP) content of green macroalgae at different irradiances. *Marine Biology* **108**, 151–155 (1991).
7. G. Blunden, B. E. Smith, M. W. Irons, M. Yang, O. G. Roch and A. V. Patel. Betaines and tertiary sulphonium compounds from 62 species of marine algae. *Biochemistry System Ecology* **20**, 373–388 (1992).
8. M. P. De Souza, Y. P. Chen and D. C. Yoch. Dimethylsulfoniopropionate lyase from the marine macroalga *Ulva curvata*: purification and characterisation of the enzyme. *Planta* **199**, 433–438 (1996).
9. R. P. Kiene. Dimethylsulfide production from dimethylsulphoniopropionate in coastal seawater samples and bacterial cultures. *Applied and Environmental Microbiology* **56**, 3292–3297 (1990).
10. R. P. Kiene and T. S. Bates. Biological removal of dimethyl sulfide from sea water. *Nature, London* **345**, 702–705 (1990).
11. J. Stefels and W. H. M. Van Boekel. Production of DMS from dissolved DMSP in axenic cultures of the marine phytoplankton species *Phaeocystis* sp. *Marine Ecology Progress Series* **97**, 11–18 (1993).

12. M. R. Diaz and B. F. Taylor. Comparison of dimethylsulfoniopropionate lyase activity in a prokaryote and a eukaryote. *Annual General Meeting of the American Society of Microbiology, Abstract* **N18**, 319 (1994).
13. J. Stefels, L. Dijkhuizen and W. W. C. Gieskes. DMSP-lyase activity in a spring phytoplankton bloom off the Dutch coast, related to *Phaeocystis* sp. abundance. *Marine Ecology Progress Series* **123**, 235–243 (1995).
14. M. Steinke and G. O. Kirst. Enymatic cleavage of dimethylsulfoniopropionate (DMSP) in cell-free extracts of the marine macroalga *Enteromorpha clathrata* (Roth) Grev. (Ulvales, Chlorophyta). *Journal of Experimental Marine Biology and Ecology* **291**, 73–85 (1996).
15. M. Steinke, C. Daniel and G. O. Kirst. DMSP lyase in marine macro- and micro-algae: interspecific differences in cleavage activity, pp. 317–324. In: R.P. Kiene *et al.*, (ed.) *Biological and Environmental Chemistry of DMSP and Related Sulfonium Compounds.* (Plenum Press, New York, 1996).
16. J. E. Lovelock, R. J. Mags and R. A. Rasmussen. Atmospheric dimethyl sulphide and the natural sulphur cycle. *Nature, London* **237**, 452–453 (1972).
17. R. J. Charlson, J. E. Lovelock, M. O. Andreae and S. G. Warren. Oceanic phytoplankton, atmospheric sulphur, cloud albedo and climate. *Nature, London* **326**, 655–661 (1987).
18. G. P. Ayers and J. L. Gras. Seasonal relationship between cloud condensation nuclei and aerosol methanesulphonate in marine air. *Nature, London* **353**, 834–835 (1991).
19. M. O. Andreae, W. Elbert and S. J. De Mora. Biogenic sulfur emissions and aerosols over the tropical South Atlantic. 3. Atmospheric dimethylsulfide, aerosols and cloud condensation nuclei. *Journal of Geophysical Research* **100** (D6), 11335–11356 (1995).
20. J. E. Lovelock. *Gaia—A New Look at Life on Earth.* (Oxford University Press, Oxford, U.K., 1979).
21. J. E. Lovelock. *The Ages of Gaia.* (Oxford University Press, Oxford, U.K., 1995).
22. K. Caldeira. Evolutionary pressures on planktonic production of atmospheric sulphur. *Nature, London* **337**, 732–734 (1989).
23. G. V. Wolfe and M. Steinke. Grazing-activated production of dimethyl sulfide (DMS) by two clones of *Emiliania huxleyi. Limnology and Oceanography* **41**, 1151–1160 (1996).
24. G. V. Wolfe, M. Steinke and G. O. Kirst. Grazing-activated chemical defence in a unicellular marine alga. *Nature, London* **387**, 894–897 (1997).
25. W. D. Hamilton and R. M. May. Dispersal in stable habitats. *Nature, London* **269**, 578–581 (1977).
26. R. H. Reed. The osmotic responses of *Polysiphonia lanosa* (L.) Tandy from marine and estuarine sites: evidence for incomplete recovery of turgor. *Journal of Expermental Marine Biology and Ecology* **68**, 169–173, (1983b).
27. J. M. Sieburth. Acrylic acid, an 'antibiotic' principle in *Phaeocystis* blooms in Antartic waters. *Science* **132**, 676–677 (1960).
28. K. L. Van Alstyne. Herbivore grazing increases polyphenolic defenses in the intertidal brown alga *Fucus distichus. Ecology* **69**, 655–663 (1988).
29. V. J. Paul and K. L. Van Alstyne. Activation of chemical defenses in the tropical green algae *Halimeda* spp. *Journal of Experimental Marine Biology and Ecology* **160**, 191–203 (1992).

30. D. C. Blanchard. The production, distribution and bacterial enrichment of the sea-salt aerosol, pp. 407–454. In: P.S. Liss and W.G.N. Slinn, (ed.) *Air-Sea Exchange of Gases and Particles*. (Reidal, Dordrecht, 1983).
31. C. E. Tanner. Chlorophyta, Life histories, pp. 218–247. In: C. S. Lobban and M. J. Wynne (ed.) *The Biology of Seaweeds*. (University of California Press, Berkeley and Los Angeles, 1981).
32. R. C. Hoxmark and Ø. Nordby. Haploid meiosis as a regular phenomenon in the life cycle of *Ulva mutabilis*. *Hereditas* **76**, 239–250 (1974).
33. E. R. Bonneau. Asexual reproduction capabilities in *Ulva lactuca* L. (Chlorophyceae). *Botanica Marina* **21**, 117–121 (1978).
34. C. D. Amsler and R. B. Searles. Vertical distribution of seaweed spores in a water column offshore of North Carolina. *Journal of Phycology* **16**, 617–619 (1980).
35. A. J. Hoffman and P. Camus. Sinking rates and viability of spores from benthic algae in central Chile. *Journal of Experimental Marine Biology and Ecology* **126**, 281–291 (1989).
36. P. Kornmann and P.-H. Sahling. Meeresalgen von Helgoland. Benthische Grün-., Braun-. und Rotalgen. *Helgol: Meeresunters* **29**, 1–29 (1977).
37. C. S. Lobban and P. J. Harrison. *Seaweed Ecology and Physiology*. (Cambridge University Press, Cambridge, U.K., 1994).
38. L. V. Evans and A. O. Christie. Studies on the ship-fouling *Enteromorpha*. I. Aspects of fine structure and biochemistry of swimming and newly settled zoospores. *Annals of Botany* **34**, 451–466 (1970).
39. P. S. Low. Molecular basis of the biological compatibilty of nature's osmolytes, pp. 469–477. In: R. Gilles and M. Gilles-Baillien, (ed.) *Transport Processes, Iono- and Osmoregulation*. (Springer-Verlag, Berlin, 1985)
40. S. N. Timasheff. Stabilization of proteins by cosolvents in aqueous media. *Cryobiology* **30**, 218–219 (1993).
41. C. L. Winzor, D. J. Winzor, L. G. Paleg, G. P. Jones and B. P. Naidu. Rationalization of the effects of compatible solutes on protein stability in terms of thermodynamic nonideality. *Archives of Biochemistry and Biophysics* **296**, 102–107 (1992).
42. M. Potts. Desiccation tolerance of prokaryotes. *Microbiological Reviews* **58**, 755–805 (1994).
43. P. Mazur. Survival of fungi after freezing and desiccation, pp. 325–394. In: G. C. Ainsworth and A. S. Sussman, (ed.) *The Fungi. Vol.* **3**, *The Fungal Population*. (Academic Press, New York, 1968)
44. I. S. Bhandal, R. M. Hauptman and J. M. Widholm. Trehalose as cryoprotectant for the freeze preservation of carrot and tobacco cells. *Plant Physiology* **78**, 430–432 (1985)
45. J. F. Carpenter, S. C. Hand, L. M. Crowe and J. H. Crowe. Cryoprotection of phosphofructokinase with organic solutes. Characterisation of enhanced protection by divalent cations. *Archives of Biochemistry and Biophysics* **250**, 505–512 (1987).
46. M. K. Nishiguchi and G. N. Somero. Temperature and concentration-dependence of compatibility of the organic osmolyte β-dimethylsulfoniopropionate. *Cryobiology* **29**, 118–124 (1992).
47. J. H. Crowe, L. M. Crowe, J. F. Carpenter and C. Aurell Wistrom. Stabilization of dry phospholipid bilayers and proteins by sugars. *Biochemical Journal* **242**, 1–10 (1987).

48. J. H. Crowe, F. A. Hoekstra and L. M. Crowe. Anhydrobiosis. *Annual Review of Physiology* **54**, 579–599 (1992).
49. S.B. Leslie, E. Israeli, B. Lighthart, J. H. Crowe and L. M. Crowe. Trehalose and sucrose protect both membranes and proteins in intact bacteria during drying. *Applied and Environmental Microbiology* **61**, 3592–3597 (1995).
50. J. L. Green and C. A. Angell. Phase relations and vitrification in saccharide-water solutions and the trehalose anomoly. *Journal of Physical Chemistry* **93**, 2880–2882 (1989).
51. C. A. L. S. Colaço, C. J. S. Smith, S. Sen, D. H. Roser, Y. Newman, S. Ring and B. Roser. Chemistry of protein stabilization by trehalose, pp. 216–240. In: J. L. Cleland and R. Langer, (ed.) *Formulation and Delivery of Proteins and Peptides.* (American Chemical Society, Washington DC, 1993).
52. F.A. Hoekstra, L.M. Crowe and J.H. Crowe. Differential desiccation sensitivity of corn and *Pennisetum* pollen linked to their sucrose content. *Plant Cell and Environment* **12**, 83–91 (1989).
53. G. Rosenberg and J. Ramus. Ecological growth strategies in the seaweeds *Gracilaria foliifera* (Rhodophyceae) and *Ulva* sp. (Chlorophyceae): the rate and timing of growth. *Botanica Marina* **24**, 583–589 (1981).
54. P. Viaroli, M. Naldi, C. Bondavalli and S. Bencivelli. Growth of the seaweed *Ulva rigida* C. Agardh in relation to biomass densities, internal nutrient pools and external nutrient supply in the Sacca di Goro lagoon (Northern Italy). *Hydrobiologia* **329**, 93–103 (1996).
55. A. Sfriso, A. Marcomini and B. Pavoni. Relationships between macroalgal biomass and nutrient concentrations in a hypertrophic area of Venice lagoon. *Marine Environment Research* **22**, 297–312 (1987).
56. A. Sfriso, B. Pavoni and A. A. Orio. Macroalgae, nutrient cycles, and pollutants in the lagoon of Venice. *Estuaries* **15**, 517–528 (1992).
57. P. Viaroli, A. Pugnetti and I. Ferrari. *Ulva rigida* growth and decomposition processes and related effects on nitrogen and phosphorus cycles in a coastal lagoon, pp. 77–84. In: G. Colombo *et al.*, (ed.) *Marine Eutrophication and Population Dynamics.* (Olsen and Olsen, Fredensborg, 1992).
58. P. G. Soulsby, D. Lowthion, M. Houston and H. A. C. Montgomery. The role of sewage effluent in the accumulation of macroalgal mats on intertidal mudflats in two basins in southern England. *Netherlands Journal of Sea Research* **19**, 257–263 (1985).
59. C. Bliding. A critical survey of European taxa in Ulvales. Part II. *Ulva, Ulvaria, Monostroma, Kornmannia. Botanical Notes* **121**, 535–629 (1963).
60. D. F. Kapraun. Field and cultural studies of *Ulva* and *Enteromorpha* in the vicinity of Port Aransas, Texas. *Contributions to Marine Science* **15**, 205–285 (1970).
61. R. C. Hoxmark. Experimental analysis of the life cycle of *Ulva mutabilis. Botanica Marina* **18**, 123–129 (1975).
62. E. D. Parker, Jr. Ecological implications of clonal diversity in parthenogenic morphospecies. *American Zoologist* **19**, 753–762 (1979).
63. R. C. Vrijenhoek. Factors effecting clonal diversity and coexistance. *American Zoologist* **19**, 787–797 (1979).
64. R. J. Jeffries and L. D. Gottlieb. Genetic variation within and between populations of the sexual plant *Pucinellia* × *phryganodes. Canadian Journal of Botany* **61**, 774–779 (1983).

65. B. Schmid. Effects of genetic diversity in experimental stands of *Solidago altissima*—evidence for the potential role of pathogens as selective agents in plant populations. *Journal of Ecology* **82**, 165–175 (1994).
66. W. R. Barnard, M. O. Andreae and R. L. Iverson. Dimethylsulfide and *Phaeocystis poucheti* in the southeastern Bering Sea. *Continental and Shelf Research* **3**, 103–113 (1984).
67. J. A. E. Gibson, R. C. Garrick, H. R. Burton and A. R. Mc Taggart. Dimethylsulfide and the alga *Phaeocystis pouchetii* in antartic coastal waters. *Marine Biology* **104**, 339–346 (1990).
68. K. M. Crocker, M. E. Ondrusek, R. L. Petty and R. C. Smith. Dimethylsulfide, algal pigments and light in an Antartic *Phaeocystis* sp. bloom. *Marine Biology* **124**, 335–340 (1995).
69. G. L. Cantoni and D. G. Anderson. Enzymatic cleavage of dimethylpropiothetin by *Polysiphonia lanosa. Journal of Biological Chemistry* **222**, 171–177 (1956).
70. M. K. Nishiguchi and L. J. Goff. Isolation, purification and characterisation of DMSP lyase (dimethylpropiothetin dethiomethylase (4.4.1.3)) from the red alga *Polysiphonia paniculata. Journal of Phycology* **31**, 567–574 (1995).
71. K. M. Ledyard, E. F. De Long and J. W. H. Dacey. Characterization of a DMSP-degrading bacterial isolate from the Sargasso sea. *Archives of Microbiology* **160**, 312–318 (1993).
72. M. P. De Souza and D. C. Yoch. Purification and characterisation of dimethylsulfoniopropionate lyase from an *Alcaligenes*-like dimethyl sulfide producing marine isolate. *Applied and Environmental Microbiology* **61**, 21–26 (1995).
73. B. F. Taylor and D. C. Gilchrist. New routes for aerobic biodegradation of dimethylsulfoniopropionate. *Applied and Environmental Microbiology* **57**, 3581–3584 (1991).
74. M. J. E. C. Van Der Maarel, P. Quist, L. Dijkhuizen and T. A. Hansen. Anaerobic degradation of dimethylsulfoniopropionate to 3-S-methylmercaptopropionate by marine sulfate-reducing bacteria. *Applied and Environmental Microbiology* **62**, 3978–3984 (1993).
75. P. T. Visscher and B. F. Taylor. Demethylation of dimethylsulfoniopropionate to 3-mercaptoproprionate by an aerobic marine bacterium. *Applied and Environmental Microbiology* **60**, 4617–4619 (1994).
76. P. Brimblecombe and D. Shooter. Photo-oxidation of dimethylsulfide in aqueous solution. *Marine Chemistry* **19**, 343–353 (1986).
77. D. P. Kelly and N. A. Smith. Organic sulfur compounds in the environment: biogeochemistry, microbiology and ecological aspects. *Advances in Microbiology and Ecology* **11**, 345–385 (1990).
78. P. T. Visscher and H. Van Gemerden. Photoautotrophic growth of *Thiocapsa roseopersicina* on dimethyl sulfide. *FEMS Microbiology Letters* **81**, 247–250 (1991).
79. S. H. Zinder and T. D. Brock. Dimethyl sulphoxide reduction by micro-organisms. *Journal of General Microbiology* **105**, 335–342 (1978).
80. H. M. Jonkers, M. J. E. C. Van Der Maarel, H. Van Gemerden and T.A. Hansen. Dimethylsulfoxide reduction by marine sulfate-reducing bacteria. *FEMS Microbiology Letters* **136**, 283–287 (1996).
81. A. F. Mohsen, A. F. Khaleafa, M. A. Hashem and A. Metwalli. Effect of different nitrogen sources on growth, reproduction, amino acid, fat and sugar contents in *Ulva fasciata* Petite. *Botanica Marina* **17**, 218–222 (1974).

82. G. Nilsen and Ø. Nordby. A sporulation-inhibiting substance from vegatative thalli of the green alga, *Ulva mutabilis* Føyn. *Planta* **125**, 127–139 (1975).
83. R. A. Niesenbaum. The ecology of sporulation by the macroalga *Ulva lactuca* L. (Chlorophyceae). *Aquatic Botany* **32**, 155–166 (1988).
84. F. L. Eisele and P. H. McMurray. Recent progress in understanding particle nucleation and growth. *Philosophical Transactions of the Royal Society of London (B)* **352**, 191–201 (1997).
85. P. Caumette. Phototrophic sulfur bacteria and sulfate-reducing bacteria causing red waters in a shallow brackish coastal lagoon (Prévost lagoon, France). *FEMS Microbiology Ecology* **38**, 113–124 (1986).
86. J. Castel, P. Caumette and R. A. Herbert. Eutrophication gradients in coastal lagoons as exemplified by the Bassin d'Arcachon and Étang du Prévost. *Hydrobiologia* **329**, IX–XXVIII (1996).

CHAPTER 16

LIFE, EVOLUTION AND DEVELOPMENT IN THE AMAZONIAN FLOODPLAIN

PETER HENDERSON

Our paper *Evolution and Diversity in Amazonian Floodplain Communities* was in part an essay on the biological wonders and inspirational diversity of the Amazon Rainforest. It was written expressly for a conference held in Cambridge, England in 1998, but it is far more than a quick paper stitched together to justify attendance at a meeting. Bill and I had come to suspect that tropical wetlands were the most important generators of biological innovation on the planet. Over a number of trips with Will Crampton to the Mamirauá reserve near Tefé, Brazil, we had noted how the particular conditions within the floodplain created novel forms of adaptation. Microscopic floating plants, sponges living in trees, carnivorous plants, floating ferns, pack hunting fish that swim through forest, fish that communicate using electricity, dolphins that hunt between trees, the list of forms with unusual behaviour or life styles was almost endless. Yet by tropical forest standards of diversity the floodplains are not species rich. It was not the vast number of species that came to be the focus of our deliberations, but the potential of this habitat to produce new forms of life. We came to believe that this was because the floodplain selects for simplicity and plasticity and as the last sentence of our paper states, 'Simplified, responsive organisms are a clay from which evolution may be moulding not so much abundance of species as novel forms.'

Once I would have described our long-term relationship as a succession of odd coincidences that kept us in contact. The events that led from our first meeting to the production of this paper now seem like a straight, inevitable path. From the moment of our first conversation at Imperial College in 1975, to our final walk in the New Forest, Hampshire in January 2000 on the eve of his final field trip to Africa, we discussed evolution and rain forests. As I considered what I should do after graduating, Bill suggested that I travel to Brazil. He thought that all evolutionists benefited from time in the tropics; 'just look at the influence the tropics had on the intellectual development of Darwin, Bates and Wallace'. True, I could not speak the language, had little money and no contacts in the Amazon. This seemed of little consequence to Bill. Brazilians are generous people and it was cheap to live in the forest. Fortunately, I did not take his advice, for it would probably have ended in disaster. Bill's scientific intuition was always superb, but his advice on career development and how to find work almost always ruinous and frequently dangerous. But he had sown the tropical seed, and when the opportunity came in 1981 to go and work in Brazil with Dr Ilse Walker in Manaus, Amazonia, I grabbed the chance.

It was on this first trip to Brazil that I met the man who later would bring Bill and me to the floodplains of the Rio Japurá, near Tefé and lead to the writing of this paper. Marcio Ayres had returned to INPA, the Brazilian research institute in Manaus in 1981 to present his masters thesis and in addition to the main dissertation, he had to write an essay on a selected topic in evolutionary biology. His chosen essay topic was the evolution of sexual reproduction. Somehow he heard that a visiting professor who worked with Bill Hamilton on the evolution of sex was at INPA, so he sought me out. Our friendship rapidly developed and Marcio, who was a native Amazonian, set about expanding my Amazonian knowledge. This included introducing me into the world of late-night scandalous gossiping, a Brazilian speciality, interspersed with in-depth scientific discussion and smoking until the sun rose, which was a Marcio speciality. I was amazed to discover that Marcio used to work for Dr Warwick Kerr who, in turn, was a friend of Bill. It seemed as if Fate had determined that Bill, Marcio and I would work together, although it would be more correct to state that mutual interests had drawn us together. When Marcio and I parted company after 3 months together we knew we would meet again in England because the following year he planned to study for his doctorate at Cambridge, England. It is painful to write that Marcio died not too long after Bill. While the Amazon and its wide range of debilitating diseases was not the cause, I feel that tropical field biology has taken too many of my friends and acquaintances.

For his doctoral studies Marcio worked on a rare monkey only found in the floodplain of the Rio Japurá. This monkey is known to Brazilians as the English monkey because, as Marcio put it, 'it has a face like a sunburnt English gin drinker'. It was while undertaking his field work that he started to formulate the idea for the Mamirauá Ecological Reserve. Later, after completing his thesis, he developed the idea for a great inundation forest reserve further and contacted Bill and me to ask us if we would work with him. The moment had arrived when we would work and travel together in Amazonia. Soon after this suggestion was made Bill and I flew to Tefé and spent two weeks travelling with Marcio in his small boat Giovotta. We travelled extensively around the area that would later become the Mamirauá reserve. With hindsight I now see that Marcio took us to selected places with wonderful views and extraordinary ecology. They were also considerably more mosquito-free than many places we later came to work in. He was keen that we should see the beauty and ecological importance of the region. The trip was wonderful and we were captivated. We were certainly not inspired by the food, as Marcio was suffering from an ulcer, and he told the cook to only prepare plain boiled chicken and serve it with white rice. Despite two weeks of extraordinarily dull food, Bill and I agreed that there could be no more inspirational place to work. I would study the fish and aquatic ecology while Bill would focus on the plants. I think Marcio wanted Bill to join the project because he felt that Bill would bring scientific weight and influence. Perhaps I was included to look after Bill, but I hope my knowledge of Amazonian aquatic ecology was also a factor. He was also aware that Henry Walter Bates, the great nineteenth-century British biologist had stayed for some years at Tefé (then called Ega) and had visited Mamirauá. We imagined ourselves as part of a great British tradition and the project needed initial financial support from Britain. We were able to obtain generous funding from both the Royal Society and the Overseas Development Agency of the British Government (now the Department for International Development). It may surprise some readers that Mrs Thatcher was keen that her government should support work on conservation in Amazonia.

It was on this first trip that Bill and I spent some time seeking out angelfish, *Pterophyllum scalare*, and discus, *Symphysodon aequifasciata*. These wonderful cichlid fish, which are adapted to swimming through roots and submerged trees, have become familiar aquarium species and both had once been abundant in the vicinity of Tefé. We were interested in the evolution of their social behaviour and parental care. While fish watching, I introduced Bill to the weird and little-known electric fish fauna. It was clear that the Rio Japurá floodplain held an exceptional assemblage of electric fish, although at this time I could only guess at the wonderful species that would later be found by Will Crampton. It was these preliminary considerations that led to the third author of this paper coming to work in Amazonia. Will was an Oxford graduate with a driving desire to work in Amazonia. He came to hear of our work in Brazil and contacted Bill to see if he would supervise him for his doctorate. I suggested the electric fish, Bill suggested angelfish, and after a depressing period trying to sample angelfish, Will undertook a superb study of the electric fish. Will followed my suggestion because, unlike Bill's, it was feasible. Bill's suggestions on interesting work to be undertaken on fish were at best difficult and frequently impossible. His background was as an entomologist and he tended to view the activities of aquatic ecologists as something of a sideshow. I could never persuade him that the wood louse was the advance guard of the coming crustacean invasion of the land. He would point out that the real evolutionary action lay with the insects and terrestrial plants.

So in the early 1990s, the three of us were working regularly together in the Mamirauá Reserve where we gradually developed a joint approach to study Amazonian floodplain animals and plants. On some trips we would mostly stay on a floating house moored in Lago Mamirauá. On other occasions Bill accompanied me on boat trips around the reserve. Irrespective of the locality or accommodation, our days were generally spent in a similar manner. While I went out with Will and our assistants to net fish and sample the life of floating meadows, Bill would go off in his own metal canoe with an outboard to study and record the life of the lakes and channels. Sometimes he would work alone, but on many trips Luisa Bozzi, his partner, or an Amazonian field guide accompanied him. On occasion he would even undertake extended trips and sleep overnight in the canoe. I would always be glad to see him at the end of such a trip. Not only was there relief that he was well, but the evening would be spent listening to his account of the places visited and the plants and animals seen. Sadly, we did not always take great pleasure in each other's company. Bill could also be a most annoying shipmate—his time keeping was often bad, he left his kit everywhere and it was most unwise to leave any organising of stores to him. He was apt to buy a wide range of unusual fruits and forget the toilet paper. He could also fill a boat with plants, spiders, snakes and even dead monkeys. I used to organise our joint expeditions and gradually developed the role of unofficial minder. However, when things went wrong, Bill was always the first to volunteer to jump over the side or take the injured man 40 miles back to Tefé in a canoe. He was extremely brave and a fine companion in a tight spot.

In late afternoon we would all return to the boat and normally the last to come back was Bill with his canoe filled with vegetation that he would heave onto the deck. This large compost heap would now be in place on the deck until after dinner. Sailors the world over are very tidy and keep all decks spotless, and Bill's specimens drove them mad. However, they never complained to him, only to me. 'Please Peter—you must help us. The Professor has now brought a snake back in his vegetation and the engineer almost sat on it . . . ' After a shower we would sit down to dinner and discuss what we had found. Afterwards, we would work through our specimens while drinking coffee. Will and I would identify, weigh and measure fish and Bill would work through his compost heap, selecting specimens to be pressed. At the end of the evening, with everything stowed and our hammocks set, Bill and I would occasionally sit in the cool on the roof of the boat and drink a glass of whisky. Below us we would hear the clatter of the crew playing dominoes while we watched the insects fly around the navigation lights. On these occasions we experimented with the latest fruits that Bill had found while collecting. We discovered some that combined wonderfully with scotch. One evening as we sat talking, a wasp attracted to the navigation lights flew up my shirt and stung me. I jumped, and complaining, threw the wasp off. With a surprisingly swift and practiced swipe Bill grabbed the wasp in flight and examined it. 'Arrgh I know this group—in fact I know the species—I believe it is called—hamiltonii', and with an embarrassed, self-conscious look he told me it was named after him. I had been stung by my friend's namesake.

So our days would go on as the boat progressed from lake to lake. A boat develops a rhythm and everyone gradually finds their spot and their storage area. Bill's domain would get gradually larger as his plant presses expanded to be joined by a collection of bogwood, bromeliads and skulls. Then, a time was reached when the crew knew we would soon be returning home. The mood started to change—there was expectancy in the air. Discussion turned to Tefé and what the town had to offer. I discussed our return with the captain and the momentum was unstoppable. The crew wanted, needed, must return. We turned for home, but the dash for civilization was curtailed because Bill had noticed a particularly interesting island that just could not be missed. I was glad because he had refocused us on scientific study. He was utterly unaware of the crew's feelings. Other considerations were secondary, for we must only leave when the science had been achieved. At this moment, island exploration was not a popular move. Some hours later we were again underway, rushing down river to try to make Tefé before dark. We approached after dark and the lights of the town filled everyone with excitement. When we left it seemed like a small rather shabby Amazonian town. Now it was the big town with beckoning lights. As we took our mooring all the crew had shaved and had found wonderfully clean shirts. Even the engineer had a clean shirt! They were reborn men. Bill sat in the corner on his chest, examining an interesting specimen with binoculars and camera at hand. He was dishevelled, in torn trousers and looking tired. But he was clearly content, but almost reluctant to leave the boat. We set to work helping Bill to offload his kit as we all wanted to be off to the bar. The first cold beer on land was a pleasure, and we sat contented, looking not unlike the last scene in the film Ice Cold in Alex. Our specimens were safe and tomorrow we would work in air conditioning.

I think the Amazon became particularly important to Bill because he escaped the desk and constant demands upon his time and so enjoyed being surrounded with active, noisy, hyper-diverse life. At Oxford, Bill had to deal with an almost impossibly large correspondence, mostly by e-mail, which he always tried to answer in depth and with care. I watched it gradually take its toll of his energy, although he never once complained. I am sure that he was at his happiest in the Amazon. I remember a particularly happy night when we camped on the banks of Lago Tefé, cooking on an open fire and watching bats, mole crickets, nightjars and a multitude of flying insects. We spent some time fishing with cotton thread for mole crickets on the beach. We lowered the threads down their vertical tunnels in the sand and when the insects grabbed our threads we pulled them from their burrows. As a teenager in the New Forest I had always wanted to see a mole cricket, but could never find this elusive and remarkable insect. Here they were plentiful and after admiring their great strength by feeling them burrowing out of our clenched fists we watched them return to their burrows. After dark we lay on our backs on the sand enjoying the cool of the night and discussing the brilliance of the Milky Way and the succession of shooting stars. It is a powerful memory, for I believe he was more at peace and content that night than at any time I can recall over our more than 20 years of working together.

Back in Oxford, we gradually wrote the text of our paper over about a 6-week period. The text is an almost total fusion of our writing styles. By this point we had spent so much time together that I would not always know if an expression was one of his or mine. But the one word in the introduction that certainly came from Bill was palimpsest. So precise and such fun to use, but how many of our readers will understand? He found my description of Amazonian waters as languid equally amusing (and I hope accurate) so that rather odd word also remains. At the start we had no idea of the structure the paper would take or the size of the finished product. We both wrote our individual bits and I bolted them together. Then we passed the text back and forth perhaps four times. During the working day we spoke little except during morning and afternoon tea breaks. In fact, we sometimes hardly moved. I had not realised how quiet and still we were until Bill was given a new suite of offices which were fitted with lights that would switch off after a set period if they did not detect movement. On occasions, as we worked we would be plunged into darkness, and I would wave an arm to activate the lights. We took pleasure in such quiet contemplation and the utter lack of need to speak.

The verbal presentation of our paper was a rerun of a number of similar disasters. Will was not present as he was still working in Brazil. I could see what would unfold, it was inevitable and as such I was not concerned. We met outside our accommodation in Cambridge and as we carried our kit in I wondered where Bill had packed his slides. A few minutes later there was a knock on the door and Bill asked if he could borrow some of my slides as he seemed to have left his at Oxford. This was anticipated and was no surprise, although I have yet to get them back and rather suspect they have now become part of the Hamilton archive at the British Library. We had a 20-minute slot for our presentation and it was agreed that Bill would start with a general introduction to our ideas on macroevolution and then talk about our thoughts on the development of novel plant forms and the transition between herbs and trees. I would then take over to discuss the fish and the physical environment. Bill's part of the presentation was captivating and the audience was interested. When I got up to speak the chairman pointed out that I had 1 minute left. I spoke for 4 minutes and decided not to push our luck further. This turnout was not really a disappointment nor annoying. Why we both tried to speak in a single 20-minute slot is a question I still cannot fully answer. It is not as if either of us desired to perform in front of an audience. In part, it was because of Bill's sense of fair play. He was well known. I was not. If I did not speak, the audience would view the work as solely Bill's. Perhaps, but it probably had more to do with our general lack of planning and our constant desire to try to do something different. The result was that nothing was ever rehearsed and slides and overheads were constructed at the last possible moment. Sometimes it worked and we both felt elated as we talked with other workers during the breaks. More often it did not.

In *Evolution and Diversity in Amazonian Floodplain Communities*, we wanted to consider large-scale major evolutionary innovations. We discussed events over millions of years and ranged over both the plant and animal kingdoms on both land and in water. We wrote about some of the greatest events in biological evolution, the conquest of land by the vertebrates and the evolution of wood so that plants could become trees and forests could form. Such great generalizations are risky and I remain unsure if we were successful. I think our combination of ecological and evolutionary thought within a single paper is interesting and unusual and reflects our combination of complementary skills. It was good to work with Will and Bill. The change in scale from my earlier work with Bill on the mathematical genetics of sex using one, two or three locus models (see Volume 2) is striking and reflects the gradual expansion of the vision and confidence of Bill in the later part of his life. Having studied and understood the fall of the apple, he was now willing to consider planets. This paper did not represent the final stage in this development and at the time of his death, we were working on a Gaian style of model for the evolution of life on planets. As Charles Darwin or Henry Bates would almost certainly have agreed—the Amazon inspires big ideas. If in this paper we rather over extended ourselves, I hope the reader will be generous and remember that it was written while the grandeur of the floodplain and the flash of sunsets over Lake Tefé were still clear in our minds.

If we had known time was short, we would have worked with more urgency. This paper was a first attempt. Bill and Luisa were planning to come and live in the New Forest so that we could continue our studies. I miss my friend.

EVOLUTION AND DIVERSITY IN AMAZONIAN FLOODPLAIN COMMUNITIES†

P.A. HENDERSON, W.D. HAMILTON AND W.G.R. CRAMPTON

Summary

1. The physical environment of the Rio Solimões and Rio Japurá floodplain confluence is shown to be highly variable both spatially and temporally at scales ranging from 0.01 to 10 000 m and 1 h to 100 years.
2. In the case of fish the floodplain probably holds > 300 species; however, most all these species can be found in other local habitats such as main river channels, black-water lakes or forest streams.
3. For both plants and fish it is argued that factors positive for gene-flow and negative for patch persistence make speciation in the floodplain uncommon.
4. Speciation is much more common under the isolation afforded by side streams and peripheral areas; floodplain species and clades are probably usually derived from these areas.
5. Lack of speciation combines with the high rate of temporal change to demand plastic adaptability in most floodplain inhabitants.
6. Low speciation is not low macroevolution; indeed much remarkable floodplain adaptation is present. 'Genetically assimilable' plasticity often precedes radical novelty.
7. Examples are given of past important macroevolutionary developments that may have originated in tropical floodplains, including the likely origins of angiosperms[1] and of tetrapods.[2] Radical innovations of the present and recent past in the Amazon floodplain are illustrated.

Introduction

The landscape of the Upper Amazon floodplain is a palimpsest written by its recent erosions and depositions of the river. Land is cut and relaid to maintain

† In D. M. Newbery, H. H. T. Prins and N. D. Brown, *Dynamics of Tropical Communities*, The 37th Symposium of The British Ecological Society, London (1998), pp. 385–419 (Blackwell Science, Oxford, 1998).

an intricate mosaic of forest, scrub, marshes, lakes and channels. Owing to the annual flood regime, all habitats flood occasionally. Most habitats flood every year, some submerging by up to 11 m. Some floating communities do not submerge but are still subject to mixing and disruption by the floods. We discuss how this restless physical setting affects its ecological and evolutionary patterns, particularly those of fish and plant life.

All habitats are dominated by the physical activities of the river. Much of the 'land' begins as ridges alternated with swamps or lakes. The ridges are created by annual floods and sedimentation at the sides of major channels. As land distances from the main channels, fine deposits in-fill this relief and poor drainage in the low-water season forms backswamps. In such interior areas the vegetation cannot progress to the high forest of levees and sandbar ridges but forms lower swamp forest *(chavascal)*. Subsequently, this land is recut by the return of the previous channel or another. Therefore, in sharp contrast to the adjacent unflooded forest no climax vegetation type is ever reached.[3] Likewise, no old waterbodies develop. Along straight transects, open-water, swamp and forest communities change abruptly, while along the more natural lines of the water channels changes are smoother but still show hierarchies of rather discrete patches at many scales.

Richly vegetated floodplains of tropical rivers were probably present in one continental site or another ever since macrophytes and large animals evolved in the early Palaeozoic. Many important episodes of fossilization are due to such riverine environments. Freshwater flood-prone forest habitats contributing to coal measures, for example, have been well studied. In spite of their importance for the geological record and for evolution, we are not aware that either the general evolutionary characteristics of the biota of such riverplains, or how these are likely to differ from those of contemporary habitats of other kinds, have been discussed. A thread of very relevant research on floodplain ecology and evolutionary outcome, however, has been extended for the single beetle family Carabidae[4] and we return to this later. General questions that arise include, how typical are such habitats of the biodiversity of their times? Are they rich or poor in the incipient radical novelties that may later come to demarcate major groups?

Based on our experience over 6 years at a nature reserve in western Brazil (Estaçäo Ecologica Mamirauá) (Fig. 16.1) we discuss ecological and evolutionary characteristics of the plants and fish of the particularly broad section of the alluvial plain of the Rio Solimões (Amazon) that surrounds its junction with a large ex-Andean tributary, the Rio Japurá. First, we describe the physical aspects of the floodplain, the template[5] upon which the diversity and life-histories of the biota are formed. Second, we illustrate the biodiversity supported using data on plants and fish. Finally, we consider the scope and evidence for species formation and macroevolution within the várzea (white-water floodplain).

Figure 16.1. Map of the floodplain of the Rios Solimões and Japurá showing the main rivers referred to in the text. The map shows that part of the floodplain designated as the focal area of Reserva Mamirauá. The networks of lakes and channels within the reserve are filled in black.

Habitat Types

The Andean rocks supply both the sediments from which the floodplain is formed and the dissolved nutrients which allow high biomass productivity. Lower productivity black-water habitats fed by nutrient-poor forest catchments exist nearby. These two types of seasonally flooded habitat occurring inside and outside of the outermost levees of the river and known respectively as várzea and Igapó are often sharply contrasted even when spatially close.[6] Igapó is not confined to the neighbourhood of the flood-plain; it is also found around the edges of the sometimes very large impeded ('black-water') lakes and tributaries (the whole lower stretch of the Rio Negro is one example) and is scattered as poorly drained and temporarily flooding patches within the matrix of non-flooding terra firme forest. In contrast, várzea habitats and vegetation are strictly confined to silty 'white-water' floodplains.

Within the general category of várzea it is useful to distinguish several habitat subtypes. These are defined in Table 16.1 (for Upper Amazon comparable habitats see ref 7). It is inevitable that all the named landscape features intergrade, and contrasting the simplicity of Table 16.1 with the complexity of Fig. 16.1 (with even the last obviously far from exhausting the habitats that could be shown by a closer mapping) reaffirms an unavoidable vagueness in the definitions. A rapid-flowing *parana* at flood season may transform to a series of lakes at low water; lakes are often elongated like *paranas* and may be not much wider than their own channels; and so on. At high water, the entire floodplain can be under water that is for the most part nearly static. At such a time only the pattern of the emergent vegetation allows the different types of waterbodies to be distinguished. Even when not all is covered, it is useful to remember that all land has been deposited by the river and therefore *várzea* includes nothing, excepting upper parts of substantial trees, that is never flooded.

LARGE-SCALE PROCESSES

All of the physical features mentioned range greatly in size, reflecting the variety of processes that form them. Those on the largest scale gain their characteristic

Table 16.1. Floodplain habitats with their equivalent local names in italics.

Mainly terrestrial	Wholly or mainly aquatic
Island and point sand bars—*praias*	Main river channels—*rios*
Levees—*restingas* (silt-heightened riverbanks and bars, on which grows tall forest)	Lesser river channels—*paranas* (Usually flowing but sometimes static during low water)
Low swampy woodland—*chavascal* (usually behind levees) White water floodplain—*várzea*	Lake channels—*canos* (as above but connecting to lakes rather than to other channels)
Black-water floodplain—*Igapó* (i.e. around lakes and streams outside the flanking levees of a *várzea* Never inundated land—*terra firme* (land flanking the floodplain)	Lakes—*lagos* (waterbodies of variable shape and size but usually fluvual in origin holding static or slow-flowing water)

forms over time scales of $>10^4$ years. The most important of these events for the Reserva Mamirauá area have been the eustatic ocean changes of the Holocene.[8] During periods of lower sea water level, rapid channel incision occurred. Over the past 4000 years a rise in sea water level has widened the floodplain and aggrading silt has slowly limited the blocked-valley ('black-water') lakes along the border of the reserve to their present shape. So-called 'black-water' of such blocked valleys and side-streams is characteristically red-tinted with forest-derived humic material. Margins of the black-water lakes support the *Igapó* flooded forest and have little floating meadow.

MEDIUM-SCALE PROCESSES

The patchy, often laminar, structure so conspicuous in radar and satellite photographs of the floodplain is created over time scales of 10^2–10^4 years. Main channel migration over this time scale results in the general pattern of lakes and channels we see. Channel migration can be rapid. Puhakka, *et al.*[9] give an average erosion rate of 115 m day^{-1} and a maximum of 300 m day^{-1} for the Amazon in Peru. Other estimates are 400 m $year^{-1}$ near the Peruvian border with Brazil[10,11] and 200 m $year^{-1}$ beach growth at the Solimões side of the Mamirauá reserve (Pedro Santos, personal communication). Within the *várzea* new waterbodies are formed in various ways. Meander migration results in a scroll-swale topography within the meander loops. The low-lying region between the meander bar and the inside bank initially forms a crescent lake which later becomes a back swamp because of overbank sedimentation. Many lakes within the reserve are crescent lakes of this kind and they offer a series of successional habitats.[12] Larger oxbow lakes are formed by the cutting of whole meanders. These vary considerably in age and size. The largest result from the abandonment (avulsion) of a main river channel and may include a number of meanders. Within the reserve it is likely that its eponymous Lago Mamirauá is an abandoned main river channel, a notion supported by its maximum depth of about 40 m which is similar to that of the adjacent main channel of the Rio Solimões.

The continual reworking of the floodplain in the manner described produces a characteristic scaling of lake size. Hamilton *et al.*[13] showed that for Amazonian floodplain lakes size distribution is self-similar over a size range from 0.74 to 1000 km^2 with an estimated fractal dimension of 1.8. Using 1 : 50000 maps for the Mamirauá reserve focal area, which is the triangle of floodplain bounded by the Rio Solimões, Rio Japurá and the Parana Aranupu, the Pareto plot in Fig. 16.2 shows that this self-similar structure extends to waterbodies as small as 10 ha. Below this size the Pareto plot is non-linear, probably reflecting census bias against smaller waterbodies; over the approximately linear part of the curve and slope is -1.27, exactly the same as Hamilton *et al.*[13] calculated for Orinoco channel lakes and confirming the fractal patchiness of water distribution at this scale.

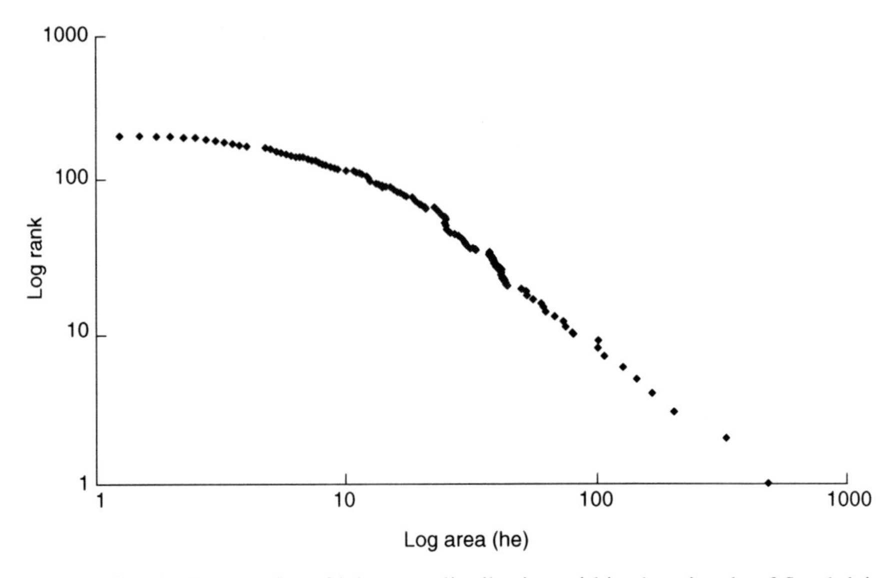

Figure 16.2. The Pareto plot of lake area distribution within the triangle of floodplain within the Rios Solimões and Japurá and the Parana Aranapu.

SHORT-TERM PROCESSES

Over time-scales of $1–10^2$ years, shear stress and sediment deposition channel structures such as undercut banks, sand bars and levees. Erosion and sedimentation create, change and destroy both aquatic and 'land' habitats. For example, bank erosion fells and sinks trees to create submerged wood habitats which are characteristic for many animal and plant species. Sedimentation changes flow patterns so that well-oxygenated flowing channels transform to languid, oxygen-poor water.

At time scales of <1 year by far the most important process is the seasonal change in river flow which creates the annual flood. Seasonal variation in water depth is shown in Fig. 16.3. The 0-m datum in the Mamirauá water-level series was originally selected as a level to which the water was likely to fall during the low-water season.

As for other rivers,[14] the course of the annual flood cycle is highly variable. Variation within the annual cycle alters both the amount and duration of habitat availability. Marked on Fig. 16.3 is the depth at which the forest surrounding Lago Mamirauá becomes completely inundated. Forest-dwelling fish must adapt to an ever-changing habitat availability. A much-longer time series 1902–86 is available for the Solimões–Negro confluence at Manaus collected by the Brazilian Port Authority PORTOBRAS. A simple conversion was possible by comparing the flood waveforms from the Mamirauá reserve in 1983–84 in Ayres[15] with the same period in the PORTOBRAS series. Using these data,

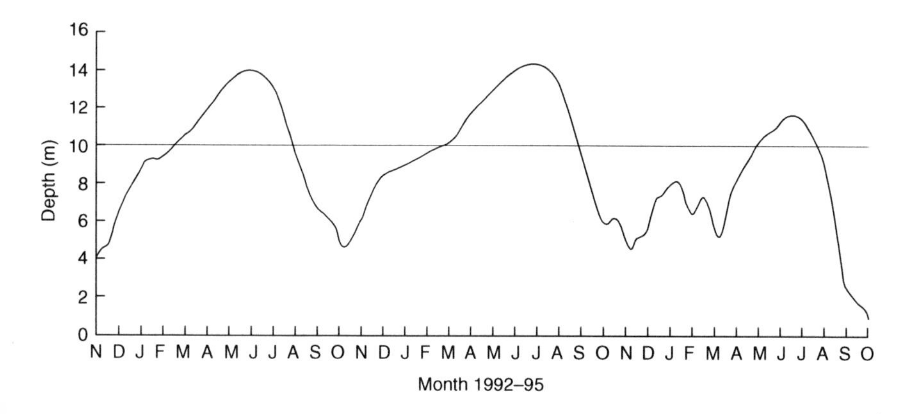

Figure 16.3. Seasonal variation in water depth for the hydrograph station in Systema Mamirauá. This is one of the main lake systems within the focal area of Reserva Mamirauá in the Rios Solimões-Japurá floodplain. The horizontal line at about ten marks the level at which all the surrounded forest is underwater. (Modified after Henderson and Crampton[16]).

Table 16.2. The frequency at which the water level declines below selected water levels during the annual low water season between 1902 and 1984. The values are given for an arbitrary 0-m datum and the frequencies were calculated using PORTOBRAS data

Water depth (m)	Frequency
−2	Once every 25 years
−1	Once every 10 years
0	Every 4–5 years
1	Every 2–3 years
2	Every 2 years
3	8 out of 10 years
4	9 out of 10 years

Table 16.2 Shows the frequency at which different water depths were experienced in the Mamirauá reserve over the twentieth century.

These data suggest that the drying up of shallow lakes which have a bed at +3 m is an almost consistently annual phenomenon, and that channels which largely have a bed at 0 m dry out on average about every 4–5 years (about 22% of low-water seasons). Throughout the *várzea* most lakes have beds at −1 to −2 m and would be expected to be dry once every 10–50 years. However, some channels in the Mamirauá reserve, such as the upper Cano do Lago Mamirauá, have stretches with a bed depth of about 15 m. These areas, of limited extent, act as refugia during the most extreme conditions.

Spatial and Temporal Variation in Water Quality

Water quality has been relatively easy to measure and we report on this as perhaps exemplifying the situation for other physical factors. As pointed out, 'terrestrial' habitats of the study area are never completely non-aquatic over time and at least comparable niche complexity will doubtless eventually be found to apply to them.

Much limnological research in Amazonia has shown the importance of dissolved oxygen, temperature, turbidity and dissolved nutrient levels to animal presence and abundance (e.g. refs 17–20). The overall nutrient content is reasonably summarized by water conductivity. As these four main variables are influenced by the physical structure of the waterbodies and by the temporal variation in flow and flooding mentioned in the previous section we can anticipate that these too must show spatial and temporal variation. However, the variables now in focus reflect new factors: they are influenced by the extracts, excretions and decay products of organisms. These themselves are intrinsically patchily distributed, so creating even more fragmented spatial and dynamic temporal patterns than would follow from physical processes alone. Below we limit the discussion to the temporal-spatial variation of these variables at scales which may influence species presence and community diversity.

Solar heat and oxygen enter water via the free surface. Thus, still water typically shows vertical gradients of both. Oxygen is also generated by photosynthesis and is consumed by decomposition. This enhances the vertical stratification as light is normally only available near the surface and most decomposition occurs within the sediments. The depth, shape, orientation and flow of lakes and channels influence temperature and dissolved oxygen resulting in large-scale spatial variation in these variables. Small-scale variation is also evident. Typical examples of vertical profiles of oxygen and temperature are shown in Fig. 16.4. Another example of the dramatic small-scale spatial variation in oxygen appears in Fig. 16.5 where oxygen levels are shown for localities under forest, floating meadow and in open water only 20 m apart. Temporal variation can be equally extreme. Because of the role of vertical mixing in determining oxygen concentration and temperature there is long-term variation seasonally and between years in line with the flood cycle. For example, flowing water in *paranas* is vertically mixed and deeper water here holds more dissolved oxygen than in lakes. The deeper *paranas* become anoxic during exceptional low water periods when they cease to flow with main river water. Oxygen availability always varies diurnally and may change dramatically even by the hour. Figure 16.6 shows the variation in dissolved oxygen and temperature in surface waters within Cano do Mamirauá during October 1995. Following an afternoon rainstorm, the water went from supersaturation during the day to almost total anoxia at night, resulting in a mass fish mortality. Similar mortalities have been seen several times during our

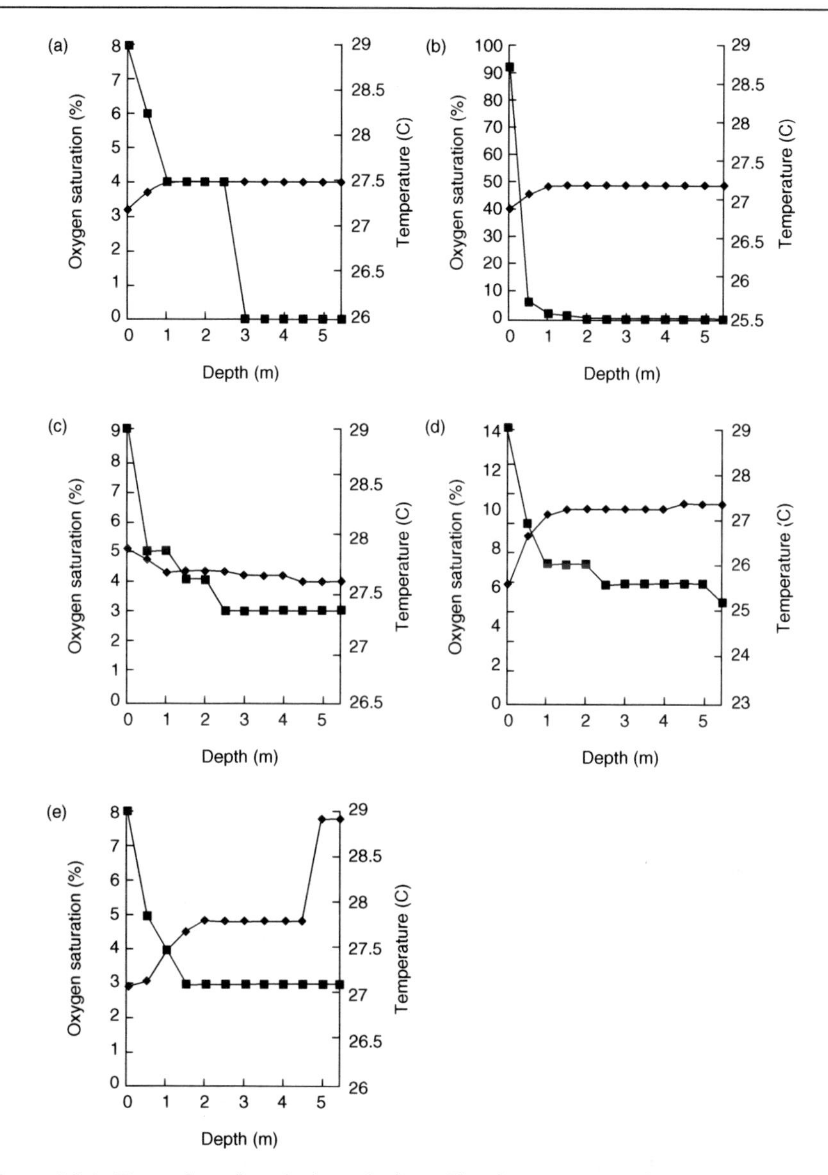

Figure 16.4. Examples of typical vertical profiles for oxygen and temperature within the Rio Japurá-Solimões floodplain waters. (a) The Apara Parana along which Rio Japurá water flows: (b) the lake Lago Mamirauá: (c) (d) and (e) different localities within the mixing zone between the Apara Parana ands ago Mamirauá Key: diamonds—temperature, squares–oxygen.

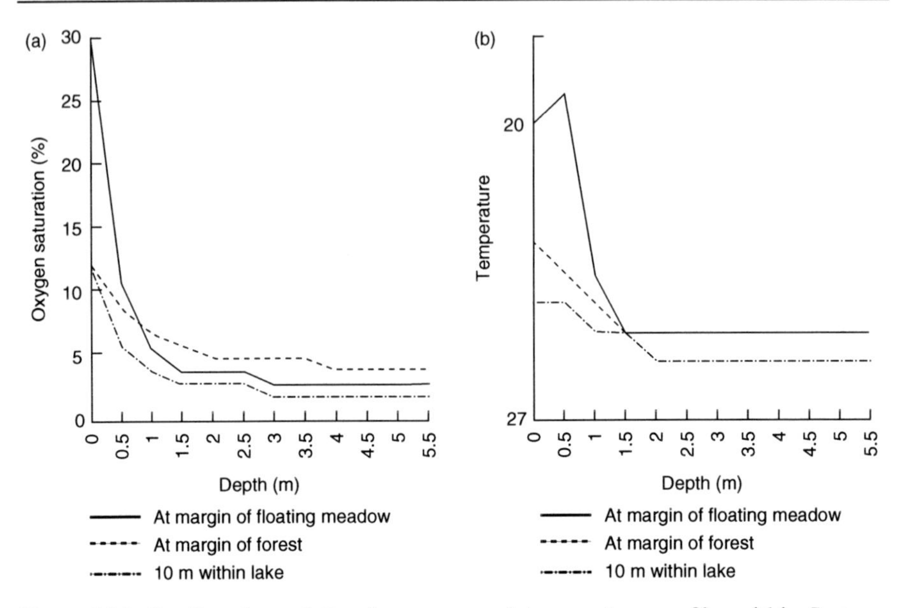

Figure 16.5. Small-scale variation in oxygen and temperature profiles within Systema Mamiraoá: (a) dissolved oxygen: (b) temperature. Data were collected between 11.35 am and 12.35 pm on 20 July 1994.

limited visits. On a slightly longer time-scale these events emphasize the high productivity generating the huge numbers of fish that are always present to be killed.

The meeting of waters of different temperature, such as when a flowing *parana* enters a lake, creates a sharp spatial discontinuity because of incomplete mixing. Such frontal systems can range in scale from a few metres to many kilometres wide at the meeting of major rivers such as the Japurá and Solimões. They offer unique sets of conditions and are often zones of enhanced productivity.

Turbidity and conductivity also vary spatially and temporally at many scales. The Rios Japurá and Solimões, because they collect waters from different geographical regions, differ in conductivity and suspended sediments. Above the Auati-Parana (the first channel via which the Japurá receives Solimões water though occasionally the flow is reversed), the Japurá and Solimões have conductivities of 8 $mS\,cm^{-1}$ and 80 $mS\,cm^{-1}$, respectively, a 10-fold difference. Thus, the floodplain has the interesting large-scale feature that it is bounded by, and receives, waters of different character. The main river inputs, however, are not the only factor determining conductivity, which varies considerably between lakes and within a lake between seasons. Some lakes during the low-water period have conductivities greater than 180 $mS\,cm^{-1}$ at 20°C. High conductivities are probably linked to both evaporative concentration

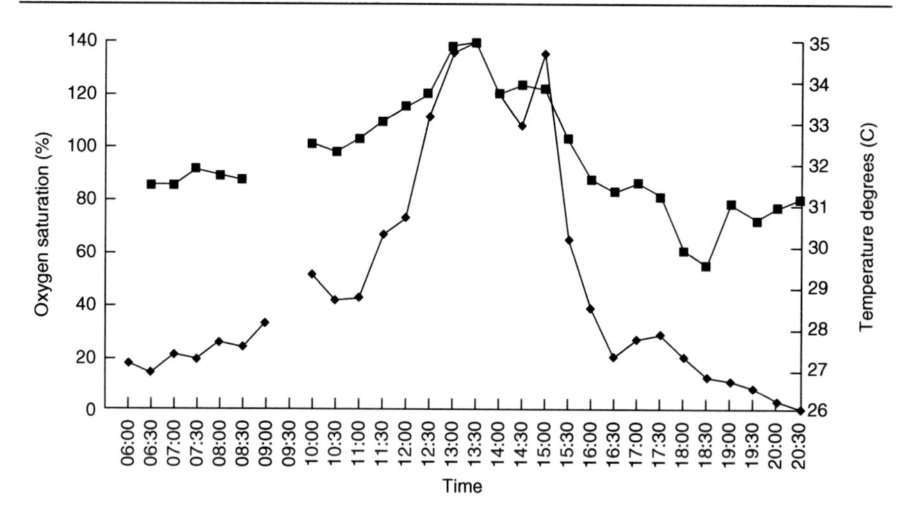

Figure 16.6. Diurnal variation in surface dissolved oxygen and temperature in Cano de Mamirauá over 24 October 1995. A light rain shower occurred at 14:00 h and heavier rain fell between 15:30 and 16:30 h. The first distressed fish were seen at the surface at 20:15 h. Key: squares–temperature, diamonds–oxygen.

and decomposition of floating meadow. Turbidity is linked to flow, wave action, bed sediments, plankton and macrophytes, all of which are continually changing.

As will by now be evident, we have sketched a set of habitats with intermittent potential for high biological productivity but which is extremely variable and unstable on all scales. For sufficiently agile species, it could provide a wealth of niches, just on its physical characteristics alone. We now turn to the life-forms present, especially plants and fish.

Some Examples of Diversity in the Floodplain

PLANTS

Several lists of plants, emphasizing the trees, are available for our region but a full list is not, herbs especially being neglected. However, it can be inferred with fair confidence from comparisons of *várzea, Igapó* and *terra firme* forests at other Amazonian sites (ref 7 and A. J. G. Ferreira, personal communication) that Mamirauá *várzea* will have a reduced tree list compared to its nearby *terra firme.* This list is probably about equal in size to that of the local *Igapó* which occupies, in this region, a much smaller area. Given the much greater fertility and productivity of *várzea* compared to *Igapó*, the result of this comparison may seem surprising but the first point to note is that it conforms with the so-called 'paradox of enrichment'—the frequent observation that nutrient enrichment can lead to reduced biodiversity.[21–23]

When one descends to the lower storeys of the forest or moves to habitats outside, the flora remains poor by absolute Amazonian standards but becomes more special. The ground flora of forests of levees and *chavascal* seems similar in floristic richness to that of *Igapó* and *terra firme*. As leaf-litter and soil emerge after the flood, a forest-floor semblance of the spring activity of herbs of north temperate forest arises in the *várzea* with the same tendency to local dominance by a few species. In *restinga* forest, a small *Calathea* for example, leafing from bulbs that a few weeks previously were underwater and had remained so by a metre or more for several months, creates an effect of the dense stands of *Endymion* or *Anemone* of a European deciduous wood, while elsewhere clumps of Cyperaceae simulate the temperate woodland grasses like *Deschampsia, Brachypodium* and *Melica* as well as sedges like *Carex pendula*.

Occasional true grasses are present in these shady sites; however, grasses are vastly more dominant in the early stages of succession outside of forest, especially in the very characteristic 'floating meadow' which grows over open shallow water and across slow-flowing channels. The largest of all grasses, of course (potentially common on river banks and even in mature *chavascal*) are two *Guadua* bamboos, while second and third in size are the giant reed *Gynerium sagittatum* and *Laziacis procerrima*. The last two are colonists respectively of high sand banks[12] and of forest edges and gaps. After these in size come the main floating grasses. 'Floating meadows' are often tens to hundreds of hectares in extent but they contain only very few species.[24,25] *Paspalum repens* and *Echinochloa polystachya* are equally dominant, but *Hymenachne amplexicaule* and *Oryza glumipatula* are also present locally. A less robust grass, *Leersia hexandra*, forms smaller floating mats in less eutrophic (but not 'black') water bodies.

Significant grass populations have existed along the Amazon for at least 9000 years[26] but it is possible that even among the few species listed above are recent additions, as the following points suggest:

1 Absy showed pollen in recent deposits currently on a rising trend and at the maximum she has found.
2 The *Oryza glumipatula* in our area is very close to *O. rufipogon*, which is regarded as a wild (or feral) species of Africa. Local opinion claims this '*arroz bravo*' to have increased greatly during living memory.
3 Pires and Prance[27] treat *Echinochloa polystachya* as absent or at least uncommon above Manaus and abundant below it until the Atlantic Ocean.

This is far from the case now: in the Tefé region *E. polystachya* now floats in dense largely monospecific and parasite-free stands over larger areas than its main rival, *Paspalum repens*, which by contrast is often severely damaged by insects and fungi. The situation with related *Echinochloa* suggests that

E. *polystachya* comes from Africa ultimately but perhaps indirectly to the Amazon via more northern neotropical areas and with hybridization and/or polyploidy involved. All the grasses so far mentioned are perennial, but at low water a half dozen or so annual species plus a similar set of annual Cyperaceae (including several *Cyperus*) flourish on open mudbanks with a modest variety of annual dicot herbs intermixed. All the species sometimes make dense uniform stands. Conspicuous among the herbs are *Sphenoclea zeylanica*, some five or so species of *Ludwigia* (Onagraceae), and three or four of *Aeschynomene* (Papillionaceae).

Floating and truly aquatic macrophytes make up the most remarkable communities and show the most unexpected diversity in *várzea* plant-life. Three types are represented. First, there are plants typically rooted on the bottom but with floating parts that for some or most of the flood-cycle photosynthesize at or above the water surface. Second, there are plants always submerged but floating unrooted. Third, there are plants floating at the surface and photosynthesizing above it.

The first class includes the large grasses already mentioned. When water is present these grow extremely fast, easily keeping pace with the rising flood and spreading rapidly laterally across the water surface at the same time. When dropped back on the mud at the end of the flood season they are greatly reduced by disease, decay and consumption by fish, but do not necessarily die or cease growing completely.[25] Only *Oryza* (later, when floods return, to vie with *Hymenachne* as the least buoyant floater) grows substantially during the 'terrestrial' phase. Less common but similarly semi-aquatic are *Polygonum acuminatum* and *Caperonia castanaeifolia*. Fully aquatic and requiring a pool that persists even at low water, is the characteristic and magnificent *várzea* water lily, *Victoria amazonica*. Dubiously also in this category fall a species of *Calliiriche* starwort and a *Riccia* liverwort. It is noteworthy that each of these very different rooted plant types seems to have only one species and this is true of most other genera. Besides the exceptions already mentioned in *Cyperus, Ludwigia* and *Aeschynomene*, however, small annuals of the genera *Bacopa* and *Lindernia* also show several species each.

The second class, that of free-floating but submerged plants, has surprisingly few representatives and again genera are mostly monospecific in our area, for example *Cerotoplyllum, Najas* and *Wolffiella*, while *Utricularia* has two species. Perhaps the paucity of submerged aquatics arises because such plants are unusually at the mercy of: (i) the currents that affect most habitats at one season or another—they are less likely to be held amongst attached vegetation than full floaters; and (ii) the abundance, diversity, specialism and hunger of the low-water fish population. Herbivorous fish such as anostomids are to be observed eating leaves and shoots of *Paspalum*. Fish may act against any fully exposed floating plants like piranhas may against ducks, which are another at first sight surprisingly under-represented group.

The third class, the surface floaters, make up the most immediately striking community. It is remarkable in the following respects:

1. While the community comprises at least 23 species, about a half dozen are so abundant that at the margins of any open area of still water they can usually be found in a few minutes' search.
2. In a 'floating lawn,' as we prefer to call an assemblage of these free-floating small species, plants often exist interspersed as a true community in which the smaller fill gaps between the larger. Thus, in descending size, may be observed: *Eichhornia* > *Pistia* > *Limnobium* > *Salvinia* > *Azolla* > *Ricciocarpus* > *Spirodela* > *Lemna*. The 'floater' genus *Wolffia* (Lemnaceae) containing the smallest of all angiosperms, is recorded from Amazonia but we have not encountered it. Lawns of single species occur but are less common than mixtures.
3. On all floating lawns phytophagous insects are abundant and cause obvious damage. Fungal and microbial diseases are also evident.
4. So far as can be judged by the presence of sexual structures, all plants of these floating lawns reproduce sexually.
5. All the species also reproduce prolifically by vegetative means.
6. While some species may have long been pantropical, several from the community have been spread by human agency from seemingly just these kinds of communities in tropical South America, where in natural waters they are virtually never pests, to other continents[28] where they quickly become pests due to their uncontrolled multiplication and the formation of dense monospecific mats.[29,30]
7. Soon after so migrating they are found to have partially or wholly lost sexuality,[29,31] either through having become inbreeders and/or almost exclusively reliant on vegetative propagation. Among the water pests to have originated in this habitat or others closely similar elsewhere in S. America arc: *Gymnocoronis, Eichhornia, Pistia, Utricularia, Salvinia* and *Azolla*.

Similar habitats occurring elsewhere in South America need co-emphasis in the above claim. In a trend which applies to water plants generally and especially to small ones (probably because propagules or pieces of these are most easily carried by migrating birds), ranges of *várzea* water plants are often very wide. In the case of Amazon water plants a particularly common range extension is into the *pantanal* and *chaco* areas of the Parana river basin to the south.

Returning more generally to the plants in our region, no local endemic has yet been found and the most local species seem not to be from other habitats than *várzea*. For example, a collection so far unidentified to species is a tree in the genus *Bucida* (Combretaceae) whose species are only otherwise from coastal forests of the western Caribbean. This particular small tree is from the *Igapó* of a small black-water lake adjacent to the reserve. While such an unexpected genus is easily noticed, our survey cannot generally resolve what might be new species from mere local varieties: our knowledge of the plants cannot match, for example, the discrimination achieved by Ayres[32] and Yonenaga-Yassuda and Chu[33] for the new squirrel monkey, *Saimiri vanzolinii*, a truly endemic species of the *várzea* of the Mamirauá area. As noted in a subsequent paper[34] major river channels in Amazonia quite commonly serve as barriers to gene flow for terrestrial large mammals which are poor swimmers: the rivers thus delimit species ranges and control speciation. Migration of islands and channel avulsions in the unstable Amazon floodplains, however, generally work against isolation by water. In the Tefé area this factor seemingly has not been sufficient to prevent speciation in the case of *Saimiri*, and has also been insufficient in the case of a social stingless bee, *Melipona seminigra*.[35] So far, we lack any parallel case for a plant.

Fish

At present, 286 species of fish have been identified within the floodplain 'reserve' (Fig. 16.1) and its immediate vicinity, an area of approximately 2×10^3 km. This is about 10% of the total species known from the Amazon basin, which has an area of $7 \times 10^6\,km^2$. Probable upper limits for species diversity within the Solimões-Japurá floodplain and the Amazon are 300 and 3000 species, respectively.[36] High as these numbers may seem, a more impressive diversity is that of form and life-style shown by floodplain fish and this is discussed further below. Floodplain diversity seems not to be based on the rapid speciation of a few ancestral forms such as is found, for example, in the remarkably species-rich cichlid communities of the African great lakes (see ref. 37 for a recent review of cichlid speciation).

So far, as with the plants, no species endemic to the Japurá-Solimões floodplain has been identified. Floating meadow and floating lawn, especially abundant and extensive in the left area and almost confined to the floodplain, might seem to provide promising habitats for endemism. The former especially has been the subject of intensive studies[16,38,39] and yet no example has been found. The importance of electric fish, a group characteristic of these habitats, has been discussed by Crampton.[39] A complete survey of the gymnotids of floating meadows, for example, revealed 14 species with seven Hypopomidae and five Gymnotidae. All of the species with confirmed identifications are known from other regions of the Amazon basin and similar forms to the unidentified species have been seen in museum collections from other areas. Thus, there is as yet no compelling evidence that any gymnotids are endemic to the immediate vicinity of Tefé.

Diversity at deeper phyletic levels is also high. Table 16.3 lists the 42 families and number of species of fish in each family within the Mamirauá reserve. The

Table 16.3. The number of fish species subdivided into families recorded from the Rios Solimões-Japurá floodplain. Data were collected between 1990 and 1996 using a wide range of fishing techniques

Family	Number of species
Potamotrygonidae	2
Clupeidae	1
Engraulidae	2
Arapaimidae	1
Osteoglossidae	1
Erythrinidae	3
Ctenolucidae	1
Crenuchidae	1
Characidiidae	6
Anostomidae	9
Hemiodidae	3
Lebiasinidae	9
Curimatidae	16
Gasteropelecidae	2
Serrasalmidae	12
Characidae	48
Sternopygidae	1
Eigenmanniidae	11
Rhamphichthyidae	6
Hypopomidae	10
Apteronotidae	12
Gymnotidae	6
Electrophoridae	1
Doradidae	16
Auchenipteridae	10
Aspredinidae	2
Pimelodidae	25
Ageneiosidae	4
Cetopsidae	1
Hypophthalmidae	3
Trichomycteridae	3
Callichthyidae	4
Loricariidae	17
Belonidae	1
Cyprinodontidae	3
Synbranchidae	3
Sciaenidae	2
Cichlidae	21
Gobiidae	1
Soleidae	1
Tetraodontidae	1
Lepidosirenidae	1
Total species number	286

dominant groups in terms of species number are the characin families Curimatidae, Serrasalmidae, Characidae, the electric fish groups Eigenmaniidae, Hypopomidae, Apteronotidae, the catfish groups Doradidae, Auchenipteridae, Pimelodidae and Loricaridae, and the Cichlidae. This list is typical of Amazonian floodplain habitats.[40] In addition, the floodplain also supports a number of species-poor families of predatory fish, the Osteoglossidae, Arapaimidae, Synbranchidae, Erthyrinidae and Electrophoridae. The following single representitive of each family, *Osteoglossum bicirrhosum, Arapaima gigas, Synbranchus marmoratus, Hoplias malabaricus* and *Electrophorus electricus*, are important members of the fauna both in number and biomass. All five occur along the full length of the lowland Amazon floodplain.

In general, a spatial and temporal patchiness fit to accommodate a large fauna is not difficult to imagine. Looking only to the most major divisions within the river corridor we recognize four principal aquatic domains: forest streams (S), river channels (R), blocked valley lakes (B) and floodplain waters (V). How fish species distribute among these major domains can be conveniently represented in a Venn diagram (Fig. 16.7). Several features are at once striking. The set $R + V$ with 239 members represents the fish of the white waters of all kinds. The great majority are found in flowing channels, $V \cap R = 236$ and $V - R = 8$. Of the set $V - R$ only two species, an undetermined *Gymnotus* knife fish and the lungfish *Lepidosiren paradoxa*, have been found to reside solely in static 'white' water and even here it is possible that the gymnotid may move outside of the *várzea* as it has not been captured at low water. Also notable is

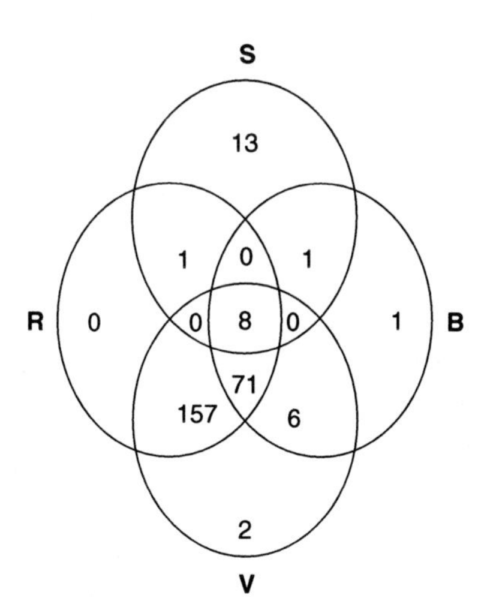

Figure 16.7. Venn diagram showing the distribution of fish species between the four principal aquatic domains in the vicinity of the Solimões-Japurá floodplain; S, forest streams: R, flowing channels: B, black-water lakes: V, floodplain (*várzea*) lakes.

that the set of fish only caught in static water $(13+V)-(C+S)=9$ is small; evidently most species prefer, or at least tolerate, flowing water at some time. In comparison, the much more poorly sampled forest stream habitat holds 13 species not found in C, V or B. Finally, the set of generalist species $R \cap V \cap C \cap S$ holds eight species and includes three of the most dominant in terms of biomass S. *marmoratus, H. malabaricus* and *E. electricus*. Within this particular Amazon habitat it seems that these fish manage to be both jacks and masters of all trades. Two important features that they share is a predatory life-style and having adaptations to cope with low dissolved oxygen.

Each of the four aquatic domains can be divided into a large number of habitats each offering patches widely differing in size. A good example is the lake size distribution already discussed. Open waters hold specialized fish communities changing with the size of the habitat. Another important example is floating meadow.[16,38] Both shelter and food amongst the roots and rhizomes are favourable for small fish including the juveniles of important commercial species.[41,42] During quantitative seine net sampling of floating meadow 79 species have been captured. The species accumulation curve shows an almost linear increase in species number up to a sampled area of 400 m^2.[216] Given the lack of asymptotic behaviour and the recorded presence of further small species using hand nets, there can be no doubt that the total species number for this habitat will exceed 100. Thus, this one habitat holds more than one-third of the total species complement for the region (presently 286 species with those of forest streams and blocked valley lakes included) and about half of all the species found within the *várzea* lakes and channels. However, no single site ever holds more than a small part of such species diversity. The floating meadows with the most species occur along the edge of flowing channels such as the Apara-Parana which carries water from the Rio Japurá into Lago Mamirauá. The maximum species number collected at one station in this channel was 23 in 10 m^2. Probably the most species-rich localities have complements of about 2.5 species m^{-2}.

Rank-abundance graphs and dominance indices can be found in Henderson and Crampton.[16] Very few species contribute more than 90% of the biomass and these dominants, *Hoplius malabarirus, Aequidens* cf. *tetramerus, Hypopomus* spp. and *Parauchenipterus sp.*, can be classified as floating-meadow generalists since they were found at almost all the stations sampled. The majority of species of floating meadow are infrequently caught and rarely abundant. It is probable that they use the habitat for only one period within the annual cycle or only when other more favoured areas such as flooded forest are unavailable.

The flood-cycle greatly affects fish density and standing crop within the meadows. At low water the fish are concentrated in the lakes. For example, a floating meadow site within the reserve sampled at low water had an average fish standing crop of 195.4 g wet weight m^{-2}, whereas at high water the value was only 17.3 gm^{-2}. Order of magnitude changes in fish density are probably typical.

From the preceding it can be seen that even where floodplain habitat may appear uniform it is exceptionally rich in aquatic vertebrate species. For comparison, 10 m^2 of a simple floating meadow habitat dominated by at most two plants (*Paspalum repens* and *Echinochloa polystachya*) had a species total of fish almost equal to the total for NW Europe. Clearly, the fish data are telling us that *várzea* niches are abundant. But this is not to say that there are as many species as there could be. Within a fluvial system the presence of large numbers of species within one area does not imply that they have been generated *in situ* to fill niche gaps nor that they have filled all of them. Indeed, we argue below that probably for all groups the floodplain is like a battery which is charged by a constant stream of forms originating in forest streams and headwaters. This battery accumulates endlessly (though presumably with some dissipation/ extinction), but very occasionally generates dramatic exterior pulses—that is, macroevolutionary changes appear capable of making important contributions elsewhere.

Details of the fish fauna offer some support for this view. It was noted above that the Japurá-Solimões floodplain held no known endemic fish species, suggesting little *in situ* speciation. The argument that temporally variable habitats do not favour fish speciation has been made before[37,43] but has been little discussed. In our case three factors seem to conspire to prevent genetic separations. First, spatially all the habitats are physically interconnected, both by channels and seasonally by the mantle of flood water. Second, all habitats are ephemeral on the various longer time scales we have discussed. Third, at low water many populations are forced out of the floodplains into the adjacent channels. Finally, there is the mass transfer of fauna in drifting islands of meadow.[38] Jointly, these factors determine that:

(i) it is almost inevitable that the lowland Solimões will show many fish species ranging over the full floodplain length; and

(ii) as predicted by the general direction of flow, adaptation and speciation will be increasingly restricted further down the river. As shown by the greater species list in Henderson and Crampton[16] for the Solimões-Japurá junction compared to that of Junk *et al.*[19] for *várzea* in the vicinity of Manaus. 800 km downstream, our present indications seem to fit this. However, given the number of large tributaries converging in the Upper Amazon Basin, the excess may (also or instead) be a further indication of the effect already suggested, that the number of tributaries rather than Amazon channel size or even local extent of floodplain is the significant factor in adding to regional species lists.

It was noted above that the long-term 'charging' of the species diversity of the floodplains may derive from a slow trickle out of the surrounding hinterland of headwaters. Two conditions are needed to fortify this suspicion. The first is that the headwaters are capable of generating new species. The second is that these species are sometimes able, perhaps only very occasionally, to invade floodplain habitats. The reality of these conditions is considered in turn.

While each stream holds a comparatively low diversity fish fauna in comparison to the *várzea*, samples from two forest streams gave 13 fish species not found in any floodplain or river habitat. In comparison, the considerably larger *várzea* community was found to hold only two species not found in either river, lake or stream, and one of these is doubtful. This simple example gives some measure of the potential for isolation and speciation within forest headwaters compared with *várzea*.

Two cases serve to suggest how fish species may 'evolve' from headwaters to floodplains. Both are based on known 'pre-adaptations'.

The first concerns the gymnotiformes. A large *várzea* floodplain subtends on each side a comb of local forest streams within each of which infaunal, cryptic, gymnotiform fish are largely isolated. Such isolation offers ideal conditions for speciation within groups such as *Gymnotus, Steatogenys* and *Hypopomidae*. In our area only two species are shared between floodplains and streams. These are the electric eel *Electrophorus electricus* and *Sternopygus macrurus;* in other areas *Gymnnotus* cf. *carapo* is also known to be shared. Interestingly, each of these three species belongs to a different, ancient, clade branching near the base of the phylogeny of the gymnotiformes. However, none of the other 58 or so gymnotiformes known from the area is shared between streams and floodplains and indeed six species are known from streams and nowhere else. Despite the restriction of many species to streams there are strong reasons to believe that they could be pre-adapted to floodplain habitats and could hence be candidates to become future invaders. Most of the gymnotiformes of the streams occur either in dense leaf-litter banks or among hanging roots. These habitats are structurally very similar to the hanging root mass of floating meadows. The stream-dwelling gymnotiformes, which live in leaf-litter banks or root masses, resemble their floodplain infaunal counterparts of the floating meadow both in being small and, by merit of their anguilliform shape and electric sense, in their admirable adaptation to manoeuvre and forage in the dense reticulate habitat of a floating meadow. Furthermore, although streams on the whole tend to be better oxygenated and lower in temperature than the floodplain waters, the leaf-litter banks are atypical being often not only hypoxic but also relatively warm[44,45] Hence, conditions in the leaf-litter bank pose similar physiological challenges to the floating meadows of the *várzea*. Experimental investigations of three common leaf-litter bank specialists: *Microsternarchus bilineatus, Hypopomidae* sp. nov, and *Gymnotus* cf. *anguillaris* showed that all were tolerant to protracted periods of anoxia (W. R. G. Crampton, unpublished observations). In summary, it would appear as if these three species have attributes pre-adapted to life in the floodplain.

An examination of species distribution within the family Curimatidae also shows the potential for speciation in isolated catchments and headwaters.

This group makes a good example as its members are some of the commonest fish within the várzea.[47] They occur in large numbers, grazing upon algae, vegetation and detritus. As detritivores they are able to exploit a food source which is available in all forest streams and is at its most plentiful within the floodplain. The taxonomy has been recently revised by[48–51] and 95 species are now recognized mostly with limited distributions. About nine species are known from the Solimões-Japurá floodplain. Vari[52] hypothesized that the present distribution could be explained by allopatric speciation within tributaries and headwaters with occasional cases of wide-scale dispersal. The present distribution of species shows a clear tendency for headwaters and tributaries to differ in their species while the main stem of the river holds a set of widely distributed abundant forms.

BEETLES

Erwin and Adis[4] claim that the beetle family Carabidae is more diverse in inundation habitats of Amazonia than in the *terra firme*. Their assessment includes, and is possibly mainly based on, *Igapó* habitats: they record *Igapó* as higher in carabid diversity than *várzea*, whereas, as noted above, the situation for plants and fish turns out more equal.

When discussing the local generation of species, they conjecture that great variation in extent of *Igapó* is likely to have accompanied world climatic (and in particular sea level) changes during the Cenozoic, and that this might, through a kind of 'mini-refuge' scenario, account for a high rates of speciation in *Igapó*. While patches of *Igapó* around mouths of side streams may indeed expand, fuse and contract during climatic changes, it is harder to see this happening for *várzea*. Continuance of the Amazon river plus a regime of floods almost guarantees, differently from *Igapó, a* minimal corridor of floodplain existing at all times along the main valley. Thus again *várzea* seems unlikely to provide sufficient barriers for allopatry, strengthening the impression given by Erwin and Adis[4] that it is *Igapó* rather than *várzea* which contributes most to floodplain carabid diversity.

Macroevolution

What next? Does nothing move out of *várzea* ? Of course 'nothing' is perhaps unlikely for any well-filled habitat but, given the above-argued lack of speciation, it is striking to find hints that those groups that very rarely do emerge into other habitats may be very important. Strong claims on these lines for the last-mentioned group, Carabidae, have been made and qualitatively they concur remarkably with our own data for plants and fish. Erwin and Adis[4] indicate that carabids of inundation forest are characterized by generalism and include species occupying broad and sometimes even several different niches.

For example, some carabids of *Igapó* alternate during the seasonal flood-cycle between tree tops and forest ground litter. Although little is said on mechanism (and that somewhat indefinitely) the authors imply that this generalism (or 'plesiotypy') creates an important evolutionary potential, with the result that on a geological time-scale floodplains generate 'taxon pulses'[53] that spread out to carry new wetland-originating carabid forms, speciating and specializing as they go, right to the ultimate limits of life of the insect biosphere in deserts, mountains, the far north, and so forth.

We argue similarly that absence of common speciation in the floodplain actually abets rather than denies the possibility that the habitat has been important for macroevolution. Being 'stretched' over several niches is conducive to polymorphism[54–56] and polymorphism in turn is conducive to radical changes.[57] Extremely rarely, developmental morphs of plastic species may transform into new and more efficient adaptations enabling descendants to invade new habitats. Examples where wing polymorphism may have been antecedent to the origin of major groups (e.g. ants, anthocorid and cimicid bugs) are shown by Hamilton[58]. (A wingless very ant-like bethylid female, seemingly somewhat communal in its social life, is common under bark of recently floating or submerged dead logs in levee forest of Mamirauá; immature stages have not been seen.) Henderson and Bamber[59] argued that the adaptations for plasticity required of fish in estuarine conditions (which are not unlike those of *várzea*) allow rapid invasion of fresh water when niches became available. Judging by the presence of typically marine fish families such as the Clupeidae, Soleidae, Sciaenidae, Belonidae or Tetraodontidae, in the channels of the study area, 2600 km from the sea, this transition has frequently been accomplished. Many similar examples of ex-estuarine invasions of rivers by fish could be given for other regions—for example, atherinids in Mexico.[60] As regards specific *várzea* potentials for novelty, it is notable that the one species of fish of the study area which seems to be a complete floodplain specialist is the lungfish *Lepidosiren Paradoxa*. The unique set of adaptations of this animal is particularly directed at survival in low oxygen, seasonal environments. Its air breathing is unlikely to open new evolutionary avenues to it now but this is because terrestrial habitats are so well occupied by what are, in effect, its remote relatives. We are not of course suggesting *Lepidosiren* here as a perfect 'frozen' ancestor of the land vertebrates: other lungfish exist in seasonally flooded habitats in Africa and Australia and are not very closely related. Combined, the three merely suggest likely characters of the common ancestor. Instead, what is significant for the present theme is that all three combine to implicate seasonally flooded freshwater habitat.

Prior occupancy of the land did not apply in the early Palaeozoic, either for animals or, as we will discuss later, for plants. While the invasion of land by vertebrates is still poorly understood, there is a general consensus that tetrapods originated from sarcopterygian fish during the late Devonian,[2] probably from shallow freshwater forms although Bray[61] suggests that coastal marine swamp habitats could have been important. By the Visean-Namurian epochs of the early Carboniferous 'most tetrapod assemblages [were in] lowland swamps or lakes on coastal plains, and are all situated in the continent of Euramerica close to the palaeoequator of the time'.[2] Tetrapod fossils from the Lower Carboniferous in Scotland are found in freshwater limestone, black shales and tuffs in association with scorpions, eurypterids, millipedes, harvestmen, ostracods and plant material. The presence of ostracods as the only aquatic group is indicative of temporary flooding. Thus, the general picture we presently have of the originating habitat of tetrapods bears similarities to the recent *várzea*.

Put more theoretically, a habitat's demands for plasticity could well accelerate macroevolution via genetic assimilation (provided—the usual difficulty for this concept—plasticity can be sufficiently maintained as the process goes on). A lungfish that has to live as an adult in an expanse of drying mud as well as living at other times in a shallow lake may find itself, via genetic assimilation, within reach of macroevolutionary slopes that are unattainable to a pure shallow-lake specialist. Its advance may then occur in ways not necessarily connected with aestivation in mud (see ref 62 for a good outline explanation of these ideas although the difficulty concerning the maintenance of plasticity has not yet been treated). Likewise for tambaqui, *Colossoma macropomum*, which varies from being a predator when young to a seed eater in sparsely populated flooded woodland as an adult.[42] As an adult this fish displays a remarkable further plasticity in switching seasonally from feeding on seeds crushed by strong teeth at high water to planktivory using gill-rakers at low water. Such an example may help in part to explain a related fish group whose origin may indeed be in the floodplain. The biting piranhas (family Serrasalmidae) are found currently in many aquatic habitats of tropical South America. They include at least one representative, the group-living, predatory *Serrasulmus nattereri* that has to be regarded as essentially a floodplain species even though its young are also found in river margins. The piranha method of feeding has its special advantages in two circumstances: first, during the overcrowded conditions of a *várzea* low-water lake when the species attacks large prey in groups (as well as certainly continuing its normal and individualistic fin and tail biting habit); second, at high water within the forest where its dentition allows it to scavenge flood-stricken victims. Plausibly ancestors of Serrasalmidae at some point specialized from the plastic repertoire of a tambaqui to become unique neotenous chunk-biting predators, thus initiating the radiation of form and behaviour that we see. The observation that most neotropical fish have an African analogue does not hold for the piranhas, and the idea of the arrival of a fertile pair of piranhas in an African river leads to interesting evolutionary speculations. These cases are only a few out of a list that could be given of thwarted or promising adaptive transitions in *várzea* fish.

In plants, parallels exist for both the situations mentioned for fish, that is both for 'frozen' ancestors and new expansions. In the first class, and perhaps like *Lepidosiren* now thwarted by the very extent of their ancestral triumphs, we find *Gnetum leyboldi* and *Ceratophyllum demersum*, one a liane dangling above the water surface and the other commonly snagged on twigs just beneath it. Both occupy very basal positions in most suggested seed-plant phylogenies. *Ceratophyllum* is so thoroughly an aquatic it is a little hard to imagine either its ancestors or descendants being anything else.

Nevertheless, in *várzea* it is regularly dumped on wet mud and so is being 'given a chance'. *Gnetum* is not quite so thoroughly a liane. Its wood is weak under bending and tension and it is not difficult to imagine a herbaceous ancestor for the Amazonian riverine species. This could have lived on channel-side mud and sprouted rather *Piper*- or *Saururus*-like jointed stems, their weakness here probably costing the plant little. Instead speed of growth was the essence and more significantly for their success we can imagine such herbs unfurling paired broad leaves with unprecedented rapidity. These leaves would be the best solution for leaf design to have evolved on earth to date. To summarize, the imagined growth style would be very similar to that of various of the recently focused 'palaeoherbs' of today which group some modern phylogenies actually count *Gnetum* to be the immediate sister. Four further points seem worth making about *Gnetum leyboldi*. First, its adaptations to seed dispersal by fish show thorough commitment to life in flooded forest.[63] Second, mature lianes in the *várzea* show capability to create long horizontal snake-like shoots at ground level suggestive (whether for past, future or both) of a rhizomatous potential (as in a *Piper* sp. of the same area): so far as we know both this feature and the fish seed dispersal are unique to the flooded-forest species. Third, as seen at a glance the floodplain *Gnetum* easily passes as a dicot liane: close attention is needed to distinguish if from various opposite-leafed vines such as *Salacia* with which the gymnosperm is often intermingled; whether by convergence or homology *Gnetum* has very angiospermous leaves. Fourth and finally, we must add a caution to all the above: Gnetum species occur pantropically and often in much less watery habitats: therefore without review of other members a too-glowing portrait of a single species calls for reserve. A similar caution needs to be applied to our discussion of *Ceratophyllum* which is far more cosmopolitan still and which so far as we know has no special features to distinguish *várzea* representatives.

Too much wind- and water-borne gene dispersal for local genetic differentiation plus competition with such angiospermous vines as *Salacia* could explain why *Gnetum* itself is held to a fairly rigid niche at present. However, a particular advanced angiosperm of the *várzea* might reveal something of the opportunities that might have been open to a *Gnetum* ancestor in the Jurassic. *Phyllanthus fluitans* is classified into a very large. almost cosmopolitan, genus (more than 600 species) that includes very numerous herbs, some bushes, and a few small trees. Yet *P. fluitans* is a tiny floating plant mingling with equally small *Azolla, Salvinia* and *Spirodela* in floating lawns. It extends only a few internodes of stem along the water surface and expands only a few of its rounded, crinkly leaves before breaking into pieces that become independent plants. When stranded it puts down shallow roots into the soil but is at best static in this situation and virtually never grows an upward stem. Its tiny white and separately male and female flowers presumably provide the characters for its generic assignment, but it is easy to imagine that the reduced flowers of such a reduced plant could lead eventually to a radical new arrangement so that if the plant ever evolved to be large again, and its flowers with it, its placement might be obscure. Some superficially radical characters are present in the genus *Phyllanthus*, at least for the untechnical observer; for example, the absence of typical euphorb latex and the curious arrangement of flowers on side shoots that makes them seem to spring from the axils of petiolules of a compound leaf. However, it does not seem that *P. fluitans* can be a basal species with respect to these developments since, while the wood of members of the subtribe including *P. fluitans* is indeed stated to be unusually uniform and distinct,[64] which is what might be expected if secondary thickening had been reinvented after a 'fluitans' phase, *P. fluitans* is placed in a section of the subtribe seemingly not treated as primitive and containing many non-aquatic members.[65] Only one other member of this huge genus is aquatic. Obviously, this favours *P. fluitans* being a derived outlier from either a herbaceous or a woody ancestor, quite the contrary to itself showing the ancestral form. A similar situation holds for the ephemeral herb *Aciotis aequatorialis*, a member of the Melastomaceae which is otherwise almost entirely woody in Amazonia as elsewhere. Flowering sometimes on a monopodial stem at about 5 cm tall, *A. aequatorialis* grows on waterside mud and along forest edges and is not aquatic, while above it in such sites may arch a *Miconia* plant, member of a huge, woody (thus far more typical) genus of the same family. Examples of floral innovation in small (new?) taxa of small plants of open habitats, with such taxa likewise allied to large genera, occur in *Aphanes* and *Sibbaldia* of the Rosaceae. The corresponding larger sister taxa of which are *Alchemilla* and *Potentilla*. As an example of a structural transposition that could be important for future flower form, in the case of *Aphanes* (formerly *Alchemilla*) the stamens are outside rather than inside the flower's nectar-secreting disk.

A similar rationale to that suggested above for *P. fluitans* and *A. aequatorialis* (as also for *Aphanes* and *Sibbaldia*) could be applied to other small floaters mentioned, including the heterosporous ferns. An actual example of novelty in very small *várzea* plants is a 'spike-rush', *Eleocharis radicans*, which forms creeping mats over wet sandy mud at low water. Inflorescence-destined stems from the small caespitose plants arch down to touch the mud at a very early stage of the initiation of their flower spikelet. Roots, one or more new shoot meristems, and more leaves then develop simultaneously with further expansion (and upward reorientation) of the inflorescence spike. Such 'inflorescence plantlets' are often seen growing detached from their parents following early decay or breakage of the connecting stem: the plant then has the paradoxical appearance of having flowered before it produced its first leaves. Effectively, this form achieves rapid vegetative spread and seed production in a single and very economical process. Touching the mud is not essential in the above process. Inflorescence plantlets are often held suspended in the mat of hair-like leaves but still emit leaves and onward-colonizing flower stalks.

Such speculations of evolutionary importance for small, short-lived forms must be admitted to be quite opposite to a long-held botanical preference which traces the major developments of higher plant evolution through woody forms and even sometimes designates aquatics specifically as end states.[66,67] Fossil findings, however, have recently reversed this belief for the ultimate origins of angiosperms and have led to designation of a set of 'palaeoherbs' that are likely never to have had woody angiospermous ancestors. The idea is currently much discussed[1] referring mainly to an environmental background in Cretaceous North America that is strongly suggestive of *várzea* (e.g. see Fig. 9.2, ref 1). This encourages a further suggestion that plant macroevolution may have involved important herbaceous phases at other points, with woodiness intermittently evolving. Besides implicating all monocotyledons as never-arborescent descendants of palaeoherbs (and thus embracing a substantial set of the *várzea* water plants already mentioned), the family Nymphaceae is often specially focused upon in the discussions. Within Nymphaceae *Victoria amazonica* is even more *várzea*-adapted than *P. fluitans*, if less currently abundant. Like *P. fluitans, V. amazonica* is unusual for its group in several ways. Some of these suggest past evolutionary states and others suggest current plasticity. In spite of its giantism (and unlike most other water lilies) *V. amazonica* can sometimes (here in a kind of parallel to *Aciotis*?) behave as an annual. Its very large spine-protected flowers are pollinated by bulky dynastine beetles.[68] On other grounds beetle pollination is suggested to be the oldest entomophilous system.[69]

Comparing *V. amazonica* to *Brasenia* or *Cabomba*, genera generally considered even more 'primitive' than any Nymphaceae, we find *Victoria* to be a pachycaul plant,[70] contrasted to small herbaceous leptocauls while at the same time we see a great increase in size. In this respect two other perhaps 'promising' *várzea* plants deserve notice as possible parallels.

First, the aroid *Montrichardia arborescens* is a pachycaulous and nearly monopodial giant herb growing up to about 5 m tall. It is a common early colonist of newly deposited sand especially along the Lower Amazon (in our area it is widespread but not abundant). Second, *Cecropia membranacea* and *C. latiloba* are pachycaul trees, the former attaining to perhaps 25 m. As is necessary for such a stature, *Cecropia* are woody, and in fact in our area they provide the first abundant tree colonists of new levees and islands. Closely related to *Cecropia* and sometimes in our area appearing on open riverbanks is a genus of mostly forest trees, *Pourouma*. Current botanical interpretation typically sees *Pourouma*, the tree, as the primitive form in a sequence: *Cecropia* by this interpretation is an intermediate while a large herb, like *Urera baccifera* again in our study area (Urticaceae, close and even possibly sister group to Cecropiaceae), could be taken to illustrate what might eventuate from the 'descent' of a *Cecropia*.[1] However, another view is possible. *Monotrichardia* is very obviously not a descending tree and it thus suggests a reversed view for *Cecropia*. The alternative sequence would have an *Urtica-like* herb at the start and *Pourouma* a 'neo-tree' at the end, in this case *entering* the forest. Strengthening this point, the characteristics of island evolution suggest woody *Urtica ferox* in New Zealand as far more likely ascended from a herb, similar to the many woody 'speedwells' *(Hebe)* and arborescent 'daisies' of the same region, rather than descended from a tree. The moderately pachycaulous *Urera baccifera* (and perhaps still better *U. carascara* of the same region) can be a model of an intermediate life-form which, in a parallel context, might have led from an *Urtica* (or a *Dorsienia*) towards the state of a *Cecropia* and then a *Pourouma*. In Africa the cecropiacean *Musanga*, showing short-lived but well-built trees, could represent a further development in this line. Remarkably in this context *Cecropia mem-branacea*, which of the two main *várzea* species favours the richer and therefore more 'nettly' habitats, actually has fragile, irritant hairs on its leaves and young stems. (Yet another parallel *Cecropia* has with many urticaceous 'nettles' is dioecy, but this is a character of all Cecropiaceae.)

Surprisingly, a theory that evolutionary enlargements and lignifications in herbs have been common could revive attention to various positive features of Corner's neglected 'Durian' theory of plant evolution,[70,71] the main difference being that the stem of angiosperm evolution would lie largely through short-lived fast-evolving herbs which, from time to time (probably usually following radical biochemical innovations, see ref 72) would send up lines to become arborescent. As already mentioned for New Zealand, there seem to be many trends in plants especially on islands (including mountains as 'islands' in land areas) where this can be seen happening today.[73] Herbs evolving to become woody do not have to proceed through a pachycaul habit, of course, as many of the island examples and other woody oddities occurring within herbaceous genera show (e.g. in *Bupleurum, Mimulus, Tabacum*) but they are perhaps especially likely to become pachycaul when there is a premium on either water conservation or on fast growth in rich soils. The last would apply to *várzea*. Here, leaves would be expected to be large and these and the few meristems producing them would need to be well protected by latex, novel poisons, stinging hairs, physical defences (as with the spines in *Victoria*), or, as in *Cecropia*, by ants. *Macaranga*, the Asian life-form and ecological parallel to *Cecropia*, extends its parallel even to the possession of ants. As a woody *Urtica* might stand in a transition towards *Cecropia*, so in the euphorb case various co-tribal genera could be considered model transitions through herbs, treelets and even lianes, as exemplified in the genera *Ricinus, Manihot, Acalypha* and *Dalechampia*.

Since palms worldwide tend to be tolerant of flooding, palms as a group might be suspected to have floodplain origins. This suspicion might find supporting evidence in other continents but it seems to gain none from Amazonian *várzea* or *Igapó*. There palms are less common and diverse in the flooded habitats than they are on adjacent *terra firme*. In SE Asia the genera *Nypa* and *Metroxylon* with their unusual rhizomatous tendencies show themselves long adapted to forest-water edges and swampy situations. This is especially true of *Nypa* which has a pollen record of over 100 million years to the Cenomanian, a record not excluding the New World. *Elaeis oleifera*, the western oil palm, may also approximate to life-styles of *Nypa* and *Metroxylon* in the places where it is native in Central America. In Amazonia it is local and perhaps only adventive in sites of pre-Colombian human introduction: however, it still always occupies 'low-lying wet areas'.[74] Apart from sometimes having air roots or stilt roots, the palms of the *várzea* and of neighbouring *Igapó* do not show obvious adaptation to floods. Spruce[75] argued long ago that biogeographic evidence suggests Amazon palms came into their present flood-prone habitats from origins in the shield-rock areas to north and south of the basin, or else from Andean foothills. Spruce's seemingly unchallenged opinion for this group obviously converges to ours for fish and to that which we hold, as yet more tentatively, for the plant life generally. One palm exception that has been found to Spruce's claim is monotypic *Barrcella odora, a* small acaulous palm of semiopen campina scrub. This species is truly endemic to central Amazonia: but it is on *terra firme*. The only local endemic palm of the study area, *Baetris tefensis* (a single member of a large Amazonian genus) is likewise from *terra firme*.[74] It seems, therefore, that the adaptations of floodplain palms were probably acquired by ancestors in wet sites near to or beyond the periphery of the Amazon basin rather than acquired near its centre. The very successful lepidocaryoid palms *Mauritia flexuosa* and *Mauritiella armata* currently grow on fringes of floodplains of the Amazon tributaries along most of their length but they are also dispersed all over the basin and also beyond it. Notably, they are common in extremely different communities along streams and in marshes of grassy savannas both to the north and south of the forest area.

As with the groups discussed earlier in this chapter, however, there seems to be at least one exception to offset what is for palms a repetition of our generally negative theme. In our area as in others along the main rivers, species of the genus *Desmoncus* are typical of forest-water edges and swampy liane forests. Phyletically *Desmoncus* occupies an undistinguished position very close to the upright bactridine palms.[74] The life-form of *Desmoncus*, however, is entirely different from the typical bactridines since all its members are lianes. In this case, biogeography and ecology seem at present consistent with a floodplain origin. (However, the existence of huge suitable nonfloodplain areas of *chavascal* and liane forest that are distant from navigable channels, and therefore botanically unexplored, should be borne in mind.) As suggested by its systematic placement and paucity of species (five in Amazonia, two others elsewhere, as stated by Henderson[74]), the evolution *of Desmoncus is* evidently recent. Nevertheless, if at least two strikingly parallel but much older unrelated developments in the palms in SE Asia are taken as guide, the emergence of *Desmoncus* in Amazonia could have importance for the future. In the eastern realm several hundred species of rattan of the palm genus *Calamus* provide forests and scrub habitats with dominating, smothering lianes of very diverse sizes and types, while yet another eastern genus of spiny liane palms, *Korthalsia*, has begun, uniquely for the whole order, an alliance with ants.

As already mentioned, the evolution of *Desmoncus* in *várzea* fits with the broken, semi-open quality of the vegetation which is very favourable to lianes and generated a long list. Among many other examples of incipient or recent scandent forms in *várzea*, two other plants deserve mention, both, like *Desmoncus*, standing out from the growth habits typical of their family. One *is Byttnera ancistrodonta, a* species from the only 'scandent' genus within Sterculiaceae (as a juvenile the plant is a bush). The other *is Scleria secans* in Cyperaceae. In this second case the climbing habit is unique within the genus, all other *Scleria* having caespitose or reedy forms that are much more typical for their 'sedge' family. *S. secans* ('tiririca' in our region, well known for its backward-cutting sharp leaves) climbs high into trees where it forms dense festoons. It is especially common in disturbed *Igapó* which is not very 'black'.

To conclude, the patchwork of ephemeral *várzea* habitats selects for colonizing attributes such as short life-cycles, rapid growth and life-style adaptability. In spite of its multiplicity of niches it is not favourable to high specialism or to *in situ* processes of speciation. In its inhabitants, simplification of body form and phenotypic plasticity are common outcomes. In both plants and fish this commonly manifests in weedy life-styles, short lives and size-change experiments both towards large and small. Many fish demonstrate striking plasticity in their ability to change their physiology, behaviour and body form in response to anoxia and factors of the floodcycle. In plants, as just one example, diverse rooted life-forms swiftly develop arenchyma and become floaters when they are flooded. Many other examples have been shown. Simplified, responsive organisms are a clay from which evolution may be moulding not so much abundance of species as novel forms.

References

1. D. W. Taylor and L. J. Hickey, Evidence for and implications of an herbaceous origin for Angiosperma. In *Flowering Plant Origin, Evolution and Phylogeny* D. W. Taylor and L. J. Hickey (ed.), pp. 232–266. Chapman & Hall, New York (1995).

2. P. E. Ahlberg and A. R. Milner, The origin or tetrapods. *Nature*, **372**, 507–514 (1994).
3. J. Salo, R. Kalliola, L. Hakkinen, Y. Makinen, P. Niemela, M. Puhakka, *et al.* River dynamics and the diversity of Amazon lowland forest. *Nature*, **322**, 254–258 (1986).
4. T. L. Erwin and J. Adis, Amazonian inundation forests. Their role as short-term refuges and generators of species richness and taxon pulses. In *Biological Diversification in the Tropics* G.T. Prance (ed.), pp. 358–371. (Columbia University Press, New York, 1982).
5. T. R. E. Southwood, Habitat, the template for ecological strategies? *Journal of Animal Ecology*, **46**, 337–365 (1977).
6. G. T. Prance, Notes on the vegetation of Amazonia I11. The terminology of Amazon forest types subject to inundation. *Brittonia*, **31**, 26–38 (1979).
7. R. Kalliola, M. Puhakka, J. Salo, H. Tuomisto, and K. Ruokolainen, The dynamics, distribution and classification of swamp vegetation in Peruvian Amazonia. *Annales Botanici Fennici*, **28**, 225–239 (1991a).
8. G. Irion, Sedimentation and sediments of Amazonian rivers and evolution of the Amazonian landscape since the Pliocene times. In *The Amazon. Limnology and Landscape Ecology of a Mighty Tropical River and its Basin* H. Sioli (ed.), pp. 201–243. Junk, Dordrecht (1984).
9. M. Puhakka, R. Kalliola, M. Rajasilta, and J. Salo, River types, site evolution and successional vegetation patterns in Peruvian Amazonia. *Journal of Biogeography*, **19**, 651–665 (1992).
10. L. A. K. Mertes, *Floodplain development and sediment transport in the Solimões–Amazon River, Brazil.* PhD thesis, University of Washington, Washington (1985).
11. R. Kalliola, J. Salo, T. Hame, M. Rasanen, R. Neller, M. Puhakka, *et al.* Upper Amazon channel migration: implications for vegetation perturbance and succession using bitemporal Landsat MSS images. *Naturwissenschaften*, **79**, 75–79 (1992).
12. R. Kalliola, J. Sato, M. Puhakka, and M. Rajasitta, New site formation and colonising vegetation in primary succession on the Western Amazon floodplains. *Journal of Ecology*, **79**, 877–901 (1991b).
13. S. K. Hamilton, J. M. Metack, M. F. Goodchild, and W. M. Lewis, Estimation of the fractal dimension of terrain from lake size distributions. In *Lowland Floodplain Rivers: Geomorphological Perspectives* P. A. Carling and G. E. Petts (ed.), pp. 145–163. Wiley, New York (1992).
14. B. B. Mandelbrot, *The Fractal Geometry of Nature.* W. H. Freeman, San Francisco (1982).
15. J. M. Ayres, As Matas de Várzea do Mamirauá. CNPq. Brasilia (1993).
16. P. A. Henderson and W.G.R. Crampton. A comparison of fish diversity and density between nutrient rich and poor lakes in the Upper Amazon. *Journal of Tropical Ecology*, **13**, 175–198 (1997).
17. G. W. Schmidt, Primary production of phytoplankton in the three types of Amazonian waters. The limnology of a tropical flood-plain lake in central Amazonia (Lago do Castanho). *Amazoniana*, **4**, 139–203 (1973).
18. E. J. Fittkau, U. Irmler, W. Junk, E. Reiss, and G. W. Schmidt, Productivity, biomass and population dynamics in Amazonian water bodies. In *Tropical Ecological Systems* F. B. Golley and E. Medina (ed.), pp. 289–311. Springer, New York (1975).

19. W. J. Junk, G. M. Soares, and F. M. Carvalho, Distribution of fish species in a lake of the Amazon river floodplain near Manaus (Lago Camaleao) with special reference to extreme oxygen conditions. *Amazoniana*, **7**, 397–431 (1983).
20. B. A. Robertson and F. R. Hardy, Zooplankton of Amazonian lakes and rivers. In *The Amazon, Limnology and Landscape Ecology of a Mighty River and its Basin* H. Sioli (ed.), pp. 57–102. Junk, Dordrecht (1984).
21. A. Rosenzweig, Paradox of enrichment: destabilisation of exploited ecosystems in ecological time. *Science*, **171**, 385–387 (1971).
22. A. Riebesell, Paradox of enrichment in competitive systems. *Ecology*, **33**, 183–187 (1974).
23. J. T. A. Verhoeven, W. Koerselman, and A.F.M. Meulman, Nitrogen- or phosphorus-limited growth in herbaceous, wet vegetation: relations with atmospheric inputs and management regimes. *Trends in Ecology and Evolution*, **111**, 494–497 (1996).
24. W. J. Junk, Investigations on the ecology and production-biology of the floating meadows (*Paspalum-Echinochloetum*) on the Middle Amazon, *Amazoniana*, **2**, 449–495 (1970).
25. N. J. Junk, Investigations on the ecology and production-biology of the floating meadows (*Paspalum-Echinochloa*) on the Middle Amazon. *Amazoniana*, **4**, 9–102 (1973).
26. M. L. Absy, Palynology of Amazonia: the history of the forests as revealed by the palynological record. In *Key Environments: Amazonia* G. T. Prancc and T. E. Lovejoy (ed.), pp. 72–82. (Pergamon Press, Oxford, 1985).
27. J. M. Pires and G. T. Prance, The vegetation types of the Brazilian Amazon. In *Key Environments: Amazonia* G. T. Prance and T. E. Lovejoy (ed.), pp. 109–145. (Pergamon Press, Oxford, 1985).
28. C. D. K. Cook, Range extensions of aquatic vascular plant species. *Journal of Aquatic Plant Management*, **23**, 1–6 (1985).
29. C. D. K. Cook, Vegetative growth and genetic mobility in some aquatic weeds. In *Differentiation Patterns in Higher Plants* K. M. Urbanska (ed.), pp. 217–225. (Academic Press, London, 1987).
30. A. H. Pieterse and K. J. Murphy, *Aquatic Weeds. The Ecology and Management of Nuisance Aquatic Vegetation*. (Oxford University Press, Oxford, 1990).
31. S. C. H. Barrett and B. C. Husband, The genetics of plant migration and colonisation. In *Plant Population Genetics: Breeding and Genetic Resources* A. H. D. Brown, M. T. Clegg, A. L. Kahler, and B. S. Weir (ed.), pp. 254–277 Sinauer, Sunderland, MA (1989).
32. J. M. Ayres, On a new species of squirrel monkey genus. Sainciri, from Brazilian Amazonia (Primates, Cebidae), *Papeis Avulsos De Zoologia*; **6**, 147–164 (1985).
33. Y. Yonenaga-Yassuda and T. H. Chu. Chromosome banding patterns of *Saimiri vanzolinii* Ayres, 1985 (Primates, Cebidae). *Papeis Avulsos de Zoologia*, **36**, 165–168 (1985).
34. J. M. Ayres and T. H. Clutton-Brock, River boundaries and species range size in Amazonian primates. *American Naturalist*, **140**, 531–537 (1992).
35. I. M. F. Camargo, Biogeographia de Meliponini (Hymenoptera. Apidae. Apinae): a fauna Amazonica. Anais do 1st Encontro sobre Abelhas, pp. 46–59 Ribeirao Preto, SP, Brazil (1994).

36. A. L. Val and V. M. F. Almeida-Val, *Fishes of the Amazon and Their Environment. Physiological and Biochemical Aspects*. (Springer, Berlin, 1995).
37. A. J. Ribbink. Biodiversity and speciation of freshwater fish with particular reference to African cichlids. In *Aquatic Ecology, Scale, Pattern and Process* P. S. Giller, A. G. Hilldrew and D.G. Raffaelli (ed.), pp. 261–288. (Blackwell Scientific Publications, Oxford, 1994).
38. P. A. Henderson, and H. E. Hamilton. Standing crop and distribution of fish in drifting and attached floating meadow within an Upper Amazonian várzea lake. *Journal of Fish Biology*, **47**, 206–276 (1995).
39. W. G. R. Crampton. Gymnotiform fish: an important component of Amazonian flood plain fish communities. *Journal of Fish Biology*, **48**, 298–301 (1996).
40. R. H. Lowe-McConnell. *Ecologicat Studies in Tropical Fish Communities*. (Cambridge University Press, Cambridge, 1987).
41. U. B. P. Saint-Paul. A situacao da pesca na Amazonia Central. *Acta Amazonica*, **9**, 109–114 (1979).
42. M. C. Goulding and M. L. Carvatho, Life history and management of the tambaqui (*Colossoma macropomum*, Characidae): an important Amazonian food fish. *Revista Brasileira de Zoologia Sao Paulo*, **l**, 107–133 (1982).
43. G. Fryer, and T. D. Iles Alternative routes to evolutionary success as exhibited by African eichlid fishes of the genus Tilapia and the species flocks of African great lakes. *Evolution*, **23**, 359–369 (1972).
44. P. A. Henderson and I. Walker On the leaf litter community of the Amazonian black water stream Taruma-Mirim. *Journal of Tropical Ecology*, **2**, 1–17 (1986).
45. P. A. Henderson and I. Walker, Spatial organisation and population density of the fish community of the litter banks within a central Amazonian blackwater stream. *Journal of Fish Biology*, **37**, 401–411 (1990).
47. C. A. R .M. Araujo-Lima, A. A. Agostinho, and N. N. Fabre Trophic aspects of fish communities in Brazilian rivers and reservoirs. In *Limnology in Brazil* J. G. Tundisi, C.E.M. Bicudo and T. Matsumura-Tundisi (ed.) 384 pp., ABC/SBL, Rio de Janeiro (1995).
48. R. P. Vari A phylogenetic study of the neotropical Characiform family Curimatidae (Pisces: Ostariophysi). *Smithsonian Contributions to Zoology*, **471**, 1–71 (1989a).
49. R. P. Vari Systematics of the neotropical genus *Psectrogaster* Eigenmann and Eigenmann (Pisces: Ostariophysi). *Smithsonian Contributions to Zoology*, **481**, 1–42 (1989b).
50. R. P. Vari Systematics of the neotropical Characiform genus *Steindachnerina* Fowler (Pisces: Ostariophysi), *Smithsonian Contributions to Zoology*, **507**, 1–118 (1991).
51. R. P. Vari Systematics of the neotropical Characiform genus *Curimatella* Eigenmann and Eigenmann (Pisces: Ostariophysi), with summary comments on the Curimatidae. *Smithsonian Contributions to Zoology*, **533**, 1–48 (1992).
52. R. P. Vari The Curimatidae, a lowland neotropical fish family (Pisces: Characiformes); distribution, endemism and phylogenetic biogeography. In *Neotropical Distribution Patterns: Proceedings of a Workshop* P. E. Vanzolini and W. R. Heyer (ed.), pp. 106–137. (Academia Brasiliera de Ciencias, Rio de Janeiro, 1988).
53. T. L. Erwin *Taxon Pulses, Vicariance and Dispersal: An Evolutionary Synthesis Illustrated by Carabid Beetles*. (Columbia University Press, New York, 1979).

54. N. A. Moran, The evolution of host plant alternation in aphids: evidence that specialisation is a dead end. *American Naturalist*, **132**, 681–706 (1988).
55. N. A. Moran, Phenotype fixation and genotypic diversity in the life cycle of the aphid *Pemphigus betae. Evolution*, **45**, 957–970 (1991).
56. N. A. Moran, The evolutionary maintenance of alternative phenotypes. *American Naturalist*, **139**, 971–989 (1992).
57. M. J. West-Eberhard, Phenotypic plasticity and the origins of diversity. *Annual Reviews of Ecology and Systematics*, **20**, 249–278 (1989).
58. W. D. Hamilton, Wingless and fighting males in fig wasps and other inseets. In *Reproductive Competition, Mate Choice and Sexual Selection in Insects* M. S. Blum and N. A. Blum (ed.), pp. 167–220. (Academic Press, London, 1979).
59. P. A. Henderson and R. N. Bamber, On the reproductive strategy of the sand smelt *Atherina Boyeri* Risso and its evolutionary potential. *Zoological Journal of the Linean Society of London*, **32**, 395–415 (1987).
60. R. N. Bamber and P. A. Henderson, Pre-adaptive plasticity in atherinids and the estuarine seat of teleost evolution. In *Fish in Estuaries* (Symposium proceedings: by P. A. Henderson and A. R. Margetts (ed.)). *Journal of Fish Biology*. **23**(A), 17–23 (1989).
61. A. A. Bray, The evolution of terrestrial vertebrates: environmental and physiological considerations. *Philisophical Transactions of the Royal Society Series* B. **309**, 289–322 (1985).
62. D. C. Dennett, *Darwin's Dangerous Idea*. Penguin, London (1995).
63. K. Kubitski. Ichthyochory in *Gnetum venosum. Anais da Academia brasileira de Ciencia*, **57**, 513–516 (1985).
64. A. M. W. Mennega, Wood anatomy of the Euphorbiaceae. In *The Euphorbiaceae. Chemistry. Taxonomy and Economic Botany* S. L. Jury, T. Reynolds, D. F. Cutler, and E. J. Evans (ed.), pp. 111–126 (Academic Press, London, 1987).
65. W. Punt, A survey of pollen morphology in the Euphorbiaceae with special reference to *Phyllanthus. Botanical Journal of the Linnean Society*, **94**, 127–142 (1987).
66. A. H. Church, *Thalassiophyta and Subaerial Transmigration*. (Clarendon Press, Oxford, 1919).
67. V. H. Heywood, *Flowering Plants of the World*. (Mayflower, New York, 1978).
68. G. T. Prance and J. R. Arias, A study of the floral biology of *Victoria amazonica* (Poepp.) Sowerby *(Nymphaceae). Acta Amazonica*, **5**, 109–139 (1975).
69. W. L. Crepet and E. M. Friis, The evolution of insect pollination in angiosperms. In *The Origins of the Angiosperms and their Biological Consequences* E. M. Friis, W. G. Chaloner, and P.R. Crane (ed.), pp. 181–202. (Cambridge University Press, Cambridge, 1987).
70. E. J. H. Corner, The evolution of tropical forest. In *Evolution as a Process* J. S. Huxley, A. C. Hardy, and E. B. Ford (ed.), pp. 34–46. (Cambridge University Press, Cambridge, 1954).
71. E. J. H. Corner, The Durian theory or the origin of the modern tree. *Annals of Botany New Series*, **13**, 367–414 (1949).
72. M. Berenbaum and D. Seigler, Biochemicals: engineering problems for natural selection. In *Insect Chemical Ecology: An evolutionary approach* B. D. Roitberg and M. B. Isman (ed.), pp. 89–121. (Chapman and Hall, New York, 1992).

73. S. Carlquist. Wood anatomy of Compositae: a summary with comments on factors controlling wood evolution. *Aliso*, **6**, 25–44 (1966).
74. A. Henderson. *The Palms of the Amazon.* (Oxford University Press, New York, 1995).
75. R. Spruce, Palmae Amazonicae. *Journal of the Proceedings of the Linnean Society, Botanical*, **4**, 58–63 (1871).
76. W. J. Junk, and M.T.F Piedade, Biomass and primary-production of herbaceous plant communities in the Amazon floodplain. *Hydrobiology*, **263**, 155–162 (1993).
77. E. Nevo, Adaptive speciation at the molecular and organismal levels and its bearing on Amazonian biodiversity. *Evolucion Bioloica*, **7**, 207–249 (1993).

CHAPTER 17

A VIEW FROM MARS

SAM P. BROWN

Autumn colours . . . can be spectacular: a Martian might easily guess them to be almost as important in the latitudes in question as the green of summer.

Bill Hamilton

By the time I encountered the name W. D. Hamilton, I had already abandoned the science of biology as being impossibly dull. As an idle second-year philosophy student, I chanced upon a dog-eared copy of Dawkins' *The Selfish Gene*[1] on a friend's bookshelf. This book knocked my socks off. I was filled with wonder at the forces of evolution, and the force of Bill's ideas.

Bill's memes certainly had a dramatic effect on my behaviour. With near missionary zeal I at last experienced a sense of calling beyond the college bar. Within weeks I was even pleading to be re-admitted to the biology programme I had so gleefully escaped only a year before. Bill had been a biology undergraduate in Cambridge 30 years before me, and had not been enamoured with the prevalent disregard for evolutionary analyses.[2] A generation later, Cambridge was enamoured with Bill. As a final year student in the Zoology Department, I learnt of evolutionary approaches driving in all directions, with Bill's ideas often leading the charge.

My new passion for biology led me to a Master's degree in 'bioscience' at Oxford. My plan was to steer my fascination with behavioural and evolutionary ecology towards the study of infectious diseases, and where better, I thought, than in the company of the legendary Professor W. D. Hamilton? The course involved two extended projects. Scanning the options I saw a range of fascinating themes, but nevertheless felt a sense of anti-climax. No Bill! Before I entirely realised what I was doing, I found myself sidling along the uppermost floors of the remote West Wing, in search of the door marked 'WDH'. My momentum carried me through that door, only to desert me as soon as I entered the room and found I had no idea what on earth I was going to say. Fortunately, Bill wasn't too put out by my stammering; he managed to draw out from me that I was a Masters student on the look-out for a project, and he listened patiently as I muttered some fairly mundane ideas about host–parasite coevolution. By now he was gazing out of the window. This gave me a welcome reprieve from my apparent seizure, and a chance to take in my surroundings. I found myself in a large rectangular office, overflowing with boxes upon boxes of files. Following Bill's gaze, I looked down onto South Parks Road, with a glimpse of the University Parks beyond. I was bowled over by the thought that the great man might be finding my ideas worthy of such seemingly deep contemplation; this was Bill Hamilton—in the flesh! His shock of white hair, his glasses and that slight stoop fulfilled all the criteria for absent-minded professordom. Absent minded? Ah, yes; perhaps he'd forgotten I was there. Hoping to convey the impression I was concentrating hard too, I settled on counting the boxes. Bill finally turned from the window and said: 'I had an idea, but I can't quite remember what it was—could you come back tomorrow?' So I promised I would, and left his office in a state of heady excitement.

The next day I duly knocked on his door (now seemingly far less remote) expecting to be challenged by some elaboration on Red Queen host/parasite dynamics. Instead, Bill welcomed me with an enthusiastic 'Ah yes: TREES!', and led me to his window. He pointed excitedly at this tree and that as I desperately tried to figure out what he was on about. Gradually he slowed down—or maybe I was catching up. Five minutes earlier I'd been looking down on a typical (if distinguished) English street in October; now I was dazzled by images of combat-trees flexing their muscles at the world. I was hooked.

Over our next few meetings Bill outlined the scope of his idea. His leap was to view autumn colours as something other than a pretty accident, and to suggest that they might be functioning as expensive 'handicap' signals that asserted the vigour of the plant. At this point he had two potential classes of receivers in mind. His preferred hypothesis was that richly coloured trees were signalling to autumn-ovipositing insects, of which aphids were likely to be especially important. In a short note, Bill summarized the signal of a brilliantly coloured tree as saying 'I'm strong: watch my potlatch of energy and pigment! Be warned; I keep plenty back and I'm going to be highly inedible next spring. Don't land on me—go somewhere else, find someone weaker.' His secondary hypothesis was that these signals might be directed at nearby plants of the same species that could be competing for light and soil with the signaller.

But how could these elaborate hypotheses be tested in a 6-month MSc project? In line with a previous landmark work of his concerning colours in nature (this time of birds[3]), he suggested a comparative analysis as an essential—and manageable—first step. And so I began an accountant's audit of the world's temperate trees. Which tree turns what colour? Where are they? How big? What size leaves? Can they move (by root suckers)? And finally: which aphids come to call? Once the data was collected, we could then look for tell-tale signs of a link between the ecological pressure of pesky aphids on a given tree species, and an evolutionary response of striking autumn colour. The data was a joy to collect. I really couldn't get over the cheekiness of Bill's approach. He suggested I look up 'works of a "gardening" slant', and use the descriptions of the foliage as the data for my scoring of autumnal colours. Thus if, say, a *Collins*' guide to trees reports that a particular species turns 'bright red' in autumn, then that species would score '3' for red, whilst another that turns 'dull yellow' would score a point for yellow, and so on.

The autumn of 1995 may well have produced only run-of-the-mill colours, but I had never experienced a more spectacular fall. Every tree entranced me. With the colours rapidly deepening all around, Bill suggested I attempt some form of observational study. 'I suggest you find a number of spindle trees (*Euonymus europea*) around Wytham and quantitatively sample *Aphis fabae* arriving on them, while at the same time assessing in some way the wave of autumn colours as it comes on—you have to act fast!'[4] Sadly, the experimental part of my project fell apart in a farcical tangle of insect-glue and absent aphids (or at least an absence of aphids willing to make contact with my makeshift traps).

Once I'd finished the comical exercise of judging characteristic species autumnal colour via the use of gardening books, we turned to the challenge of filling in the data on aphid involvement. How on earth were we to categorize the enormous diversity of aphids on the world's trees? Bill wandered down to one of his favourite Oxford libraries, the cavernous Blackwells Bookshop, and was delighted to find the answer in the form of a monstrous doorstop entitled no less than *Aphids on the World's Trees*![5]

Armed with this treasure-trove of beautifully organized information, our aphids columns were quickly filled up.

With the data-trawl complete, the statistical analyses could finally be run. Under the tutelage of Jeremy John I discovered the joys of Unix, GLIM and PHYLO.GLM in one fell swoop. Our results were positive and the excitement was enough to carry me through the sleepless nights of writing up my thesis as the March 1996 deadline approached.

I saw less of Bill during the second half of my MSc, as I was largely occupied by a lab project on malaria immunology. However, our occasional chats were given added spice by the prospect of a conference paper in Florence that August. If I hadn't seen Bill for a while, I would sometimes feel an attack of nerves on approaching his office, classic imposter syndrome in the face of Bill's awesome reputation. Yet once inside, these worries would soon subside in the company of a wonderfully gentle, shy and considerate man. I remember his office as an oasis of calm, a shelter from the headlong rush of my Masters programme. Together we would while away hours discussing endless questions thrown up by the notion of signalling leaves. How might a signal of defence originate? What of the perceptual limitations of the receivers? How might signal/receiver coevolution relate to other colourful plant structures such as fruits and flowers? Out of these discussions grew Bill's first draft of a manuscript. It was in this draft that Bill asked what might a Martian make of all this autumnal colour? I have a feeling that assuming an extraterrestrial perspective was a useful tool in more than one of Bill's contributions to Earthly biology. Compared to the terse note [next section] that finally appeared in print in 2001, this draft was more a monograph, discussing many of the intriguing avenues we had explored during our discussions. To give a picture of the broader scope, here is Bill's wonderful introductory paragraph:

Cut or saw open a large animal or plant and usually in its interior you will see little colour. Blood in a vertebrate and dependent colours may appear an exception to this claim but blood is so liable to exposure on the surface of its possessor that a signalling function (at least to its possessor) could be suspected. The uncommon wild roots and woods that are coloured never rival the vividness of blood, still less rival that of flowers. Deep in soil and in other unlighted places even the outsides of organisms are usually colourless, and this is in strong contrast to exteriors in the light. Combined, such facts suggest that it rarely 'just happens' that the molecules of functional organic chemistry are coloured. If chance colouring is rare then colours must usually be intended to be seen—thus plant colours, like animal colours, probably usually communicate.[6]

In August 1996 we left Oxford for a week in Florence at the XXth International Congress of Entomology; my first visit to Italy, and my first scientific conference! The event itself largely passed me by. I felt quite lost among 2000 strangers discussing even stranger tales of insects, and all in a manner that at the time seemed quite unrelated to my budding interests in ecology and evolution. But Bill was at hand. He would rescue me from my plainly bemused state with slices of pizza and discussions on how we should go about informing the world of our subversive ideas. Faced with the prospect of presenting a 15-minute summary, I took a red pen to the combined text of MSc thesis and draft manuscript, arriving at a simple outline of Bill's autumn colours hypothesis, the data we collected, and supporting hints from tree and aphid biology. Bill agreed that this editing was probably a wise step, but he was clearly frustrated that we weren't going to present our increasingly wacky ideas on the origins of colourful pigmentation, and the subsequent diversification of pigment uses into aversive (e.g. autumn colours) and attractive (fruits and flowers) syndromes. Bill's terrifying solution, announced only days from show time, was that we should split our 15-minute slot into two sections. For the first '7 or 8 minutes', I should do my dry run-down of the autumn colours hypothesis and our supporting case, then in the final half Bill would hold forth on a big-picture contextualization of our findings. Having finally rejected the 'talk twice as fast to get it all in' solution, I whittled down my already sparse résumé into a 7-minute skeleton. This skeleton lives on in the structure of our final publication.

Feeling rather like the straight-man in a double-act, I got up to run through my precisely timed, tightly scripted rendition. Coming in at a little over 8 minutes, I put Bill in a tight situation. Bill responded with a glorious sequence of slides outlining unexpected colours in strange corners of the plant world, accompanied with drifting anecdotes and fascinating diversions; and, not for the first time in his lecturing career, he ran out of time. I think most of the audience were aphid systematists, and not necessarily familiar with Bill's name. A kindly spectator reported that one eminent systematist surmised Bill to be completely bonkers; some lunatic from Oxford, who ought not to be let out at conferences!

The conference proceedings afforded us a publication of sorts, in the form of a tiny and obscure abstract,[7] followed and enlivened by a couple of newspaper reports.[8] Sheltered by this tiny note of precedence, we each took turns to be distracted by other projects, new and old. Bill wrote in early 1997 'I admit there is a highly anomalous situation at the moment where our idea has been published by *La Stampa* and *The Economist*[8] ... and there is no proper paper in print'.[9] Unfortunately, our 'proper paper' took another four years to appear.

By 1997 I was already heavily distracted by a PhD back in Cambridge (I had failed to secure funding to continue studying with Bill). Bill wrote that by Autumn he might have had some time to devote to colours; until then he would be snowed under writing the introductory chapters for *Narrow Roads II*, before spending a month in Brazil. Autumn came and Bill reported some fascinating observations on the colours of the Brazilian *cerrado*, adding: 'I have to confess ... I have in fact allowed myself to be distracted ...'[10] ... What by this time? Clouds! Rising to the challenge of Lovelock's *Gaia* hypothesis, he was exploring how cumulus cloud formation could be in part a bacterial mechanism of dispersal,[11] and had naughtily prioritized this work above *Narrow Roads*, Autumn Trees, and everything else. I wasn't doing any better in turning our work into a paper either. Despite Bill's continuing insistence that I should go ahead and write a concise paper of our key findings, my own (very part-time) work in this area consisted solely of developing what I saw as a partner publication on the origins and diversification of colourful signals.

Finally, in early 1998, feeling more secure in the progress of my PhD, I wrote a concise summary of our project, with a view to submission to *Nature*. We submitted in Autumn, naturally, and despite the encouragement of the unfurling seasonal show, our reviewers mirrored the range of reactions with which we had become familiar: everything from fascination to complete incredulity—'bonkers' being a not-infrequent response! So, another autumn gone, and still no paper. Progress picked up again in 1999, by which time the most critical of the referees' comments had thrown up some new references that were actually rather helpful for our theme. So we began to consider a resubmission to *Nature*.

My last visit to see Bill was late that summer. We'd arranged to meet in his office to thrash out the final details before resubmission, but he wasn't there. Next day I phoned, and he apologized and told me he was extremely busy preparing 'to take advantage of the cease-fire in the Congo to go and collect monkey droppings'. Once again Bill's inquisitive adventurousness had leapt ahead onto new challenges.[12] I left Oxford, and put 'Trees' on hold as the end of my thesis loomed.

Bill was planning what turned out to be his final trip to the Congo. When I heard of Bill's death in March 2000 I was letting new opportunities get the better of my PhD, enjoying a 3-month stay in Montpellier, France. I hadn't spoken to Bill since the autumn of 1999, and mingled in my sadness was a terrible guilt that I had once again been distracted from prioritizing the resubmission of our revised manuscript. So it was that when I got back to Cambridge, 'Trees' took precedence over my thesis, and our manuscript was soon resubmitted to *Nature*. Our changes weren't enough to sway the reviewers, so I took the manuscript to the *Proceedings of the Royal Society* where it was rapidly accepted.

The paper created quite a media stir. I was interviewed by all manner of journalists from all over the world, largely (no doubt) because of the appeal of Bill's crazy idea, but also because of its news value as the posthumous work of a scientific legend. The wave of media interest was followed by an attempt at a damning scientific critique,[13] since when, I'm happy to say, there has been a growing wave of meticulous and largely supportive studies of specific insect–tree interactions.[14–18]

Working with Bill has had a profound influence upon my development as a scientist. Sometimes while we were chatting, Bill would pause for several minutes before answering a question, or before commenting on an observation. These pondering pauses remain in my memory in noble contrast to the rapid and harried discussions that all too often comprise scientific exchange. Rather than work from the assumption that he knew the answer to anything, Bill was more likely to start from scratch, as if from the perspective of a visitor from Mars. I believe this otherworldly perspective, this almost naive and seemingly ponderous reflection, contributed to Bill's uncanny ability to capture the everyday and make it seem strange; to see with other eyes; to refocus the mind. Through his company and through his work, Bill showed me the essential importance of imagination in understanding the world.

References

1. R. Dawkins, *The Selfish Gene* (Oxford University Press, Oxford, 1976).
2. W. D. Hamilton, *Narrow Roads of Gene Land*. Volume **1**, pp. 21–22 (WH Freeman, Oxford, 1996).
3. W. D. Hamilton and M. Zuk, Heritable true fitness and bright birds: a role for parasites? *Science* **218**, 384–387 (1982).
4. Letter from W. D. Hamilton, 15 October (1995).

5. R. L. Blackman and V. F. Eastop, *Aphids on the World's Trees*. CABI, (Wallingford, UK 1995).
6. Draft manuscript entitled 'Autumn colours and "larch roses": roles for insects?', July (1995).
7. W. D. Hamilton and S. P. Brown, Autumn colours: a role for aphids? *Proc. Int. Congr. Entomol.* **20**, 232 (1996).
8. *The Economist* (5 October 1996) and *La Stampa* (30 August 1996).
9. E-mail from W. D. Hamilton, 30 April (1997).
10. E-mail from W. D. Hamilton, 10 September (1997).
11. W. D. Hamilton and T. M. Lenton, Spora and Gaia: How microbes fly with their clouds *Ethol. Ecol & Evol.* **10**, 1–16—Section 15.3 (1998).
12. E. Hooper and W. D. Hamilton, 1959 Manchester case of syndrome resembling AIDS *Lancet* **348**, 1363–1365 (Section 14.2) (1996).
13. D. M. Wilkinson, T. N. Sherratt, D. M. Phillip, S. D. Wratten, A. F. G. Dixon, and A. J. Young, The adaptive significance of autumn leaf colours. *Oikos* **99**, 402–407 (2002).
14. S. B. Hagen, I. Folstad, and S. W. Jakobsen. Autumn colouration and herbivore resistance in mountain birch (*Betula pubescens*) *Ecology Letters* **6**, 807–811 (2003).
15. S. B. Hagen, S. Debeausse, N. G. Yoccoz, and I. Folstad (2004). Autumn colouration as a signal of tree condition. *Proc. R. Soc. Lond.* B **271**, 5184–5185.
16. M. Archetti and S. P. Brown (2004). The coevolution theory of autumn colours. *Proc. R. Soc. Lond.* B **271**, 1219–1223.
17. M. Archetti and S. Leather (2005). Testing the coevolution theory of autumn colours. *Oikos* 110, 339.
18. Marco Archetti (see previous two notes) was a visiting student in the Zoology Department in the spring of 1996, and worked with Bill on a theoretical model of insect–tree coevolution (M. Archetti 2000. The origin of autumn colours by coevolution. *J. Theor. Biol.* **205**, 625–630). Despite only meeting for a total of perhaps three minutes, Marco and I experienced a remarkably parallel path of working with Bill. Marco's memories of Bill are available at http://www.evolution.unibas.ch/hamilton/archetti.htm

AUTUMN TREE COLOURS AS A HANDICAP SIGNAL†

W. D. HAMILTON AND S. P. BROWN

Abstract

Many species of deciduous trees display striking colour changes in autumn. Here, we present a functional hypothesis: bright autumn colouration serves as an honest signal of defensive commitment against autumn colonizing insect pests. According to this hypothesis, individuals within a signalling species show variation in the expression of autumn colouration, with defensively committed trees producing a more intense display. Insects are expected to be averse to the brightest tree individuals and, hence, preferentially colonize the least defensive hosts. We predicted that tree species suffering greater insect damage would, on average, invest more in autumn-colour signalling than less troubled species. Here, we show that autumn colouration is stronger in species facing a high diversity of damaging specialist aphids. Aphids are likely to be an important group of signal receivers because they are choosy, damaging and use colour cues in host selection. In the light of further aspects of insect and tree biology, these results support the notion that bright autumn colours are expensive handicap signals revealing the defensive commitment of individual trees to autumn colonizing insect pests.

Introduction

The diversity and expense of pigment synthesis in leaves about to be shed has, apart from biochemical study, attracted little academic attention. The most commonly encountered explanation of spectacular autumn colours is that they are incidental side-effects of a controlled senescence. For instance, when discussing the synthesis of new pigments in leaves about to be shed, Matile[1] concludes that 'such biochemical extravagancies associated with leaf senescence

---

† *Proc. R. Soc. Lond.* B **268**, 1489–1493 (2001).

may have evolved in the absence of selection pressure'. To our knowledge only two functional hypotheses exist in the literature. Note, however, that the fruit-flag hypothesis (colourful leaves serve to highlight fruits; Stiles[2]) is necessarily limited to fruit-bearing trees, and the ultraviolet-screen hypothesis (yellow carotenoid pigments serve to protect leaves from photo-oxidative damage prior to abscission; Merzlyak and Gittleson[3]) is more relevant to leaf construction. Importantly, while neither contradicts the signalling hypothesis, unlike the signalling hypothesis, neither can account for the striking variation in autumn colouration both within and between species of deciduous trees.

The signalling hypothesis makes predictions on two interrelated levels. On an interspecific level, tree species suffering greater insect attack should invest more in defence and defensive signalling. On an intraspecific level within signalling species, the most defensively committed individuals should produce the most intense displays, increasing the likelihood that the specialist pests driving the signal evolution will land on another individual of the same species. The inter- and intraspecific predictions are separate but interrelated, mirroring an earlier study on parasite-driven signals in birds.[4] Here, we concentrate on examining interspecific trends, to provide an empirical framework for future experimental research on individual signalling species.

Methods

We investigated the signalling hypothesis by examining the association between aphid diversity and degree of autumn colouration across 262 north-temperate tree species. The data-set is available from S. P. B. on request. We focused on aphid—deciduous-tree interactions because aphids are, at present, an important and well-documented pest of trees[5] and have a long proven history of association.[6] Furthermore, many aphids show a peak in migratory behaviour in autumn,[7] just as the tree leaves are changing colour. Note, however, that the proposed signalling system is likely to hold between other plant groups and their most significant insect pests.

Aphid diversity was used as the best available comparative estimate of insect damage, leading to the prediction that trees beset by many aphid species would, on average, invest more in autumn-colour signalling than less troubled trees. Furthermore, we expected specialist (single winter host species) aphids to be most important in driving the evolution of autumn signalling because specialist pests tend to be more damaging.[8,9]

The set of 262 tree species was constructed from two field guides,[10,11] subject to a number of exclusions. Hybrids, clones and other cultivated species were excluded, in order to focus on naturally selected colour traits. To minimize likely false zeros in the aphid data, tree genera for which no aphid data were available were also excluded. Using strict rules based on the wording in the guides, we gave each tree species a score for the development of yellow colouration and another score for red. The scores corresponded to the following field-guide wordings relating to the degree of autumnal coloration: 0, colour change not mentioned; 1, dull (or brown-) yellow or red; 2, yellow or red; 3, bright (or striking, intense, etc.) yellow or red. A mention of brown autumn colouration scored 1 for both colour variables.

Estimates of the control variables (leaf size, tree size, fruit and flower colours, climate and geographical range) were obtained, where possible, from the same guides, based on simple information on geographical location, typical leaf and tree dimensions and fruit and flower colours. Fruit ($n = 135$) and flower ($n = 132$) redness were of interest in assessing the possibility of pleiotropy in leaf, fruit and flower development, and were recorded as either present or absent. Species range ($n = 117$) was recorded on a four-point scale: 0 = <130 000 km^2; 1 = 130 000–520 000 km^2; 2 = 520 000–2 600 000 km^2; 3 = >2 600 000 km^2. Climate ($n = 117$) was quantified categorically on a three-point scale: dry, warm temperate, and cool temperate. An estimate of leaf size ($n = 135$) was obtained by multiplying leaf length by leaf width. An index of tree size ($n = 135$) was obtained by multiplying tree height by the square of the trunk diameter at waist height.

After completing colour and control scoring, we tabulated the number of autumn colonizing aphid species associated with each tree species using a recent monograph of world tree aphids.[5] Only aphids with winged autumnal morphs were included in the analysis. Five counts were recorded for each tree species, giving the numbers of aphid species specific to the tree at the species level, and at the levels of infesting just one, two, three, or more than three genera of trees. Aphid—tree associations recorded as very recent, when either partner has been introduced into a new area by human activity, were excluded.

The strength of association between autumnal colouration and the diversity of aphid attack was quantified using both cross-species correlation and phylogenetic regression.[12,13] Phylogenetic regression allows the incorporation of phylogenetic information into the statistical analysis of comparative datasets, in order to identify correlated character changes that occur repeatedly in evolutionary lineages. As with other correlated-divergence methods, each radiation in the phytogeny, rather than each present-day species, contributes one data point to the regression analysis.[14]

A phylogenetic tree linking the 262 tree species was constructed using the taxonomic information captured in the two field guides, with branch lengths specified using Grafen's[12] default 'figure 2' method. Phylogenetic regression was implemented using PHYLO.GLM v. 1.03, running in GLIM 3.77.[15]

Results

The phylogenetically controlled analysis revealed that the degree of yellow colouration in autumn leaves was significantly correlated with the number of aphid species, regardless of host specificity (Table 17.1). Repeating the analysis

Table 17.1. Phylogenetic regression[12] of autumn colour against aphid diversity

(The columns present statistics for a range of aphid—host specificities, from extreme specialists (one winter host species) to extreme generalists (more than three genera of winter hosts). The first and second rows are *F*-values and the third row shows the corresponding degrees of freedom.)

	aphid diversity, categorized by winter-hostspecificity					
Colour	All aphids	One species	One genus	Two genera	Three genera	More than three genera
Redness	0.96	5.97*	0.49	2.93	1.45	0.77
Yellowness	7.72**	17.28***	2.89	0.01	0.00	0.05
d.f.	1, 49	1, 40	1, 39	1, 29	1, 10	1, 33

*$p < 0.05$, **$p < 0.01$ and ***$p < 0.001$.

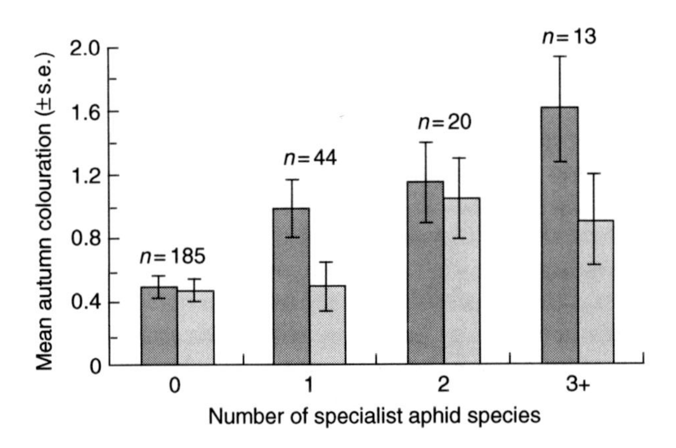

Figure 17.1. Autumn colour (±s.e.m.) as a function of specialist aphid diversity. Dark and light bars record mean yellow and red scores, respectively. For yellow, Spearman's rank correlation coefficient, $r_s = 0.30$, $n = 262$, $p < 0.005$; for red, $r_s = 0.14$, $n = 262$, $p < 0.01$.

for specialist aphids alone, the positive association with yellowness was strengthened, and an independent significant correlation with redness emerged in both the raw (cross-species) and the phylogenetically controlled analyses (Table 17.1 and Fig. 17.1).

The analyses of both the raw and the phylogenetically controlled data show statistical significance, hence the evidence for a non-chance relationship between strength of autumn colouration and diversity of specialist aphids can be regarded as secure. Strong autumn colouration is over-represented in tree species that are hosts to specific aphids (i.e. aphids that are their most particular enemies). Note that the correlation between aphid diversity and autumn redness

Table 17.2. Phylogenetic regressions[12] of autumn colour against specialist aphid diversity, given a range of controls

(The columns present phylogenetic regressions of autumn colour on specific aphid diversity, given different controls. The first and second rows are *F*-values and the third row shows the corresponding degrees of freedom.)

Colour	No control	Leaf size	Tree size	Climate	Geographical range	Autumnal redness	Autumnal yellowness	Fruit redness	Flower redness
red	5.97*	7.47*	6.91*	0.51	2.22	—	5.62*	5.90*	5.53*
Yellow	17.28***	11.40**	8.36**	9.91**	12.97**	17.96***	—	8.94**	9.34**
d.f.	1, 40	1, 36	1, 37	1, 28	1, 27	1, 43	1, 42	1, 24	1, 25

*$p < 0.05$, *$p < 0.01$ and ***$p < 0.001$.

strengthens when phylogenetic control is imposed, suggesting numerous convergences to autumnal redness accompanied by an aversion to red on the part of the appropriate aphids. That yellow autumn colours might need to have re-evolved on fewer occasions fits with the relative ease of manifesting yellowish pigments. These are mainly derivatives of β-carotene and related compounds, which are universal accessories of photosynthesis.

To investigate the potential for a causal relationship between aphid diversity and autumn colouration, the above analyses were repeated for specific aphids while controlling for a number of relevant and potentially confounding variables (climate, geographical range, leaf size, tree size, fruit colour and flower colour; Table 17.2). In all cases, the significant correlation between yellowness and specific aphid diversity remained, regardless of the control used. The climate and geographical-range controls resulted in the redness correlation losing significance; however, further tests indicated that the loss of significance in these two cases was a consequence of a reduced data-set (geographical information was available for only 117 species) rather than an effect of the controls. Leaf yellowness was found to be positively and significantly correlated with both range size (phylogenetic regression, $F_{1,18} = 7.39$, $p < 0.05$) and tree size (phylogenetic regression, $F_{1,22} = 11.44$, $p < 0.01$). These results may reflect a bias in field knowledge: big, widespread trees will generally be better described. None of the remaining variables showed any significant relationship with either the redness or the yellowness of autumnal leaves. Overall, the use of varied controls strengthens the argument that the specific-aphid—autumn-colour link is causal.

Biological Context

In addition to the association between parasite burden and elaboration of autumn colour, several aspects of aphid and tree biology make the case for autumn signalling more compelling.

AUTUMN COLORATION IS EXPENSIVE

An essential component of a handicap signal is cost.[16,17] If the signal is not costly then a weak signaller could pretend to be strong and thus undermine signal reliability. A number of potential honesty-ensuring costs are associated with leaf-pigment signalling.

The carotenoid yellows of many autumn displays follow the selective resorption of chlorophyll.[18] The highly selective resorption of leaf pigments leads to a significant and avoidable loss of energy; for instance in *Acer platanoides* no more than 50% of carotenoids are recovered before leaf fall, whereas around 95% of chlorophylls are resorbed.[3] The cost to the tree is compounded by the loss of the lipids in which the carotenoids are suspended. A further cost of autumn signalling may be incurred due to the loss of primary production following an early cessation of photosynthesis.

In addition to differential pigment decomposition, colour change also results from the synthesis of new compounds. A number of fluorescent compounds and optical brighteners have been recorded only in autumn leaves.[19,20] The case against a simple economy is even stronger for the anthocyanin reds and purples because they are manufactured in often massive quantities in autumn leaves,[21,22] suggesting a major energetic cost to red autumnal colour.

The range of potential costs associated with the synthesis and loss of pigments indicates that there is the potential for handicap signalling in autumn trees. The variability in autumn colours within species (see below) may indeed honestly reflect the variation in defensive commitment among trees. The expense, waste and scattered phylogenetic distribution of autumn displays suggest that autumn colours are adaptive. The following three points show that their most probable function is to serve as conspicuous signals: in short, they are 'pick on someone else' signals to specialist autumn-flying insects.

APHIDS ARE DAMAGING

An expensive signalling strategy will only evolve if the recipients' behaviour is economically important to the signaller. Aphid damage is often high: if *Drepanosiphum* aphids were eliminated from sycamore trees, it has been estimated that wood deposition would increase 2.8-fold.[23] Infested saplings of lime and oak often weigh less at the end of a year than they did at the start.[24] Even fleeting contact with aphids can cause serious damage through the transmission of viruses[25]; hence, signalled deterrents that act before the insect samples the plant (anti-xenosis) carry an important benefit over deterrents (e.g. toxins) that act at the point of sampling (antibiosis).

The aphid record back to the Carboniferous[6] suggests that damage has occurred throughout the entire history of the angiosperms, the most autumn-coloured group. Maples (*Acer*) are particularly troubled by specific aphids, as this quote from Blackman and Eastop[5] reveals: 'There are several aphid genera which are virtually specific to *Acer*, and within these genera there is a high degree of monophagy, although a few species are able to colonize several *Acer* species.' We note that maples are well known across north-temperate regions for their brilliant autumn colors.

APHIDS ARE DISCRIMINATING

Individuals within a plant population commonly show extensive variation in quality as hosts for their insect pests.[26–28] For example, tree-specific survivorship of autumn migrants of *Pemphigus betae* colonizing *Populus angustifolia* varies from 0 to 76%.[26]

In the face of such high fitness differentials between hosts, colonizing insects will experience a strong selection pressure to develop mechanisms of adaptive discrimination between good- and poor-quality hosts. When herbivores are selecting existing tissues, direct assessment of a critical fitness determinant, say leaf size, is generally possible. In the case of autumn-ovipositing insects, the situation is more complex due to the time lag between host selection in autumn and host exploitation by progeny the following spring. Adaptive discrimination is constrained by the availability of host features detectable at colonization and correlated with host quality at the time of exploitation. Despite these constraints, autumn migrants of *P. betae* colonizing cottonwood trees show preferential colonization of more favourable hosts.[27]

The potential role of autumn colouration in the discriminatory process of autumn migrants is supported by the existence of considerable variation in both the timing and the degree of colour change among individuals of colourful tree species at a single site.[29,30] Autumn migrants of *Periphyllus californiensis* colonizing *Acer palmatum* in Japan were recorded to colonize yellow-orange *A.palmatum* individuals preferentially, leaving the most intensely red individuals almost aphid free.[29,30] Note that the few aphid individuals colonizing the reddest trees suffered reduced fitness as measured by the fecundity of their progeny in spring.[30]

APHIDS USE COLOUR IN HOST SELECTION

For a number of well-studied aphids, yellow-green light has been shown to be the most attractive,[31,32] though for the great majority of species the precise preferences are unknown. Interestingly for the present hypothesis, however, it has been shown for three species of yellow-seeking aphids that an undiluted hue of the most attractive wavelength attracts fewer aphids than the same hue diluted with white.[33] We suggest that this preference for impure yellow will result in avoidance of trees showing maximal signalling in favour of others managing to produce only impure tints. We argue that the important cue is relative intensity: individual hosts are competing to produce the most intense colour and so avoid colonization. An analogy can be made with the handicap-signalling theory of stotting in gazelles, in which gazelles perform energetic stotting displays to dissuade cheetahs from giving chase.[34] While a cheetah may select a metaphorically 'off-colour' gazelle, aphids may literally prefer off-colour trees.

Among aphids with known visual sensitivities, reflected red light is thought to be at or just beyond the perception of the majority.[31] However, a number of studies suggest that reds can play an important role in aphid antixenosis.[35–37] Our hypothesis predicts that specialist aphids of red autumnal trees will have red-sensitive vision and show increasing aversion to increasing red coloration.

Discussion

The diversity of autumnal coloration has been remarked upon many times in comparisons of trees both across[1,22] and within[1,29] species. Simple observation easily reinforces this claim. Here, we suggest that this diversity of coloration reflects, in part, a signalling interaction between certain tree species and their most aggressive pests. The comparative data on autumn coloration and aphid diversity point to a suggestive pattern of association. Evidently, direct experimental investigations are now required to test these ideas on characteristically bright tree species and their autumnal pests.

The use of comparative data to investigate the signalling hypothesis is complicated by a lack of comparative data on the damage sustained by specific tree species as a result of autumnal infestation with aphids. Ideally, we would present data on an 'impact factor' of aphids on each individual host species, but, as is often the case in comparative studies, the limits of the available data force the adoption of alternative measures. As a result of this limitation, we used instead the diversity of aphids per tree species as an indicator of the insect damage sustained per tree species. Interestingly, the diversity of specialist aphids emerged as the key correlate with autumn coloration, which, when viewed in conjunction with the signalling hypothesis, reinforces the independent suggestion that specialist pests tend to be more damaging to their plant hosts.[8,9]

From an experimental perspective, the major outstanding prediction of the handicap-signalling hypothesis is the intraspecific association between signal intensity and insect attack, within a signalling species. The importance of relative signal intensity in intraspecific host selection needs to be investigated both in observational field studies and in simple choice experiments, focusing on colourful tree species and their specialist autumn colonizing pests.

The causes of within-species variation in signal intensity are likely to be diverse, though we have focused primarily on the degree of defensive commitment. Note that defensive commitment is not necessarily correlated with plant resource holding or vigour; indeed the inverse may even be true.[38] Experimental investigations on tree clones raised under differing resource conditions could offer important insights into the relationships between resource holding, defensive expenditure and autumnal coloration.

Broadening the perspective on insect pests beyond aphids is likely to be most fruitful for the red-signalling trees. That the colour-signalling system may have evolved against many insects in addition to aphids is suggested by the weaker, yet still positive, correlation found for red.

In summary, we suggest that tree species suffering increased insect damage are more likely to evolve signals of defensive commitment perceptible to their insect pests. Autumn coloration is suggested to be a handicap signal of such commitment: the intensity of coloration honestly indicates defensive commitment through the costs of pigment synthesis, resource loss and primary-production loss. Aphids are likely to be an important group of receivers because they are choosy, damaging and have colour-sensitive vision.

Acknowledgements

We thank Jeremy John for help with the phylogenetic regression, and Pej Rohani and David Earn for comments on the manuscript. Support from The Royal Society (to W. D. H.) and the Biotechnology and Biological Sciences Research Council (to S. P. B.) is gratefully acknowledged.

References

1. P. Matile, Biochemistry of Indian summer: physiology of autumnal leaf coloration. *Exp. Geront.* **35**, 145–158 (2000).
2. E. W. Stiles, Fruit flags: two hypotheses. *Am. Nat.* **120**, 500–509 (1982).
3. W. N. Merzlyak and A. Gittleson, Why and what for the leaves are yellow in autumn? On the interpretation of optical spectra of senescing leaves (*Acer platanoides* L.). *J. Plant Physiol.* **145**, 315–320 (1995).
4. W. D. Hamilton and M. Zuk, Heritable true fitness and bright birds: a role for parasites? *Science* **218**, 384–387 (1982).
5. R. L. Blackman and V. F. Eastop, *Aphids on the World's Trees.* (CABI, Wallingford, UK, 1994).
6. O. E. Heie, Why are there so few aphid species in the temperate areas of the southern hemisphere? *Eur. J. Entomol.* **91**, 127–133 (1994).
7. A. F. G. Dixon, Population dynamics of the sycamore aphid *Drepanosiphum platanoides* (Schr.) (Hempiptera: Aphididae): migratory and trivial flight activity. *J. Anim. Ecol.* **38**, 585–606 (1969).
8. P. D. Coley and J. A. Barone, Herbivory and plant defence in tropical forests. *A. Rev. Ecol. Syst.* **27**, 305–335 (1996).
9. A. Mackenzie, A trade-off for host plant utilisation in the black bean aphid, *Aphis fabae. Evolution* **50**, 155–162 (1996).

10. A. Mitchel, *A Field Guide to the Trees of Britain and Northern Europe*. (Collins, London, 1974).
11. E. L. Little, *The Audubon Society Field Guide to North American Trees, Western Region*. (Knopf, New York, 1980).
12. A. Grafen, The phylogenetic regression. *Phil. Trans. R. Soc. Lond.* B **326**, 119–157 (1989).
13. A. Grafen and M. Ridley, Statistical tests for discrete cross-species data. *J. Theor. Biol.* **183**, 255–267 (1996).
14. T. Price, Correlated evolution and independent contrasts. *Phil. Trans. R. Soc. Lond.* B **352**, 519–529 (1997).
15. Royal Statistical Society. GLIM (general linear interactive model), version 3.77. (London, Royal Statistical Society 1985).
16. A. Zahavi, Mate selection—a selection for a handicap. *J. Theor. Biol.* **53**, 205–214 (1975).
17. A. Grafen, Biological signals as handicaps. *J. Theor. Biol.* **144**, 517–546 (1990).
18. J. E. Sanger, Quantitative investigations of leaf pigments from their inception in buds through autumn coloration to decomposition in falling leaves. *Ecology* **52**, 1075–1089 (1971).
19. T. Duggelin, K. Bortlik, H. Gut, P. Matile, and H. Thomas, Leaf senescence in *Festuca pratensis*: accumulation of lipofuscin-like compounds. *Physiologica Plantarum* **74**, 131–136 (1988).
20. P. Matile, B. M. P. Flach, and B. M. Eller, Autumn leaves of *Ginkgo biloba* L: optical properties, pigments and optical brighteners. *Bot. Acta* **105**, 13–17 (1992).
21. M. Boyer, J. Miller, M. Belanger, E. Hare, and J. Wu, Senescence and spectral reflectance in leaves of northern pin oak (*Quercus palustris* Muench.). *Remote Sens. Environ.* **25**, 71–87 (1988).
22. S. B. Ji, M. Yokoi, N. Saito, and L. S. Mao, Distribution of anthocyanins in Aceraceae leaves. *Biochem. Syst. Ecol.* **20**, 771–781 (1992).
23. A. F. G. Dixon, The role of aphids in wood formation. I. The effect of the sycamore aphid, *Drepanosiphum platanoides* (Schr.) (Aphididae) on the growth of the sycamore, *Acer pseudoplatanus* (L.). *J. Appl. Ecol.* **8**, 165–179 (1971*a*).
24. A. F. G. Dixon, The role of aphids in wood formation. II. The effect of the lime aphid *Eucallipterus tiliae* L. (Aphididae) on the growth of the lime *Tilia x vulgaris* Hayne. *J. Appl. Ecol.* **8**, 393–9 (1971*b*).
25. K. F. Harris and K. Maramorosch, *Aphids as Virus Vectors*. (Academic Press, New York, 1977).
26. T. G. Whitam, Host manipulation of parasites: within plant variation as a defence against rapidly evolving pests. In *Variable plants and herbivores in natural and managed systems* R. F. Denno and M. S. McClure (ed.), pp. 15–41 (Academic Press, New York, 1983).
27. N. A. Moran and T. G. Whitam, Differential colonisation of resistant and susceptible host plants: *Pemphigus* and *Populus*. *Ecology* **71**, 1059–1067 (1990).
28. R. S. Fritz, Direct and indirect effects of plant genetic variation on enemy impact. *Ecol. Entomol.* **20**, 18–26 (1995).
29. K. Furuta, Host preference and population dynamics in an autumnal population of the maple aphid, *Periphyllus californiensis* Shinji (Homoptera, Aphididae). *J. Appl. Entomol.* **102**, 93–100 (1986).

30. K. Furuta, Early budding of *Acer palmatum* caused by the shade; intraspecific heterogeneity of the host for the maple aphid. *Bull. Tokyo University Forests* **82**, 137–145 (1990).
31. R. J. Prokopy and E. D. Owens, Visual detection of plants by herbivorous insects. *A. Rev. Entomol.* **28**, 337–364 (1983).
32. J. Hardie, Spectral specificity for targeted flight in the black bean aphid, *Aphis fabae. J. Insect Physiol.* **35**, 619–626 (1989).
33. V. Moericke, Host-plant specific colour behaviour by *Hyalopterus pruni* (Aphididae). *Entomol. Exp. Appl.* **12**, 524–534 (1969).
34. C. D. Fitzgibbon and J. H. Fanshaw, Stotting in Thompson's gazelles: an honest signal of condition. *Behav. Ecol. Sociobiol.* **23**, 69–74 (1988).
35. P. R. Ellis and J. A. Hardman, Investigations of the resistance of cabbage cultivars and breeders lines to insect pests at Wellesbourne. In *Progress on Pest Management in Field Vegetables*. C. Cavalloro and C. Peleprents (ed.), pp. 99–105. (Balkema Press, The Netherlands, Rotterdam, 1988).
36. R. Singh and P. R. Ellis, Sources, mechanisms and bases of resistance in Cruciferae to the cabbage aphid, *Brevicoryne brassicae. Int. Org. Biol. Control/West Palaearctic Regional Section Bull* **16**, 21–35 (1993).
37. P. R. Ellis, R. Singh, D. A. C. Pink, J. R. Lynn, and P. L. Saw, Resistance to *Brevicoryne brassicae* in horticultural brassicas. *Euphytica* **88**, 85–96 (1995).
38. P. Mutikainen, M. Walls, J. Ovaska, M. Keinanen, R. Julkunen-Tiitto, and E. Vapaavuori, Herbivore resistance in *Betula pendula*: effect of fertilisation, defoliation and plant genotype. *Ecology* **81**, 49–65 (2000).

CHAPTER 18

TOMATO ATTRACTORS ON THE WALL OF AN ABANDONED CHURCH

AKIRA SASAKI

It was early July of 1999 when I arrived at Oxford to resume the collaborative research on the desynchronization in host–parasite metapopulation dynamics with Bill Hamilton, which we had started in the previous summer. He had just returned from a journey to Congo. The journey was not to test the Red Queen hypothesis for the evolution of sex or the Hamilton–Zuk hypothesis about sexual selection. He went to collect chimpanzee feces. And the feces must have been collected from chimpanzee troops living around the former Stanleyville in the Belgian Congo. It was to examine the OPV/AIDS hypothesis on the origin of HIV, in defiance of scientific authority that blacked out the hypothesis (see Chapter 14 of this book). When I asked him what was inside the bunch of test tubes thrown out on the desk of his office, he briefly explained his expedition and the reason why he was undertaking it. I had been impressed by his strong and steady will for testing a hypothesis that lies beyond the frontiers of the academic world. I of course did not know that 6 months later he would set off again for the Congo and, as a consequence, lose his life. When I think of Bill, I immediately think of this image, of his talking to me, choosing his words, with those test tubes in his hands.

I have to wind the clock back several years to begin this introduction. My memory of Bill can be traced back to a scene of his walking to and fro along a trail of Unzen volcano in western Kyushu island. It was just a few years before its major volcanic eruption in 1991. He disappeared into the bush, inspecting leaves of seemingly unimportant shrubs, paying no attention to the old temples and Buddhist statues along the trail that the host of the tour wanted to show him. It was an excursion of a scientific meeting, and we had to stop every five minutes to look around the places where this world-famous evolutionary ecologist had disappeared. Bill went in for this kind of side-stepping wherever he was invited in Japan, and at least in some cases, he found treasures during it. Yoshiaki Ito described to me how amazed Bill was when he first found color polymorphism of aphids within a leaf of a shrub while he was walking at the foot of the Japanese Alps. Professor Ito said that Bill stood still with admiration and seemed to have started building a new theory of host–parasite coevolution. As a graduate student studying mathematical biology, I had already read some of his famous papers at that time. His character, which was more that of a naturalist than a theoretician, impressed me and many other Japanese ecologists and mathematical biologists as well. Shortly after this visit to Japan, Bill contributed a long article to a popular Japanese entomological journal, which focused on the close relationship between his theoretical works and the insects he loves. With this Japanese translated article (see Chapter 3 of this book) Bill has become one of the best known and beloved scientists in Japan.

I had an opportunity to visit Oxford for a couple of weeks in February 1997. As he did for many others, Bill kindly offered me his college flat for my stay, and after this kind offer in his e-mail dated on February 11 1997, Bill asked me the following questions:

[Suppose that a] metapopulation has host–parasite cycles occurring in all its demes. If migration rates are zero, the phase of cycling across the metapopulation stays the way it is started—synchronised if started in phase or random if started randomly. But if low average migration rates are brought in, what happens? Steven Frank has stated that when migration is low the populations desynchronise and only when the rates are very high do they synchronise. This fits with my own intuition but I would like to know more:

1. If they desynchronize, presumably there is still correlation from one deme to a neighbour. So is there a pattern and if so what? Perhaps spiral waves in the 2-D case like those on the cover of the recent TREE that had the article on this?
2. If we had 'island model' migration instead of local, what happens? For low migration, do we again when there is low migration have neutrally random phases of cycles across the population, as in the case of no migration? (At very

high migration there must be synchrony, I suppose, whether the case is 'island model' or 'stepping stone'.)

3. Are there *maxima* over the range of average migration in any tendency to desynchronise, and if so what can be said about the parameter states associated with these maxima?
4. (i) How does *finite deme population* as opposed to infinite affect all this?
   (ii) How does making the parameters of the host-parasite interactions in the different populations *slightly different*, instead of all exactly the same, affect all the above results?
   (iii) How do disparate migration rates, as between host and parasite, affect all the above?

That's far more questions than I meant to ask you when I started writing them! [. . .]

In preparing this introduction, I found this message in my old e-mail log, and was surprised by how exactly Bill envisaged the future research plan that we later 'followed' in September 1998, though both of us had already forgotten the e-mail and its four questions. I would call them 'W.D. Hamilton's 4 unresolved problems in the Red Queen theory', after Hilbert's 23 unsolved problems in pure mathematics. Reading my reply to Bill's e-mail, it seems that I thought at that time that the topics had already been actively studied by many people, because I had heard too many talks on spirals and metapopulations, and was a bit surfeited with them, and thought there would be nothing left for us. This was, however, not true, as I realized when I actually started it. Restrospectively, Bill's 2nd question led me to the finding of pacemaker phenomena, which is the main topic of the paper following this introduction.

Fortunately, I had an opportunity of another and a longer visit to Oxford in the summer of 1998. During July and August, however, Bill was busy in writing the introductions to Volume 2 of Narrow Roads, and I focused on my work on the arms race of host resistance and parasitoid's countermeasure in collaboration with Charles Godfray at Silwood Park, around 50 miles from Oxford. Bill and I finally started our work together in September after Bill came back from summer vacation in Italy. Bill wanted me to extend Judson's work[1] on the evolution of sex in HAMAX-type metapopulations (HAMAX stands for the Hamilton–Axelrod model for host–parasite matching genotype dynamics[2]). He was interested in the spatial correlation and patterns generated in coupled host–parasite dynamics, and how they are related to the advantage for sex. I looked at Judson's paper, reread the HAMAX paper, and thought that we'd better first simplify the model to clarify what's going on in these simulations. The original HAMAX model was individual-based (and hence is a finite population model with demographic stochasticity), assumed multilocus interaction between host and parasite, and allowed different mean generation times of host and parasite by adopting different thresholds of the matching score in truncation selection. These assumptions made any analytical treatment extremely difficult. Although the HAMAX model generated a secure advantage for sex, it was not clear what factor of the HAMAX model was responsible for such simulation results. Several hypotheses have been proposed about what gives rise to the evolutionary advantage for sex and recombination. Some of them, for example, the breaking down of Hill–Robertson fitness interference among loci, and the reverse winding of the Muller's ratchet (irreversible accumulation of deleterious mutations), essentially rely on random genetic drift due to finite population size. At that time, it was not clear whether or not the use of finite populations led to the success of the HAMAX model. Both of us conjectured it is not, but to confirm this we need to let the population be infinite (i.e. make it deterministic) and see if the same advantage for sex is obtained. We later revealed in our pacemaker paper that our conjecture was right.

Bill's earlier model published in Oikos in 1980[3] had a much simpler structure than did the HAMAX model. The criticisms on the robustness of producing the advantage of sex in the Oikos-type model,[4] led Bill to consider more realistic but complex models. He found that obtaining robust advantage for sex is easier with more loci participated in host–parasite recognition, and with more difference in the strength of truncation selections and in the generation times between host and parasite. I, however, believed that these criticisms, based on a particular association between epidemiology and the frequency-dependent selection in host genotypic dynamics, miss the mark and wondered why Bill did not focus more on a simpler version.

Another criticism of Bill's matching genotype model was that real gene-for-gene interaction empirically observed in plants and fungal pathogens is extremely asymmetric in contrast with what Bill assumed in the Oikos paper, and this asymmetry collapses the genotypic cycles that favour sex and recombination.[5] One of my contributions to this topic was to show that even with empirically common asymmetric gene-for-gene interaction between multiple resistance loci of host and avirulence loci in parasite, complex cycles of the frequencies of the resistance allele in host and the virulence allele in parasite is retained, and their cycles are desynchronized between loci, allowing the recombination to produce genotypic diversity in resistance.[6] I had almost finished this work, and that was why I planned to start a collaboration with Bill.

On September 3 1998 I finished programing a deterministic HAMAX model, and the liquid crystal display of my notebook started to produce interesting patterns. Quite often Bill stood by and watched the display before we went downstairs to have tea. We sat down at the tea lounge of Zoology department and discussed for a long time the results we had just obtained. I removed the individual-based features of the HAMAX model—the fitness of a host individual, for example, became a function of the mean matching score for a given parasite genotype distribution. Bill called this a 'multiple bite model', and called the original HAMAX 'single bite'. It was now clear that whenever the HAMAX model demonstrates an advantage for sex, the prominent pair of host genotypes in a given time interval are genetically far distant from each other, and the prominent pairs flip to another distant pair when the parasite catches up the host genotype (see Figure 18.3 of the following paper). By the recombination between the prominent pair, the host can produce diverse genotypes, and hence can escape the parasite that undergoes evolutionary chase of the matching host, the host it can exploit. Asexually reproducing parasite chases the host by point mutations, and hence quite often loses its sight when the host jumps over by recombination.

One afternoon (the time stamp of the program source file indicates that it was September 11, 1998) I was playing around with a simulation of gene-for-gene host–parasite metapopulation model. If the demes are connected locally as stepping stone model, it produced sprials and target patterns (which Bill called 'tomato attractors'), which was quite exciting to see but, we knew, not quite new. Then I just decided to take a look when the migration is global (the 'island model') to answer Bill's second problem. What we found was a strange pacemaker phenomenon. I found desynchronization in globally coupled gene-for-gene dynamics where every pair of subpopulations exchange migrants equally often with each other. With this global connection, it was clear that there is a quite robust and strong tendency towards global synchronization of cycles, but we found that a small fraction of demes remain unmoved by this sea of synchrony. The phase of the cycle in this small fraction of demes keeps ahead of the other demes, and regulates the period and amplitude of the whole population. We called this the pacemaker phenomenon. Bill immediately recognized its importance, and said he had never seen anything like it. Bill could not resist looking more, and asked me to port my program to his G3 desktop computer in his house in Wytham village.

Bill was attracted by the very rich behaviors of the simulations played on the computer screen. They seemed to inspire his thought. I was happy to think that the way he was attracted to the patterns my program generated might look like the way he was charmed by a bunch of aphids in different colours on a leaf when he found them in Nagano. I remember many scenes in the study of Bill's house, with Bill observing every detail of what was going on in the display, and whispering about their implications. To my surprise, he even started modifying my program written in C. I knew Bill was quite efficient in writing Fortran programs—I once looked at his original program source file of the HAMAX model, which was a well organized program of the good old days, but I was sure that he never knew C language. Here is Bill's artwork from the snap shot of the simulation of our model (see Fig. 18.1).

I should be pleased if I could give some clue of how Bill was excited with our findings from the paper. Just after I returned to Japan, our paper was almost finished. But Bill wrote in November 1998 in his e-mail that in the last part of our manuscript he was 'distracted by exciting theme'. He wrote: 'Looking up the vast publication list of Paul Hebert, who has worked a lot on microcrustacea (and other invertebrates) both around Ontario and in the Canadian Arctic I found that there is really a lot of data on the clone mixtures found in arctic lakes...' Bill thought that adding statistical

Figure 18.1. Bill's artwork: growing ring patterns in host–parasite metapopulation model.

analysis of these empirical data was far better than just finishing the manuscript as a theoretical paper. Bill later introduced to me his 'brilliant' student, Francisco Ubeda, who had just arrived to work with Bill and was interested in this project. We asked him to do some statistical analyses on the time series data of Daphnia clone mixture, and later on the time series data of bacteria clones in Bruce Levin's intestine that exhibits 'revolutionary change in clone frequencies in every two weeks with a bit of discomfort and looseness', which came to fruition in the last part of the paper.

In the summer of 1999, I and my family joined Bill, Luisa, Mary and their family at Land's End in Cornwall to see the total eclipse of the sun. Not surprisingly for a British summer, the sun was totally behind the thick cloud and none of us could see annular eclipse. The sky became pitch dark in the noon and we all knew exactly when the annular eclipse must be seen above the clouds. A few days after the day of eclipse, we went on a walk along natural trail and near the tip of the cape we found an abandoned church. Bill and Luisa beckoned me to look at the stone wall of the church. At the place Bill pointed, I found lichen forming 'tomato attractors' and spirals that might be produced by instability by lateral inhibition in the circular growth of a lichen colony.

References

1. O. P. Judson, Preserving genes—a model of the maintenance of genetic-variation in a metapopulation under frequency-dependent selection. *Genet. Res.* **65**, 175–191 (1995).

2. W. D. Hamilton, R. Axelrod, and R. Tanese. Sexual reproduction as an adaptation to resist parasites (a review). *Proc. Natl. Acad. Sci. USA* **87**, 3566–3573 (1990).

3. W. D. Hamilton, Sex vs non-sex vs parasite. *Oikos* **35**, 282–290 (1980).

4. R. M. May, and R. M. Anderson, Epidemiology and genetics in the coevolution of parasites and hosts. *Proc. R. Soc. Lond. Ser. B* **219**, 281–313 (1983).

5. M. A. Parker, Pathogens and sex in plants. *Evol. Ecol.* **8**, 560–584 (1994).

6. A. Sasaki, Host-parasite coevolution in multilocus gene-for-gene system. *Proc. R. Soc. Lond. Ser. B* **267**, 2183–2188 (2000).

CLONE MIXTURES AND A PACEMAKER: NEW FACETS OF RED-QUEEN THEORY AND ECOLOGY[†]

A. SASAKI, W. D. HAMILTON AND F. UBEDA

Abstract

Host–parasite antagonistic interaction has been proposed as a potential agent to promote genetic polymorphism and to favour sex against asex, despite its twofold cost in reproduction. However, the host–parasite gene-for-gene dynamics often produce unstable cycles that tend to destroy genetic diversity. Here, we examine such diversity destroying coevolutionary dynamics of host and parasite, which is coupled through local or global migration, or both, between demes in a metapopulation structure. We show that, with global migration in the island model, peculiar out-of-phase islands spontaneously arise in the cluster of islands converging to a global synchrony. Such asynchrony induced by the 'pacemaker islands' serves to restore genetic variation. With increasing fraction of local migration, spots of asynchrony are converted into loci or foci of spiral and target patterns, whose rotating arms then cover the majority of demes. A multi-locus analogue of the model reproduces the same tendency toward asynchrony, and the condition arises for an advantage of asexual clones over their sexual counterpart when enough genetic diversity is maintained through metapopulation storage—migration serves as a cheap alternative to sex.

Introduction

Suppose a species in an archipelago is synchronously repeating a series of changes in gene and genotype frequencies. One may think either that cyclica

[†] *Proc. R. Soc. Lond.* B **269**, 761–772 (2002).

climatic factors underlie the variation—sunspot, 'El Niño' or similar effects[1]—or else that some predator or parasite cycle may be in progress, with an inter-island migration rate high enough to ensure synchrony.[2] But suppose that a single island in the group is found in cyclical change with the *same period* as the rest (or nearly) but it changes with a *lesser amplitude and is permanently out of phase with all of the others*. Suppose further that nothing in the environment, either physical or biotic and excepting its different oscillation, distinguishes the odd island—it is not, for example, peripheral, it emits and receives exactly as much migration as all the others, and so on. What mechanism can account for such an odd behaviour?

In Fig. 18.2(a) we illustrate that such a paradoxical stable out-of-phase island may spontaneously arise in so-called 'island' models of population structure[3] (1970) that manifest oscillations of host and parasite genotypes. Two facts, one new (with its analysis to be published later) and one old, are noteworthy. First, such a spontaneous 'pacemaker deme', as we entitle the anomalous island in these situations, is more, not less, likely to arise when the cluster of like islands is more numerous. Second, when the odd island fails to arise (which most often occurs for small groups of islands), then globally synchronized expansive ('heteroclinic') fluctuations in gene and genotype frequencies of host and parasites occur and ultimately destroy variation through gene fixations. When a pacemaker island does arise, however, both that island and the main set quickly settle into two *different*, mutually stabilizing limit cycles.

We show in this paper how such peculiar asynchrony of gene and genotype frequency fluctuation develops in a globally and locally coupled metapopulation under standard dynamics of host–parasite coevolution, the gene-for-gene (in § 2) and the multi-locus matching allele dynamics (in § 3). In § 2, a spontaneous emergence of a small fraction of peculiar phase outlier deme (the pacemaker) in globally coupled gene-for-gene metapopulation dynamics is detailed, together with its effect on the maintenance of genetic diversity in the whole population and the relationship with foci and loci of spirals and target patterns that arise in locally coupled metapopulation. In § 3, the multi-locus matching genotype dynamics between host and parasite (HAMAX model) is analysed in metapopulation structure. The factors that favour sex over asex mixture under the host–parasite interaction in a metapopulation are clarified. The characteristic relationship between the abundance (and its temporal fluctuation patterns) of asex clones and the genetic distance observed in this system are detailed. The implications of these findings to classical geographical parthenogenesis is discussed. In § 4, our findings are compared with the spatial and temporal fluctuation patterns of the snail clones in New Zealand lakes. Furthermore, statistical analyses are conducted for the quantitative relationship between the genetic distances and the abundance of clones using the time series data of boreal zooplankton *Daphnia pulex* and of the intestinal flora of *Escherichia coli*. The statistical analysis is aimed to test the predictions of the Red Queen

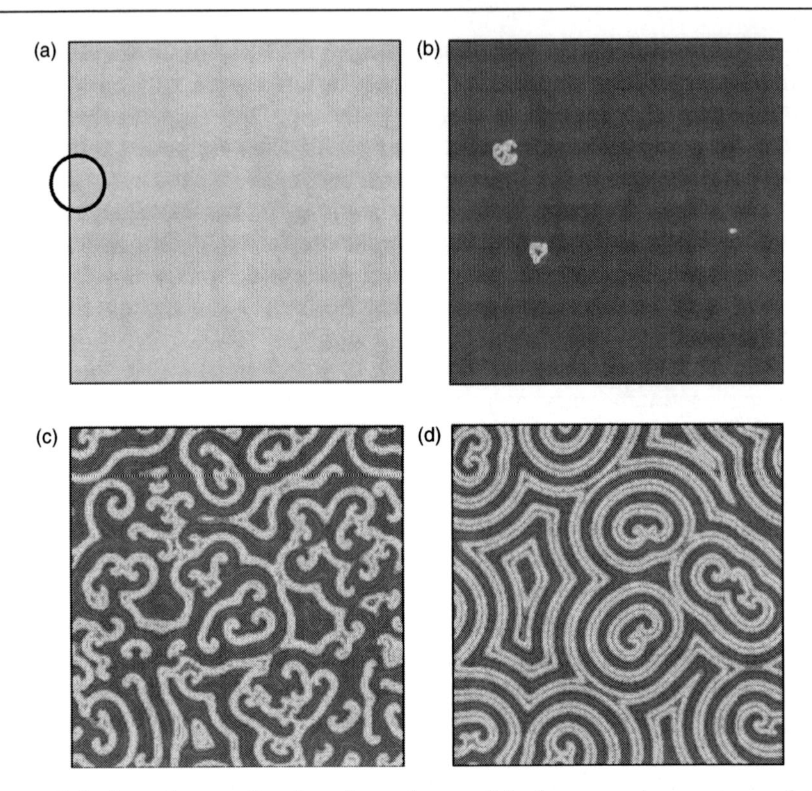

Figure 18.2. Asynchrony of cycles of a resistant allele frequency in a metapopulation model of classic host–parasite gene-for-gene-like interaction. Two host genotypes, susceptible (S) and resistant (R) and two parasite genotypes, avirulent (A) and virulent (V), are segregating respectively with frequencies $1-x$ and x in host and $1-p$ and p in parasite in each of 100×100 demes. In each deme, the host and parasite allele frequencies change according to the dynamics of equations 2.1–2.3. After the change by selection, the host and parasite migrate with migration rates m_{H} and m_{P}, where the fraction ν of migrants comes from one of z $(=4)$ nearest-neighbour demes (local migration) and the fraction $1-\nu$ from one of n $(=100 \times 100)$ demes in the whole population (global migration).

Panels show the different ways by which, depending on the fraction of local migration, the asynchrony developed in metapopulation prevents monomorphism and leads to stable limit cycles in each deme. (a) A single deme out of 10 000 demes refuses to join the global synchronized cycles of the rest. The migration is totally global ($\nu = 0$). (b) Snapshot when a small fraction (20%) of local migration is introduced ($\nu = 0.2$). A few spots of differently beating demes originate and release circular waves into 'the sea of synchrony' surrounding them. (c) Spots develop as the foci of spirals when local migration becomes dominating ($\nu = 0.85$). (d) For further increased local migration ($\nu = 1$), spirals give way mostly to target patterns with double spirals in their centre.

Parameters: $\beta_{\mathrm{H}} = \beta_{\mathrm{P}} = 5$, $c_{\mathrm{H}} = c_{\mathrm{P}} = 2$, $m_{\mathrm{H}} = m_{\mathrm{P}} = 0.2$. Boundaries of the metapopulation are periodic. All snapshots are taken about 250 generations after the simulations started from randomly chosen initial frequencies in each deme. All sources of asynchrony (pacemaker demes and foci of spirals, targets—not shown—and double spirals) are stationary, once formed.

hypothesis for the evolution of sex, with the deleterious mutation hypothesis as an alternative.

Gene-For-Gene Dynamics in a Metapopulation

To see how the asynchrony is developed in the host–parasite frequency cycles in a metapopulation, we first consider the host–parasite allele frequency dynamics of classic gene-for-gene interaction. Two host genotypes, susceptible (*S*) and resistant (*R*) and two parasite genotypes, avirulent (*A*) and virulent (*V*), are segregating with frequencies $1-x$ and x in host and $1-p$ and p in parasite in each deme. The fitness of each genotype is:

$$\begin{aligned} &\text{susceptible host:} w_s = \exp(-\beta_H), \\ &\text{resistant host:} w_R = \exp(-c_H - \beta_H p), \\ &\text{avirulent parasite:} w_A = \exp(\beta_P(1-x)), \\ &\text{virulent parasite:} w_V = \exp(-c_P + \beta_P), \end{aligned} \tag{1}$$

where β_H and β_P are the loss in a host and the gain in a parasite of their fitness due to a successful infection, respectively. An underlying assumption is that a resistant host is infected only by virulent parasites and an avirulent parasite can infect only susceptible hosts. The costs of resistance and virulence are denoted by c_H and c_P. The frequencies of the resistant host and the virulent parasite after selection are:

$$\begin{aligned} x* &= F(x,p) = w_R x/(w_R x + w_S(1-x)), \\ p* &= G(x,p) = w_V p/(w_V p + w_A(1-p)). \end{aligned} \tag{2}$$

The gene frequencies in the metapopulation are then coupled by migration:

$$\begin{aligned} x_i' =&(1-m_H)F(x_i,p_i) + v\frac{m_H}{z}\sum\nolimits_{|i-j|=1} F(x_j,p_j) \\ &+ (1-v)\frac{m_H}{n}\sum\nolimits_j F(x_j,p_j), \\ p_i' =&(1-m_P)G(x_i,p_i) + v\frac{m_P}{z}\sum\nolimits_{|i-j|=1} G(x_j,p_j) \\ &+ (1-v)\frac{m_P}{n}\sum\nolimits_j G(x_j,p_j), \end{aligned} \tag{3}$$

where x_i and p_i are the gene frequencies in the *i*-th deme, m_H and m_P are the migration probabilities of host and parasite, where the fraction v of migrants come from one of z (=4) nearest-neighbour demes (local migration) and the fraction $1-v$ from one of n (=100 × 100) demes in the whole population (global migration). The case $v=0$ corresponds to the Wrightian 'island' model where all demes are 'equally distant' from each other; $\nu = 1$ to the 'stepping stone' model where migration occurs only between the nearest neighbours; and $0 < \nu < 1$ to the mixture of the migration modes. Without metapopulation structure and migration, the genetic dynamic described by eqn (2) is invariably unstable and the frequencies of the resistant host and the virulent parasite fluctuate with increasing amplitudes, spending increasing periods of time near monomorphic vertices of genotype space. Increasing extremity in these cycles makes eventual allele extinctions inevitable for finite populations.

ASYNCHRONY UNDER GLOBAL AND LOCAL MIGRATIONS

Metapopulation structure, however, drastically changes the situation. Asynchrony in gene frequency fluctuation develops in either globally or locally coupled metapopulations and serves to maintain genetic diversity that is inevitably depleted in a single population dynamics. The asynchronous fluctuation pattern observed with globally coupled metapopulation is particularly interesting (Fig. 18.2(a)), in which a single deme out of 10 000 demes refuses to join the globally synchronized cycles of the rest. This unique deme rules the period of the whole system and leads to 'stable' polymorphism, even in the case of totally global migration, as here ($\nu = 0$). Mathematical analysis in a subsequent paper will show that this form of asynchrony with a tiny fraction of out-of-phase demes is much more likely in coupled gene-for-gene systems than an asynchrony with different phases assigned to more equal numbers of demes, and is also more likely to arise as metapopulation size is increased.

When a small fraction (20%) of local migration is introduced, a few spots of differently beating demes originate and release circular waves into 'the sea of synchrony' surrounding them (Fig. 18.2(b)). The waves dissipate into the sea but still affect it enough to prevent occupation of a monomorphic vertex of gene frequencies to which the whole system otherwise converges. Spots develop as the foci of spirals when local migration becomes dominating ($\nu = 0.85$) and ever-outward-moving arms now mostly cover the whole lattice, although regions of global fluctuation still remain (Fig. 18.2(c)). For further increased local migration ($\nu = 1$), spirals give way (in this case) mostly to double spirals in which one spiral pairs with (or helps to create) another opposite centre close by (Fig. 18.2(d)). A wave generated at an oscillating centre or a focus of rotation travels undiminished until it meets another and then vanishes. In the genotype space, the total frequencies of genotypes stay nearly constant and central.

THE PACEMAKER ISLAND

Most of the time (in some cases, all of the time), the phase of the odd island is advanced relative to the rest, hence our chosen term 'pacemaker'. That a minority set of pacemaker islands 'pulling' the rest, however large, is essential in maintaining such stabilizing asynchronous limit cycles. In doing so, it balances the continual tendency of the rest towards overshoot and expansion. If the whole system was divided into similarly sized phase groups, as opposed to the extremely asymmetric division shown in Fig. 18.2, the advanced islands send too many migrants to allow the followers to catch up, resulting in a diversity destroying global synchrony. In its spontaneous appearance during the random start of our systems, the pacemaker is different from the imagined or actual externally imposed, forcing oscillators or 'pacemakers' sometimes discussed in dynamical and physiological contexts.[5] On the other hand, the effect seems to be indeed closely related to the still mysterious spontaneous 'pacemakers' concept of pulsing tissue systems, such as the vertebrate heart and the pheromonally unified aggregations of cellular slime moulds. Generally, the spontaneous pacemaker with its associated wave features arising under local migration (described below) appear generic and are certainly not restricted to the classic 'gene-for-gene' model used for Fig. 18.2. For example, a more simplistic gene-for-gene model[6] can reproduce all of the above results, though showing a lesser variety of types of oscillator that can appear under local migration. Below, we also show that multi-locus models with quite different schemes of host-parasite fitness interdependence (approaching truncation selection) produce similar features.

Especially under the panmictic conditions of migration so far described (the 'island model'; ref. 3), our pacemaker deme is usually unique: all other islands, including the phase-amplitude outliers that might have seemed equal initial candidates to become pacemakers, eventually converge to the majority rhythm—that is, all converge but one. In the past, the classic gene-for-gene model underlying Fig. 18.2, in the case shown using a single-locus interaction of host with parasite[7] and with costs of virulence and resistance also assumed, has been treated as a system unlikely to cycle permanently because, if the costs are low, states of fixation are reached directly and if they are high enough to induce cycling, the system is unstable and permanently escalates, as already described, ending in allele fixation.[8–10] For a natural metapopulation, however, our demonstration of control by a spontaneous pacemaker or other self-differentiating set(s) shows this conclusion to be questionable.

SPIRALS AND TARGET PATTERNS UNDER LOCAL MIGRATION

When, under conditions otherwise identical, migration is predominantly local instead of global, there is even less threat to continued variety in the metapopulation. Our models show how structured systems of cyclical wave-like variation readily occur (Fig. 18.2(a) and (d); and see Fig. 18.4) and, reinforcing previous results,[11] imply strong asynchrony on the metapopulation scale. In our illustrated case, waves spreading from spontaneous centres (in which sites seem picked from the phase outliers in the random starts that would be the candidate or actual pacemakers under the Wrightian 'island' migration pattern) generate 'spiral' and/or 'target' patterns similar to those already known (either by models or real demonstration) in chemical, biological and populational systems that constitute 'excitable spatial media'.[4,5,12–15] Our diagrams only illustrate the more common patterns that we have observed, notably 'targets', spirals and what may be called 'detached curl-ended waves'. In many regions of parameter space, especially under moderate or/and multi-locus selection, all these well-formed patterns (among a variety of others) dissolve into flickering, patchy, irregular designs of varying grain, amplitude and period. As far as any of the above dynamic patterns exist in nature, or, correspondingly, there exist other still more complex and irregular alternatives such as may be imposed by real land forms and ecology, they may help to explain not only some puzzling actual cases of out-of-phase population dynamics[16] but also, as in our next topic, help explain various long-known spatial and temporal phenomena in the ecology of sex and of clone mixtures.

Sex Versus Clonal Mixtures in a Metapopulation

It has been recognized for some years that spontaneously asynchronous variation in metapopulations potentially stores variation both within species, as genetic variation,[11] and at the community level, as species diversity.[17–19] The evidence is already substantial for some wild situations.[16,18,20] As stated, the importance of multi-locus varietal asynchrony in this picture has also been recognized.[11,21,22] However, the possibility that clone mixtures in a metapopulation might achieve storage of 'anti-parasite' adaptations *better* than can sex itself has been addressed so far only in peripheral comments. In our present models, by contrast, we have been forced to recognize broad parameter swathes where *mixtures of asexuals* effect the storage. By the fact of particular widely divergent pairs being in prominent frequencies at any one time, such mixtures greatly lower infestation (and thereby damage) from the coevolving parasites.

We first summarize the characteristics of multi-locus matching-genotype dynamics in a single locus.[23] Figure 18.3 shows cycles of host genotypic frequencies in a single deme host–parasite multi-locus matching-genotype dynamics (HAMAX) model in which soft truncation selection based on the matching score of host loci and corresponding parasite loci is underway[23]—see Appendix A. Striking features of these trajectories are: (i) that in a short time span, the most distant pair of host genotypes ('prominent pairs') is common; (ii) that they alternate in taking the highest abundance; and (iii) that in a longer time span, the prominent pair is switched to another prominent pair when the parasite genotypes catch up with the currently common host genotypes. Figure 18.4 demonstrates a snapshot of spatio-temporal waves of host genotype frequencies when the above single deme model is extended into a metapopulation structure where demes are connected by local migration between nearest neighbours (see the legend to Fig. 18.4 for detail).

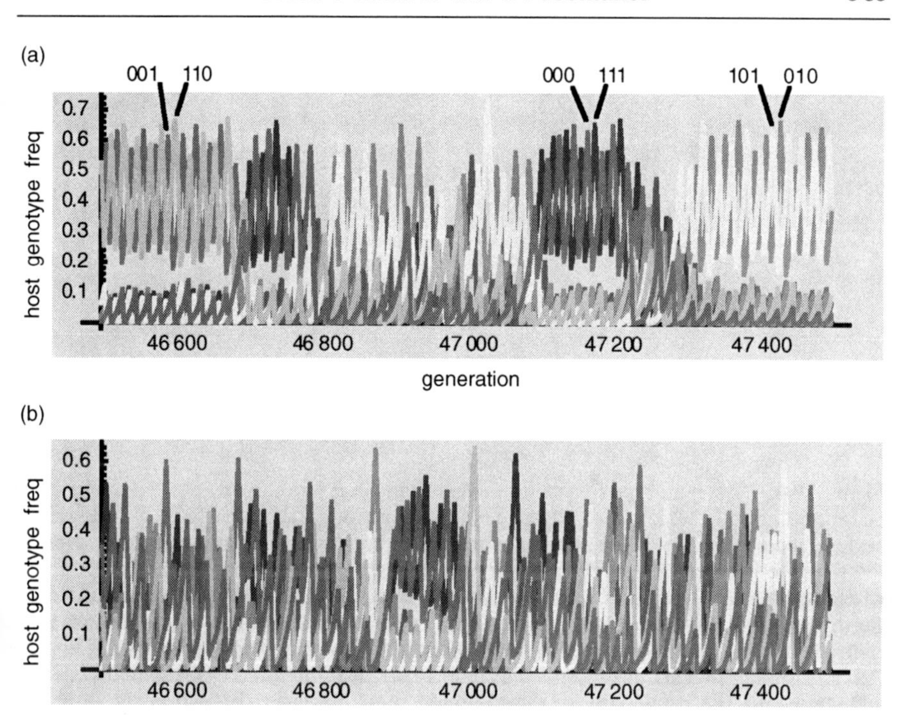

Figure 18.3. Multi-locus haploid genotype frequencies in a single population under soft truncation selection based on host–parasite matching of genotypes. L of resistance loci of hosts correspond with virulence loci of parasite, each with two alleles 0 and 1. (a) A trajectory of host genotype ($L=3$, $\theta_H=0.1$, $\theta_P=0.9$, $r=0.2$, $\mu_H=10^{-4}$, $\mu_P=0.01$) shown here can be divided into the periods in which one of the pairs of most distant genotypes are common ('prominent') and alternate their frequencies. (b) When recombination rate is further increased ($r=0.5$, the other parameters remaining unchanged), host genotype frequencies fluctuate in more chaotic fashion, though a tendency of most distant pairs to be prominent remains. In both panels, the trajectory for genotype 110 is marked with dots to indicate the amount of change between generations.

TRAJECTORIES IN HYPERCUBE

The pairs that we call 'prominent' in this context have: (i) genotypes as opposite as are available; and (ii) frequencies at a given time above those of all others. As we illustrate for a single population in Fig. 18.3, prominent pairs of genotypes alternate in taking highest frequency. As both morphs are common, they provide a continuously difficult problem to the parasite population, which at best can match only one-half of the hosts, even in an ideal situation that the parasite trails just behind the host by switching its own maximally distant pairs.

When the number of loci involved in host–parasite interaction is small, as illustrated in Fig. 18.3, such systems successfully exclude sex when this has its

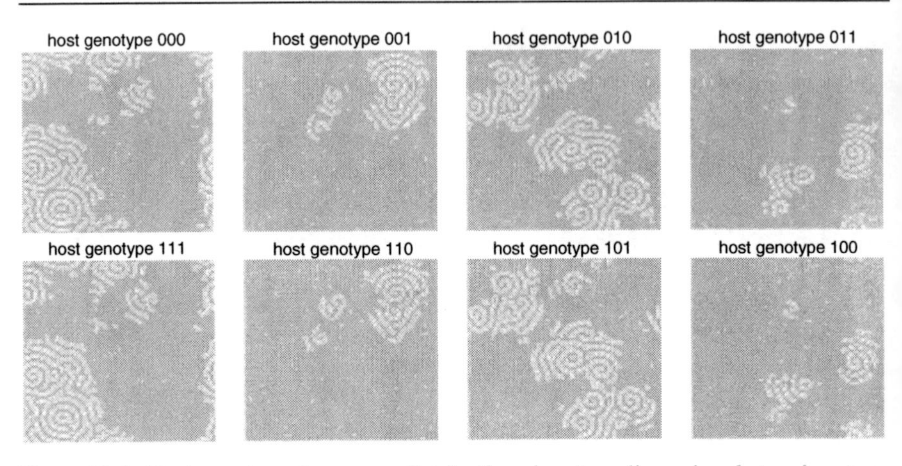

Figure 18.4. Host genotype frequency distributions in a two-dimensional stepping stone metapopulation in which host–parasite matching genotype dynamics occur as described in the legend of Fig. 18.3 The number of resistance loci in the host and of virulence loci in the parasite is three, giving eight genotypes in each species. After a random start, the whole space quickly divides into four regions in each of which a pair of most distant genotypes (e.g. 001 and 110) is 'prominent'. These regions form mutually exclusive centred patterns (in this case spirals) that compete at their margins. Under conditions like that shown, long-term quasi-permanent limit cycles in all demes are attained after a slow adjustment in the total pattern; development may, however, be extremely slow: for example the closer-spaced waves of spirals may, once they form, slowly 'eat back' the slower wave patterns from some types of target centre (for the rationale see Boerlijst *et al.*[4]). Parameters: $L=3$, $r=0.5$, $m_H=m_P=0.25$, $\theta_H=\theta_P=0.5$, $\mu_H=0$, $\mu_P=10^{-5}$.

In addition to the wave patterns shown we have also observed a wide variety of parameter states generating more irregular and non-repetitive dynamical patterns as well as other types of centre.

characteristic halved efficiency due to unproductive maleness. In the cases in Fig. 18.3, the situation is sexual with no asexual competitors present; if they are brought in, they very quickly replace sex and a permanent symmetrical alternation of just two asexual opposites follows (the period is about 9 and the amplitude about 0.5 in the case chosen). In this unrealistic infinite population model all possible asexual genotypes are present, albeit some only at extremely low frequencies.

The threat to sex of mimetic and more efficient prominent pair-cycles persists if, from the state in Fig. 18.3, we simply raise the number L of loci even though (with $L=8$, for example) successful invasion by an opposite pair becomes very slow. However, as the selection on host is increased (from host culling fraction $\theta_H=0.1$ in Fig. 18.3 to about $\theta_H=0.16$—see Appendix A for the definitions of θ_H and θ_P), the stability of a prominent pair cycle is lost and thereafter opposite pairs have only brief tenures, even in pure asexual populations. Now recombination plays a part in assisting transitions from one pair vibration to another.

At an even stronger selection on host, $\theta_H = 0.2$, we find as few loci as $L = 5$ gives full security to sex once it is established; asex may make occasional incursions but is always repelled as parasites adapt to each new common genotype. In metapopulations with local migration and with consequent spreading waves of local synchrony, as in Fig. 18.2, escalation is prevented by local migration and particular pairs persist (with alternation) without difficulty, and asex, without wasteful segregation, normally wins.

A similar but variant outcome is sometimes seen in three-locus asexual metapopulations but this again emphasizes the importance of 'distant' mixtures. Some parameter conditions give rise not to pairs but to prominent 'tetrahedral' quadruples (e.g. $000 \rightarrow 011 \rightarrow 110 \rightarrow 101 \rightarrow 000$), i.e. sets of four genotypes vibrating in frequencies above those of others and with each member differing in two of its loci from all others in its set. For example, the case of Fig. 18.3 put into a stepping-stone lattice metapopulation, with the local migration rates set at 0.25, gives this result. The tetrahedral oscillation that appears is interrupted from time to time by the rise of an opposite pair (one member being contributed from the non-dominant tetrahedral set) but shortly it is always the same tetrahedral set as before that resumes abundance, thus revealing a strong and exclusive dominance to the alternative tetrahedron in spite of high irregularity at all times in the vibration that reigns.

FACTORS FAVOURING SEX IN METAPOPULATION

In metapopulations, any tendency *away* from finely striped (Fig. 18.2(c) and (d)) or speckled (not shown) dynamics towards broader patterns that show local or global synchrony (Fig. 18.2(a) and (b)) is relatively favourable to sex. Recombination rather than local migration then becomes the principal producer of the genotypes able to escape current parasite waves. In summary, while it is true that spatially asynchronous dynamic states are favourable to the preservation of variability,[11] they turn out to be more favourable to its preservation as asexual strain mixtures than as sexual variation by Mendelian segregation and recombination. This is not to say that sex cannot be superior in the metapopulation conditions that have only local migration; this is untrue in the model and also abundantly contradicted in nature. Rather, the point to be made is that any conditions that facilitate steady alternation of opposites, and that make transitions from one prominent pair (or fleeting rise of opposites) to another pair uncommon (Fig. 18.3(b) compared with 18.3(a)), must be seen as relatively favourable to asex. Our prediction, therefore, is that old parthenogenetic populations in nature will always, apart from brief interludes, consist of distant mixtures.

Our models assume asexual haploid parasites tracking hosts by mutation only. This may be unrealistic in nature, both as regards haploidy and 'mutation only'. It is especially unrealistic for larger parasites that are slower breeding and inevitably more 'visible' to a host's defence system. Restricted ploidy and limitation to facultative change are, however, common for small parasites, especially for viruses and bacteria.[25] The advantage of a change in axis of oscillation of unlike types for the host population in such cases, where there are two or more loci involved in resistance, is then obvious. In our models, we in fact find that sex is rapidly better maintained against competing asex as:

(i) the number of resistance/virulence loci in the model is increased;

(ii) conditions creating the 'broad patchy' chaotic regimes (or generally more synchrony as induced by global migration) supervene;

(iii) demes are large and thus little subject to drift.[23]

In those multi-locus cases, where orderly spirals or target patterns arise in lattices, we find that different regions of the spreading waves often involve different pairs of 'opposites'. In Fig. 18.4, we show this effect as occurring, in this case, in a HAMAX-like[23] host–parasite fitness system. This looks at all matches at three loci and severely disadvantages a fixed fraction of the most-matched tail. When such selection (near to a truncation) occurs under sexuality and free recombination, with only two alleles per locus, the mating of opposite genotypes produces potentially all possible genotypes; however, in 'spreading wave' conditions, and subject to the current position of the parasite genotype swarm, such sexual variants are commonly outcompeted by the more efficient low matching being achieved in the pairs of maximally 'opposite' clones.

Models of the type that produced Fig. 18.4 are also capable of showing pacemakers if the migration is made global. However, the outlier demes in this case are much more changeable and often also remain more numerous. Situations with a single pacemaker (or two in course of a transition) are apt to be interrupted by periods showing several pacemakers, or even by interludes of seeming random asynchrony. Permanent single pacemakers still appear, however, for interactions based on three loci, for example, albeit only through use of different and smoother fitness functions than that of Fig. 18.4. Truncation selection as actually used in Fig. 18.4 seems adverse to pacemaker phenomena. In general, the complex dynamic polymorphisms of the several to multi-locus cases (sometimes referred to as 'tangled wool' polymorphisms due to their appearance in the gene frequency phase space;[23,26] seem to limit dynamic extremity by their more general asynchrony and they do not need a rhythmic input from a pacemaker to prevent escalation. So far, we have found no situation that can be described as a permanent single pacemaker for four loci, although temporary flickerings of distinct out-of-phase, multiple-though-minority sets lasting through the periods of dynamic near synchrony, are commonly seen.

METAPOPULATION STORAGE AND GEOGRAPHICAL PARTHENOGENESIS

Omission of few-locus trials in sex models in the past may explain the relatively limited conditions for success of asex that have been highlighted in other studies on metapopulation models in the past, including previous studies of the HAMAX type.[11,22] As indicated in Hamilton *et al*,.[23] high culling fractions in hosts tend to cause simultaneous loss of all bearers of an allele and thus they initiate an accelerating disaster in the competition with asex. Obviously, as variability is lost at one locus after another there comes to be less and less that recombination can achieve. At the other extreme, culling fractions that are too low give insufficient selection to overcome the efficiency of parthenogenesis. Failure of sex therefore occurs at both high and low culling rates and this happens even though sexuals are always maintaining a greater mean match distance from parasites than are asexuals. Besides the strong effect of number of polymorphic defence loci and of culling rates, other conditions detected as favourable to sex in the present study, mostly confirming previous work[23] (and additional to § 3b(i)–(iii) already mentioned above), include: (iv) low host migration; (v) high parasite migration; (vi) large deme size; and (vii) low host mutation. In our study, synchrony was favourable to sex and local dynamical differentiation is favourable to asex. Synchronization induced in HAMAX-type multi-locus truncation models by low rates of global migration may be accompanied by differentiation of one or more pacemaker demes; but even when 'pacemakers' are absent, escalating cycles seem not to occur when the number of match loci is greater than three. Thus, in these states migration plays a role closely similar to a high reversible mutation rate and makes loss of alleles very unlikely. Pacemakers are, in this sense, not needed. Retrospectively, we rationalize our unexpected finding: that increase in the number of demes and the formation of a local-migration metapopulation is often more favourable to clone mixtures than it is to sex, on the grounds that the metapopulation is providing an alternative mode for the storage and release of species anti-parasite technology compared with that provided by recombination.[21]

The notion hinted by others[22,27–29] and supported generally in the models above, that a metapopulation structure may be specially conducive to clone mixtures as an alternative to sexuality and provide a way of reducing parasitism, can be strengthened by three kinds of data taken from biological literature. But first we make points about the actual distribution of parthenogenesis.

It has long been known that parthenogenesis is more frequent on the margins than in the centre of a species range. This 'geographical parthenogenesis' (GP) has been variously explained.[30–32] As follows simply from physical adaptation, all versions tend to accept, however, that species ranges can often be treated as three roughly concentric zones[33]: (a) the largest continuous populations are in the centre of the species range; (c) at the opposite extreme, where the species has physical difficulty in surviving at all, populations tend to be temporary (see especially Peck *et al.*);[34] and in between and less extreme, (b) there is a broad third zone where the species exists as a metapopulation—numerous long-lived local concentrations in suitable habitats and those exchanging migrants with each other. In accordance with the parasite Red-Queen (PRQ) theory of sexuality, the central species range typically has the highest rate of sexuality and it also has the highest prevalence and diversity of parasites. The multi-species diversity of parasitism falls off especially in marginal populations. Much needs to be done to explain the many conspicuous exceptions to this triple correlation (sex–GP–parasitism); reindeer[35,36] and large-flowered self-incompatible Arctic–Alpine plants, for example, seem to flout the rule. Are these species in some way *more* parasitized at their margins? PRQ predicts this, but only in a few cases is there some evidence.[35,36] Overall, however, the trend in sexuality and in parasitism accords with the theory. In the present study we are suggesting a new intermediate form of anti-parasite diversity that participates as a part of the spectrum of GP and also at the same time reveals a new facet of PRQ. We now outline some preliminary data in support of our hypothesis.

Applications

MULTI-LINE EFFECT

The first evidence is a concept that entered agricultural science in the 1950s and has persisted: multi-lines. This concept invokes the advantage of mixtures of varieties in crop stands in conferring partial protection against pests and diseases. The varieties are not asexual but, through the ways they are bred, they are often effectively uniform. That genotype mixtures could impede the spread *and also adaptation* of parasites was conjectured and proven early on,[37] but in practice unfortunately the benefits of mixtures were (and remain) generally less attractive to farmers than control by spraying.

GEOGRAPHICAL PARTHENOGENESIS IN SNAILS

Second, clone diversity in local populations does often involve, just as our model predicts, the coexistence of deeply divergent clones[20,38–46] The exceptions are seemingly confined to cases of recent invasions. Such divergent coexistence is quite different from the expectation if GP is described in an extension of a 'deleterious mutation hypothesis' (DMH) about of the function of sexuality according to the DMH, the differences within a deme should be slight and only due to recent mutation. However, of course, a more powerful test will be whether the coexistence of clones is static or dynamic (see § 4c). The DMH suggests no reason for highly diverse clones to be together, still less why they should be dynamic. Apart from the references already given, that clone associations are generally indeed distant seems particularly well illustrated in Ostracoda and Cladocera of lakes of boreal and formerly glaciated terrain.[43,47–49] For some cases, as we shall see, there is evidence that specializing and resistable predators can play a part in the observed dynamics.[50–52] and various variation-protective aspects of the metapopulation structure have already been noted.[53] In the temperate zone, *Daphnia* species naturally suffer a wide array of damaging parasites,[54–56] *Daphnia magna* have been shown to have the highest susceptibility to parasite varieties in their own pond and to show asymptotic fall for more distant ones, although some hints of random and maladaptive rises in virulence in cases of exposure to very distant parasites have also been noted.[54] These facts, taken together with the modern theory of virulence,[57,58] generally accord well with the trend in our models. In some cases, even more detailed correspondence can be found.

For some years, C. Lively and co-workers have studied parthenogenesis and parasites in a metapopulation of lake snails in New Zealand that show both geographical[44,59] and what may be called 'extremity', or more exactly 'depth', parthenogenesis.[45,60–62] Intermediate populations, studied on both the local ecological and geographical scales, show many resemblances to the assumptions of present models and to HAMAX studies,[11,23] including lake-to-lake transferral by water birds of both snails and parasites (but more of the latter; see ref. 63 and in the snails being iteroparous (while also, speeding all micro-evolution, they have 2–3 generations per year). Intensive study of the mixed population of Lake Alexandrina (South Island, New Zealand) over 5 years has revealed that sexual snails occur in the shallow water habitat[60,61] and in that zone are much parasitized by trematodes.[60,64] In the next zone by depth the proportion of clones is higher, showing a few common clones[61] and 40 or so rare ones. In this zone, parasitism is still substantial.[60,64] A third, deeper zone has a few sexuals, many clones and lower risk of infection.[61,64] This is equivalent to what happens in zone (C) in the GP scenario above. In the all-clonal population of Lake Peorua (South Island, New Zealand), common clones appear to take turns at being extremely common and then reduced under the lagged rises of genetically coadapted parasite varieties.[65,66] Although time has not allowed more than a few clone replacements to be observed and no repeat abundance of any ultra-common clone has yet to be seen, the overall patterns, including high sexuality in the more parasitized shallows, resembling the evidence for other systems,[20] are suggestive of the model situations we describe.

CLONE DISTANCE AND TEMPORAL VARIATION

Boreal zooplankton

Third, there is evidence regarding alternation and 'distance' of clones. In the above New Zealand lake study, there is as yet nothing to support the idea of particular distant clones repeatedly exchanging positions when common; but, returning to published studies on the boreal cyclic zooplankton, we have found signs of this. It needs to be remembered here that the two clearest predictions of our model are that, first, in the intermediate zone of GP, common co-clones in a locality should not be each other's closest clones and, secondly, no frequencies should be static. (Exceptions to the former prediction, however, have to be expected in cases of recent unique introductions—refs 67, 68—such populations having had time only a few mutations of any kind.) Consequently we have sought other data on clonal organisms that would combine sufficient separate habitats sampled and loci assessed to give varied genetic distances and show the abundance variation of this over real space and time.

We have selected two studies on *Daphnia pulex* in North America.[24,48] In one, 20 loci were assessed in the melanic *D. pulex* in Canadian high arctic ponds, serving to distinguish 13 non-rare clones. For each sampling site in turn, we differentiate the two most common clones, i.e. the most common pair, from all other clone pairs in the sampling site (these other clone pairs include pairings involving the top two). Again, for each sampling site we calculated Nei's genetic distance between clone pairs based on the 20 loci. Pooling such distances we compared the distance values of the 'most common pairs' group with the distance values of the 'not most common pairs' group using a Wilcoxon rank sum test. We calculated signed rank sums w for these two groups and, because the distances are not independent from each other, we 'bootstrapped' 400 w values by repeatedly randomly reassigning pairs' pertainance to each group ('most common' versus 'not most common') and evaluating w for every new group allocation. By comparison of the original w with the distribution so formed, we obtained $p = 0.02$ (one-tailed test) for the chance occurrence of a set of Nei's distances between two clones that was as large as we found. We conclude that the two most common *Daphnia pulex* clones, in the sampling sites surveyed, are significantly more genetically distant from each other than would be clones picked at random. For purely parthenogenetic organisms, it seems hard to suggest any explanation for this that does not involve frequency-dependent selection by at least one antagonistic coadapting enemy species. Of course, the loci and alleles involved in the survey are probably not directly involved in parasite defence but, given the known wide dispersal of resistance alleles in genomes generally,[69–71] it is likely that they are statistically associated with other genes that are involved, having become so due to founder effects and similar.

In a second approach, we looked at the time series of frequencies of the five most common clones in a particular pond in Illinois over a period of almost 2.5 years, as sampled by L. J. Weider. On this time-scale, as seen in Fig. 18.5(a), the changes in relative frequency are clearly too great to be random drift. This is further highlighted in Fig. 18.5(b), where we show the Nei's distances for the two most common clones at half-month intervals[24] together with the unweighted mean of distances within all six non-rare pairs (Weider identifies six clones but does not provide temporal data for the rarest). During 24 out of 29 months of data available, the Nei's distance for the most common pair exceeds the mean. Further correspondence with the pattern in our Fig. 18.3(b) lies in the numerous immediate or delayed reversals within the temporary top pairs: it is

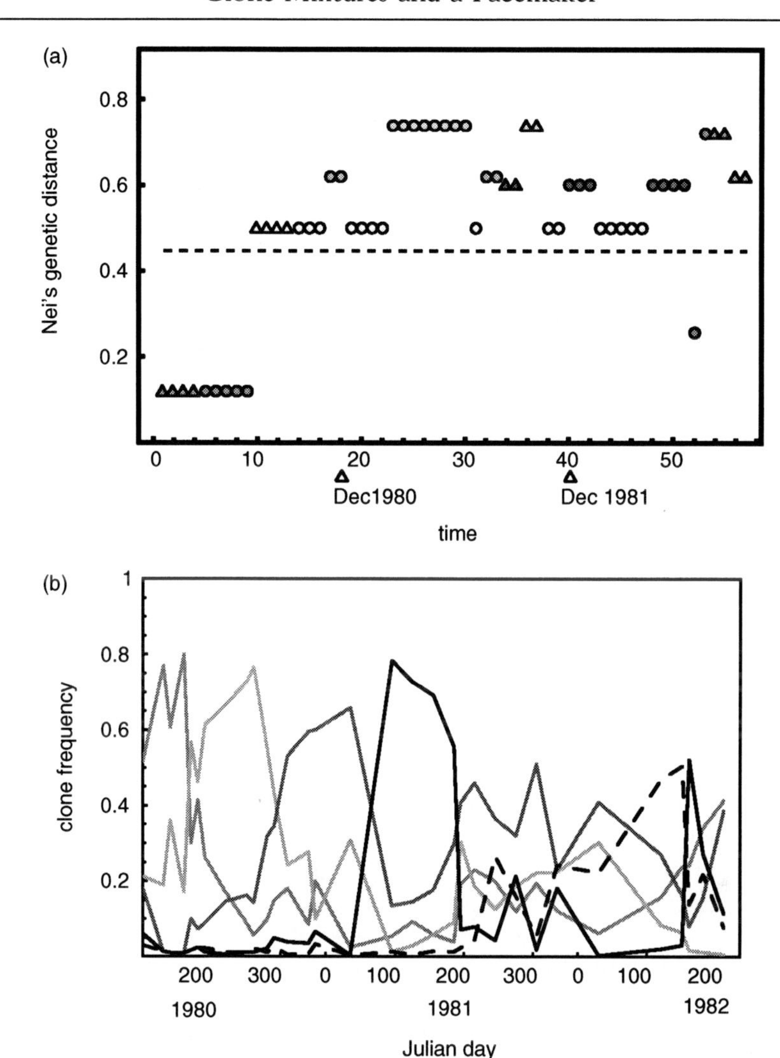

Figure 18.5. Clone and clone pair abundances of *Daphnia pulex* in an Illinois lake,[24] and genetic distances, over time. Each time unit represents half a month. (a) Colours identify different clones amongst the six overall most common clones. Most abundant pairs: red triangles (1, 2), red circles (2, 1); yellow triangles (2, 3), yellow circles (3, 2); green triangles (1, 3), green circles (3, 1); light blue triangles (3, 4), light blue circles (4, 3); dark blue triangles (3, 5), dark blue circles (5, 3); purple triangles (1, 5), purple circles (5, 1); grey triangles (1, 4), grey circles (4, 1). (b) Nei's distance in most common clone pairs against time. Colours now identify different common clone pairs. Symbols show which clone is the most frequent within a pair (*i, j*): triangle for 'direct' order (*i* more abundant than *j*) and square for 'reverse' order (*j* more abundant than *i*). Red line, clone 1; green line, clone 2; blue line, clone 3; black line, clone 4; dashed line, clone 5. Dashed line indicates the average genetic distance between clones.

noteworthy that by the end of the study period every one of the five most common clones had taken at least a brief turn at being most abundant clone.

The literature of pond plankton studies has references to 'checkerboard distributions', a term implying a distribution of frequencies among ponds that are both highly variable and spatially over-dispersed, while at the same time seeming to show no connection of the clones or alleles to physical factors. Comparing the very numerous 'spotty' patterns observed in our simulations (highly clumped peaks emerging and disappearing chaotically, which we have not illustrated) also, noting the fact that in the ponds we are not seeing the primary resistance genes but other genes linked to them, we see possible interpretation of these 'checkerboards', again, as parallels to the fine-grain restless asynchrony of our models. In one area, a species said to display the 'checkerboard' type of distribution for two clones, Wilson and Hebert found that the differences were due at least partly to a differential clone-specific susceptibility to a predatory pond copepod, itself patchily distributed.[51,52]

Intestinal flora

Intestinal flora of *Escherichia coli* populations in human guts seems to provide further qualitative support. The *E. coli* population in a human gut consists of a mixture of asexual clones, which would be subject to antagonistic selection pressure due to bacteriophages or host immune system. Biotic interaction between *E. coli* and phages/immune system would simulate the coevolution of host and parasite changing in comparable time-scales. This view of human gut as a hostile biotic environment against *E. coli* seems reasonable if we take into consideration the potential danger of this bacteria when the immune response is for some reason weakened. It is therefore probable that some sort of control from the immune system is taking place all the time. Making an analogy with *Daphnia* populations, each human host would play the role of a 'pond', *E. coli* as a 'host' and a certain molecular species of antibodies secreted by the immune system as a 'parasite'.

Such a system would meet all of the conceptual requirements of our model, hence our prediction that a genetic distance between the most common pair of clones would be higher than that expected between randomly chosen pairs and that the temporal substitution of the most frequent pair by another prominent pair would take place. That is exactly what we found in a study conducted on the genetic diversity in the *E. coli* population of a human host.[72]

From the data of Caugant *et al.*,[72] 13 loci were assessed serving to distinguish 53 non-rare clones in 22 samples collected over a four-month period of time. Following the same kind of analysis applied to *Daphnia* data in the previous section, we calculated Nei's distance for all possible pairs of clones. Ignoring the sampling times, the probability of getting, at random, a genetic distance equal to or bigger than the one corresponding to the prominent pair (0.572) is equal to

0.024. Clustering the frequencies according to the month in which the sample was taken, we used a Wilcoxon rank sum test together with a bootstrap method to compare the distance values of the 'most common pairs' group with the distance values of the 'not most common pairs' group, as described in the previous section. The only methodological difference in this case is that instead of using a geographical criterion for clustering, we are using a temporal criterion. The result obtained ($w = 3.22327$; $p < 0.001$; one-tailed test; bootstrap of 400 replications) indicates that the two most common clones are significantly more distant than expected if picked at random over the period of time considered. This analysis is viable because in each of the four months' clusters the prominent pairs were different; in fact, only two of the 53 clones were present in more than one month's cluster. This change of prominent pairs every month was described by one of the authors as a 'revolution every four weeks accompanied by a bit of looseness and discomfort'. The change is consistent with the idea that the most abundant types of clones in a period of time are matched by the immune system and reduced to frequency levels that are difficult to detect in the next sampling period. This result, coming from evolutionary microbiology, provides additional evidence for the presence in natural systems of some of the features described in our model.

Independently of these hints of parallels to real distributions, host–parasite interactions, admittedly treated in our models in realistic genetic detail rather than demographically,[73,74] reveal many simple and unexpected dynamic patterns. The patterns are unexpected in the cases we have displayed in aspects both of their antisymmetry (as in the genetic oppositeness of oscillating clone pairs) and their asymmetry (as of the pacemaker—one emergent different deme 'controlling' an unlimited number of others). Recognized in theory, such 'Red-Queen' patterns can be searched for carefully in nature and, insofar as have been found, give credence to the theory.

Acknowledgements

The authors thank Dr L. Weider for providing the data on the time-series of *Daphnia*, Dr J. Jokela for useful comments on New Zealand snail data and Dr N. Pierce for her kind suggestions on the manuscript.

APPENDIX A: MULTI-LOCUS MATCHING GENOTYPE DYNAMICS AND EVOLUTION OF SEX

We here introduce the multi-locus haploid genotype frequency dynamics in a single population under soft truncation selection based on host–parasite matching of genotypes. L of resistance loci of hosts correspond with virulence loci of parasite, each with two alleles 0 and 1. The mean distance of host genotypes (expressed as binary sequences for their allelic states) $\mathrm{s}=s_1s_2\ldots s_L(s_i \in \{0,1\})$ from parasites and that of parasite genotypes $\mathrm{t}=t_1t_2\ldots t_L(t_i \in \{0,1\})$ from hosts, respectively, are

$$\begin{aligned} \bar{d}_H(\mathrm{s}) &= \sum\nolimits_{\mathrm{t}} d(\mathrm{s},\mathrm{t})p(\mathrm{t}) \\ \bar{d}_P(\mathrm{t}) &= \sum\nolimits_{\mathrm{s}} d(\mathrm{s},\mathrm{t})x(\mathrm{s}) \end{aligned} \qquad (A1)$$

where $x(\mathrm{s})$ and $p(\mathrm{t})$ are the frequencies of host genotype s and parasite genotype t and $d(\mathrm{s},\mathrm{t})$ is the Hamming distance of s and t, i.e. the number of different allelic states in corresponding loci of host genotype s and parasite genotype t. Population sizes are infinite. Truncation selection based on the mean Hamming distance (A 1) of genotypes is assumed for both host and parasite, in which a constant fraction θ_{H} of hosts having the smallest mean distance from parasites and a constant fraction θ_{P} of parasites having the largest mean distance from hosts are killed in each year. Vacancies opened by mortality are filled by reproduction of survivors. Both species are therefore iteroparous. Parasites are asexual, transmitting exact copies of their genotypes. Sexual hosts randomly mate and recombine their genotypes before providing the progeny. If both sexual and asexual hosts are present, the fecundity of a sexual host is halved compared with asexuals (the twofold advantage of asexuals due to no investment in males or male functions). The recombination rates between adjacent loci are the same, denoted by r $(0 \leq r \leq 0.5)$. Mutation rate per locus is μ_{H} in host and μ_{P} in parasite. The model is simplified from the described model HAMAX[23] in two ways. (i) The populations are infinite and thus the model deterministic, so that any implicit effect of finite population size that may favour sex is absent. (ii) An individual always starts reproduction after the first time unit and there is no menopause.

By having usually a distant pair of genotypes more common than others, the host population provides a situation that, when the two prominent genotypes are in mid-frequency range, cannot be matched to a level >0.5 by the parasite. If sexual a prominent opposite pair also generates diverse recombinants at all times and thus provides opportunity to switch to other prominent opposite pairs of genotypes as parasite genotypes optimize their frequencies for the current pair. However, in the particular cases shown (Figs. 18.3(a) and (b)), selection is too weak and the genotype space too restricted for clones arriving by mutation to be resisted. Indeed, at the selection strength used $(\theta_{\mathrm{H}}=0.1)$, sex seemingly fails to compete even for $L=8$ although its defeat becomes very slow at this locus level. With stronger selection $(\theta_{\mathrm{H}}=0.2)$, however, only two more loci than in Fig. 18.3($L=5$) are enough to protect sex. If there is very low or zero recombination, the host population loses most of its genotypic diversity and becomes trapped into a single unchanging mode of fluctuation of opposites (case not shown), the particular pair (in our symmetrical model) arising according to the starting conditions. Asexuals can achieve the last-mentioned advantage but have no chance of switching to other pair prominences and thus of exploiting the extensive mismatching during prominence changeovers. If the host stays in one fixed mode of fluctuation, the match score of the host is drastically decreased since the parasite, by its shorter generation time plus higher mutation rate, can completely catch up and track the currently commonest genotypes in the host cycle. This indicates that sexual hosts outcompete asexual counterparts in spite of a twofold efficiency advantage of the latter when the number of loci is increased: the sexual host jumps by recombination to completely different opposite prominent pairs of genotypes in the high-dimensional hypercube of genotypes. The parasite must find each new prominences step by step by mutations (see also Hamilton *et al.*).[23]

References

1. B. T. Grenfell, K. Wilson, B. F. Finkenstadt, T. N. Coulson, S. Murray, S. D. Albon, J. M. Pemberton, T. H. Clutton-Brock, and M. J. Crawley, Noise and determinism in synchronized sheep dynamics. *Nature* **394**, 674–677 (1998).
2. F. R. Adler, Migration alone can produce persistence of host–parasitoid models. *Am. Nat.* **141**, 642–650 (1993).
3. J. Crow and M. Kimura, *An Introduction to Population Genetics Theory*. (Harper and Row, New York, 1970).
4. M. C. Boerlijst, M. E. Lamers and P. Hogeweg, Evolutionary consequences of spiral waves in a host parasitoid system, *Proc. R. Soc. Lond.* B **253**, 15–18 (1993).
5. J. D. Murray, *Mathematical biology*, 2nd edn. (Springer, Berlin, 1989).
6. W. D. Hamilton, Sex vs non-sex vs parasite. *Oikos* **35**, 282–290 (1980).
7. I. R. Crute, E. B. Holub, and J. J. Burdon. *The Gene-for-Gene Relationship in Plant-Parasite Interactions*. (CAB International, Wallingford, UK, 1997).
8. J. Hofbauer and K. Sigmund, *The Theory of Evolution and Dynamical Systems*. (Cambridge University Press, 1984).
9. M. A. Parker, Pathogens and sex in plants. *Evol. Ecol.* **8**, 560–584 (1994).
10. S. A. Frank. Problems inferring the specificity of plant-pathogen genetics—reply. *Evol. Ecol.* **10**, 323–325 (1996*b*).
11. O. P. Judson, Preserving genes—a model of the maintenance of genetic-variation in a metapopulation under frequency-dependent selection. *Genet. Res.* **65**, 175–191 (1995).
12. A. Winfree, *The Geometry of Biological Time*. (Springer, Berlin, 1980).
13. H. N. Comins, M. P. Hassell, and R. M. May, The spatial dynamics of host parasitoid systems. *J. Anim. Ecol.* **61**, 735–748 (1992).
14. A. White, M. Begon, and R. G. Bowers, Host-pathogen systems in a spatially patchy environment. *Proc. R. Soc. Lond.* B **263**, 325–332 (1996).
15. P. Rohani, T. J. Lewis, D. Grunbaum, and G. D. Ruxton, Spatial self-organization in ecology: pretty patterns or robust reality? *Trends Ecol. Evol.* **12**, 70–74 (1997).
16. C. D. Thomas, Spatial and temporal variability in a butterfly population. *Oecologia* **87**, 577–580 (1991).
17. B. P. Zeigler, Persistence and patchiness of predator-prey systems induced by discrete event population exchange. *J. Theor. Biol.* **67**, 687–713 (1977).
18. I. Hanski, Metapopulation dynamics: does it help to have more of the same? *Trends Ecol. Evol.* **4**, 113–114 (1989).
19. G. D. Ruxton, Low-levels of immigration between chaotic populations can reduce system extinctions by inducing asynchronous regular cycles. *Proc. R. Soc. Lond.* B **256**, 189–193 (1994).
20. B. Schmid, Effects of genetic diversity in experimental stands of *Solidago altissima*—evidence for the potential role of pathogens as selective agents in plant-populations. *J. Ecol.* **82**, 165–175 (1994).

21. W. D. Hamilton, Pathogens as causes of genetic diversity in their host populations. In *Population biology of infectious disease* R. M. Anderson and R. M. May (ed.), pp. 269–296. (Springer, New York, 1982).
22. R. J. Ladle, R. A. Johnstone, and O. P. Judson, Coevolutionary dynamics of sex in a metapopulation—escaping the red queen. *Proc. R. Soc. Lond.* B **253**, 155–160 (1993).
23. W. D. Hamilton, R. Axelrod, and R. Tanese, Sexual reproduction as an adaptation to resist parasites (a review). *Proc. Natl Acad. Sci. USA* **87**, 3566–3573 (1990).
24. L. J. Weider, Spatial and temporal heterogeneity in a natural *Daphnia* population. *J. Plankton Res.* **7**, 101–123 (1985).
25. M. Tibayrenc, F. Kjellberg, J. Arnaud, B. Oury, S. F. Breniere, M. L. Darde, and F. J. Ayala, Are eukaryotic microorganisms clonal or sexual—a population-genetics vantage. *Proc. Natl Acad. Sci. USA* **88**, 5129–5133 (1991).
26. W. D. Hamilton, Haploid dynamics polymorphism in a host with matching parasites: effects of mutation/subdivision, linkage, and patterns of selection. *J. Heredity* **84**, 328–338 (1993).
27. O. P. Judson and B. B. Normark, Ancient asexual scandals. *Trends Ecol. Evol.* **11**, A41–A46 (1996*a*).
28. O. P. Judson and B. R. Normark. Ancient asexuals: scandal or artifact? Reply. *Trends Ecol. Evol.* **11**, 297 (1996*b*).
29. O. P. Judson, A model of asexuality and clonal diversity: cloning the red queen. *J. Theor. Biol.* **186**, 33–40 (1997).
30. A. Vandel, La spanandrie, la parthenogenenie geographique et la polyploidie chez les curculionides. *Bull. Soc. Ent. France* **37**, 255–256 (1932).
31. G. Bell, *The Masterpiece of Nature: The Evolution and Genetics of Sexuality.* (University of California Press, Berkeley, 1982).
32. P. Bierzychudek. Patterns in plant parthenogenesis, *Experientia* **41**, 1255–1264 (1985).
33. P. F. Brussard, Geographic patterns and environmental gradients: the central-marginal model in *Drosophila* revisited. *Ann. Rev. Ecol. Syst.* **15**, 25–64 (1984).
34. J. R. Peck, J. M. Yearsley, and D. Waxman, Explaining the geographical distributions of sexual and asexual population. *Nature* **391**, 889–892 (1998).
35. K. Bye and O. Halvorsen, Abomasal nematodes of the Svalbard reindeer. *J. Wildlife Dis.* **19**, 10–15 (1983).
36. O. Halvorsen, Epidemiology of reindeer parasites. *Parasitol. Today* **2**, 334–339 (1986).
37. C. A. Suneson, Genetic diversity—a protection against diseases and insects. *Agron. J.* **52**, 319–321 (1960).
38. E. D. Parker Jr, Ecological implications of clonal diversity in parthenogenetic morphospecies. *Am. Zool.* **19**, 753–762 (1979).
39. R. C. Vrijenhoek, Factors affecting clonal diversity and coexistence. *Am. Zool.* **19**, 787–797 (1979).
40. R. A. Angus, Geographical dispersal and clonal diversity in unisexual fish populations. *Am. Nat*, **115**, 531–550 (1980).
41. R. J. Jeffries and L. D. Gottlieb, Genetic variation within and between populations of the sexual plant *Pucinellia* × *phryganodes. Can. J. Bot.* **61**, 774–779 (1983).

42. N. C. Ellstrand and M. L. Roose, Patterns of genotypic diversity in clonal plant species. *Am. J. Bot.* **74**, 123–131 (1987).
43. P. D. N. Hebert, Genotypic characteristics of cyclic parthenogens and their obligately asexual derivatives. In *The evolution of sex and its consequences* S. C. Stearns (ed.), pp. 175–195. (Birkhauser, Basel, Switzerland, 1987).
44. M. F. Dybdahl and C. M. Lively, Diverse, endemic and polyphyletic clones in mixed populations of a fresh-water snail (*Potamopyrgus antipodarum*). *J. Evol. Biol.* **8**, 385–398 (1995).
45. J. Jokela, C. M. Lively, J. A. Fox, and M. F. Dybdahl, Flat reaction norms and 'frozen' phenotypic variation in clonal snails (*Potamopyrgus antipodarum*). *Evolution* **51**, 1120–1129 (1997).
46. R. C. Vrijenhoek and E. Pfeiler, Differential survival of sexual and asexual *Poeciliopsis* during environmental stress. *Evolution* **51**, 1593–1600 (1997).
47. P. D. N. Hebert and D. B. McWalter, Cuticular pigmentation in arctic *Daphnia*: adaptive diversification of asexual lineages. *Am. Nat.* **122**, 286–291 (1983).
48. L. J. Weider and P. D. N. Hebert, Microgeographic genetic heterogeneity of melanic *Daphnia pulex* at a low-arctic site. *Heredity* **58**, 391–399 (1987).
49. M. J. Beaton and P. D. N. Hebert, Geographic parthenogenesis and polyploidy in *Daphnia pulex*. *Am. Nat.* **131**, 837–845 (1988).
50. P. D. N. Hebert and T. Crease, Clonal diversity in populations of *Daphinia pulex* reproducing by obligate parthenogenesis. *Heredity* **51**, 353–369 (1983).
51. C. C. Wilson and P. D. N. Hebert, The maintenance of taxon diversity in an asexual assemblage—an experimental analysis. *Ecology* **73**, 1462–1472 (1992).
52. C. C. Wilson and P. D. N. Hebert, Impact of copepod predation on distribution patterns of *Daphnia pulex* clones. *Limnol. Oceanogr.* **38**, 1304–1310 (1993).
53. M. Lynch, K. Spitze, and T. Crease, The distribution of life-history variation in the *Daphnia pulex* complex. *Evolution* **43**, 1724–1736 (1989).
54. D. Ebert. Virulence and local adaptation of a horizontally transmitted parasite. *Science* **265**, 1084–1086 (1994).
55. D. Ebert, C. D. Zschokke-Rohringer, and H. J. Carius, Within- and between-population variation for resistance of *Daphnia magna* to the bacterial endoparasite *Pasteuria ramose*. *Proc. R. Soc. Lond.* B **265**, 2127–2134 (1998). (DOI 10.1098/rspb.1998.0549.)
56. T. J. Little and D. Ebert, Associations between parasitism and host genotype in natural populations of *Daphnia* (Crustacea: Cladocera). *J. Anim. Ecol.* **68**, 134–149 (1999).
57. D. Ebert and E. A. Herre, The evolution of parasitic diseases. *Parasitol. Today* **12**, 96–101 (1996).
58. S. A. Frank, Models of parasite virulence. *Q. Rev. Biol.* **71**, 37–78 (1996*a*).
59. C. M. Lively, Evidence from a New Zealand snail for the maintenance of sex by parasitism. *Nature* **328**, 519–521 (1987).
60. J. Jokela and C. M. Lively, Parasites, sex and early reproduction in a mixed population of freshwater snails. *Evolution* **49**, 1268–1271 (1995*a*).
61. J. A. Fox, M. F. Dybdahl, J. Jokela, and C. M. Lively, Genetic structure of coexisting sexual and clonal subpopulations in a freshwater snail (*Potamopyrgus antipodarum*). *Evolution* **50**, 1541–1548 (1996).

62. J. Jokela, M. F. Dybdahl, and C. M. Lively, Habitat-specific variation in life history traits, clonal population structure and parasitism in a freshwater snail (*Potamopyrgus antipodarum*). *J. Evol. Biol.* **12**, 350–360 (1999).
63. M. F. Dybdahl and C. M. Lively, The geography of coevolution: comparative population structures for a snail and its trematode parasite. *Evolution* **50**, 2264–2275 (1996).
64. J. Jokela and C. M. Lively, Spatial variation in infection by digenetic trematodes in a population of freshwater snails (*Potamopyrgus antipodarum*). *Oecologia* **103**, 509–517 (1995*b*).
65. M. F. Dybdahl and C. M. Lively, Host–parasite coevolution: evidence for rare advantage and time-lagged selection in a natural population. *Evolution* **52**, 1057–1066 (1998).
66. C. M. Lively and M. F. Dybdahl, Parasite adaptation to locally common host genotypes. *Nature* **405**, 679–681 (2000).
67. L. Hauser, G. R. Carvalho, R. N. Hughes, and R. E. Carter, Clonal structure of the introduced fresh-water snail *Potamopyrgus antipodarum* (Prosobranchia, Hydrobiidae), as revealed by DNA fingerprinting. *Proc. R. Soc. Lond.* B **249**, 19–25 (1992).
68. T. J. Little, R. Demelo, D. J. Taylor, and P. D. N. Hebert, Genetic characterization of an arctic zooplankter: insights into geographical polyploidy. *Proc. R. Soc. Lond.* B **264**, 1363–1370 (1997). (DOI 10.1098/rspb.1997.0189.)
69. S. J. O'Brien and J. F. Evermann, Interactive influences of infectious disease and genetic diversity in natural populations. *Trends Ecol. Evol.* **3**, 254–259 (1988).
70. B. N. Kunkel, A useful weed put to work: genetic analysis of disease resistance in *Arabidopsis thaliana*. *Trends Genet.* **12**, 63–69 (1996).
71. E. B. Holub, *The Gene-for-Gene Relationship in Plant Parasite Relationships* I. R. Crute, E. B. Holub and J. J. Burdon (ed.), pp. 5–43. (CAB International, Wallingford, UK, 1997).
72. D. Caugant, B. Levin, and R. Selander, Genetic diversity and temporal variation in the *E. coli* population of a human host. *Genetics* **98**, 467–490 (1981).
73. A. B. Korol, V. M. Kirzhner, Y. I. Ronin, and E. Nevo, Cyclical environmental changes as a factor maintaining genetic polymorphism. 2. Diploid selection for an additive trait. *Evolution* **50**, 1432–1441 (1996).
74. V. M. Kirzhner, A. B. Korol, and E. Nevo, Complex limiting behaviour of multilocus genetic systems in cyclical environments. *J. Theor. Biol.* **190**, 215–225 (1998).

CHAPTER 19

BECAUSE TOPICS OFTEN FADE

Letters, essays, notes, digital manuscripts and other unpublished works

JEREMY LEIGHTON JOHN[1]

Introduction

I still have the note of a telephone message written by a flatmate in my absence. The words that stood out from the paper were: '. . . Prof. Hamilton[2] is seriously ill . . .' With that brief pausing of the heart familiar to all who have lived long enough to know sorrow, concern turned to growing sadness and ultimately grief: a visit to Middlesex Hospital in London, telephoned, written and printed news of his death, a funeral in the Oxfordshire village of Wytham where as a student I had lived for four years with Bill Hamilton and family, a beautiful memorial service in New College Chapel with the music that had been played for Darwin (the musical score having been obtained from the British Library), the despair of a coroner's inquiry (a mundane proximal explanation for a desperate death), all these followed in tragically inexorable succession.

I also still have the annotated e-mail printout that I found lying on my desk in the bioacoustics section of the Sound Archive of the British Library where I worked at this time. A subsequent meeting soon confirmed that Bill's papers had been offered to the British Library and that the Chairman had personally accepted them. It was both a humbling privilege and a relief to be able to act, to participate in coming to terms with the loss for myself, his family, friends and colleagues. So it was that almost five months later in August 2000, I found myself ensconced at a different desk, across the way, installed as Curator with over 200 large boxes of WDH's materials sitting on the shelves in the nearby repository of the Department of Manuscripts.

Of course, Bill had said more than once that the British Library might suit me. What neither he nor I would have known was that it would be under these circumstances, and that I would be working in this department with its historical roots in the library of the British Museum, with its papyri, vellum, charter rolls, seals, quarto leaves and flourishing script in browned ink, and its historic gems from scientists such as Newton, Leibniz, Halley, Linnaeus, von Humboldt, Babbage, Darwin, Wallace and Fleming. I say that Bill would not have known, but occasionally I come across brief explanatory comments in pencil besides his earlier writings, and in my imagination it feels as if he is quietly helping me sort out his papers.

The W. D. Hamilton Archive: Bill's Last Great Work

'. . . because topics in science often fade for no good reasons–certainly not because of error—it may sometimes help inspiration to be reminded of forgotten facts or to see new ones from an old point of view'[3]

W. D. Hamilton

In early June 2000 the bulk of what was to become the W. D. Hamilton Archive was collected from Oxford by the British Library in a large removal van supported by professionals with special expertise in handling archival materials and museum exhibits: his office in the Department of Zoology comprising two rooms; his college rooms in Saville House, an annexe of New College situated on Mansfield Road; and his home, at 28 Wytham Village in the countryside close to Wytham Woods, famous to field researchers the world over. Some other materials later came from Oaklea, his childhood home in Kent that remains in the family. There are today 21 boxes, in part but not purely because the contents of some of the very large and unwieldy original boxes were transferred to smaller ones. With donations of manuscripts still coming in, the total number of boxes continues to increase.

Bill's family kindly made available almost all of his academic and personal papers, including those from his childhood and youth.

These materials—including a selection from outside his family—range from the scripted to the photographic, from flimsy paper to rigid card, from dense printed directory of names to dynamic audio-visual records of outdoor interviews. Although some additional items will be mentioned in passing, I shall not attempt to discuss all types of items, nor shall I elaborate on the envelopes containing significant amounts of foreign currency. The diversity is illustrated by the list in the accompanying box.

List of some of the items in the W. D. Hamilton Archive

audio cassette tapes — bank records and research account cheque stubs — bibliographic reference cards — black & white and colour prints — books — certificates awarded — computer hardware — computer printouts of data, analyses and graphics including colour — computer program workings — conference proceedings — correspondence—personal, scientific and administrative — data record cards — day-to-day lists of things to do — defunct passports retained — displays — drawings — draft articles and annotated off-prints and photocopies — e-mail printouts — examinations taken — expedition journals and travel notebooks — field record books and cards — floppy disks and other digital storage media with electronic files and software — graphs — illustrations — index cards — leaflets and other printed ephemera — manuals, technical and computer — marking notes — mathematical calculations — newsletters — notes from university lectures attended — overhead transparencies — paintings — photographic slides — pocket diaries — posters — receipts — referee reports — research proposals — scientific and, to a lesser extent, literary periodicals — seminar programmes with notes — study proposals — testimonials — theses — thoughts and sketches of ideas — travel itineraries and tickets — university telephone directories — various paper manuscripts of unpublished exposition such as essays and lectures prepared — VHS video tapes

For the first 18 months this archival project,[4] the sorting and describing of Bill's papers, was supported primarily by generous donations to the W. D. Hamilton Memorial Fund, administered by New College, Oxford. Later, the British Library funded me on its own with a remit that included scientific manuscripts generally, and in 2003 I became a full curator in the Department of Manuscripts with responsibility for digital and scientific manuscripts. This means that a significant proportion of my time and effort is, and has been for some time, directed towards other archival work, but it has had, and continues to have, the benefit that it has enabled the W. D. Hamilton Project to continue on a much firmer foundation.[5]

Besides the diversity of the materials, requiring very different approaches, three aspects of the archival work will indicate its time-consuming nature.

Figure 19.1. Wasp Data Cards

A reference card box, labelled 'POLISTINAE dissections', with 5″ × 8″ cards containing data entries. The cards record the sex of the insect and the status of its ovaries, spermatheca and fat body; most date from 1964 and many derive from various localities within Brazil. Species represented include *Polybia sericea*, *Apoica pallida*, *Bombus atratus* and *Polistes versicolor*.

The first and most obvious is the size. There are hundreds of thousands of items including, for example, thousands of photographic slides. The second is that because of the highly contemporary nature of many materials, numerous decisions have to be reached about when these can be made available. For example, there are many laser-printed manuscripts of articles by other scientists, potentially containing ideas or results that the originator might not yet have published.

The third aspect arises out of a wish to capture and preserve the layout and organization of Bill's papers and other materials. Often, though by no means invariably, papers arriving at an archive are subject to sorting and selection before the cataloguing, numbering and description takes place. In these cases, information about the original organization of the papers and their physical context may be lost. The combination of WDH's tendency to keep much material, of having a large office in which to do so, and the unexpected timing and nature of his death, means that there is an opportunity to record the layout and organisation of a working scientist's papers at the close of the twentieth Century.[6]

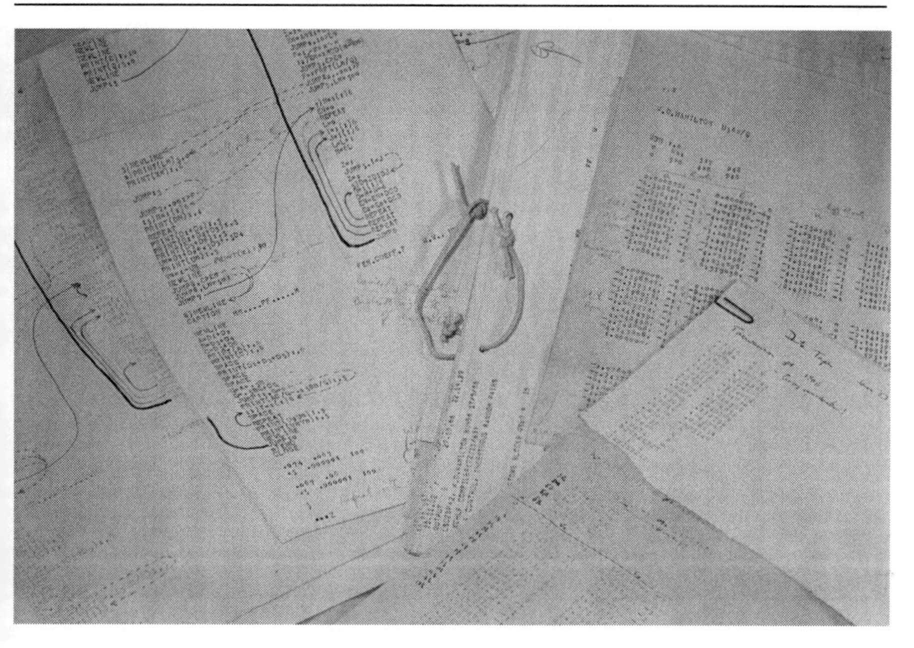

Figure 19.2. Atlas Computer Printouts

A series of computer printouts dating from the 1960s and produced at the University of London Computer Unit, mainly on the Atlas computer using Autocode and the CHLF3 compiler and later the EXCHLF version. The printouts shown include both the output and the programs (heavily annotated in some cases), some of which date from 1963 when W. D. Hamilton was at the London School of Economics and Political Science; these 1963 printouts (right) are believed to be derived from the Mercury computer.

Thus a hierarchical scheme has been adopted for registering numerically the placement of letters within envelopes within folders within a filing cabinet drawer, and so on, recording as completely as possible the items as they existed in his work places, yielding ultimately a detailed inventory mapping of item to location.[7]

A series of programmes and projects have been put in place to process these materials.[8] These and the diverse nature of manuscripts and historical archives can be illustrated with two of these activities.

Digitization of Annotations in Books

The DAB ***hand*** Project entails the following activities: recording the titles of all the books in WDH's personal library, locating all annotations, and creating digital images of these scripted markings. Besides the obvious

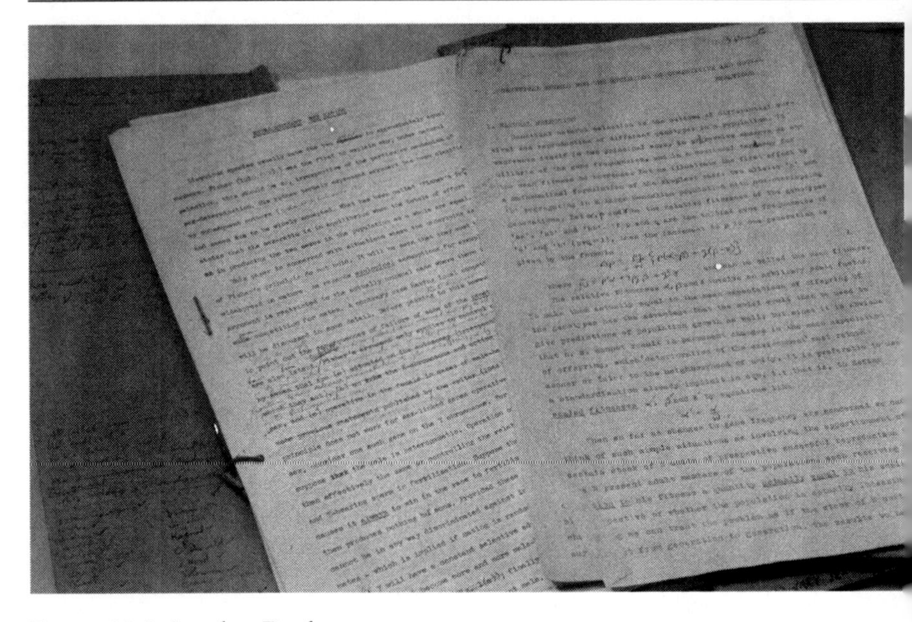

Figure 19.3. London Drafts

The typescript unpublished drafts of W. D. Hamilton's most celebrated early papers. The first (right) is entitled 'Genetical models for the evolution of competitive and social behaviour', and it is a rudimentary form of the 1964 publication, and the second (left) is relatively close to the version finally published with the title 'Extraordinary sex ratios' in 1967. On looking at the early manuscripts that WDH worked, one sees the aspects of social behaviour that interested him—altruism, competition, relatedness and sex ratio—intermingled in different writings and typings, with a returning to the same themes as time passes. The card wallet folder as a manuscript is also demonstrated (left) as is the use of treasury tags.

interest of a pencil comment by WDH on a portion of text within well-known books such as Mandelbrot's classic on fractals, there are often dates of acquisition at the front of the book, and in the case of a field guide to flora (for example) records of the date and place of observations of particular species.

The recording of annotations is a long-standing and respectable archival activity, carried out, for example, with the books owned by one of the founders of the Royal Society, John Evelyn,[9] and also with those of Newton.[10] Apart from the candid nature of some remarks, these provide evidence that a book was read. For example, annotations by Mendel show that he read 'The Origin of Species'.[11] The list of books in a personal library combined with dates of acquisition can give a hint of how patterns of reading interest change over the years as the scientist develops and matures.

Figure 19.4. Marginalia

Some of the books in the personal library of W. D. Hamilton, with one open to show some annotations in the margins. E. B. Ford's book on 'Butterflies' may have stimulated the young Bill Hamilton profoundly but his few annotations in 'Ecological Genetics', Ford's principal text, indicate some differences in detailed opinion. It is in the nature of annotations in books that they include remarks that question and evaluate what is being read, but there are also approving comments too. For example, a sentence in G. C. Williams' classic book 'Adaptation and Natural Selection' is met with the pithy: 'Excellent'. The open book is Benoit B. Mandelbrot's (1977) 'Fractals. Form, Chance and Dimension', with WDH's name in the front in his hand, dated 26 August 1978. It is open at pages 122–123; the note in pencil on the left says 'Summarises main idea of book'. WDH's annotation on the right concerns Mandelbrot's presentation of Hoyle's model of stellar clusters. In the collection there is another copy of the book, a revised edition (1983), apparently with Mandelbrot's own signature, dated 1991.

Whereas Newton folded pages right down precisely to the phrase of interest,[12] WDH would draw a diagonal pencil line at the top outer corner so that a flick through the pages quickly identifies the noteworthy ones.

Insertions: Paper and Organic

This project follows on from the DAB ***hand*** Project, and involves the compiling of materials found inserted between the pages of Bill's textbooks and notebooks. These include organic specimens, mostly plant leaves. This

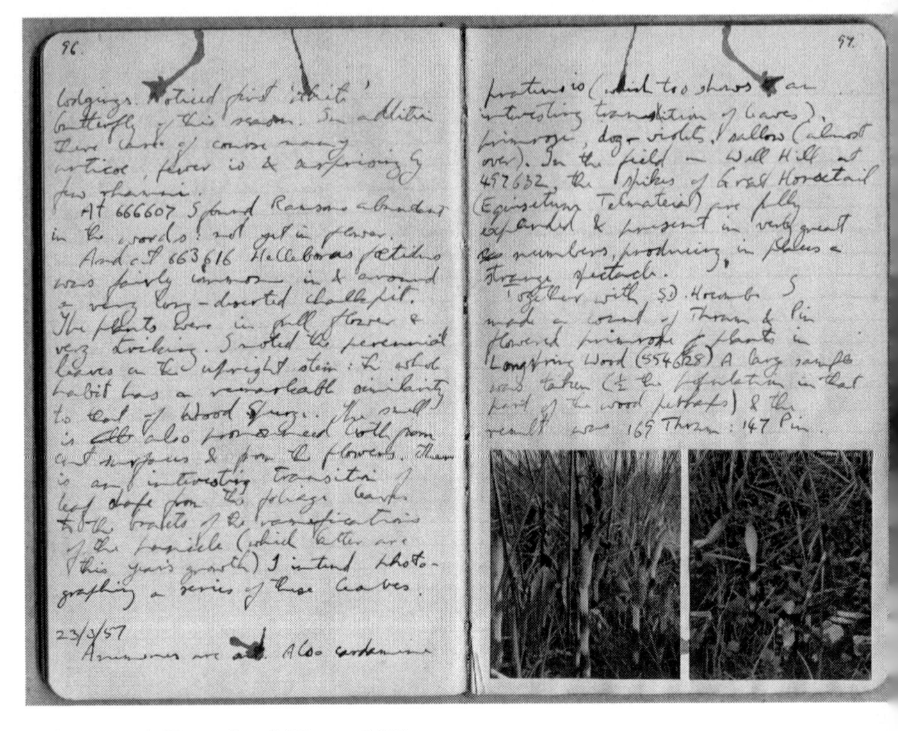
96

lodgings. Noticed first 'white' butterfly of this season. In addition there were of course many [illegible], fever [illegible] & surprisingly few [illegible].

At 666607 I found Ransoms abundant in the woods: not yet in flower.

And at 663616 Helleborus foetidus was fairly common in & around a very long-deserted chalk pit. The plants were in full flower & very striking. I noted the perennial leaves on the upright stem: the whole habit has a remarkable similarity to that of Wood Spurge. The small [illegible] is also [illegible] from [illegible] & from the flowers. There is an interesting transition of leaf shape from the foliage leaves to the bracts of the ramifications of the panicle (which latter are this year's growth) I intend photographing a series of these leaves.

23/3/57

Anemones are [illegible]. Also cardamine

97

pratensis (which too shows an interesting transition of leaves), primrose, dog-violets, sallow (almost over). In the field on Well Hill at 497632, the spikes of Great Horsetail (Equisetum Telmateia) are fully expanded & present in very great numbers, producing, in places a strange spectacle.

Together with [illegible] I made a count of Thrum & Pin flowered primrose plants in [illegible] Wood (554628). A large sample was taken (½ the population in that part of the wood perhaps) & the result was 169 Thrum : 147 Pin.

Figure 19.5. Downland Natural History

Pocket notebook, a personal journal of natural history observations by Bill Hamilton, 1956-1957. Pages 96 and 97 (according to original pagination) are shown. Page 96 comprises mos of the notebook's entry for the 22 March 1957. The entry for '22/3/57' on the previous page is 'Cycled to Aylesbury in search of . . .', continuing at the top of page 96. The two lines at th bottom of page 96 along with page 97 represent the entry for 23 March 1957. As with th books, organic materials (plants and insects) are found inserted in various notebooks. I addition, there are occasional diagrams, sketches or photographs, as shown in this case A partial transcription shows that the notebook records places (with Ordnance Survey ref erence numbers) as well as the identities of plants, insects and other wildlife, their apparen abundance and state of development: 'In the field on Well Hill at 497632, the spikes of Grea Horsetail (Equisetum Telmateia) are fully expanded & present in very great numbers, pro ducing in places a strange spectacle'. This is *Equisetum telmateia*, of course. Other plant recorded on these pages include: 'Ransoms' (wild garlic, *Allium ursinum*) and '*Helleboru foetidus*' (Stinking Hellebore). Entries elsewhere in this notebook mention, for example, th location and timing of the flowering of orchids such as *Orchis purpurea*, and the presence the 'coridon', the Chalkhill Blue Butterfly, *Lysandra coridon*.

is a somewhat haphazard academic resource, but original specimens of thi kind have yielded great historical and biological benefits, as with sample found attached to the manuscripts of van Leeuwenhoek.[13] By recording th location of a plant specimen within a particular notebook, some possibl pointers to context can be obtained, as with WDH's field notebooks from Brazil. The pile of fragments and pressings emanating from over 1000 books and manuals from WDH is not far from 30 cm in height.

Lessons and Examples: Paper Manuscripts

WDH's papers will be a rich resource for historians and scientists interested in evolutionary and theoretical biology in the late twentieth Century. To a significant degree, this is because of his correspondence with other scientists. There are, moreover, his qualitative and quantitative biological observations recorded on cards and in notebooks. But in addition, there is a glorious profusion of printed ephemera and personal letter and diary writing that inform about many other aspects of life during these times: everyday life, growing up, academic experience generally, and the natural, social and commercial environments that surround an individual. His papers have a value far beyond science and the history of science.

Archival pearls include personal diaries from his adolescence with tales of silently leaving the house at night without his parents knowing, and making a fire in the cold woods, and of an experimental meal of squirrel shot with a gun, of records of travelling across Europe to Belgrade and to Turkey; there are accounts of his experience of National Service, airmail letters and hardback notebooks that tell of his travels to and within Brazil in the 1960s that can be compared with those written by him later in the same country, in the 1970s; there is a long sequence of correspondence with one of his earliest friends J. C. Hudson, the botanical foray into southern Chile in the 1970s with maps and travel ephemera, observations in Bangalore in the late 1980s, and the poignant collection of materials that were in his possession on his return from the last Congo expedition. There is also an unpublished novel written by Bill, its scripted evolution preserved in successive drafts: a fantastic adventure, provisionally entitled 'The Dark of the Stars'.

There is much of palaeographic interest too, preserving a historical record of the changing nature of correspondence and note taking during these decades as well as Bill's own idiosyncratic styles.

One of the most characteristic forms of paper manuscript of WDH is his grouping of sheets together by means of one or two treasury tags or string passing through one or two punched holes.[14] The folios will usually be written in small handwriting in pencil. Numerous details, such as these, help

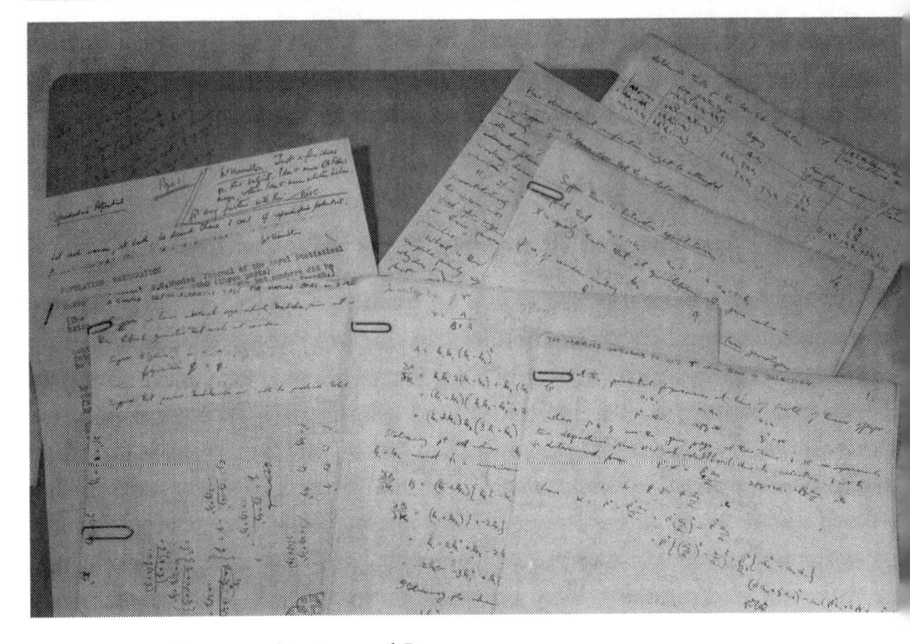

Figure 19.6. Norman Carrier and Lectures

Lecture notes by WDH (with exercises) from Norman Carrier's course covering human demography and population mathematics at the London School of Economics and Political Science. There is also a sheet of foolscap paper about 'Reproductive Potential' written by Carrier himself (left). It begins with a note in the corner: 'Mr Hamilton. Just a few ideas on this subject. I don't know R. A. Fisher's work, & thus I don't know whether he has got any further with this'. A collection of early handwritten mathematical workings by WDH is also displayed (right). The corner of the corresponding card wallet folder is also visible, characteristically bearing a poem (top left); the folder also has written on it WDH's name and address, 14 Hadley Gardens, Chiswick, London.

to enable the authentication of manuscripts that emerge at a later date, one of the key roles of archives. Similarly, characteristic changes in the style and formatting of writing and writing materials provide a chronological guide for use with undated manuscripts.[15]

Some readers will have seen examples of Bill's postcards where he writes across the surface in more than one orientation, sometimes with different colours of ink. Another typical type of early correspondence is represented by his densely written aerogrammes, his minute writing covering the entire face of the sheet up to the edges.

Traditionally one of the first things an archivist looks for in a collection is the personal catalogue compiled by the originator.[16] Bill does not disappoint

Although less rigorous in his later years, there are such lists; for example, a series of 3″ × 5″ note cards held together by two elastic bands, 'Index of 1963–64 Brazil Note Books', will be consulted for years to come.

Even these few examples begin to reveal that the way manuscripts are studied and used is extremely diverse in academic research.[17] As a means of further illustrating the scholarly value of manuscripts, two themes may be considered: the importance of youth to science as witnessed by manuscripts, and the importance of travel to the recording of history and place.

Useful youthful thoughts

It is easy to temporarily forget that eminent discoverers and researchers often do some of their most important work when still quite young. Many people (especially those who did not know the person in his or her younger days) find it difficult to imagine the distinguished Professor as a younger scientist.

Photographs from when Bill was a relatively young man, around the time of his early publications in the 1960s, for example, are one way to make

Figure 19.7. Bill in the Sixties

A photograph showing Bill Hamilton with Christine A. Friess (Hamilton), bearing a manuscript of some kind, possibly a letter, at 17 New North Street, London, 1967 (according to WDH's own note, confirmed by personal communications, Dr Janet Hamilton, Dr Christine Hamilton and Dr Mary Bliss, November 2004).

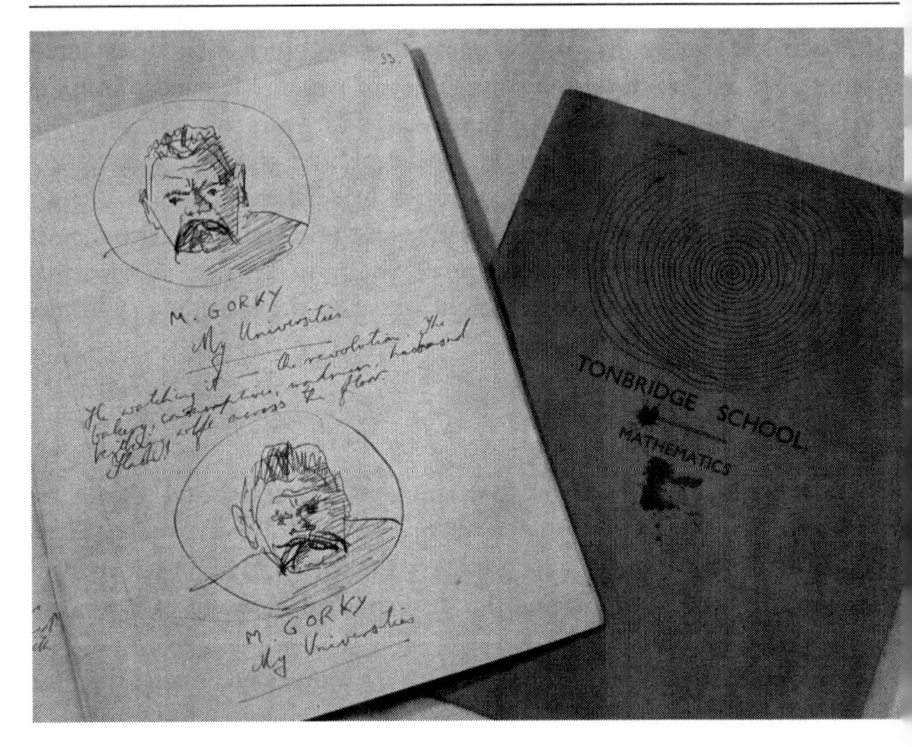

Figure 19.8. 1954

The 1954 diary opened, and lying on another, similar, exercise book, showing two rough sketches intended to be of the Russian writer, Maxim Gorky. The cover of the 1954 diary itself is entitled 'Rough Notes'. It seems likely that the sketches are based on a photograph of Gorky to be found on the front of the book jackets of a series of publications of his work, obtained by WDH in 1954: Soviet Literature, Foreign Languages Publishing House, Moscow. By all accounts Bill Hamilton was at this time entranced by Gorky's writings, which led to his fascination with Anton Chekhov.

The sketches shown are preceded in the diary by the brief entry: '9/9/54, Thursday, 10.15 p.m. "Christ or Darwin"—these two. One or the other ~~See My Univ~~'. After the Gorky sketches shown in this Figure, the diary continues: '26/9/54, Sunday, Morning. But for that crazy last entry, which I would tear out but for my resolve never to alter anything once it is in. I have written nothing for months. I have acquired a dislike of this diary, God knows why, regarding it as childish futile, something of the past best forgotten. However, I still have faith that it ought to turn out useful & so here I go again . . . ' The crossed out 'My Univ.' refers to Gorky's autobiographical book 'My Universities', in which there is a mention of a sermonising speech by a visiting Tolstoian that concludes: 'And so I ask: do you follow Christ, or Darwin?'. Thus the teenaged Bill Hamilton, like many pensive young people, reflects on the relationship between science and religion.

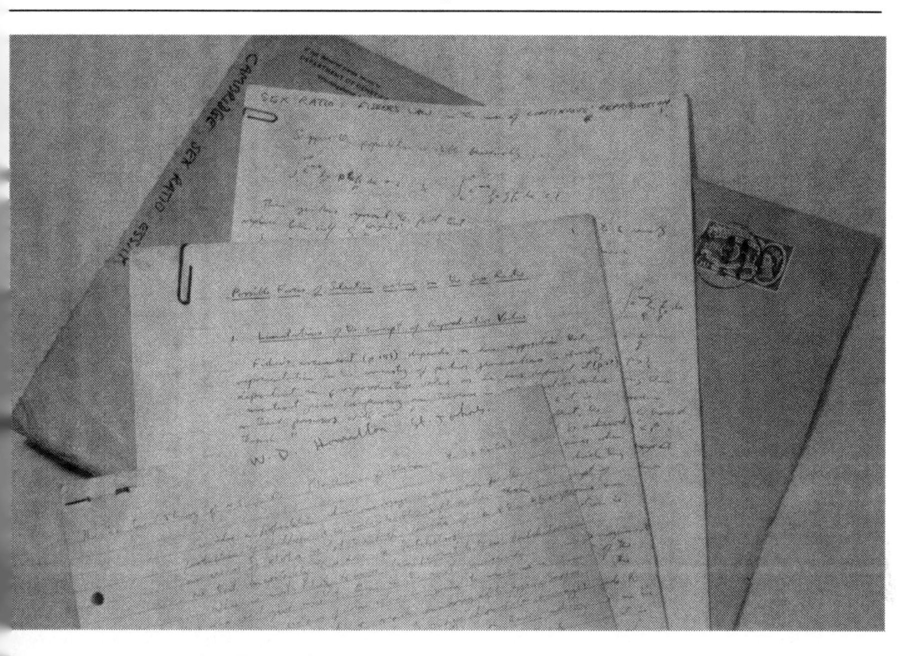

Figure 19.9. Cambridge and Fisher

A review of Fisher's discussion of sex ratio in his book 'The genetical theory of natural selection'. The two folios held together by a pin (bottom), originate from Hamilton's time at the University of Cambridge, and quite possibly represent an essay written for a college tutorial. Thus the date would be around 1960 or earlier. One sees in it a growing germplasm of interest and thought that later produced longer accounts of more adventurous supposition and enquiry. Two other early essays and workings are also shown; both were found in the envelope labelled by WDH 'Cambridge sex ratio essay' (seen in the Figure). Also in the archive are notes of lecture courses WDH attended at Cambridge, importantly a series of courses in the field of genetics. The British Library's manuscript collections have long incorporated scientific and natural philosophy lectures scribed by students and other attendees, including, for example, notes of the lectures of Joseph Black (1728–1799), Professor of Chemistry at Edinburgh University (BL: Add. MS 52495 & 59843).

this point. Nonetheless, writings can have the same effect, as in the case of Bill's mother writing to him expressing concern as he is about to embark on further travels within the Brazilian interior, and his father sending him £200 to help prevent a serious misadventure.[18]

Indeed, very young people can find it hard to imagine that an older person was ever young. Manuscripts from his childhood and adolescence bring home his youth in a very evocative way, and make the sometime gruff elder instantly both human and vulnerable. One of Bill's most moving is his 1954 diary written in a Tonbridge School mathematics exercise book that will

~~[illegible]~~ passed neither 'Fazenda', or at least, there
was one mighty ~~made~~ old house & a small group of
very primitive ones. Down below in the valley the stream
ran through a series of large ponds & around them
was native woodland. On the other side of the valley
however, beside this woodland there were some bad
patches of soil erosion, & indeed less recent soil
erosion may have been much more prevalent than
[illegible]. ~~After this the road~~

After this the road was green & grassy with
only the faintest wheel ~~ruts~~ marks showing, & soil
drifting onto it from the hillside above made it
slope rather precariously for the jeep. In a side way
ravine we passed through some fine native woodland, &
in the crook of the ravine there was a quite gigantic
tree, the biggest I have yet seen in Brazil I should think,
the trunk at ground level being 6 ft through at least.
One or two thin lianas ascended against the rough bark
of the trunk but on the whole the canopy of rather
fine smallish [illegible]-like ~~[illegible]~~ branches & leaves was
unusually free from lianas & epiphytes.

After this we seemed to ~~[illegible]~~ the road altogether
& wandered at random through ~~[illegible]~~ [illegible]
~~[illegible]~~ then the not-quite wholly wild contents,

258.

following fence-lines etc until by a very steep descent
we arrived at Fazenda Cuscuzeiro.

Thence by a small sandy road through the cerrado
we rejoined the Corumbataí–Analândia road [illegible]
to just a little beyond the point which I already
reached when the clutch gave out (25/4/64, 1226)

Analândia is after all a not-so-dusty township:
it impressed me as being more wealthy & more modern
than Corumbataí. In part it looked a slightly newer
town, & the 'praça' had small younger trees.

The Pedra of Cuscuzeiro turned out to be [illegible]
well up to expectations as a spectacle. It is of
red-rock (basalt I suppose) & rises like a 'stack'
at the top of a small hill covered with native
woodland. The rock looks as if it would be extremely
difficult to climb, at least on the side which we
saw. The cliffs must be 150 ft, & there would
be room on top for a few small trees & a palm.
The other side of the road going to Fazenda Boa
Vista is another very steep rocky hill with,
in this case, an easily climbable pinnacle. Both
looked very beautiful in the rays of the setting
sun & all members of our party showed the
animation which sights like this usually produces.

Figure 19.10. A Diary from Latin America

Two pages—numbers 257 (verso, i.e. the back) and 258, according to original pagination—showing the middle of the entry for 17th May 1964. A partial transcription will illustrate the description, written in blue ink, of an outing by jeep with three children and two women, local Brazilians, to Analândia in Sao Paulo State. After describing inevitable diversions towards natural history including a fine nest of *Trigona spinipes*, a species of stingless bee, the account on Page 257 goes on: 'Anyway, . . . the road became smaller & smaller & the scenery more & more attractive. On the site of a steep hill we', continuing at the top of Page 257 (verso), as shown, 'passed another "Fazenda", or at least, there was one nicely-made old house & a small group of very primitive ones. Down below in the valley the stream ran through a series of large ponds & around these was native woodland. On the other side of the valley however, beside this woodland, there were some bad patches of soil erosion, & indeed less recent soil erosion may have been much more prevalent than was obvious. After this the road was green & grassy with only the faintest wheel marks showing, & soil drifting onto it from the hillside above made it slope rather precariously for the jeep. In a side ravine we [*slightly uncertain transcription*] passed through some fine native woodland, & in the crook of the ravine there was a gigantic tree, the biggest I have yet [*added later in a slightly different colour, probably because WDH would, or would expect to, see larger trees in Brazil*] seen in Brazil I should think, the trunk at ground level being 6 ft through at least. One or two thin lianes [sic] ascended against the rough bark of the trunk but on the whole the canopy of rather fine eucalyptus-like branches and lianes [sic] was unusually free from leaves & epiphytes. After this we seemed to lose the road altogether & wandered at random through the not-quite wholly wild cerrado, following fence-lines etc until by a very steep descent we arrived at Fazenda Cuscuzeiro. Thence by a small sandy road through the cerrado we joined the Corumbatai-Analandia road just a little beyond the point which I reached when the clutch gave out (25/4/64 . . .)' 'Analandia is after all a not-so-dusty township: it impressed me as having more wealth & movement than Corumbatai. In fact it looked a slightly newer town, & the 'praça' had younger trees.' 'The Pedra of Cuscuzeiro turned out to be well up to expectations as a spectacle. It is of red-rock (basalt I suppose) & rises like a 'stack' at the top of a small hill curved with native wood. The rock looks as it would be extremely difficult to climb, at least on the side which we saw. Its cliffs must be 150 ft, & there is room on the top for a few small trees & a palm. The other side of the road going to Fazenda Boa Vista, is another very steep rocky hill, with, in this case, an easily climbable pinnacle. Both looked very beautiful in the rays of the setting sun & all members of our party shared the animation which a sight like this usually produces.' On the other side of the page there are further natural history observations, of passing through country with very numerous armadillo burrows, and of stopping to watch an animal caught by the headlights of the jeep, an opossum or some other marsupial. The locality is well known to Brazilians and is favoured for outdoor activities including climbing.

become a classic manuscript that both delights and surprises. Here are words from the first page:

'The difficulty that has been worrying me for some time is how to conceal it [the diary]. If I could think of some perfectly safe way I could be perfectly frank, which would be a good thing—if not, however, I had better be cautious'.
'The trouble is I cannot get out of my head the idea that someone else is not sometime going to read this, and I don't suppose I shall ever be able to get rid of this idea. It means that I write for the benefit of this unknown person whoever it may be, whereas I ought to write purely for myself, because unconsciously when addressing someone else we all adopt some sort of pose & indeavour [sic] to appear what we are not. Perhaps we even fool ourselves a little but I pride myself that in my case it is only a little. However, I will try in this diary to be as honest & unpretentious as I possibly can, & probably as I get used to it I will improve.'

On the more scientific side, it will be interesting to watch future researchers shedding light on the emergence of Bill from the shadow of R. A. Fisher, discerning the manner and timing of his growing confidence. At what time did he finally feel that his own thinking was original and beyond that of his hero?

Archives, science and travel

One of the most striking things about making one's way through Bill's papers is how international is scientific communication, not least among evolutionary biologists, and also how much Bill himself travels, always taking the opportunity to sample local natural history.

Academic life with its conferences, visiting lectures and seminars and moves to new universities involves, for some, considerable and frequent travel. Indeed, scholarship and travel have long been intertwined, and this is a very good thing for history. People who travel are likely to write about the place, about what they find—whereas residents often neglect to do so. But in the age of the telephone before e-mails, distant if not continental travel was often necessary to motivate writing. Bill's travels to, and residence in, North and South America, in particular, have yielded a veritable jewel box of information in the form of letters and diaries.

Travel provides exposure to new and unfamiliar places and customs and i good for writing and for science, but the other side is true too; letter writing exposes one to the rest of the world, inducing alliances and friendships tha provide moral support and crucial sources of scientific information, and this long before the advent of the internet. Both Darwin and Linnaeus gleaned crucial information from networks of communicating allies. Bill's correspondence shows a similar flow of news, ideas and data from all around the world, mostly in letters and postcards, but later in the increasingly prevalent e-mails.

Evolutionary Capture, Preservation and Conservation of Information: Digital Manuscripts

There is in the archive a very wide range of materials derived from computing. These date from the 1960s and provide an insight into the history of the *use* of computers by scientists. Items include 5-hole paper tape, 80-column punched cards, 0.5″ 9-track magnetic tape, a long magnetic strip for the programmable calculator, floppy disks of various sizes and kinds, data cartridge tapes, optical media such as CDRs, and also a series of computers and hard drives, including Silicon Graphics workstations. The phrase digital manuscripts or eMANUSCRIPTS has been used to refer to unpublished writings and works derived from these kinds of sources. (The term 'manuscript' long ago lost its purely literal meaning, of course.)

In keeping with the perspective of the paper manuscript, digital manuscripts originating on electronic media present three principal requirements for archivists and archives: (i) to guarantee authenticity (and to continue to be able to do so); (ii) to ensure complete confidentiality where necessary; (iii) to retain the styles and formatting of the manuscripts—in this case electronic documents. It is useful to draw a distinction between the initial capture of the information from often obsolete computer media, and the long-term preservation of this information.

These concerns have been exercising the British Library and its sister institutions. No one who knew Bill or knows his work will be in any doubt what he would have suggested in these circumstances: an evolutionary approach. It might seem that the last thing one wants when one is striving to preserve something is evolution, but of course, there are living fossils that have existed relatively unchanged morphologically for millions of years. Even more significantly, there is a considerable amount of conservative DNA (the ultimate digital information) that has existed for many hundreds and thousands of millions of years, ancestral biochemical pathways being

Figure 19.11. Digital Manuscripts

Various digital media are shown including 5-hole paper tape for an Atlas computer (left), 80-column IBM cards (right), 0.5″ 9-track magnetic tape on a reel (top), and some floppy disks, 5.25″ and 3.5″ (bottom). There is also a Hewlett-Packard magnetic program card strip (middle).

shared by many forms of life. This offers an important source of optimism—deep digital preservation is certainly possible even if not always easy to direct. It also indicates that the fact that digital preservation is a new challenge, one that presents a great deal of uncertainty, means that there is a profound need for a variety of strategies.

In the public mind, the focus of natural selection tends to be on the selection. A corresponding need for variation is often overlooked. There are several possible reasons for the phenomenal variation that exists in nature, and continues to exist in populations for generation after generation. The principal answer is unpredictability, for it reflects the existence of multiple strategies: diversity in the face of unpredictability. It is, of course, a matter of folk wisdom that one should not put all one's eggs in one basket. In other circumstances it appears that the variation reflects damage, aberration, the need for repair. Either way, the evolutionary quasi-metaphor hints at specific strategies and tactics, and a source of insights for digital repair and for system emulation and replication.

An immediately practical manifestation (not always explicitly understood) is the policy of making duplicate copies, and making them on different batches of a particular type of media such as a CDR. In the case of the WDH archive, an early decision was made to use both optical and magnetic media combined with regular and monitored refreshing of the storage media.[19]

My thoughts went in this direction in the summer of 2000. Recently, this year (2004), I came across a letter written by Bill in 1991, from which the following is extracted:

'For these types of damage an intact homologue can provide a template for repair. This is the aspect that does not require outbreeding: indeed in principle polyteny with inbreeding would seem the better course for maintaining highly faithful replication. It is, of course, the principle which microcomputer users work with their disks: keep multiple copies'.

Conclusion

That Bill's manuscripts are now being sorted and described in the national library seems most apt. He long had an empathy for any publicly spirited source of information, most notably, accessible research libraries, as is plainly evident in both the volumes that preceded the present one.[20] WDH repeatedly referred in the same breath to libraries and fields and woods as sources of information. That this outlook was not contrived for his autobiographical sketches can be demonstrated by recalling that WDH was as likely to be taking his undergraduate and postgraduate students to the many second-hand bookshops of far off Hay-on-Wye as showing them the wonders of the bark of dead trees in Wytham Woods or the sphagnum of Tregaron Bog. Further evidence of Bill's psychological juxtaposition of bibliospace and biosphere is provided by the following letter:

'Now for the exciting new addition to my menagerie of social adaptations, I think I may have found something close to being another eusocial insect—rather in the line of Aoki's aphids—in a non-hymenopteran haplodiploid group. I found it in a library (almost needless to say at this time of year), rather than in the field . . .'

WDH, letter 1978.[21]

Although WDH obtained his information primarily through reading and field observation, he had considerable admiration for the deftly manipulative touch in experimentalists, and he was not above a little intervention himself. It was not uncommon to be sitting in a comfortable chair reading in his house, and to see in the periphery of one's vision a beetle scurrying by with an identifying spot painted on its dorsal surface. Still, Bill was no avid practitioner of the experimental approach; clearly his liking was for gleaning information from the field or from scientific books and periodicals.

One feels that WDH certainly understood the value of unpublished information too (ideas, data, creative writing, and records of happenings)—indeed his fairly careful retention of papers and photographs points to this view.[22] His discussion in Volume 2 of Edward Adrian Wilson, the ornithologist and polar explorer, reflects a longing for more information about this scientist's unpublished thoughts.[23] It is also apparent in the sharing of information and ideas displayed in his correspondence.

Not unexpectedly, the process of sorting out Bill's collection—replete with unpublished happenstance, contemplation and erudition—is proving to be an education in science, in social and scientific history, in the multifarious nature of modern manuscripts, and in life itself. In his archive of hoarded and marvellously diverse textual and figurative sources held in both paper and digital media, WDH has left future generations with much to

Figure 19.12. Chronological Sequencing

Pocket diaries, passports (right), receipts (left) and a self-made calendar roll (top) opened to show dates of travels. Tickets (e.g. for international flights) are also an important resource for chronological information.

think about, much to explore, much to absorb. Without doubt, this collection is destined to be one of enormous importance, due to its totality, quality and multiplicity.

Of course, no collection exists in solitude, no collection can shed light on all matters, and so equally important is the need to ensure that different collections speak to each other; in this way, one can fill gaps left in one collection and can see the same historic events recorded in different ways. One of the long-term outcomes I most look forward to seeing is a comprehensive chronological map of WDH's travels and correspondence within countries and across the world in his search for crucial information from libraries, colleagues and nature herself, joined in time by matching maps for other scientists.

With a legacy of scientific publications and autobiographical sketches, and despite an untimely death, Bill achieved more during his lifetime than most of us can hope to emulate. But even beyond his life he left an archive of unpublished material that will stimulate for decades to come. He has left some of the best until last.

Acknowledgements

So many of Bill's family and friends have helped in so many ways that it is difficult to compile a list of their names that would actually be read; but Christine Hamilton, Mary Bliss, Luisa Bozzi, and Janet Hamilton have helped me enormously, not only with information, but also with their kind and warm encouragement. Any repeated biographical and historical returning to the doings and writings of a loved one is bound to be tinged with great sadness as well as the pleasure of memory. Each of them has responded to my enquiries with unfailing geniality and dignity. These four people, together with the whole association of family, friends and close collegues, have made my work feel even more worthwhile. As many readers will know, Luisa died with desolating suddenness in May 2004. I would be pleased if this article would be a remembering tribute to her support and aspirations for the archive. I would also like to thank the organisers of and various donors to the W. D. Hamilton Memorial Fund that made possible the initiation of this archival project. The following people have graciously read various drafts: Mary Bliss, Christine Hamilton, Janet Hamilton, Elaine Ireland, Katja Küll, Christiane Ohland, and Richard Ranft. I am particularly grateful to Julia Sheppard of the Wellcome Library for her thoughtful comments. Mark Ridley's influence has been a lesson in that editorial art of encouragement, patience and gentle prodding; he also promoted the inclusion of photographs. I would also like to record my profound gratitude to the British Library[24] for enabling me to work with the W. D. Hamilton collection. For archival insights, I have enjoyed discussions over the years with Frances Harris, Rachel Stockdale and Anne Summers. Along with Christopher Wright, now Keeper of Manuscripts at the British Library, they have welcomed into their midst the scientist as curator with warmth and friendliness.

References

1. The author was a DPhil student of Professor William D. Hamilton during the 1990s. Some of the resulting work was discussed in Volume 2.
2. To avoid repeatedly referring to Professor Bill Hamilton in the same way, I frequently adopt the more familiar forms of 'WDH' and 'Bill'.
3. Hamilton, W. D. *Narrow Roads of Gene Land. The Collected Papers of W. D. Hamilton. Volume 1. Evolution of Social Behaviour.* (Spektrum, W. H. Freeman, Oxford, 1996) page 229.
4. Officially, 'W. D. Hamilton Project' at the British Library.
5. The preparation of the archive is still in process and much remains to be done.
6. Recall, for instance, that Bill Hamilton in his last years was still intellectually active, and very much so, producing at least two innovative ideas for testing, on autumn colours and on Gaia theory.
7. With new ways of doing academic research emerging rapidly and with even the entirely paperless office already existing in the commercial world—a more detailed approach to preserving the origin of materials was justified. Knowing where individual items come from is also important in helping to work out the circumstances and context of the manuscripts. It should be noted, however, that before the British Library was officially involved, the various rooms had been visited, and so the contents were not *exactly* as they were when Professor Hamilton left for the Congo.
8. For further details, see Summers, Anne and Jeremy Leighton John The W. D. Hamilton Archive at the British Library. *Ethology, Ecology & Evolution* **13**, 373–384 (2001).
9. An example is discussed on page 84 in Hunter, Michael *The British Library and the Library of John Evelyn*, pages 82–102, in *John Evelyn in The British Library* (British Library, London, 1995). See also Harris, Frances and Michael Hunter (ed.) *John Evelyn and his Milieu.* (British Library, London 2003).
10. For example, page 119 in Mandelbrote, Scott *Footprints of the Lion: Isaac Newton at Work*, (Cambridge University Library, Cambridge, 2001).
11. Consult pages 121–122 (see also pp 152–153) in Henig, Robin Marant *A Monk and Two Peas. The Story of Gregor Mendel and the Discovery of Genetics* (Weidenfeld & Nicolson, London, 2000).
12. See pages 42 & 111 in Mandelbrote, Scott *Footprints of the Lion: Isaac Newton at Work*, (Cambridge University Library, Cambridge, 2001).
13. Ford, B. J. *Hidden Secrets in the Royal Society Archive*, pp 27–42 in *Biological Collections and Biodiversity*, B. S. Rushton, P. Hackney and C. R. Tyrie (ed.

joint symposium of the Linnean Society of London and the Royal Horticultural Society held at the Ulster Museum, Belfast, (Westbury, Otley, 2001).

14. See Fig. 19.3, for example. I recently learned that Winston Churchill also preferred documents to be punched and tagged before he read them—indeed insisted on it; see page 23 in IWM *The Cabinet War Rooms*. A guide; with an Introduction by Philip Reed. (The Imperial War Museum, London, 2003).
15. A simple example of observations that can at least hint at the chronological sequencing of undated notes, is the tendency for WDH's writing to be on yellow paper, US Legal size, when derived from his time in the USA, and on greying cream paper when from his time in the UK, foolscap becoming A4.
16. Among the most celebrated are the lists of Sir Hans Sloane PRS, whose collections along with those of Sir Robert Bruce Cotton and Robert and Edward Harley (the 1st and 2nd Earls of Oxford) founded the British Library's world-renowned manuscript collections.
17. For a brief elaboration see Summers, Anne and Jeremy Leighton John *The W. D. Hamilton Archive at the British Library*. *Ethology, Ecology & Evolution* **13**, 373–384 (2001).
18. Extract from air mail letter (aerogramme) by Dr A. M. Hamilton (Bettina) to WDH: 'Home, 2.5.64 [.] Dear Bill, Even after all this time I am not sure how long airmail takes to reach you. Yours to us seem very variable. But I think this one is pretty certain to be in Rio Claro before you leave on the 21st... What a stupendous trip that is going to be. Fairly simple you say as far as Brasilia... And after Brazilia [sic] the jungle, at least it seems you intend to go & look for it... So I'd advise you to use that £200 Dad sent out after all [.] I thought I'd persuaded him to wait till you asked for it but he said no, it must be there if he wants it. He's worth an awful lot more to us than that so he must have a 100% safe motor car. So please see that it is & that you have all the spares you can possibly need on the road'. The reader may like to turn to Fig. 19.10, which shows WDH's diary entry for 17th May 1964, just over two weeks later.
19. Systems for storing and managing digital objects are continuously being expanded and developed by the British Library and its partners.
20. See pages 139–141 in Volume 1 (Hamilton, W. D. *Narrow Roads of Gene Land. The Collected Papers of W. D. Hamilton. Volume 1. Evolution of Social Behaviour.* (Spektrum, W. H. Freeman, Oxford, 1996)), and page 8 in Volume 2 (Hamilton, W. D. *Narrow Roads of Gene Land. The Collected Papers of W. D. Hamilton. Volume 2. Evolution of Sex.* (Oxford University Press, Oxford, 2001)).
21. Of course, this quotation also provides an example of communication of ideas and thoughts in correspondence.

22. A manuscript entitled 'The nature of isolation' has, in addition to early amendments and notes, occasional yellow Post-it notes attached, with much more recent remarks by Professor W. D. Hamilton such as 'I don't think I'd be ready to use this analogy now . . .'. It is worth remembering that WDH would have consulted his early papers when preparing his essays for Volumes 1 and 2.
23. Hamilton, W. D. *Narrow Roads of Gene Land. The Collected Papers of W. D. Hamilton. Volume 2. Evolution of Sex.* (Oxford University Press, Oxford, 2001).
24. The observations made and opinions expressed in this chapter are those of the author, and do not necessarily represent the conclusions or policy of the British Library.

Illustrations: origins and copyright

I am further indebted to Dr Christine A. Hamilton for her gracious permission to reproduce wholly or partly the materials in Figures 19.1 to 19.6 and in Figures 19.8 to 19.12 Dr Janet M. I. Hamilton kindly granted permission to copy the photograph of Bill and Christine Hamilton in Figure 19.7. This photograph was taken either by Janet Hamilton herself or by her (and Bill's) brother Alex, who died as a very young man not long afterwards. This image and those found in Figures 19.5 and 19.10 were reproduced using a scanner. The photographs of Figures 19.1 to 19.4, 19.6, 19.8, 19.9, 19.11 and 19.12 were taken with a British Library digital camera by the author—myself—and any pretensions to photographic copyright I relinquish (with thanks) to the British Library. Accordingly, copyright is asserted for all the illustrations in this chapter on behalf of the Estate of W. D. Hamilton, individual members of the family of W. D. Hamilton, and the British Library Board is appropriate.

CHAPTER 20

WILLIAM DONALD HAMILTON

BY ALAN GRAFEN

Introduction

William Donald Hamilton was born in 1936 in Cairo to New Zealander parents, and was brought up for the most part in a rural and wooded part of Kent, England. He described his childhood as idyllic, full of freedom to roam, and of maternal inspiration and encouragement, and himself as a great burrower. He was fascinated by insects from an early age. A great-aunt gave him her insect collection, whose cases he used for his own (later ruing his discarding of the insects themselves), and also lent him a translation from Fabre, the great French naturalist and one of the first to study behaviour scientifically. A birthday present from his parents was a much-coveted copy of E. B. Ford's *Butterflies* in the Collins' New Naturalist Series, which introduced the 12-year old to genetics, to a scientific snobbery that looked down on 'mere collecting', to mathematical biology in the shape of Mendelian segregation ratios, and to the modern study of Evolution. After reading Ford, he asked for a copy of Darwin's *Origin of Species* as a school prize. To have inspired this one young biologist would by itself justify Ford's efforts in writing *Butterflies*.

Another childhood interest was bombs. Around the time of *Butterflies*, he unearthed cases of materials belonging to his father, connected with wartime research on grenades, which had been hidden in a rabbit hole for safety. Retrieving them from a further hiding place, the young Hamilton proceeded to cause a near-fatal explosion. A thoracotomy in King's College Hospital saved his life, but some fingers were shortened, and brass remained implanted in his chest.

Hamilton itched to travel, and did visit France after staying on at Tonbridge School for an extra term to take Cambridge entrance. Before university were two years of compulsory National Service in 1955–1957 during which his early bomb injuries prevented an overseas posting Ironically, he was commissioned as a drafting officer in the Corps of Roya Engineers, sending others abroad. Bill once told me a revealing story about a formal dinner. He broke a rule that no-one must leave before the most senior officer present, and escaped to the countryside, pursued by a search party and succeeded in his aim of remaining free until he could give himself up in the morning light. It is a characteristic Bill who found a formal social even tiresome, and fled to huddle under bushes for the night, close to the natura world he loved, hiding from the agents of authority, playing and winning a game of his own making, for his own satisfaction.

Primed in many ways for biological study, particularly study combining mathematics, genetics and natural selection, Hamilton went up to St John' College, Cambridge in 1957 with a State Scholarship to study Natura Sciences.

Cambridge, Fisher and Theoretical Biology

Cambridge was a frustrating place for Hamilton, as the undergraduate course was old-fashioned in its approach to natural selection, and contained littl mathematical biology. Hamilton escaped the confines of his teachers and discovered in his college library the book that would set his course in som detail until the mid-1970s, and its general direction for a lifetime, *Th Genetical Theory of Natural Selection* by R. A. Fisher. Then, to his surprise he found Sir Ronald himself, still in post as Arthur Balfour Professor c Genetics.

To appreciate the influence of the book, read Hamilton's endorsemer on the back cover of the 1999 Variorum edition of the *Genetical Theor* written at a time when his 'ultimate graduation' was tragically and unfor eseeably near:

> This is a book which, as a student, I weighed as of equal importance to the enti rest of my undergraduate Cambridge BA course and, through the time I spent on i

I think it notched down my degree. Most chapters took me weeks, some months; even Kafka whom I read at the same time couldn't depress me like Fisher could on, say, the subject of charity, nor excite me like his theory of civilisation. Terrify was even the word in some topics and it still is, so deep has been the change from all I was thinking before. And little modified even by molecular genetics, Fisher's logic and ideas still underpin most of the ever broadening paths by which Darwinism continues its invasion of human thought.

...

Unlike in 1958, natural selection has become part of the syllabus of our intellectual life and the topic is certainly included in every decent course in biology. By the time of my ultimate graduation, will I have understood all that is true in this book, and will I get a First? I doubt it. In some ways some of us have overtaken Fisher; in many, however, this brilliant, daring man is still far in front.

The pages of the *Genetical Theory* do take weeks and months to work through, and biology is richer for every student who takes the trouble. Hamilton's investment paid off a hundredfold, as we shall see. What might not be obvious today is how little influence the *Genetical Theory* had at that time (Edwards 2000). It was disregarded by Hamilton's teachers who, amazingly as it now seems, viewed Fisher as only a statistician, lacking standing in biology.

Much time at Cambridge was spent not on the formal teaching, but on his own work, and he spent his third year attached to the Department of Genetics, in which the retired Fisher still reigned before the appointment of his successor. Hamilton reports getting on well with Fisher. His own work seems to have been in large part a 'theory of ethics', of which I can find no details, based on new understandings gleaned from Fisher's book, and that may have been a forerunner of his work on altruism. Hamilton was a prolific postcard writer, frequently cross-writing in different colours. A card sent to his sister Mary, now Mary Bliss, suggests that he had worked out at least part of the sex ratio theory of his 1967 paper (5)* in February 1960, a few months before his final examinations. He had been assigned to help A. W. F. Edwards, then a graduate student of Fisher's and later his distinguished successor, with an experiment intended to test Fisher's sex-ratio theory, and this may have provided the stimulus for his own work.

Hamilton continued in that postcard, 'I begin to think that my ambition to be a theoretical biologist can be more than a dream in spite of my poor mathematical ability'. Such percipience and self-knowledge are remarkable

* Numbers in this form refer to the bibliography at the end of the text.

in an undergraduate. As we shall see, Hamilton's achievements with his 'poor mathematical ability' put to shame those of us who practise biology with greater mathematical skills but to lesser biological effect.

Inclusive Fitness

Hamilton's graduate student life is the period of his greatest scientific work. Rebuffed nearly everywhere he applied for his topic of genetics and altruism, he eventually enrolled for an MSc in Human Demography at the LSE, and was initially supervised for research work by Norman Carrier, who crucially also secured for him a Leverhulme Research Studentship for one year, and then a Medical Research Council Scholarship. He transferred as the work became more mathematical to John Hajnal of the LSE, and as it became more genetical, to a joint supervision by Cedric Smith of UCL. In *Narrow Roads of Gene Land I* (14 page 4), Hamilton reports a general suspicion in the institutions to which he belonged that he might have been 'a sinister new sucker budding from the recently felled tree of Fascism' simply through using words like 'gene' and 'behaviour' in the same sentence.

A vital influence for the project was an appreciation of Fisher's 'Fundamental Theorem of Natural Selection', which more or less states that natural selection should result in individuals that are well designed to produce as many offspring as possible given all the circumstances of their lives. The power and generality of the theorem greatly impressed Hamilton, but there were difficulties in applying it to social behaviour. Darwin had noticed in the *Origin* the difficulty in explaining the evolution of honey-bee worker's structures and habits (which incidentally to Darwin's point, but germane to Hamilton's, are social traits) through natural selection. The derivation of the fundamental theorem assumed that an individual's number of offspring depended on its own genotype, and not on the genotypes of others. There was fundamental work to do to incorporate social behaviour into the best contemporary Darwinian theory, and Hamilton took up that challenge.

Hamilton's situation may be hard to understand for a modern biologist because Fisher's result is not widely seen today as very interesting or important. Indeed, the situation was probably rather similar in 1960. Perhaps only Fisher and Hamilton, who so far as we know never discussed it, viewed the Fundamental Theorem as 'holding the supreme position among the biological sciences' (Fisher 1930, page 37).

The initial idea was that sharing of genes altered calculations. Selection could quite easily favour helping siblings, but less often second cousins. Hamilton built up many models of special cases, each with a different genealogical link between actor and recipient. These were unsatisfactory because clearly there was some more general phenomenon going on than these scattered separate cases, particularly as in each of them the gene frequency more or less magically dropped out from the condition for spread. More deeply, they could not aspire to provide the generalization of Fisher's fundamental theorem that was so necessary on conceptual grounds. It would be deeply unsatisfying to have a correct theory of altruism, which would disprove Fisher's fundamental theorem and throw evolutionary biology (at least as understood by Hamilton and Fisher) back into disorder. It is important to understand that references to 'the classical theory' both in the 1964 paper (4) and in *Narrow Roads of Gene Land I* (14, for example on page 27) are in fact references to the fundamental theorem, perhaps so much taken for granted that it need not be given a name, perhaps strategically veiled to deflect the reader from recalling the attacks on the theorem's truth.

At some point, Hamilton saw how to produce a general model. It employed Wright's coefficient of relatedness, which was a correlational measure of closeness of kinship. More significantly, the model encompassed a broad range of kinds of social interaction, and involved a maximization principle. Thus, essentially the whole of social behaviour had been embraced by a generalization of Fisher's Fundamental Theorem. According to that theorem, individuals should be expected to maximize their reproductive success, or 'fitness'. In Hamilton's model, the quantity that individuals were expected to act as if maximizing was named 'inclusive fitness'.

Inclusive fitness has been much admired but also much misunderstood (Grafen 1982). It was, in the words of his 1964 paper (4), 'its production of adult offspring . . . stripped of all components which can be considered as due to an individual's social environment . . . then augmented by certain fractions of the harm and benefit which the individual himself causes to the fitnesses of his neighbours. The fractions in question are simply the coefficients of relationship.' Mathematically, we assume an additive representation such that the number of offspring of individual i is the sum of contributions a_{ij} made by individuals j, formally $1 + \sum_j a_{ij}$. The inclusive fitness is then defined as $1 + \sum_i r_{ij} a_{ij}$, where r_{ij} is the relatedness between individuals i and j. The conceptual transformation is that number of offspring contains all the offspring an individual has, while inclusive fitness contains those offspring an individual causes to exist. Inclusive fitness accounts offspring by causation and not by parenthood.

The fundamental theorem provided a mathematical and conceptual underpinning for Darwinian natural selection and clarified what 'fitness' was. Inclusive fitness, erected on the theorem, went further and extended the very concept of natural selection, so that it now satisfactorily accounted for the worker honey bee that had puzzled Darwin, and stood ready for the assault on understanding social behaviour in general, which continues today.

There are many remarkable aspects of the papers reporting this work. A short *American Naturalist* (3) paper was published first, but written second. The paper submitted to *Journal of Theoretical Biology* (4) was split into two on the advice of a referee, but Hamilton consistently referred to them as Part 1 and Part 2 of a single paper. The short paper is a lucid verbal explanation of what has become known as 'Hamilton's Rule', with some applications, and it also offers the gene-centred view of selection that Hamilton always viewed as an informal aid to thinking requiring the backing of a population genetics model. In this case, the backing was in the 1964 paper (4). This gene-centred view was later developed by Dawkins (1976, 1982) into his powerful conceptual unification of Darwinian theories.

Part 1 contains the model, set perfectly in context, with special cases worked out, objections to the model raised and discussed, and the conceptual implications thought through. This part has an undeserved reputation for obscurity. The main message and content are lucid when read today, and the notation and mathematical argument present no serious obstacle for those with mathematical training.

Following the theoretical triumph of Part 1, Part 2 is an extraordinary *tour de force* of synthesis. Hamilton shows, across the range of biology, how his theory transforms the study of social traits. His logical application of the ideas combined with his deep immersion in biology led to many discussions that today look highly prophetic. The early standard examples are all given, such as distastefulness in insects, alarm calls, mutual preening and grooming. He outlines the logic of kin recognition, today a whole subject of its own, quite extensively. He discusses multicellularity in terms of the relatedness of the cells, and collaboration in other kinds of colonies, and he solves what would now be called games between relatives. Moving on to social insects, he first discusses the famous 3/4 principle, which is the only argument many later commenters noticed. It was the social insects to which his theory had given the first evolutionary key, as most of their significant behaviour is clearly social, and he gives accordingly an in-depth discussion encompassing multiple mating, multiple insemination, pleometrosis, aggression in relation to sterility, nest usurpation, and the origins of eusociality. He also discusses the possibility that females should mate only once in order to reduce the conflict between their offspring. These cases are not merely mentioned. Rather they are discussed with relevant facts, and considering apparent exceptions. The final section is on anomalies to the theory as a whole.

The extraordinary coherent power of the theory is nowhere more apparent than in his taxonomically-distant analogy between ant queens collaborating to build a nest, and sporelings of a branching red alga, as competing and potentially related. Only a strong theory can abstract from all the details of ant biology and phycology in this way, but Hamilton was clearly in no doubt about the scale of his *magnum opus*.

It is worth pausing to see Hamilton's methods, for they vary remarkably little throughout his career. The problem must be evolutionary, and the formal solution is a population genetics model. Very extensive reading of relevant literatures was a preparation. The range of facts, in terms both of species and type, is wide, and most of them were not collected for Hamilton's purpose. Thus the facts often require sophisticated handling and defy easy interpretation, closer to history or astronomy, indeed to Darwin's own arguments, than to the more straightforward experimental methodology that became more and more dominant during the twentieth century. A crucial role is usually played by anthropomorphic thinking. He would often make remarks like 'Now if I were the Ebola virus . . .', going on to explain a cunning strategy by which it could increase its spread. The problem itself is usually felt important by Hamilton because of a perceived hiatus between theory and facts, and it was his utter confidence that one should be able to explain the other that drove him forward to seek solutions.

Just how revolutionary was this work? There were frequent accusations of lack of originality, which were found hurtful. However, the view cannot be sustained that Hamilton's work was merely an elaboration of an idea that should be credited to Haldane or Fisher. They had both previously published some of the ingredients, but neither had seen or even partly understood the magnitude and significance of the problem Hamilton identified. Fisher never pointed out that his argument about aposematism contradicted the assumptions of his fundamental theorem. Haldane published his 24-line passage in a paper (Haldane 1955) in a semi-popular journal. Neither indicated that there was a general problem about how natural selection acted on social behaviour, and so naturally neither claimed to be tackling it. Neither gave the impression they attached more importance to these arguments than to the others on adjacent pages. In the large, then, there is no question of pre-emption. Even with a more detailed point, it is easy to overvalue the earlier work. A modern audience reads Haldane's use of the chances of sharing a gene and, being familiar with inclusive fitness, immediately thinks of the coefficient of relatedness, and is then tempted to credit Haldane with the importance of relatedness in social behaviour. But there is no evidence Haldane ever made that leap from gene sharing with particular relatives to relatedness in general. There was, however, a definite moment at which Hamilton realised the significance of the coefficient of relatedness, and found that, in place of separate models making special assumptions, he could instead construct a single general model in which relatedness played a crucial role. Finally, the idea that individuals acted as if maximizing some quantity that generalized Darwinian fitness was not even hinted at by Haldane or Fisher. Indeed so novel was this idea that I know of no passage in their work in which such a hint could even have found a natural place. Yet it is precisely this aspect of inclusive fitness that forms the centre-piece of Hamilton's achievement.

Thus inclusive fitness was a major conceptual advance in biology, wholly original with Hamilton. We have seen how Hamilton viewed the problem and worked on it. We turn to how other biologists reacted to this masterly reshaping of social biology.

The Reception of Inclusive Fitness

One strand in the reception of inclusive fitness was very positive. Whole areas of biology now depend on it, and it has survived essentially unchallenged as the evolutionary theory underlying social behaviour. Hamilton received many honours and awards in recognition of his achievement. Another strand, located in the population genetics tradition, has been negative and dismissive, and sometimes worse. These strands have interplayed in important ways.

It has often been remarked that Hamilton's 1964 paper are much more frequently cited than read. Four influential books seem to have been mainly responsible for publicizing inclusive fitness and conveying its message to biologists. There is a long and serious exposition and discussion on pages 328–334 in *The Insect Societies* (Wilson 1971), which took it as given that inclusive fitness theory is right as a general theory of social behaviour, and then focused on whether this theory can account for the greater incidence of eusociality in the Hymenoptera. (Incidentally, Wilson anticipated that further work would suggest it can't, still a respectable point of view.) Also in 1971, the collection of papers *Group Selection*, edited by G. C. Williams, reprinted the 1964 papers, making them more available and drawing attention to their relevance. In 1975, Wilson published the widely read and highly controversial *Sociobiology*, and mentions inclusive fitness briefly. By 1976, Dawkins could still refer in his powerful and influential *The Selfish Gene* to the papers as 'among the most important contributions to social ethology ever written' and yet 'so neglected by ethologists'. It was *The Selfish Gene* that synthesized a unified evolutionary understanding from inclusive fitness and other current ideas. The history of this fascinating period is expertly illuminated and well documented by Segerstråle (2000), who is currently writing Hamilton's biography (Segerstråle, forthcoming).

Generations of biologists have learnt about inclusive fitness through secondary sources, and today, textbooks in animal behaviour, behavioural ecology and evolution introduce inclusive fitness as an uncontroversial central principle. In succeeding decades since publication, the 1964 papers have received an annual average citation rate of less than ten (1964–73), about 60 (1974–83), about 100 (1984–93) and about 160 (1994–2002). Their total citations between 1981 and 2002 are greater than those of comparably iconic biological works (all books) of the same era such as Williams (1966), Macarthur and Wilson (1967), Wilson (1971) and Lack (1966, 1968). Inclusive fitness became a major, dominating theme in biology within 15 years of its publication.

Honours and prizes were important to Hamilton, as recognition that inclusive fitness really was accepted, and for the opportunity to travel that some prizes offered. His medals include the Scientific Medal of the Zoological Society of London (1975), the Darwin Medal of the Royal Society (1988), the Scientific Medal of the Linnean Society (1989), and the Frink Medal of the Zoological Society of London (1991). His memberships and fellowships of academies, sometimes Foreign or Corresponding memberships, include the American Academy of Arts and Sciences (1978), Royal

Society (1980), Royal Society of Sciences of Uppsala (1987), Brasilian Academy of Sciences (1993), Academy of Finland (1997) and American Philosophical Society (1999). Major Prizes included the Albert Wander Prize (1992), Crafoord Prize (1993), Kyoto Prize (1993), and Fyssen Prize (1996). The Crafoord is the closest biology comes to the Nobel: the Swedish Royal Academy awards a biological prize (frequently shared, in Hamilton's case with Seymour Benzer) usually every three years. The transformation of biology for which Hamilton was responsible was fully recognized by the scientific establishments of the world.

Such was Hamilton's standing by the time of his death that his papers were taken to join the unsurpassed collection of manuscripts of the British Library, most fittingly including those of Darwin and Wallace. Over 200 boxes of Hamilton's materials are in the process of being catalogued. Many of Hamilton's 'papers' are unpublished computer documents (now termed eMSS), bearing witness to the use of computers in science from near the beginning in the 1960s up to 2000. The library is devising new ways of archiving, most especially concerning how to capture, preserve and make available the information on 80 column punch cards, paper tape, and floppy and hard disks of many generations. Hamilton is as path-breaking in death as in life.

Against this background, the existence of the strand that never accepted inclusive fitness looks very puzzling, but is readily understandable in its historical context. Inclusive fitness theory provided a maximizing principle namely that natural selection causes organisms to act so as to maximize their inclusive fitness. It was a generalization of Fisher's Fundamental Theorem of Natural Selection.

The reception of Fisher's theorem is described by Edwards (1994). It had mainly been ignored, but at just the time Hamilton was developing and extending it, population geneticists were beginning to pay attention to it and they found it wanting. Failing to understand the derivation, they also misunderstood the statement of the theorem, taking it to imply that mean fitness must always increase. A succession of models then proved that mean fitness does not always increase. It was conclusively established that population genetic systems did not in general *have* a quantity that would always be maximized.

Believing the fundamental theorem to be false in a simple situation without social interactions, population geneticists had little patience with inclusive fitness, which claimed to establish a maximization principle in the presence of social interactions. There are two recent lines of wor hat may retrospectively unpick this negative reception and finally convince opulation geneticists. The fundamental theorem itself is now understood much better (Price 1972a; Ewens 1989; Edwards 1994): Fisher's lerivation was intelligible and correct after all, and crucially the theorem loes not imply that mean fitness always increases. Thus, population genetics egains a theorem and loses an Aunt Sally. And I myself have undertaken a roject to represent in formal mathematical terms the link between population genetics and fitness maximization principles (Grafen 1999, 2000, 002).

Based though it may have been on misunderstandings, the presence of this egative strand has been important. Lacking support from mathematical opulation geneticists, Hamilton's derivation of inclusive fitness has been atronized and overlooked. There are many rederivations of inclusive fitess, presented with the clear implication that the original derivation was omehow dubious. A curious literature results, in which inferior derivations re presented as improvements, and as avoiding special assumptions that vere never made in the first place. The 1964 paper acknowledged all the imitations of the analysis presented, and gave excellent justifications for he assumptions made.

The first derivation to improve on the original is Hamilton's own rederivation of 1970 (6), inspired by the covariance selection mathematics of rice (1970, 1972b; and see Frank 1995). George Price himself was a emarkable figure, and Hamilton's involvement with him is described in *Narrow Roads of Gene Land I* (14) and by Schwartz (2000), who shows that rice had the central insight that led to the rederivation, but turned down he co-authorship Hamilton offered. The only further paper to improve on ither of Hamilton's derivations is Taylor (1990, see also Taylor 1996), vhose brief appendix provides a still somewhat cryptic expansion of the 970 argument. It is utterly remarkable that the approaches of Hamilton's wo derivations have been so neglected, while the concept of inclusive tness has become so pervasive in biology and elsewhere, and won such enown and recognition for its inventor.

The lack of support from population geneticists for inclusive fitness also ed to an overemphasis on the informal arguments that make Hamilton's ule, a simple and convenient version of the inclusive fitness conclusion, so lausible. The rule is that if a social action adds b to the recipient's number f offspring, and subtracts c from the actor's, then it will be favoured by election if $rb - c > 0$, where r is the coefficient of relatedness. Interpreting r as the fraction of genes shared between actor and recipient, *rb–c* is simply the net effect of the action on the number of copies in the next generation of an allele in the actor.

This argument is so simple and appealing that it became the justification for inclusive fitness to generations of biologists. It brings to mind T. H. Huxley's response to natural selection itself ('How stupid not to have thought of that'), though, tellingly, Darwin expressed privately the view that Huxley didn't actually explain it very well (Burkhardt *et al.* 1993, page 84) and so perhaps didn't understand it very well either. In fact, the real argument for inclusive fitness is more complex, as the original papers make perfectly clear. Hamilton himself developed methods further in the 1970 paper already referred to (6), in his 1975 paper (7), and in his 1980 paper (26) with Richard Michod on coefficients of relatedness.

Even though most biological students of social behaviour fully accepted inclusive fitness eventually, typically for Hamilton's work it took about fifteen years to become mainstream, having been regarded initially almost as crackpot science. The consensus among practitioners that is such an important element in the workings of science does act against new and original ideas, and we will see that Hamilton spent his later years trying to encourage more open-mindedness. Radical ideas are hard to assimilate, however elegantly or clearly they are expressed, because readers start from what they (think they) know. Now, most challenges to orthodoxy are not works of genius, but trivial mistakes or confusions in intellectual terms, and often self-promoting works in careerist terms. They are appropriately resisted by most scientists. However, the resistance appropriate to the majority seems, in retrospect, scandalous when applied to the few ultimately successful challenges. Hamilton's prophet-like status persisted throughout his working life, so that his current work, whatever it may have been, was frequently regarded as odd and obscure and often rather embarrassing. Although he did complain, for example in *Narrow Roads of Gene Land I* (15, page 18), about lack of interest in his present as against his well established ideas, it must be said that the position of dissident seemed quite natural to him.

Inclusive fitness has had far-reaching effects in biology, reflecting its standing as the only significant extension to Darwinism of the twentieth century. The concepts of inclusive fitness, relatedness, altruism, spite and selfishness are firmly embedded in biology, and will remain central to the study of social behaviour.

Completing a Unified Theoretical Framework for Social Behaviour

We return to 1963, after the inclusive fitness work had been done, but before all the papers had been published. No one else appreciated the significance of Hamilton's work, and there was little sign they would. A new side of his life began with his visit to Brasil, to work with Dr W. E. Kerr in Rio Claro. His purpose was to study social insects further, with a view to collecting data relevant to the predictions of this new theory, and indeed some made it into the 1964 paper. On the day of publication of the 1964 papers, he was on the road from Brasilia to Belém, on his way to Canada and then Britain.

This section will look at Hamilton's work after the 1964 papers, and before the sex project that occupied, roughly, the 1980s. Hamilton began this period as an unknown and unappreciated graduate student, and ended it with considerable fame. He took up a lectureship at Imperial College in 1964, and was based at the field station at Silwood Park. In 1967, he married Christine Ann Friess, who was training as a dentist. Bill and Christine had three daughters. Helen is now a senior ecologist for an environmental firm, Ruth has recently completed a Ph.D. in parasitology at Cambridge, and Rowena graduated with a first class degree in Fine Art. Rowena was born in the United States, after Bill had left Silwood Park in 1977, spent time as a Visiting Professor in Harvard and then joined the University of Michigan as a Professor in 1978, persuaded there by one of the leading American champions of inclusive fitness in animal and human affairs, Richard Alexander.

The 1963 expedition was the first of many. He returned with Christine to Brasil in 1968 with the Royal Society and Royal Geographical Society expedition to central Brasil. The official report (Smith 1971) has an extraordinary picture of each taken by the other, after being attacked by wasps whose tree they had just felled, as well as a hair-raising account of Bill's driving that was immediately recognizable to one more recent passenger. He later made many further visits to Brasil, particularly the Amazon. The immediate intellectual fruits of these expeditions seem few and minor, but they clearly fed Bill's imagination, and fulfilled a deep need. Perhaps the jungle offered a respite from etiquette and compromise. During his final decade, he was engaged in research on Amazonian plant communities, and was actively involved in the Mamirauá Project based in Tefé. There is now a 'Centro Itinerante de Educação Ambiental e Científica Bill Hamilton' there, a floating school for scientific and envir-onmental education. Hamilton might well, had he lived, have gone on to make significant contributions to plant ecology.

His and others' recollections of expeditions are vivid. Bill prided himself in the friendliest possible way, on his physical fitness and ability to cope Mike Worobey, who accompanied him on the second Congo expedition himself a Judo black belt and fireman, and thirty years younger, describes Bill as the toughest man he's ever known. He also relates that in the Congo, Bill did not purchase a machete from the posh shop with gleaming blades, but in the backstreets. Once in the jungle, the others' tools were soon discarded, and Bill was doing all the hacking away of creeper and undergrowth with a blade built for business not beauty. Bill enjoyed relating how he swam under a hull on the Amazon to plug a hole, knowing full well that his explanation ('the dangers of piranhas were much over-rated') would impress much of his audience. Bill was a self-constructed hero from a boyhood comic, relishing putting his skills and his practical knowledge to the test, and succeeding He had a great love for improvisation and originality in practical and intellectual affairs.

Returning, then, to Silwood Park in 1964, he ruefully described being one of two applicants for this job, and not the first to be offered it. Students routinely complained to the director, first O. W. Richards and later Richard Southwood, asking that in view of the nature of the teaching, the marks from Hamilton's course should not be counted towards their degrees.

Silwood Park was the 'entomological Mecca of Britain' (*Narrow Roads* I p. 92) and Hamilton developed a characteristic niche as a resident genius and was regularly to be found engaged in evolutionary interaction with graduate students and others in the conservatory where morning and afternoon coffee was held. Managing an intellectually brilliant but pedagogically poor lecturer is difficult. Richard Southwood later moved to Oxford, and secured a Royal Society Research Professorship there for Bill—solution unfortunately not available to him while at Silwood.

Scope for recognizing and accommodating exceptional individuals has been diminishing in British universities ever since. Hamilton published relatively few papers, in generally low-status journals, and gained only handful of grants much later in life. Bureaucratic measures of performance are increasingly important and judge the impact of an article only by the journal it is published in. This seriously undervalues radical originality which, though extremely rare, is utterly vital to science. It is disturbing tha a young Bill Hamilton today would likely find an academic career even more difficult to pursue.

Hamilton's general research direction in these Silwood years was to explore the selection of social behaviour. I will mainly consider only four papers that together represent conceptual extension and consolidation of the inclusive fitness approach. They together develop the justification for a regulated anthropomorphism, remove unnecessary assumptions, and uncover theoretical questions that still concern us.

The first, 'Extraordinary Sex Ratios' in 1967 (5), develops a theme of Fisher's, but transforms it into new metal. Fisher's sex ratio conclusion was that investment in the two sexes should be equal at the end of the period of parental investment, but his argument was verbal, and too modern in that it viewed parents as maximizing agents. Shaw and Mohler (1953), Bodmer and Edwards (1960) and Kolman (1960) had provided mathematical versions of Fisher's argument in more conventional terms, while Verner (1965) had made the interesting point that with very small populations, there would be selection for reduced variance.

Hamilton showed his deep understanding of Fisher's argument by exploring the consequences of varying the central assumption of panmixia, whose consequence was that each additional son gained on average the same number of extra grandchildren for the parent. Hamilton considered the case in which the sons of a small number of mothers are competing for the daughters. In this case there are diminishing returns to sons as increasingly they compete among themselves. This variation produces female-biased sex ratios, matching what Hamilton showed was a common pattern in nature. This striking confirmation of a model put theory to an empirical test that was unusually stringent, especially for the 1960s. The resulting literature still provides the most quantitative tests in modern Darwinian biology (Hardy 2002).

What this paper goes on to do is of deeper significance for the whole approach of viewing organisms as maximizing agents. Hamilton did not follow Fisher in using reproductive values, but employed his usual population genetics models, with one locus determining sex ratio in a species with XY sex determination. The key step he did take was to consider the possibilities that this locus might be on either an autosome, or on an X chromosome, or on a Y chromosome. He concluded that different parts of the genome 'wanted' different sex ratios, and so discovered what is now called 'intragenomic conflict' (Haig 1997, Burt and Trivers in preparation). He explained it as resulting from the different relatednesses in the different parts of the genome. Thus while the rest of biology still had no inkling of the importance of relatednesses, or of the formal possibilities of modelling individuals as maximizing agents, Hamilton was continuing to encounter and resolve fundamental questions in his new subject, still wholly his in 1967, and in terms of its basic theoretical framework.

The 'unbeatable strategy' is an equilibrium concept introduced in the paper, a formalization of Fisher's procedure in the sex ratio argument. It is often compared to the ESS, announced four years later by Maynard Smith and Price (1971). Hamilton's concept lacked the final formal separation from dynamics that the ESS achieved, but it explicitly embraced all the rest of the bundle of ideas that launched game theory so productively and spectacularly into biology.

This short paper embodies Hamilton's work of the time: energetic, powerful developments of Fisherian themes, which simultaneously embrace empirical findings much more intimately than Fisher ever could, and lay out theoretical paths for the decades ahead. The logical structure turns out to be an extension of inclusive fitness, in which relatednesses and maximization play central roles.

The 1970 *Nature* paper (6) has already been discussed. It provides a new an more general derivation of inclusive fitness using Price's covariance selectio mathematics, and introduces 'regression coefficients of relatedness' instead o the assumption of weak selection. Along with the 1971 addendum to the reprinting of the papers in *Group Selection* (Williams 1971), it shows a furthe characteristic Hamiltonian trait. In the 1964 paper (4), Hamilton had made one outright mistake (on sex ratio in haplodiploids under maternal control) and one unnecessary approximation (Wright's coefficient of relatedness give the chance of gene sharing only under weak selection). While many author claim to improve or correct when nothing of the kind is required, it wa almost always Hamilton himself who tackled his genuine mistakes. There is a distinct note of regret in discovering that the golden ratio does not after al play a central role in haplodiploid sex ratios, but the introduction of regressio coefficients of relatedness has proved of vital significance.

The third paper in the theoretical core of this period is 'Innate Socia Aptitudes of Man' (7) from 1975, and is in part a homage to George Price who had died by his own hand in January of that year. In technical terms, hierarchical expansion of the Price equation allowed the extension c inclusive fitness to structured populations. This permitted inclusive fitnes to embrace and reinterpret existing theories of group selection, and mor generally exhibited the power of inclusive fitness to apply when panmixia failed and, by extension, to apply to real population structures. Hamilton chose to employ this theory in relation to the evolution of humans, making an implicit claim that the new extension of natural selection allowed biology to explain the essence of humanness. This radical suggestion met with extreme rebuttals from the distinguished anthropologist Sherwood Washburn, among others. Hamilton enjoyed and adopted Robert Trivers' reference to this as his 'fascist paper', as a mockery of one school of objections to it.

The final core paper is a joint publication with the political scientist Robert Axelrod in 1981, 'The evolution of cooperation' (1). Hamilton believed that cooperation, along with the existing work on relatedness and population structure, completed the selective forces relevant to social behaviour. The Iterated Prisoner's Dilemma is defended as a paradigm of social interaction, and previous findings of Axelrod (1980a,b; 1984) about Tit-for-Tat are extended using the logic of evolutionarily stable strategies. The key evolutionary ideas are that how well a strategy does against itself is centrally important (following the ideas of ESS); and that relatedness between interactants can increase the effective concentration of a new mutant strategy, so that the initial hurdle of meeting mainly an incumbent type can be easier to overcome. These central points hold, despite technical difficulties with the precise conclusions of the paper. Hamilton discussed in *Narrow Roads of Gene Land II* (16, page 132) the fact that, contrary to the paper, there is no ESS in the Iterated Prisoner's Dilemma (see Lorberbaum 1994), and there remain various serious technical issues surrounding the game in general. But whether these are ultimately resolved by fastening on a new equilibrium concept, or by studying a perhaps more realistic game with fewer technical ambivalences, the key points of this paper will undoubtedly remain.

There are two more papers from this period that should be mentioned, because of their great influence in ecology. 'Geometry for the Selfish Herd' (6) and 'Dispersal in Stable Habitats' (22, jointly with Robert May) bring an essential evolutionary individualist view to spatial distribution. Animals aggregate on plains because each individual is safer if it places conspecifics between itself and predators: no species advantage is required, and neither mystic nor cuddly communitarianism. Similarly, considerable dispersal is favoured even in the face of very high extra mortality, simply because a stay-at-home strategy can never colonize other sites and so can never increase its representation, while a strategy with some probability of dispersal has the chance to spread, and is unthreatened on home sites unless there is significant dispersal in the population as a whole. The remarkable finding of Hamilton and May in one simple model was that no matter how high the extra mortality, the evolutionarily stable strategy implied that at least half the offspring should disperse. The intellectual angle of both papers shows the importance of individual selfishness in aggregate behaviour, parallel to the significance of genetic selfishness in meiotic drive a level lower. More generally, these papers show a characteristic combination of apparent quirkiness with profound and original insight. This memoir has to focus on a central narrative with a few themes and so regrettably cannot cover all Hamilton's papers, and many biologists may find their favourite paper omitted.

Returning to the core work of this period, the current authoritative mathematical source for inclusive fitness theory comprises two papers of Peter Taylor (1990, 1996). Hamilton greatly approved of a systemization by Frank (1998) of ideas and methods in the natural selection of social behaviour, building on and extending these mathematical developments, even though at a technical level it emphasized personal rather than inclusive fitness. The key point was that Frank's work allows optimization techniques to be applied to social evolution, where Hamilton himself always had to return to population genetics models for his formal arguments.

These four core papers set out a framework, based on relatedness and agents maximizing inclusive fitness, for the evolutionary study of social behaviour. There are literatures that employ this framework, but there have been no significant additions to the framework itself. In issues such as group selection, some authors choose not to use the tools provided, but it is of enormous significance for biology that the study of social behaviour can all be treated within a single coherent theoretical structure, which is Hamilton's greatest legacy.

Sex and Sexual Selection

By 1980, Hamilton had begun his last major research project, to solve the outstanding problem in Darwinian biology: sexual reproduction. An important and popular subject, it was receiving serious attention in books by Williams (1975), Maynard Smith (1978), and Bell (1982), but none of these appeared to Hamilton to provide a satisfactory resolution. Looking for a single explanation, and unwilling to accept that lack of asexual mutants played a major role Hamilton needed a model that explained the hardest case. The culminating paper (18, Hamilton, Axelrod and Tanese 1990) claims to provide a simulation model that meets the 'challenging conditions' of 'very low fecundity, realistic patterns of genotype fitness and changing environment, and frequent mutation to parthenogenesis, even while sex pays the full twofold cost'.

The starting point was that an advantage to sex required strong continuing selection, despite outward appearances of stability of phenotypes. There was in 1980 a consensus among biologists that heritability of fitness was bound to be low, as advantageous alleles would rapidly spread to fixation. Hamilton rejected inanimate selective forces, because they would not retain variation. Eventually, drift would eliminate or fix an allele. The two classic biological situations in which variability would be positively retained *and* there would be continual change were those producing cycling, namely predator–prey models, and host–parasite models. They were ecological models, and the cycling quantities were densities of species rather than genotypes, but Hamilton thought that sauce for the ecological goose might be sauce for the evolutionary gander. He selected the host–parasite interactions for further development because all species partake in them—viruses acting as parasites to the smallest species; and because they are more species-specific, which encourages cycling.

There followed a series of six papers in which this idea was extended and developed. The methodological difficulty was that useful current methods in population genetics allowed only an analysis of stable states in simple models. Hamilton wanted a multispecies model, in which each species had varying loci, and in which one of the species had more than two loci, and he wanted to show that sexual reproduction was maintained by simultaneous (possibly irregular) cycling of gene frequencies at many loci. There was no precedent for studying such a model. Hamilton felt he had the makings of an answer in a general sense, but that models were absolutely required, not only to persuade himself and other scientists, but also to discover more particularly the significance of the different ingredients. How, though, to proceed?

Through the six papers, Hamilton moves from a simple, parable-like model to a much more realistic and richly detailed model, which consummated the project for him. Hamilton, Henderson and Moran (20) presented an analytical treatment of a species with different genotypes subject to environmentally fluctuating fitnesses, and then Hamilton (9) made the environmental fluctuations implicitly due to parasites by introducing genotype-specific frequency dependence. The next two papers were side-shoots from the main line. Eshel and Hamilton (2) studied heritabilities of

fitness under slowly fluctuating selection pressures, and Hamilton (10
investigated how ecological competition between host species affecte
parasite-induced cycling. So far, the methods had all been analytic, wit
equations sometimes solved numerically. Returning to the central theme
Seger and Hamilton (28) first continued this approach and applie
numerical methods to deterministic equations to render precise man
intuitions about host–parasite systems. They went on to employ individual
based modeling, in which each individual host was separately represented i
the computer model, to look at a multilocus system, although the parasite
remained implicit in a host-genotype-specific frequency dependence. In th
final paper, Hamilton, Axelrod and Tanese (18) applied individual-base
modelling to the host species and up to seven species of parasite. Thi
completely individual-based approach made it possible to apply vario
assumptions that rendered the model more realistic.

It is characteristic of individual-based models that they are so comple
that they need to be studied phenomologically themselves, and this leave
room for doubt about the generality of the conclusions. But it is fair to sa
that the model shows there are circumstances in which host–parasit
interactions can create a robust advantage for sexual reproduction in th
face of the 'challenging conditions' referred to above.

For Hamilton, this model completed his task, though he did publish abou
ten later papers on related matters. The application of the ultimately cruci
methodology, individual-based models, was conducted with co-authors wh
in addition to their other contributions, actually carried out the comput
tions. Thus Hamilton's central role in the whole project was to possess an
refine the key biological insights, and to encourage others to collaborat
when he needed extra technical resources.

The unity of the papers is striking, as the long-term goal is always clea
and the provisional nature of all but the final paper is fully acknowledge
One offshoot of this subject is not only important in itself but illuminat
how Hamilton felt his way through the complex of ideas. The first two pape
had involved fitness regimes in which the fitness of a given genotype swu
violently up and down in succeeding generations. Hamilton had a 'eurek
moment' in which he asked himself what an individual in the species shou
be trying to do. The perverse answer was to mate with the sickest individual,
that will provide the genes that do best in the next generation. Simultaneous
this showed Hamilton two things simultaneously: first, that sexual selecti
and mate choice were linked with the question of sex, and fluctuations

tness would affect both, and second, that he should be looking for slower tness cycles to support sex, as a mating preference for sickness seemed nlikely to be realistically advantageous. This second point led on to the joint aper with Eshel already cited, while the first led on to the paper by Hamilton nd Zuk (23). It is worth noting some features of this turning point in the sex heory. Like so much of Hamilton's work, it was inspired by an optimizing gent view. It was a simple idea, and yet no-one but Hamilton could have had t at that time, as it made sense only within a conceptual world into which ther biologists had not, yet, entered.

Hamilton's attention to sexual selection transformed the subject, and the dea of choosy females as conducting a medical examination of potential nates has been extremely influential. There was a long and animated debate ver the theoretical predictions of and empirical support for Hamilton and uk's ideas, continued with great energy in the 'Appendix to Chapter Six' in *Jarrow Roads of Gene Land II* (16). The emphasis on 'sosigonic selection', nd the theoretical importance given to extra-pair copulations, are two mportant legacies, whatever the final verdict on the particular predictions.

In 1984, after a year's hesitation caused by the property slump in Ann Arbor, Hamilton accepted a Royal Society Research Professorship, then rguably the best job in science. He was persuaded by Richard Southwood, ormerly his head of department at Imperial, to take it up in his new lepartment, the Zoology Department of Oxford University, with a Pro-essorial Fellowship at New College, where Hamilton remained for the rest f his career. With Christine and their daughters Helen, Ruth and Rowena, e moved to a university house in the nearby village of Wytham, whose vood is an active research base. Bill conducted his own projects there, ncluding the (illegal) introduction of exotic burying beetles, using a bad-er's skull protected by wire mesh.

Autobiography, Eugenics and Science

)uring the last ten years, Bill's personal life changed dramatically. Christine noved to Orkney to pursue her career as a dentist. He met Luisa Bozzi, an talian science journalist, who became his partner for what would prove to e only the last six years of his life. As well as continuing to publish some lozens of papers continuing the project on sex, sexual selection and para-ites, he opened up two new activities. He used his by then considerable prestige to support theories that he viewed as unjustly rejected by fellow scientists through prejudice and narrow-mindedness. The second activity was autobiography.

It was not in Hamilton's nature to build an institutional empire, as most similarly distinguished scientists would have done by this point in their careers, and he continued to work as a solitary scholar. He retained throughout his life a youthful political innocence, and a disdain for authority that led to various conflicts with senior figures. He gave at least as much courtesy and attention to the views of a graduate student as to those of a Fellow of the Royal Society, and inspired enormous personal affection among junior colleagues throughout his academic life, not least those who accompanied him on expeditions to the Amazon and the Congo.

The scientific output of the final decade comprises about thirty papers One thread (including Hamilton 12, Sasaki, Hamilton and Ubeda 27) continues the sex theme, pursuing multilocus models, and contains the discovery of new phenomena as well as the development of methods of attacking necessarily complex simulations in which a purely mathematica approach had so far been of little use. Further papers introduce interesting ideas. Hurst and Hamilton (25) argue that the basic logic of the sexes is tha male gametes delete cytoplasmic DNA while female gametes do not Hamilton and Lenton (21) and Welsh, Viaroli, Hamilton and Lenton (30 argue that clouds may be dispersal mechanisms for marine micro-organisms an idea used in the extremely moving words spoken by Luisa Bozzi over Bill' open grave (quoted in Richard Dawkins' preface to the second volume o *Narrow Roads*, and now inscribed on a bench near the grave). Hamilton an Brown (19) suggest that autumn colours may result from signalling by tree to aphids. The autumn leaves paper is already well cited given its recen publication, and Atkinson (2001) referred to 'a tantalizing idea that will n doubt spark an explosion of new research. Autumn will never be the sam again.' Amazonian plant communities were a major theme in his thinking i his final years, but unfortunately no major publication emerged before hi death.

Hamilton also used his prestige to support ideas that he felt were unfairl rejected by scientists. He was providing the kind of support that he fel his own early theory should have received from the famous scientists c the 1960s. Edward Hooper was helped by Hamilton to promote his poli vaccine theory of the origin of HIV, with a co-authored paper in *The Lanc* (24) and a preface by Hamilton to *The River* (Hooper 1999). It was hard t publish on the theory, and powerful interests were at work. If the theory were true, medicine would have killed more people than it had saved in its entire history, early experiments on black Africans by white European scientists would have been the cause, and pharmaceutical companies and individual scientists might have been liable in court. It was idealistic to hope to discuss such a theory with scientific dispassion, but Hamilton deeply resented what he saw as the suppression of debate by vested interests, and thought that important implications, for example for xenotransplantation, were being missed. It now seems likely HIV did not arise through the polio vaccines, though other iatrogenic involvement in the spread of HIV is still discussed. Hamilton's support for the theory included two expeditions to the Congo to collect samples of chimpanzee faeces, to obtain relevant evidence. During the second expedition, which determined to find samples from adult wild, not juvenile urban, chimpanzees, Bill contracted malaria in the Congo, collapsed with haemorrhaging soon after returning to London, and died without fully recovering consciousness a few weeks later, at the age of only 63.

In the course of supporting this theory, Hamilton persuaded the Royal Society, with some difficulty, to hold a Discussion Meeting on an urgent topic at short notice. The meeting was held posthumously (reported in Phil. Trans. R. Soc. Lond. B, **356**, 2001) and recognized as an important precedent: procedures now allow such meetings to be proposed and held rapidly. This substantial political success suggests Hamilton may not have been quite as unworldly as he appeared. The report includes a tribute by Lord May, President of the Society (May 2001).

The other major theory Hamilton tried to rescue from unjustified scientific neglect was Lovelock's Gaia. There are sympathetic remarks in a couple of the late papers, but always accompanied by his logical objections. Hamilton was determined to remain open to strange ideas, and to reject the fossilization of intellect that comes with intolerance. But in the case of Gaia, unlike the polio vaccine theory, this openness did not bring him even close to conversion.

Hamilton's hero Fisher had published five volumes of collected papers with short notes by Fisher introducing a few of them. Perhaps inspired by his example, Hamilton created a new form of scientific autobiography in his collected papers, *Narrow Roads of Gene Land*. Michael Rodgers, who had published the *Selfish Gene* and knew this part of biology well, first approached Hamilton in 1980 about publishing the collected papers, and there is a reference to 'short, linking commentaries' in the resulting publishing proposal. After renewed negotiations in 1993, the commentaries had become 'prefaces', and the first volume was published in 1996 (15). Each paper is preceded by an essay in which Hamilton relates his personal life around the time of writing the paper, the intellectual issues of the time, and contemporary reflections on the paper and the issues. The second volume was almost complete when Hamilton died, and was published with some further light editing in 2001 (16). Hamilton had written no introductions for the third volume, which will appear, under the editorship of Mark Ridley, with introductions written by his co-authors (17). The project had caught up with itself, in the sense that the period of the papers in the third volume was mainly the period in which *Narrow Roads* itself was being written.

Narrow Roads is an extraordinarily useful publication for the original papers, as Hamilton published in a wide variety of places, including multi-author volumes that can be hard to get hold of. The essays in *Narrow Roads* provide a vivid testimony to the life of a great scientist and a remarkable man. They contain evocations of childhood and a bygone age, compelling glimpses of inner thoughts and feelings, vivid depictions of jungle scenes, marvellous portrayals of animals and plants, true-to-life pen portraits, all done with extensive though unoppressive literary references, and more. There are other essentially autobiographical essays in this period: 'Inbreeding in Egypt and in this book; a childish view' (11), 'On first looking into a British treasure' (13), and 'My intended burial and why' (15). Bill also wrote an unpublished novel from which an excerpt was read at his funeral.

The figure who emerges is a romantic Englishman, a great reader and lover of literature. Russian novels were one passion and poetry another, including Flecker, Housman and Lafcadio Hearn. A solitary rebel proud of his physical fitness, loving nature and proud of his knowledge of and ability to cope in it, a restless thinker ill-at-ease with politeness and decorum, he might in a past age have colonized new lands, but instead conquered intellectual territories. Shy and diffident, he was nevertheless a marvellous story-teller and unparalleled natural history guide. Jeremy John saw a resemblance to a prospector whose wild appearance disturbs sedate townsfolk, but who has, in fact, struck gold. The unexamined life, according to Socrates, is not worth living: in the abundant autobiographical writing of this period, Bill shares with us an extended self-examination. Bill's technical writing is sometimes described as difficult to understand—in fact even this is grossly unfair as the style is literary and approachable—but reading the *Narrow Roads* and other essays is a delight and a privilege.

Reading the essays provides an experience very similar to talking to Bill. The diversions and footnotes, the footnotes to the diversions, and then the diversions within the footnotes, show a multilayered restless intellect at work, able and eager to move along narrow paths indeed from viewpoint to vista in an intensely wrought and highly productive mental landscape. Biology, literature and the lessons of life fuse together for the listener and reader as they did for Bill. We owe to Robert Trivers the remark that while most of us think in single notes, Bill thought in chords.

One non-scientific and even political theme recurs throughout the autobiographical essays, and it is one that troubles many readers. Hamilton develops a eugenic argument, which was deeply felt and persistently argued. It was not a conventional eugenic argument: its recommendation of inter-racial marriage would hardly gain the support of the political right wing. On the other hand, it shocks many readers when Hamilton advocates infanticide; suggests the denial or in more emollient mode the strict regulation of fertility treatment; and worries about the long-term effects of saving the lives of mother and neonate with a Caesarian section.

The first context into which these views must be placed is the final five chapters of the *Genetical Theory*, in which Fisher lays out an evolutionist's prescription for avoiding the downfall of modern civilization. These five chapters, virtually ignored by biologists, seem over seventy years after they were written to emerge from the particular social position of the writer, articulated though they may be with exactly the same language and approach as the first eight, seminal, chapters. They were unquestionably written out of social conscience, and may in a private calculus have compensated for Fisher's inability to enlist in the Great War. Both scientists had wide interests in history, and deeply believed that their subjects provided insights of practical assistance in matters of the highest public importance. Hamilton's determination to express eugenic enthusiasm later may have been fuelled by his early experience with the Royal Society Population Study Group in the mid-1960s, when he reports being rebuked by the chair, Lord Florey, for suggesting the group should study interracial differences in reproductive rate in relation to warfare (*Narrow Roads of Gene Land II*, 16, page *xxxvi*). It was the final five chapters of the *Genetical Theory* that

Hamilton enthused over in a postcard to his sister Mary, written on the ver day he discovered the book.

It has to be said that the scientific basis of Hamilton's eugenic views is no established, though he provided interesting possible hypotheses. In prin ciple, the selective forces he identified could be at work, but there are n measurements, and modern conditions may for all we know have introduce other counteracting forces. A start on an intellectual engagement wit Hamilton's views has been made in considered remarks of Haig (2003).

A fitting memorial to Hamilton would be a collaborative effort betwee evolutionary and medical sciences keeping a watchful eye on the select ive effects of modern medicine. The evolutionary, as opposed to medica perspective emphasises a long timespan, so that we should be concerne about ten and a hundred and more generations into the future, and simu taneously about the possibility that new challenges will emerge, such as return to stone-age conditions for one reason or another. It goes withou saying that those engaged in the collaboration will need to emplo more political sensitivity than Bill possessed if they are to do their jo effectively.

It is interesting also to view this line of Bill's thinking in a second contex that of his autobiographical work. Two of his brothers died young, one soo after birth because of a handicap, and the other at the age of nineteen in mountaineering accident. In a remarkable passage in *Narrow Roads of Gen Land II* (16, pages 477–483), the first death is compared to that of a do belonging to his own family many years later, and Bill reflects on what make creatures morally deserving, pursuing the logic through to its difficult con clusion in his own personal case. Bill cycled helmetless at high spee through Oxford traffic on the bike belonging to the second brother wh died, maintaining decades later a memorial love of thrill and disdain fo danger. His eugenic views may have been a way to make sense of these tw deaths, of his own near-fatal childhood accident, and his and his family reactions to them. These views were the public face of private conviction Bill often said he didn't want to grow old, and he took many risks. Risks lik those in his brother's mountaineering accident, and in his own expedition t the Congo, were all part of the Hamilton family ethos. Bill's distinguishe achievements, along with those of his siblings, must be put on the positi side of that balance.

Perhaps his most extraordinary autobiographical piece is concerned wi his death, or rather, 'My Intended Burial and Why' published in English

2000 (15), after originally appearing in 1991 in the Japanese entomological journal 'The Insectarium'. In this romantic flight of fantasy, Bill sweeps far and wide over his childhood, his intellectual development, the Amazon and, finally, the habits of beetles.

A Worthy Successor to Darwin and Fisher

Hamilton's great contributions to biology relied on an essential admixture of mathematics and modelling. Darwin's original arguments were purely verbal, and mathematics was first brought in by the heroic figures of population genetics, Haldane, Fisher and Wright. Among these, it was Fisher who paid particular attention to Darwinian aspects of population genetics, most notably in his microscope argument, his use of reproductive value and optimization ideas, particularly in his sex-ratio argument, and the encapsulation of natural selection in his fundamental theorem. Hamilton pursued this line in a way that was too mathematical for a Darwin, and too biological for a Fisher. His contributions have enormously enhanced Darwinian areas of biology, influencing its practitioners as well as those in the increasing spread of subjects in which Darwinism plays an important part, such as demography, economics, and anthropology. The widespread assumption that organisms maximize their inclusive fitness, valuing the reproduction of relatives according to coefficients of relatedness, is of far-reaching importance, and close enough to the truth.

The combination of intelligence, dedication and grit that Hamilton brought to bear in his work was truly outstanding, and it is futile to plan for more 'Hamiltons'. On the other hand, the nature of his achievements leaves no doubt how valuable is the combination in one individual of deep biological knowledge and commitment with mathematical skills.

When the history of natural selection comes to be written, Darwin will take pride of place. The first half of the twentieth century will see Fisher reconciling Darwinism with Mendelism, and representing in mathematical terms the force of natural selection within the maelstrom of population genetics. Hamilton will emerge as the major figure of the second half of the twentieth century, the first to extend the principle of natural selection, and the architect of the biological theory of social behaviour.

Acknowledgments

I am grateful to Mary Bliss for permission to quote excerpts from postcards from Bill, tc Michael Rodgers for providing copies of notes relating to the publication plans for *Narrou Roads of Gene Land*, and to Jim Schwartz for providing copies of correspondence involving Hamilton and George Price. I am also grateful to Mary Bliss, Luisa Bozzi, Richard Dawkins, Steve Frank, David Haig, Christine Hamilton, Jeremy John, Lord May, Naomi Pierce, Mark Ridley, Michael Rodgers, Jim Schwartz, Ullica Segerstråle, and Mike Worobey, for comments on earlier drafts. During the final stages of preparation of this memoir, on the 11th May 2004 Maria Luisa Bozzi died suddenly at her home in Italy.

Hamilton Papers Cited in the Memoir

1. R. Axelrod and W. D. Hamilton, The evolution of cooperation. *Science* **211** 1390–1396 (1981).
2. I. Eshel and W. D. Hamilton, Parent-offspring correlation in fitness unde fluctuating selection. *Proceedings of the Royal Society B* **222**, 1–14 (1984).
3. W. D. Hamilton, The evolution of altruistic behaviour. *American Naturalist* **97** 354–6 (1963).
4. W. D. Hamilton, The genetical evolution of social behaviour. *Journal c Theoretical Biology* **7**, 1–52 (1964).
5. W. D. Hamilton, Extraordinary sex ratios. *Science* **156**, 477–88 (1967).
6. W. D. Hamilton, Geometry for the selfish herd. *Journal of Theoretical Biology* **31** 295–311 (1971).
7. W. D. Hamilton, Selfish and spiteful behaviour in an evolutionary model *Nature* **228**, 1218–20 (1970).
8. W. D. Hamilton, Innate social aptitudes of man: an approach fron evolutionary genetics. In R. Fox (ed.) ASA Studies 4, *Biosocial Anthropology* pp 133–53 (Malaby Press, London, 1975.).
9. W. D. Hamilton, Sex versus nonsex versus parasites. *Oikos* **35**, 282–29 (1980).
10. W. D. Hamilton, Instability and cycling of two competing hosts with tw parasites. In S. Karlin and E. Nevo (eds.) *Evolutionary Processes and Theor* pp 645–668 (Academic Press, New York, 1986).
11. W. D. Hamilton, Inbreeding in Egypt and in this book: a childish view. I N. W. Thornhill and W. M. Shields (eds.) *The Natural History of Inbreeding an Outbreeding*, pp 429–450 (University of Chicago Press, 1993a).

12. W. D. Hamilton, Haploid dynamic polymorphism in a host with matching parasites—effects of mutation/ subdivision, linkage, and patterns of selection, *Journal of Heredity* **84**, 328–338 (1993b).

13. W. D. Hamilton, On first looking into a British treasure—a celebration of the Collins New-Naturalist Series, *Times Literary Supplement* (Aug 12), pp 13–14 (1994).

14. W. D. Hamilton, *Narrow Roads of Gene Land*, Volume 1, *Evolution of Social Behaviour*. (Oxford University Press, 1996).

15. W. D. Hamilton, My intended burial and why. *Ethology Ecology and Evolution* **12**, 111–122 (2000).

16. W. D. Hamilton, *Narrow Roads of Gene Land*, Volume 2, *Evolution of Sex*. (Oxford University Press, 2001).

17. W. D. Hamilton, *Narrow Roads of Gene Land*, Volume 3, *Last Words* (with essays by coauthors, edited by Mark Ridley) (Oxford University Press, 2004).

18. W. D. Hamilton, R. Axelrod, and R. Tanese, Sexual reproduction as an adaptation to resist parasites (a review). *Proceedings of the National Academy of Sciences USA* **87**, 3566–73 (1990).

19. W. D. Hamilton and S. P. Brown, Autumn tree colours as a handicap signal. *Proceedings of the Royal Society B* **268**, 1489–1493 (2001).

20. W. D. Hamilton, P. A. Henderson, and N. A. Moran, Fluctuation of environment and coevolved anatagonistic polymorphism as factors in the maintenance of sex. In R. D. Alexander and D. W. Tinkle (eds.) *Natural Selection and Social Behavior*, pp 363–381 (Chiron Pres, New York, 1981).

21. W. D. Hamilton and T. M. Lenton, Spora and Gaia: How microbes fly with their clouds. *Ethology Ecology and Evolution* **10**, 1–16 (1998).

22. W. D. Hamilton and R. M. May, Dispersal in stable habitats. *Nature* **269**, 578–581 (1977).

23. W. D. Hamilton and M. Zuk, Heritable true fitness and bright birds—a role for parasites. *Science* **218**, 384–387 (1982).

24. E. Hooper and W. D. Hamilton, 1959 Manchester case of syndrome resembling AIDS. *Lancet* **348**, 1363–1365 (1996).

25. L. D. Hurst and W. D. Hamilton, Cytoplasmic fusion and the nature of the sexes. *Proceedings of the Royal Society B* **247**, 189–194 (1992).

26. R. E. Michod and W. D. Hamilton, Coefficients of relatedness in sociobiology. *Nature* **288**, 694–7 (1980).

27. A. Sasaki, W. D. Hamilton, and F. Ubeda, Clone mixtures and a pacemaker: new facets of Red Queen theory and ecology. *Proceedings of the Royal Society B* **269**, 761–772 (2002).

28. J. Seger and W. D. Hamilton, Parasites and sex. Chapter 11 in R. E. Michod and B. R. Levin (eds.) *The Evolution of Sex: an Examination of Current Ideas*, pp 176–193 (Sinauer Associates, Sunderland MA, 1988).

29. R. Stouthamer, R. F. Luck, and W. D. Hamilton, Antibiotics cause parthenogenetic Trichogramma (Hymenoptera, Trichogrammatidae) to revert to sex. *Proceedings of the National Academy of Sciences USA* **87**, 2424–2427 (1990).

30. D. T. Welsh, P. Viaroli, W. D. Hamilton, and T. M. Lenton, Is DMSP synthesis in chlorophycean macro-algae linked to aerial dispersal? *Ethology Ecology anc Evolution* **11**, 265–278 (1999).

Hamilton Papers Uncited in the Memoir

P. Becker, L. W. Lee, E. D. Rothman, and W. D. Hamilton, Seed predation anc the coexistence of species—Hubbell's models revisited. *Oikos* **44**, 382–390 (1985)

H. N. Comins, W. D. Hamilton, and R. M. May, Evolutionarily stable dispersa strategies. *Journal of Theoretical Biology* **82**, 205–230 (1980).

D. Ebert and W. D. Hamilton, Sex against virulence: the coevolution o parasitic diseases. *Trends in Ecology and Evolution* **11**, 79–82 (1996).

W. D. Hamilton, The moulding of senescence by natural selection. *Journal o Theoretical Biology* **12**, 12–45 (1966).

W. D. Hamilton, Population Control, New Scientist 30 October (1969).

W. D. Hamilton, Addendum to 1964 papers on calculation of relatednesses unde haplodiploidy. In G. C. Williams (ed.) *Group Selection*, pp 87–89 (Aldine-Atherton, Chicago, 1971).

W. D. Hamilton, Selection of selfish and altruistic behavior in some extrem models, Chapter 2 in J. F. Eisenberg and W. S. Dillon (eds.) *Man and Beast Comparative Social Behavior*, pp 57–91 (Smithsonian Press, Washington DC 1971).

W. D. Hamilton, Altruism and related phenomena, mainly in social insects, *Annu Review of Ecology and Systematics* **3**, 193–232 (1972).

W. D. Hamilton, Gamblers since life began. Quarterly *Review of Biology* **5(** 175–180 (1975).

W. D. Hamilton, Evolution and diversity under bark, Chapter 10 in L. A. Mound and N. Waloff (eds.), *Diversity in Insect Fauna*, pp 154–175, (Blackwell Scientific, Oxford, 1978).

W. D. Hamilton, Wingless and fighting males in fig wasps and other insects, in M. S. Blum and N. A. Blum (eds.) *Reproductive Competition, Mate Choice and Sexual Selection in Insects*, pp 167–220 (Academic Press, New York and London, 1979).

W. D. Hamilton, Fluctuation of environment and coevolved antagonist polymorphism as factors in the maintenance of sex, in R. D. Alexander and D. W. Tinkle (eds.), *Natural Selection and Social Behavior*, pp 363–381 (Chiron Press, New York, 1980).

W. D. Hamilton, Pathogens as causes of genetic diversity in their host populations In R. M. Anderson and R. M. May (eds.) *Population Biology of Infectious Diseases*, Dahlem Conference, pp 269–296 (Springer, Berlin, 1982).

W. D. Hamilton, Mate Choice—Bateson P. *Nature* **304**, 563–564 (1983).

W. D. Hamilton, Kinship, recognition, disease, and intelligence: constraints of social evolution. In Y. Ito, J. L. Brown, and J. Kikkawa (eds.) *Animal Societies, Theories and Facts, pp 81–102 (Japan Scientific Societies Press, Tokyo, 1987)*.

W. D. Hamilton, Discriminating nepotism: expectable, common, overlooked. Chapter 13 in D. J. C. Fletcher and C. D. Michener (eds.), *Kin Recognition in Animals*, pp 417–437 (Wiley, Chichester, 1987).

W. D. Hamilton, Citation Classic—The Genetical Evolution of Social Behavior 1, Current Contents (3 different subfields) **40**, 16–16 (1988).

W. D. Hamilton, Sex and disease. Chapter 4 in G. Stevens and R. Bellig (eds.), *Nobel Conference XXIII: the Evolution of Sex*, pp 65–95 (Harper and Row, San Francisco, 1988).

W. D. Hamilton, Memes of Haldane and Jayakar in a theory of sex. *Journal of Genetics* **69**, 17–32 (1990).

W. D. Hamilton, Mate choice near or far. *American Zoologist* **30**, 341–352 (1990).

W. D. Hamilton, Seething genetics of health and the evolution of sex. In S. Osawa, T. Honjo (eds.) *Evolution of Life, Fossils, Molecules and Culture*, pp 229–251 (Springer, Berlin, 1991).

W. D. Hamilton, Oedipal mating, Sciences—New York, **32**, 5–5 (1992).

W. D. Hamilton, Recurrent viruses and theories of sex. *Trends in Ecology and Evolution* **7**, 277–278 (1992).

W. D. Hamilton, Ecology in the large: Gaia and Ghengis Khan. *Journal of Applied Ecology* **32**, 451–453 (1995).

W. D. Hamilton, Gaia's benefits. New Scientist (dated 27 July 1996), **151**, 62–6 (1996).

W. D. Hamilton, Fables of cyberspace: Tapeworms, horses, and mountains Advances in Artificial Life. *Proceedings: Lecture Notes in Artificial Intelligenc* **1674**, 5–6 (1999).

W. D. Hamilton, Of vaccines and viruses: an African legacy (Reprinted from th Foreword to The River: A journey back to the source of HIV and AIDS, 1999 *South African Journal of Science* **96**, 263–265 (2000).

W. D. Hamilton and M. Zuk, Parasites and sexual selection—Reply. *Nature* **341** 289–290 (1989).

P. A. Henderson, W. D. Hamilton, and W. G. R. Crampton, Evolution and diversity i Amazonian floodplain communities. In D. M. Newbery, H. H. T. Prins, and N D. Brown (eds.), *Dynamics of Tropical Communities, 37th Symposium of the Britis Ecological Society*, pp 385–419, (Blackwell Science, Oxford, 1998).

A. I. Houston and W. D. Hamilton, Selfishness reexamined—no man is an islan *Behavioral and Brain Sciences* **12**, 709–709 (1989).

L. D. Hurst, W. D. Hamilton, and R. J. Ladle, Covert sex. *Trends in Ecology an Evolution* **7**, 144–145 (1992).

J. Kathirithamby and W. D. Hamilton, More covert sex: the elusive femal of Myrmecolacidae. *Trends in Ecology and Evolution* **7**, 349–351 (1992).

J. Kathirithamby and W. D. Hamilton, Exotic pests and parasites. *Nature* **374**, 769 770 (1995).

N. Moran and W. D. Hamilton, Low nutritive quality as defence against herbivore *Journal of Theoretical Biology* **86**, 247–254 (1980).

F. Roes and W. D. Hamilton, In his own words. *Natural History*, **109**, 46–47 (2000

B. Sumida and W. D. Hamilton, Both Wrightian and 'parasite' peak shifts enhan genetic algorithm performance in the travelling salesman problem. Chapter 16 i R. Paton (ed.) *Computing with Biological Metaphors*, pp 264–279 (Chapman ar Hall, 1994).

B. H. Sumida, A. I. Houston, J. M. McNamara, and W. D. Hamilton, Genet algorithms and evolution. *Journal of Theoretical Biology* **147**, 59–84 (1990).

J. P. Vartanian, P. Pineau, M. Henry, W. D. Hamilton, M. N. Mulle R. W. Wrangham, and S. Wain-Hobson, Identification of a hepatitis B vir genome in wild chimpanzees (*Pan troglodytes schweinfurthi*) from East Afri indicates a wide geographical dispersion among equatorial African primate *Journal of Virology* **76**, 11155–11158 (2002).

Other References

N. Atkinson, *Out on a Limb, or a new branch of signalling theory?—Trends in Ecology and Evolution* **16**, 603 (2001).

R. Axelrod, Effective choice in the prisoner's dilemma. *Journal of Conflict Resolution* **24**, 3–25 (1980a).

R. Axelrod, More effective choice in the prisoner's dilemma, *Journal of Conflict Resolution* **24**, 379–403 (1980b).

R. Axelrod, The evolution of cooperation. Basic Books. New York (1984).

G. Bell, The Masterpiece of Nature—The Evolution of Genetics and Sexuality. (University of California Press, Berkeley, 1982).

Benford. Fisher's theory of the sex ratio applied to the social Hymenoptera. *Journal of Theoretical Biology* **72**, 701–727 (1978).

W. F. Bodmer and A. W. F. Edwards, Natural selection and the sex ratio. *Annals of Human Genetics* **24**, 239–244 (1960).

A. Burt and R. L. Trivers (forthcoming) *Genes in conflict (Belknap Press of Harvard University Press, Cambridge, Mass.).*

R. Dawkins (1976) The Selfish Gene. Oxford University Press. Also second edition (1989).

R. Dawkins (1999) The Extended Phenotype. Oxford University Press. Also second edition (1982).

A. W. F Edwards, The fundamental theorem of natural selection. *Biological Reviews* **69**, 443–474 (1994).

A. W. F. Edwards, The genetical theory of natural selection. *Genetics* **154**, 1419–1426 (2000).

W. J. Ewens, An interpretation and proof of the Fundamental Theorem of Natural Selection. *Theor. Pop. Biol.* **36**, 167–180 (1989).

R. A. Fisher, The Genetical Theory of Natural Selection. Oxford University Press. Also 1958 second edition (Dover), and 1999 variorum edition, J. H. Bennet (ed.) (Oxford University Press, 1930).

R. A. Fisher, Smoking: the Cancer Controversy. (Oliver and Boyd. Edinburgh, 1959).

E. B. Ford, Butterflies. New Naturalist Series. (Collins, London, 1945).

S. A. Frank, George Price's contributions to evolutionary genetics. *Journal of Theoretical Biology* **175**, 373–388 (1995).

S. A. Frank, The Price Equation, Fisher's Fundamental Theorem, Kin Selection and Causal Analysis. *Evolution* **51**, 1712–1729 (1997).

S. A. Frank, Foundations of Social Evolution. (Princeton University Press, Princeton, New Jersey, 1998).

M. T. Ghiselin, The triumph of the Darwinian Method. (University of California Press, Berkeley, 1969).

A. Grafen, How not to measure inclusive fitness. *Nature*, **298**, 425–426 (1982).

A. Grafen, A geometric view of relatedness. *Oxford Surveys in Evolutionary Biology* **2**, 28–90 (1985).

A. Grafen, Formal Darwinism, the individual-as-maximising-agent analogy, and bet-hedging. *Proceedings of the Royal Society (London) B*, **266**, 799–803 (1999).

A. Grafen, Developments of the Price equation and natural selection under uncertainty. *Proceedings of the Royal Society (London) B*, **267**, 1223–1227 (2000).

A. Grafen, A first formal link between the price equation and an optimisation program. *Journal of Theoretical Biology* **217**, 75–91 (2002).

D. Haig, The Social Gene. Chapter 12 in J. R. Krebs and N. B. Davies (eds.) *Behavioral Ecology 4th edition*, pp 284–306 (Blackwell Science, Oxford, 1997).

D. Haig, Genomic imprinting and kinship. (Rutgers University Press, NJ, 2002).

D. Haig, The science that dare not speak its name. (Review of Second Volume of Narrow Roads of Gene Land.) *Quarterly Review of Biology* **78**, 327–335 (2003)

J. B. S. Haldane, Population Genetics. *New Biology* **18**, 34–51. Penguin (1955).

I. C. W. Hardy (ed.). Sex Ratios: Concepts and Research Methods. (Cambridge University Press, Cambridge, 2002).

E. Hooper, The River: A Journey Back to the Source of HIV and AIDS. (Penguin, Harmondsworth and Little, Brown, Boston, 1999).

D. Lack, Population Studies of Birds. (Clarendon Press, Oxford, 1966).

D. Lack, Ecological Adaptations for Breeding in Birds. (Methuen, London, 1968).

J. Lorberbaum, No strategy is evolutionarily stable in the repeated prisoners dilemma. *Journal of Theoretical Biology* **168**, 117–130 (1994).

R. H. Macarthur and E. O. Wilson, The Theory of Island Biogeography. (Princeton University Press, Princeton NJ, 1967).

R. M. May, Memorial to Bill Hamilton, *Philosophical Transaction of the Royal Society of London, Series B*. **365**, 785–787 (2001).

J. Maynard Smith, The Evolution of Sex. (Cambridge University Press, Cambridge, 1978).

J. Maynard Smith and G.R. Price. The logic of animal conflict. *Nature* **246**, 15–1 (1973).

G. R. Price, Selection and covariance. *Nature* **227**, 520–521 (1970).

G. R. Price, Fisher's 'fundamental theorem' made clear. *Annals of Human Genetics* **36**, 129–140 (1972a).

G. R. Price, Extension of covariance selection mathematics. *Annals of Human Genetics* **35**, 485–490 (1972b).

J. Schwartz, Death of an altruist. *Lingua Franca* **10**, 51–61 (2000).

U. Segerstråle Defenders of the Truth: The Battle for Science in the Sociobiology Debate and Beyond. (Oxford University Press, Oxford, 2000).

U. Segerstråle (forthcoming) Nature's Oracle: an Intellectual Biography of Evolutionist W. D. Hamilton. (Oxford University Press, Oxford).

R. F. Shaw and J. D. Mohler, The selective significance of the sex ratio. *American Naturalist* **87**, 337–342 (1953).

A. Smith, Mato Grosso: last virgin land : an account of the Mato Grosso, based on the Royal Society and Royal Geographical Society expedition to central Brasil, 1967–9. (Michael Joseph, London, 1971).

P. D. Taylor, Allele-frequency change in a class-structured population. *Journal of theoretical Biology* **135**, 95–106 (1990).

P. D. Taylor, Inclusive fitness arguments in genetic models of behavior. *Journal of mathematical Biology* **34**, 654–674 (1996).

J. Verner, Selection for the sex ratio. *American Naturalist* **99**, 419–21 (1965).

G. C. Williams, Adaptation and Natural Selection. (Princeton University Press, Princeton NJ, 1966).

G. C. Williams (ed.). Group Selection. (Aldine-Atherton, Chicago, 1971).

G. C. Williams, Sex and Evolution. (Princeton University Press, Princeton NJ, 1975).

E. O. Wilson, The Insect Societies. (Belknap Press/Harvard University Press, Cambridge MA, 1971).

E. O. Wilson, Sociobiology. (Harvard University Press, Cambridge MA, 1975).

Name Index

Note: names of co-authors are included, as well as names referred to in the text.
Biographical references to W. D. Hamilton are located in the Subject Index.

Subject Index

Note: biographical references to W. D. Hamilton are located in this index.

Lightning Source UK Ltd.
Milton Keynes UK
173836UK00001B/86/P

9 780198 566